D1338514

Reducing carbon dioxide (CO_2) emissions is imperative to stabilizing our future climate. Our ability to reduce these emissions, combined with an understanding of how much fossil fuel–derived CO_2 the oceans and plants can absorb, is central to mitigating climate change.

In *The Carbon Cycle,* leading scientists examine how atmospheric carbon dioxide concentrations have changed in the past and how this may affect the concentrations in the future. They look at the carbon budget and the "missing sink" for carbon dioxide. They offer approaches to modeling the carbon cycle, providing mathematical tools for predicting future levels of carbon dioxide.

This comprehensive text incorporates findings from the recent IPCC reports. New insights, and a convergence of ideas and views across several disciplines, make this book an important contribution to the global change literature. It is an invaluable resource for students and researchers working in the field.

The Carbon Cycle

Office for Interdisciplinary Earth Studies
Global Change Institute Volume 6
Series Editors: Tom M. L. Wigley and Carol Rasmussen

Produced through support from the Committee on Earth and
Environmental Sciences Subcommittee on Global Change.

The Carbon Cycle

Edited by

T. M. L. Wigley
National Center for
Atmospheric Research

D. S. Schimel
National Center for
Atmospheric Research

PUBLISHED BY THE PRESS SYNDICATE OF THE UNIVERSITY OF CAMBRIDGE
The Pitt Building, Trumpington Street, Cambridge, United Kingdom

CAMBRIDGE UNIVERSITY PRESS
The Edinburgh Building, Cambridge CB2 2RU, UK http://www.cup.cam.ac.uk
40 West 20th Street, New York, NY 10011-4211, USA http://www.cup.org
10 Stamford Road, Oakleigh, Melbourne 3166, Australia
Ruiz de Alarcón 13, 28014 Madrid, Spain

First published 2000

Printed in the United States of America

Typeface Times Roman 10/12 pt. *System* Quark 3.3 [CS]

A catalog record for this book is available from the British Library

Library of Congress Cataloging in Publication Data available.

ISBN 0 521 58337 3 hardback

Contents

Preface

Since 1988, the Office for Interdisciplinary Earth Studies of the University Corporation for Atmospheric Research (UCAR) has run a series of annual Global Change Institutes (GCIs) on a range of topics under the broad theme of global environmental change. Participants in each GCI have come from a wide range of disciplines, including those peripheral to the main topic, in order to stimulate discussion and to ensure a multidisciplinary perspective. All GCIs have been highly successful and have led to important and useful proceedings volumes.

The sixth annual Global Change Institute was held over July 18–30, 1993, in Snowmass, Colorado. The topic of the institute was the carbon cycle. As a unique feature, the GCI focused on the practical problems of projecting future concentration changes for given emissions, estimating emissions for prescribed concentration profiles, and assessing the uncertainties in these calculations. Much of the discussion still involved processes, but the viewpoint fostered was as much that of the user of carbon cycle model output as of the "pure" scientist. Over the past decade, there has been a trend toward applied or socially relevant science. With the concern over future climatic change resulting from increasing greenhouse gas concentrations, and with the central role that CO_2 plays in this problem, there is no area of science in which cognizance of the social and policy implications is more important than in carbon cycle research.

In spite of this, the carbon cycle community had not made a coordinated effort prior to the GCI to address the issue of future concentration and emissions changes and their uncertainties. The primary aim of the 1993 GCI was to take a step toward providing realistic information on these problems. The goal was not to produce quantitative answers, since we did not (and still do not, in some cases) even know the right questions to ask. Carbon cycle modeling is somewhat behind climate modeling in this regard – in 1993 there had been no comprehensive intermodel comparisons and few attempts to assess and prioritize the sources of uncertainty. The 1993 GCI provided a forum to discuss these and related issues.

The institute followed its successful format of previous years, holding an initial plenary session during which about 50% of the participants made formal presentations. This was followed by separate working group sessions, a "regrouping" plenary to ensure interaction between the working groups, a breakup into working groups again, and a final plenary.

As the GCI took place prior to the 1994 and 1995 Intergovernmental Panel on Climate Change (IPCC) reports, the working groups provided a forum for the world's leading carbon cycle experts to examine three crucial topics in the IPCC context: the missing carbon sink (how big is it, how can it be explained, and how important is it?); past CO_2 variations (what is the past record, how can it be explained, and how important is it in estimating future changes?); and modeling strategies (what tools are available, what are their respective merits and weaknesses, and how are they best employed to predict future changes?). The ideas emanating from the GCI discussions and working groups provided the foundation for the carbon cycle chapters in the 1994 and 1995 IPCC reports on the carbon cycle, excerpts from which are reproduced in this book (see Chapter 2).

Acknowledgments

The 1993 Global Change Institute and the publication of this volume were made possible by the continuing and generous support from the Subcommittee on Global Change Research of the Interagency Committee on Earth and Environmental Sciences. This subcommittee is supported by the agencies who are members of the U.S. Global Change Research Program.

We are proud to give credit to John A. Eddy as the originator of the Global Change Institutes. Additional thanks are due to the many individuals who gave a special impetus to this project: GCI co-directors Tom Wigley, Robert Watson, and Ulrich Siegenthaler; series co-editors Carol Rasmussen and Tom Wigley; and the members of the GCI's Organizing Committee: William R. Emanuel, Inez Fung, Michael B. McElroy, Berrien Moore III, David S. Schimel, and William H. Schlesinger. Most important, we thank the Office for Interdisciplinary Earth Studies (OIES) staff, whose organizational, logistical, and on-site support made this institute a success: Lisa Butler, Diane Ehret, Stacy Long, and Paula Robinson. Special thanks are due to Diane Ehret, Lisa Butler, and Christina Tidd for their work in managing the production of this book.

Finally, we are indebted to all who came to Snowmass, especially to the rapporteurs for spending countless hours pulling together the material for the working group summaries, which fed directly into the 1994 and 1995 IPCC reports, and to the moderators of the various sessions for guiding the discussions toward the purposes assigned. In addition, we wish to thank the scientists who have taken time to review the manuscripts contained in this volume. The scientific worth of this volume is founded equally on their efforts and the papers they reviewed.

Contributors to the 1993 Global Change Institute

Below are the names and affiliations of the contributors to this book and the participants in the Office for Interdisciplinary Earth Studies (OIES) 1993 Global Change Institute. Each person on this list provided valuable input to this volume either through a specific chapter or by participating in discussions at the Global Change Institute. The editors would like to extend their sincere thanks to these people.

Robert J. Andres
Institute of Northern Engineering
University of Alaska
Box 755910
Fairbanks, AK 99775-5910
USA

Peter Bakwin
Environmental Research Laboratory
National Oceanic and Atmospheric Administration
R/E/CG1, 325 Broadway
Boulder, CO 80303
USA

Harriet Barker
Corporate Affairs
University Corporation for Atmospheric Research
P.O. Box 3000
Boulder, CO 80307-3000
USA

Diana Barnes
Department of Earth and Planetary Sciences
Harvard University
100B Pierce Hall
29 Oxford Street
Cambridge, MA 02138
USA

Damian J. Barrett
CSIRO, Division of Plant Industry
GPO Box 1600
Canberra, ACT 2601
Australia

Steve Bischof
Connecticut College
New London, CT 06320-4196
USA

Thomas A. Boden
Carbon Dioxide Information Analysis Center
Environmental Sciences Division
Oak Ridge National Laboratory
P.O. Box 2008, MS 6335
Oak Ridge, TN 37831-6335
USA

Bert Bolin
IPCC Chairman
Stockholm University
Kvarnásvägen 6
18451 Österskär
Sweden

Georges Bonani
Institut für Mittelenergiephysik
ETH Zurich
CH-8093 Zurich
Switzerland

Rob Braswell
Complex Systems Research Center
University of New Hampshire
462 Morse Hall
Durham, NH 03801
USA

Thomas F. Braziunas
Quaternary Research Center
University of Washington
Box 351360
Seattle, WA 98195-1360
USA

Radford Byerly, Jr.
3870 Birchwood Drive
Boulder, CO 80304
USA

Kenneth G. Caldeira
Climate System Modeling Group
Lawrence Livermore National Laboratory
7000 East Avenue, L-103
Livermore, CA 94550
USA

Stephen Craig
Department of Meteorology
Stockholm University
Arrhenius Laboratory
S-106 91 Stockholm
Sweden

Roger Dahlman
Office of Energy Research
U.S. Department of Energy
Environmental Sciences Division
ER-74
Washington, DC 20585
USA

Robert Dixon
U.S. Department of Energy
1000 Independence Avenue, SW
Mail Stop PO-60
Washington, DC 20585
USA

Bert Drake
Smithsonian Environmental Research Center
P.O. Box 28
Edgewater, MD 21037
USA

James A. Edmonds
Global Change Division
Battelle, Pacific Northwest Laboratory
901 D Street, SW, Suite 900
Washington, DC 20024
USA

William R. Emanuel
Department of Environmental Sciences
University of Virginia
Clark Hall
Charlottesville, VA 22903
USA

Ian G. Enting
CSIRO, Division of Atmospheric Research
Cooperative Research Centre for Southern Hemisphere
Meteorology
Private Mail Bag No. 1
Aspendale, Victoria 3195
Australia

Pierre Friedlingstein
Belgian Institute for Space Aeronomy
3 Avenue Circulaire
B-1180 Brussels
Belgium

Inez Y. Fung
School of Earth and Ocean Sciences
University of Victoria
P.O. Box 1700, MS 4015
Victoria, BC V8W 2Y2
Canada

Roger M. Gifford
CSIRO, Division of Plant Industry
GPO Box 1600
Canberra, ACT 2601
Australia

David O. Hall
Division of Life Sciences
King's College London
University of London
Campden Hill Road
London W8 7AH
United Kingdom

Jennifer W. Harden
U.S. Geological Survey
345 Middlefield Road, MS 962
Menlo Park, CA 94025
USA

Kevin Harrison
Department of Geology and Geophysics
Devlin Hall, Room 213
Boston College
Chesnut Hill, MA 02167-3814
USA

L. D. Danny Harvey
Department of Geography
University of Toronto
100 St. George Street
Toronto, ON M55 3G3
Canada

Ezra Hausman
Department of Earth and Planetary Sciences
Harvard University
Pierce Hall G3G
Cambridge, MA 02138
USA

Martin Heimann
Max-Planck-Institut für Meteorologie
Bundesstrasse 55
D-20146 Hamburg
Germany

Martin I. Hoffert
New York University
Meyer Hall of Physics, Room 503
Mail Code 1013
4 Washington Place
New York, NY 10003-6621
USA

Elisabeth Holland
Atmospheric Chemistry Division
National Center for Atmospheric Research
P.O. Box 3000
Boulder, CO 80307-3000
USA

Richard A. Houghton
Woods Hole Research Center
P.O. Box 296
Woods Hole, MA 02543
USA

Atul K. Jain
Department of Atmospheric Sciences
University of Illinois at Urbana-Champaign
105 S. Gregory Avenue
Urbana, IL 61801
USA

Charles D. Keeling
Geological Research Division
Scripps Institution of Oceanography
University of California at San Diego
Mail Code 0220
La Jolla, CA 92093-0220
USA

Ralph Keeling
Marine Research Division
Scripps Institution of Oceanography
University of California at San Diego
Mail Code 0236
La Jolla, CA 92093-0236
USA

Robin S. Keir
GEOMAR
Research Center for Marine Geosciences
Wischhofstraβe 1-3, Gebäude 12
D-24148 Kiel
Germany

Daniel Lashof
Natural Resources Defense Council
1200 New York Avenue, NW, Suite 400
Washington, DC 20005
USA

Peter S. Liss
School of Environmental Sciences
University of East Anglia
Norwich NR4 7TJ
United Kingdom

Ariel Lugo
USDA Forest Service
Call Box 25000
Rio Piedras, PR 00928-2500
USA

Jason L. Lutze
Ecosystem Dynamics Group
Australian National University
Canberra ACT 0200
Australia

Jean Lynch-Stieglitz
Lamont Doherty Earth Observatory
Columbia University
Route 9W
Palisades, NY 10964
USA

Gregg Marland
Environmental Sciences Division
Oak Ridge National Laboratory
P.O. Box 2008, MS 6335
Oak Ridge, TN 37831-6335
USA

J. Patrick Megonigal
Botany Department
Duke University
Durham, NC 27708
USA

Berrien Moore III
Institute for the Study of Earth, Oceans, and Space
University of New Hampshire
Morse Hall, Room 305
39 College Road
Durham, NH 03824-3525
USA

Dennis S. Ojima
Natural Resource Ecology Laboratory
Colorado State University
Fort Collins, CO 80523
USA

Bradley Opdyke
Department of Geology, The Australian National University
Canberra, ACT 0200
Australia

William J. Parton
Rangeland Ecosystem Science
Colorado State University
Fort Collins, CO 80523
USA

Tsung-Hung Peng
NOAA/AOML Ocean Chemistry Division
4301 Rickenbacker Causeway
Miami, FL 33149-1026
USA

Paul D. Quay
School of Oceanography
University of Washington
Box 357940
Seattle, WA 98195-7940
USA

Dominique Raynaud
Laboratoire de Glaciologie et Géophysique de
l'Environnement
Centre National de la Recherche Scientifique, BP 96
F-38042 Saint-Martin-d'Hères Cedex
France

Richard Richels
Environment Division
Electric Power Research Institute
P.O. Box 10412
3412 Hillview Avenue
Palo Alto, CA 94303
USA

A. B. Samarakoon
CSIRO, Division of Plant Industry
GPO Box 1600
Canberra, ACT 2601
Australia

Jorge L. Sarmiento
Atmospheric and Oceanic Sciences Program
Princeton University
Princeton, NJ 08544-0710
USA

David S. Schimel
Climate and Global Dynamics Division
National Center for Atmospheric Research
P.O. Box 3000
Boulder, CO 80307-3000
USA

William H. Schlesinger
Departments of Botany and Geology
Duke University
P.O. Box 90340
Durham, NC 27708-0340
USA

Tim Schowalter
Entomology Department
Oregon State University
Cord 2032
Corvallis, OR 97331-2032
USA

J. M. O. Scurlock
Environmental Sciences Division
King's College London
Oak Ridge National Laboratory
P.O. Box 2008, MS 6407
Oak Ridge, TN 37831-6407
USA

Albert J. Semtner, Jr.
Department of Oceanography
U.S. Naval Postgraduate School
833 Dyre Road
Monterey, CA 93943-5100
USA

Jeffery P. Severinghaus
Graduate School of Oceanography
University of Rhode Island
Narragansett, RI 02881
USA

Eileen Shea
Institute of Global Environment and Society, Inc.
4041 Powder Mill Road, Suite 302
Calverton, MD 20705
USA

Ulrich Siegenthaler (deceased)
Physics Institute
University of Bern
Sidlerstrasse 5
CH-3012 Bern
Switzerland

Lowell Smith
U.S. Environmental Protection Agency
Room 619 A West Tower, RD682
401 M Street SW
Washington, DC 20460
USA

Thomas M. Smith
Department of Environmental Sciences
University of Virginia
Clark Hall
Charlottesville, VA 22903
USA

Allen Solomon
The Ecological Society of America
U.S. Environmental Protection Agency
200 SW 35th Street
Corvallis, OR 97333
USA

Minze Stuiver
Quaternary Isotope Laboratory
University of Washington
Box 351310
Seattle, WA 98195-1310
USA

Eric T. Sundquist
U.S. Geological Survey
384 Woods Hole Road
Woods Hole, MA 02543-1598
USA

Taro Takahashi
Division of Climate, Ocean and Environment
Lamont Doherty Earth Observatory
Columbia University
Route 9W
Palisades, NY 10964
USA

Leah May B. Ver
Department of Oceanography
University of Hawaii
School of Ocean and Earth Science and Technology
1000 Pope Road, MSB513
Honolulu, HI 96822
USA

Tyler Volk
Department of Applied Science
New York University
34 Stuyvesant Street, Suite 508
New York, NY 10003
USA

Andrew Watson
Plymouth Marine Laboratory
Prospect Place
Plymouth PL1 3DH
United Kingdom

Tom M. L. Wigley
Climate and Global Dynamics Division
National Center for Atmospheric Research
P.O. Box 3000
Boulder, CO 80307-3000
USA

Julie Palmer Winkler
Department of Botany
Duke University
Durham, NC 27708
USA

Marshall A. Wise
Global Change Division
Battelle, Pacific Northwest Laboratories
901 D Street SW, Suite 900
Washington, DC 20024-2115
USA

F. I. Woodward
Department of Animal and Plant Sciences
University of Sheffield
P.O. Box 601
Sheffield S10 2UQ
United Kingdom

Part I

Introduction

Introduction

TOM M. L. WIGLEY AND DAVID S. SCHIMEL

The sixth annual Global Change Institute (GCI) was held in 1993 in Snowmass, Colorado, to evaluate the state of knowledge of the global carbon cycle. As in previous GCIs, an overarching goal was to increase the interdisciplinary communication between scientists in different disciplines. The 1993 GCI focused on those studying the various facets of the carbon cycle, including emissions of carbon dioxide, carbon in the oceans, the role of terrestrial ecosystems and land use, and measurements of carbon dioxide buildup in the atmosphere.

The goal of the institute was in part scientific, and in part to support the then-ongoing assessment of the carbon cycle by the Intergovernmental Panel on Climate Change (IPCC) (Schimel et al., 1995, 1996; Melillo et al., 1996). The IPCC had assessed the state of knowledge concerning the carbon cycle in its 1990 and 1992 reports (Watson et al., 1990, 1992); however, its 1994 and 1995 reports required a more in-depth analysis. The need for greater depth was driven by the 1992 United Nations Framework Convention on Climate Change (FCCC). Article 2 of the FCCC states as a primary objective that countries should seek to stabilize the concentrations of greenhouse gases in the atmosphere in order to stabilize future climate (within the limits of natural variability). Because of the principal role of carbon dioxide (CO_2) in the radiative forcing of climate, the IPCC chose to conduct a more thorough investigation of the relationship between CO_2 emissions and resultant atmospheric concentration changes during a transition to a stabilized atmosphere.

The need for a more extensive analysis of the relationship between emissions and concentrations motivated the IPCC to commission an international modeling exercise to analyze this relationship (this work is reported in Enting et al., 1994 – also known as the KOALA Report). This exercise allowed for more detailed and quantitative scientific analysis of processes and mechanisms regulating the carbon cycle than had been possible in earlier IPCC reports. In this assessment process, the 1993 GCI served a vital role in bringing together many of the principal workers in the field. The sophisticated dialogue that transpired between scientists with widely divergent backgrounds led to substantial synthesis. The chapters in this volume are but a subset of the papers presented at or initiated by that meeting, and many others have already appeared in the literature (e.g., Friedlingstein et al., 1995; Townsend et al., 1996; Schimel et al., 1994; Dai and Fung, 1993; Wigley et al., 1996).

The papers compiled in this book are direct contributions from the 1993 GCI. They, and the discussions they led to, contributed to the IPCC review chapters published in 1995 and 1996. These IPCC chapters are reproduced here both to bring the present volume up to date and to provide a more extensive overview of the literature in general. The science, however, continues to advance at a rapid rate across a broad range of disciplines. In this Introduction, we note several issues discussed at the GCI that were highly relevant to the development of the IPCC reports on the carbon cycle and that remain at the cutting edge of the science. This brief overview is from the editors' perspective; it does not do full justice to the breadth of the field, or even to the conversations that occurred during the GCI.

Both at the GCI and throughout the development of the IPCC reports, there was an exhaustive review of the use of inverse modeling (including related mathematical techniques) to deduce the magnitudes of sources and sinks from atmospheric concentrations (Enting, Chapter 8, this volume; Fung and Takahashi, Chapter 9, this volume). The indication from earlier analyses that the Northern Hemisphere terrestrial ecosystems must be a net sink for CO_2 (Tans et al., 1990; Enting and Mansbridge, 1991) was discussed extensively and evaluated relative to direct inventory assessments of grassland and forest ecosystems (Hall et al., Chapter 7, this volume; Houghton, Chapter 4, this volume; Schlesinger et al., Chapter 6, this volume). This line of work has since seen a major expansion recently, with analyses based on sophisticated general circulation model (GCM) treatments of transport (Denning et al., 1995), carbon isotopes (Ciais et al., 1995a, 1995b), and oxygen concentration (Keeling et al., 1996). The improved understanding of the signals in atmospheric measurements has led to a need for more detailed observational data, reflected in recent proposals for enhanced observing networks, focusing particularly on integration with flux measurements.

A central theme at the GCI was the balance of the carbon budget, and the so-called missing sink. At the time of the 1990 and 1992 IPCC reports, there was known to be an apparent imbalance between the primary sources of CO_2 (industrial and land-use-related emissions; Andres et al., Chapter 3, this volume), the ocean sink (Caldeira et al., Chapter 16, this volume; Stuiver et al., Chapter 12, this volume), and the rate of atmospheric buildup. An additional sink, of order 1–3 GtC/yr, was required to give a balance. The sink was generally attributed to a terrestrial process resulting primarily from CO_2 fertilization of land photosynthesis (Gifford et al., Chapter 5, this volume). However, major uncertainties surrounded (and continue to surround) both the fertilization effect and the other components of the budget.

A significant change occurred in the understanding of the terrestrial carbon cycle in the early 1990s. CO_2 exchange between the land biosphere and the atmosphere was considered to have three components. First was the fertilization effect (Woodward and Smith, Chapter 20, this volume; Gifford et al., Chapter 5, this volume). Second was the "land-use source" term in the budget: the emissions of CO_2 to the atmosphere following land-use conversions such as forest to pasture (Houghton, Chapter 4, this volume). In 1990 and 1992, IPCC assigned the land-use term a best estimate value of 1.6 GtC/yr (averaged over the 1980s). The third component of the annual CO_2 exchange combined the compensating processes of photosynthesis (A) and respiration (R). These processes individually dwarf the land-use and CO_2 fertilization fluxes, each of order 120 GtC. In past analyses, these two fluxes generally had been assumed to balance (e.g., Fung et al., 1987) and thus to have no effect on atmospheric concentrations on annual and longer time scales (although seasonal imbalances create the dramatic seasonal cycle of CO_2 evident in observations).

At the GCI, Inez Fung discussed modeling results indicating that interannual and decadal patterns of climate variability could create interannual and decadal imbalances in the A and R fluxes, leading to important climate-forced source/sink effects over the past 70 or so years (Dai and Fung, 1993). This suggestion has proven to be seminal, leading to a number of new global (Keeling et al., 1995, 1996) and site-specific

(Goulden et al., 1996) analyses and modeling studies (Schimel et al., 1996) of climate-driven interannual variations in terrestrial carbon exchanges. Virtually conclusive evidence now exists for interannual variations in terrestrial carbon storage on the order of a gigaton resulting from climate-driven changes in A and R. Recent long-term records from high-resolution ice cores (Etheridge et al., 1996) and firn (Battle et al., 1996) are beginning to provide a broader perspective on these interactions. These observed interannual changes in terrestrial metabolism are providing important new insights into the mechanisms that may govern the long-term responses of ecosystems to climate (e.g., Schimel et al., 1996; Woodward and Smith, Chapter 20, this volume).

The issue of CO_2 fertilization relative to other terrestrial sinks was central at the GCI (Harvey, Chapter 19, this volume). Although simple empirical fits of the missing sink to atmospheric CO_2 result in estimates of the sensitivity of global carbon uptake to CO_2 that are plausible physiologically, there is evidence for both ecosystem and climatic modulation of this sensitivity (VEMAP Participants, 1995; Gifford et al., Chapter 5, this volume; Woodward and Smith, Chapter 20, this volume). This fact raises two questions. First, what ecosystem processes modulate the well-known leaf-level response of photosynthesis to CO_2 to produce generally lower ecosystem-level carbon storage changes? This problem area was discussed vigorously at the GCI, and important papers on nitrogen feedbacks (VEMAP Participants, 1995), plant population dynamics (Bolker et al., 1995), and the analysis of measurements (Norby, 1996) have since appeared. Second, if, as it appears, ecosystem processes may on balance reduce the CO_2 fertilization effects on carbon storage, what other processes can help to account for the terrestrial uptake estimated from budget studies (Schimel, 1995)? There are two main candidates: the effects of prior land use, now leading to regrowth (Dixon et al., 1994; Houghton, Chapter 4, this volume), and the effects of anthropogenic nitrogen deposition (Schindler and Bayley, 1993; Holland et al., 1995; Townsend et al., 1996). Again, since the GCI, there has been substantial progress in defining and quantifying the roles of these processes.

This process-oriented work paves the way for calculations regarding the stabilization of atmospheric CO_2. At the time of the GCI, initial analyses had been completed by a number of groups. The main features of the emissions changes required to achieve concentration stabilization were beginning to be apparent, but the results had not been explored in depth at that time. Both physical/biological and social science (economics) concepts that impinge on requirements for stabilization were discussed at the GCI in wide-ranging and highly interdisciplinary conversations (see Edmonds et al., Chapter 14, this volume; Andres et al., Chapter 3, this volume, Wigley, Chapter 21, this volume). These conversations have resulted in a number of papers that better define biophysical (Sarmiento et al., 1995; Schimel, 1995) and economic (Edmonds et al., 1995; Manne and Richels, 1995; Richels and Edmonds, 1995; Hourcade et al., 1996; Wigley et al., 1996) questions and prospects. Two Technical Papers of the IPCC have recently been produced that define and explain these stabilization issues for the negotiators in the climate convention process (Schimel et al., Wigley et al., 1997). Clearly, over the next few years, assess-ments of the requirements for and mechanisms to achieve stabilization at a range of concentration levels will be a key factor in determining whether and how the goals of the FCCC can be met (Wigley, Chapter 21, this volume).

The 1993 GCI brought together an unprecedented range of scientists in the area of carbon cycle research, leaving them buoyant mood and eager to tackle the many challenges that were raised, yet the aftermath of the meeting was imbued with sadness: Uli Siegenthaler, our highly regarded colleague, died not long after the workshop. Uli, who attended the meeting in the midst of a long struggle with disease, was one of its most energetic and creative participants. His work over many years has been central to the understanding of the carbon cycle and to this important facet of the IPCC process. His influence continues now and will do so in the future through the legacy of his insightful scientific publications and through the widespread use of his "Bern" carbon cycle model. He is sorely missed.

References

Battle, M., M. Bender, T. Sowers, P. P. Tans, J. H. Butler, J. W. Elkins, J. T. Ellis, T. Conway, N. Zhang, P. Lang, and A. D. Clarke. 1996. Atmospheric gas concentrations over the past century measured in air from firn at the South Pole. *Nature 383*, 231–236.

Bolker, B. M., S. W. Pacala, F. A. Bazzaz, C. D. Canham, and S. A. Levin. 1995. Species diversity and ecosystem response to carbon dioxide fertilization: Conclusions from a temperate forest model. *Global Change Biology 1*, 373–381.

Ciais, P., P. P. Tans, M. Trolier, J. W. C. White, and R. J. Francey. 1995b. A large northern hemisphere terrestrial CO_2 sink indicated by $^{13}C/^{12}C$ of atmospheric CO_2. *Science 269*, 1098–1102.

Ciais, P., P. Tans, J. W. C. White, M. Trolier, R. J. Francey, J. Berry, D. Randall, P. Sellers, J. G. Collatz, and D. S. Schimel. 1995a. Partitioning of ocean and land uptake of CO_2 as inferred by $\delta^{13}C$ measurements from the NOAA/Climate Monitoring and Diagnostics Laboratory Global Air Sampling Network. *Journal of Geophysical Research 100D*, 5051–5070.

Dai, A., and I. Y. Fung. 1993. Can climate variability contribute to the "missing" CO_2 sink? *Global Biogeochemical Cycles 7*, 599–609.

Denning, A. S., I. Y. Fung, and D. Randall. 1995. Latitudinal gradient of atmospheric CO_2 due to seasonal exchange with land biota. *Nature 376*, 240–243.

Dixon, R. K., S. A. Brown, R. A. Houghton, A. M. Solomon, M. C. Trexler, and J. Wisniewski. 1994. Carbon pools and flux of global forest ecosystems. *Science 263*, 185–190.

Edmonds, J., D. Barns, M. Wise, and M. Ton. 1995. Carbon coalitions: The cost and effectiveness of energy agreements to alter trajectories of atmospheric carbon dioxide emissions. *Energy Policy 23*, 309–336.

Enting, I. G., and J. V. Mansbridge. 1991. Latitudinal distribution of sources and sinks of CO_2: Results of an inversion study. *Tellus 43B*, 156–170.

Enting, I. G., T. M. L. Wigley, and M. Heimann. 1994. *Future Emissions and Concentrations of Carbon Dioxide: Key Ocean/Atmosphere/Land Analyses*. Technical Paper 31, CSIRO Division of Atmospheric Research, Mordialloc, Australia, 120 pp.

Etheridge, D. M., L. P. Steele, R. L. Langenfelds, R. J. Francey, J.-M. Barnola, and V. I. Morgan. 1996. Natural and anthropogenic changes in atmospheric CO_2 over the last 1000 years from air in Antarctic ice and firn. *Journal of Geophysical Research 101*, 4115.

Friedlingstein, P., I. Fung, E. A. Holland, J. John, G. Brasseur, D. Erickson, and D. S. Schimel. 1995. On the contribution of CO_2 fertilization to the missing biospheric sink. *Global Biogeochemical Cycles 9*, 541–556.

Fung, I. Y., C. J. Tucker, and K. C. Prentice. 1987. Application of advanced very high resolution radiometer to study atmosphere-biosphere exchange of CO_2. *Journal of Geophysical Research 92*, 2999–3015.

Goulden, M. L., J. W. Munger, S.-M. Fan, B. C. Daube, and S. C. Wofsy. 1996. Exchange of carbon dioxide by a deciduous forest: response to interannual climate variability. *Science 271*, 1576–1578.

Holland, E. A., A. R. Townsend, and P. M. Vitousek. 1995. Variability in temperature regulation of CO_2 fluxes and N mineralization from five Hawaiian soils: Implication for a changing climate. *Global Change Biology 1*, 115–123.

Hourcade, J. C., K. Halsnaes, M. Jaccard, W. D. Montgomery, R. Richels, J. Robinson, P. R. Shukla, P. Sturm, W. Chandler, O. Davidson, J. Edmonds, D. Finon, K. Hogan, F. Krause, A. Kolesov, E. La Rovere, P. Nastari, A. Pegov, K. Richards, L. Schrattenholzer, R. Shackleton, Y. Sokona, A. Tudini, and J. Weyant. 1996. A review of mitigation cost studies. In *Climate Change 1995: Economic and Social Dimensions of Climate Change, Contribution of Working Group III to the Second Assessment Report of the Intergovernmental Panel on Climate Change*, J. P. Bruce, H. Lee, and E. F. Haites, eds., Cambridge University Press, Cambridge, England, 297–366.

Keeling, C. D., J. F. S. Chin, and T. P. Whorf. 1996. Increased activity of northern vegetation inferred from atmospheric CO_2 measurements. *Nature 382*, 146–149.

Keeling, C. D., T. P. Whorf, M. Wahlen, and J. van der Plicht. 1995. Interannual extremes in the rate of rise of atmospheric carbon dioxide since 1980. *Nature 375*, 666–670.

Keeling, C. D., J. F. S. Chin, and T. P. Whorf. 1996. Increased activity of northern vegetation inferred from atmospheric CO_2 measurements. *Nature 382*, 146–149.

Manne, A., and R. Richels. 1995. *The Greenhouse Debate – Economic Efficiency, Burden Sharing and Hedging Strategies*. Working paper. Stanford University, Stanford, California.

Melillo, J. M., I. C. Prentice, G. D. Farquhar, E.-D. Schulze, and O. E. Sala. 1996. Terrestrial biotic responses to environmental change and feedbacks to climate. In *Climate Change 1995: The Science of Climate Change, Contribution of Working Group I to the Second Assessment Report of the Intergovernmental Panel on Climate Change*, J. T. Houghton, L. G. Meira Filho, B. A. Callander, N. Harris, A. Kattenberg, and K. Maskell, eds., Cambridge University Press, New York, pp. 444–481.

Norby, R. J. 1996. Forest canopy productivity index. *Nature 381* (letter), 564.

Richels, R., and J. Edmonds. 1995. The economics of stabilizing atmospheric CO_2 concentrations. *Energy Policy 23*, 373–379.

Sarmiento, L. J., C. Le Quere, and S. W. Pacala. 1995. Limiting future atmospheric carbon dioxide. *Global Biogeochemical Cycles 9*, 121–137.

Schimel, D. S. 1995. Terrestrial ecosystems and the carbon cycle. *Global Change Biology 1*, 77–91.

Schimel, D. S., D. Alves, I. G. Entng, M. Heimann, F. Joos, D. Raynaud, and T. M. L. Wigley. 1996. CO_2 and the carbon cycle. In *Climate Change 1995: The Science of Climate Change, Contribution*

of Working Group I to the Second Assessment Report of the Intergovernmental Panel on Climate Change, J. T. Houghton, L. G. Meira Filho, B. A. Callander, N. Harris, A. Kattenberg, and K. Maskell, eds., Cambridge University Press, New York, pp. 65–86.

Schimel, D. S., B. H. Braswell Jr., E. A. Holland, R. McKeown, D. S. Ojima, T. H. Painter, W. J. Parton, and A. R. Townsend. 1994. Climatic, edaphic, and biotic controls over carbon and turnover of carbon in soils. *Global Biogeochemical Cycles 8*, 279–293.

Schimel, D. S., B. H. Braswell, R. McKeown, D. S. Ojima, W. J. Parton, W. Pulliam. 1996. Climate and nitrogen controls on the geography and time scales of terrestrial biogeochemical cycling. *Global Biogeochemical Cycles*, pp. 677–692.

Schimel, D., I. G. Enting, M. Heimann, T. M. L. Wigley, D. Raynaud, D. Alves, and U. Siegenthaler. 1995. CO_2 and the carbon cycle. In *Climate Change 1994: Radiative Forcing of Climate Change and an Evaluation of the IPCC IS92 Emission Scenarios* J. T. Houghton, L. G. Meira Filho, J. Bruce, H. Lee, B. A. Callander, E. Haites, N. Harris, and K. Maskell, eds., Cambridge University Press, Cambridge, England, 35–71.

Schimel, D. S., M. Grubb, F. Joos, R. K. Kaufmann, R. Moss, W. Ogana, R. Richels, and T. M. L. Wigley. 1997. *Stabilization of Atmospheric Greenhouse Gases: Physical, Biological and Socio-economic Implications: IPCC Technical Paper 3*, J. T. Houghton, L. G. Meira Filho, D. J. Griggs, and M. Noguer, eds. Intergovernmental Panel on Climate Change, Geneva, Switzerland, 52 pp.

Schindler, D. W., and S. E. Bayley. 1993. The biosphere as an increasing sink for atmospheric carbon: Estimates from increased nitrogen deposition. *Global Biogeochemical Cycles 7*, 717–734.

Tans, P. P., I. Y. Fung, and T. Takahashi. 1990. Observational constraints on the global atmospheric CO_2 budget. *Science 247*, 1431–1438.

Townsend, A. R., B. H. Braswell, E. A. Holland, and J. E. Penner. 1996. Spatial and temporal patterns in terrestrial carbon storage due to deposition of fossil fuel nitrogen. *Ecological Applications 6*, 806–814.

VEMAP Participants. 1995. Vegetation/Ecosystem modeling and analysis project (VEMAP): Comparing biogeography and biogeochemistry models in a continental-scale study of terrestrial ecosystem responses to climate change and CO_2 doubling. *Global Biogeochemical Cycles 9*, 407–438.

Watson, R. T., L. G. Meira Filho, E. Sanhueza, and A. Janetos. 1992. Sources and sinks. In *Climate Change 1992: The Supplementary Report to the IPCC Scientific Assessment*, J. T. Houghton, B. A. Callander, and S. K. Varney, eds., Cambridge University Press, Cambridge, England, 25–46.

Watson, R. T., H. Rodhe, H. Oeschger, and U. Siegenthaler. 1990. Greenhouse gases and aerosols. In *Climate Change: The IPCC Scientific Assessment* (J. T. Houghton, G. J. Jenkins, and J. J. Ephraums, eds.), Cambridge University Press, Cambridge, U. K., 1–40.

Wigley, T. M. L., A. Jain, F. Joos, P. R. Shukla, and B. S. Nyenzi. 1997. *Implications of Proposed CO_2 Emissions Limitations: IPCC Technical Paper 4*, J. T. Houghton, L. G. Meira Filho, D. J. Griggs, and M. Noguer, eds. Intergovernmental Panel on Climate Change, Geneva, Switzerland, 41 pp.

Wigley, T. M. L., R. Richels, and J. A. Edmonds. 1996. Economic and environmental choices in the stabilization of atmospheric CO_2 concentrations. *Nature 379*, 240–243.

1

CO_2 and the Carbon Cycle

(Extracted from the Intergovernmental Panel on Climate Change (IPCC) Report, "Climate Change, 1994")

D. SCHIMEL, I. G. ENTING, M. HEIMANN, T. M. L. WIGLEY,
D. RAYNAUD, D. ALVES, U. SIEGENTHALER

Contributors:
S. Brown, W. Emanuel, M. Fasham, C. Field, P. Friedlingstein, R. Gifford,
R. Houghton, A. Janetos, S. Kempe, R. Leemans, E. Maier-Reimer, G. Marland,
R. McMurtrie, J. Melillo, J.-F. Minster, P. Monfray, M. Mousseau, D. Ojima,
D. Peel, D. Skole, E. Sulzman, P. Tans, I. Totterdell, P. Vitousek

Modellers:
J. Alcamo, B. H. Braswell, B. C. Cohen, W. R. Emanuel, I. G. Enting,
G. D. Farquhar, R. A. Goldstein, L. D. D. Harvey, M. Heimann, A. Jain, F. Joos,
J. Kaduk, A. A. Keller, M. Krol, K. Kurz, K. R. Lassey, C. Le Quere, J. Lloyd,
E. Meier-Reimer, B. Moore III, J. Orr, T. H. Peng, J. Sarmiento, U. Siegenthaler,
J. A. Taylor, J. Viecelli, T. M. L. Wigley, D. Wuebbles.

Summary

Interest in the carbon cycle has increased because of the observed increase in levels of atmospheric CO_2 (from ~280 ppmv in 1800 to ~315 ppmv in 1957 to ~356 ppmv in 1993) and because the signing of the UN Framework Convention on Climate Change has forced nations to assess their contributions to sources and sinks of CO_2, and to evaluate the processes that control CO_2 accumulation in the atmosphere. Over the last few years, our knowledge of the carbon cycle has increased, particularly in the quantification and identification of mechanisms for terrestrial exchanges, and in the preliminary quantification of feedbacks.

The Increase in Atmospheric CO_2 Concentration Since Pre-Industrial Times

Atmospheric levels of CO_2 have been measured directly since 1957. The concentration and isotope records prior to that time consist of evidence from ice cores, moss cores, packrat middens, tree rings, and the isotopic measurements of planktonic and benthic foraminifera. Ice cores serve as the primary data source because they provide a fairly direct and continuous record of past atmospheric composition. The ice cores indicate that an increase in CO_2 level of about 80 ppmv paralleled the last interglacial warming. There is uncertainty over whether changes in CO_2 levels as rapid as those of the 20th century have occurred in the past. However, there is essentially no uncertainty that for approximately the last 18,000 years, CO_2 concentrations in the atmosphere have fluctuated around 280 ppmv, and that the recent increase to a concentration of ~356 ppmv, with a current rate of increase of ~1.5 ppmv/yr, is due to combustion of fossil fuel, cement production, and land use conversion.

The Carbon Budget

The major components of the anthropogenic perturbation to the atmospheric carbon budget are anthropogenic emissions, the atmospheric increase, ocean exchanges, and terrestrial exchanges. Emissions from fossil fuels and cement production averaged 5.5 ± 0.5 GtC/yr over the decade of the 1980s (estimated statistically). The measured average annual rate of atmospheric increase in the 1980s was 3.2 ± 0.2 GtC/yr. Average ocean uptake during the decade has been estimated by a combination of modelling and isotopic measurements to be 2.0 ± 0.8 GtC/yr.

Averaged over the 1980s, terrestrial exchanges include a tropical source of 1.6 ± 1.0 GtC/yr from ongoing changes in land use, based on land clearing rates, biomass inventories, and modelled forest regrowth. Recent satellite data have reduced uncertainties in the rate of deforestation for the Amazon, but rates for the rest of the tropics remain poorly quantified. For the tropics as a whole, there is incomplete information on initial biomass and rates of regrowth. Potential terrestrial sinks may be the result of several processes, including the regrowth of mid-latitude and high latitude Northern Hemisphere forests (0.5 ± 0.5 GtC/yr), enhanced forest growth due to CO_2 fertilisation (0.5–2.0 GtC/yr) and nitrogen deposition (0.2–1.0 GtC/yr),

and, possibly, response to climatic anomalies (0–1.0 GtC/yr). Partitioning the sink among these processes is difficult, but it is likely that all components are involved. While the CO_2 fertilisation effect is the most commonly cited terrestrial uptake mechanism, existing model studies indicate that the magnitude of contributions from each process is comparable, within large ranges of uncertainty. For example, some model-based evidence suggests that the magnitude of the CO_2 fertilisation effect is limited by interactions with nutrients and other ecological processes. Experimental confirmation from ecosystem-level studies, however, is lacking. As a result, the role of the terrestrial biosphere in controlling future atmospheric CO_2 concentrations is difficult to predict.

Future Atmospheric CO_2 Concentrations

Modelling groups from many countries were asked to use published carbon cycle models to evaluate the degree to which CO_2 concentrations in the atmosphere might be expected to change over the next several centuries, given a standard set of emission scenarios (including changes in land use). Models were constrained to balance the carbon budget and match the atmospheric record of the 1980s via CO_2 fertilisation of the terrestrial biosphere.

Stabilisation of Atmospheric CO_2 Concentrations

Modelling groups carried out stabilisation analyses to explore the relationships between anthropogenic emissions and atmospheric concentrations. The analyses assumed arbitrary concentration profiles (i.e., routes to stabilisation) and final stable CO_2 concentration; the models were then used to perform a series of inverse calculations (i.e., to derive CO_2 emissions given CO_2 concentrations). These calculations did three things: (1) assessed the total amount of fossil carbon that has been released (because land use was prescribed), (2) determined the partitioning of this carbon between the ocean and the terrestrial biosphere, and (3) ascertained what the time course of carbon emissions from fossil fuel combustion must have been to arrive at the selected arbitrary atmospheric CO_2 concentrations while still matching the atmospheric record through the 1980s.

Results suggest that in order for atmospheric concentrations to stabilise below 750 ppmv, anthropogenic emissions must eventually decline relative to today's levels. All emissions curves derived from the inverse calculations show periods of increasing anthropogenic emissions, followed by reductions to about a third of today's levels (i.e., to ~2 GtC/yr) for stabilisation at 450 ppmv by the year 2100, and to about half of current levels (i.e., ~3 GtC/yr) for stabilisation at 650 ppmv by the year 2200. Additionally, the models indicated that if anthropogenic emissions are held constant at 1990 levels, modelled atmospheric concentrations of CO_2 will continue to increase over the next century.

Among models the range of emission levels that were estimated to result in the hypothesised stabilisation levels is about 30%. In addition, the range of uncertainty associated with the parametrization of CO_2 fertilisation (evaluated with one of the models) varied between ±10% for low stabilisation values and ±15% for higher stabilisation values. The use of CO_2 fertilisation to control terrestrial carbon storage, when in fact other ecological mechanisms are likely involved, results in an underestimate of concentrations (for given emissions) of 5 to 10% or an overestimate of emissions by a similar amount (for given concentrations).

Feedbacks to the Carbon Cycle

Climate and other feedbacks via the oceans and terrestrial biosphere have the potential to be significant in the future. The effects of temperature on chemical and biological processes in the ocean are thought to be small (tens of ppmv changes in the atmosphere), but the effects of climate on ocean circulation could be larger, with possible repercussions for atmospheric concentration of ±100–200 ppmv. Effects of changing precipitation, temperature, and atmospheric CO_2 can also have effects on the terrestrial biosphere, resulting in feedbacks to the atmosphere. Models suggest transient losses of about 200 GtC from terres-

trial ecosystems as temperatures warm, with a potential for long-term increases in carbon storage above present levels by a few hundred gigatons. Patterns of changing land use will have a substantial effect on terrestrial carbon storage and decrease the potential of terrestrial systems to store carbon in response to CO_2 and climate. Representation of feedbacks on the carbon cycle through oceanic and terrestrial mechanisms needs to be improved in subsequent analyses of future changes to CO_2.

1.1 Description of the Carbon Cycle

Atmospheric CO_2 provides a link between biological, physical, and anthropogenic processes. Carbon is exchanged between the atmosphere, the oceans, the terrestrial biosphere, and, more slowly, with sediments and sedimentary rocks. The faster components of the cycle are shown in Figure 1.1. In the absence of anthropogenic CO_2 inputs, the carbon cycle had pe-

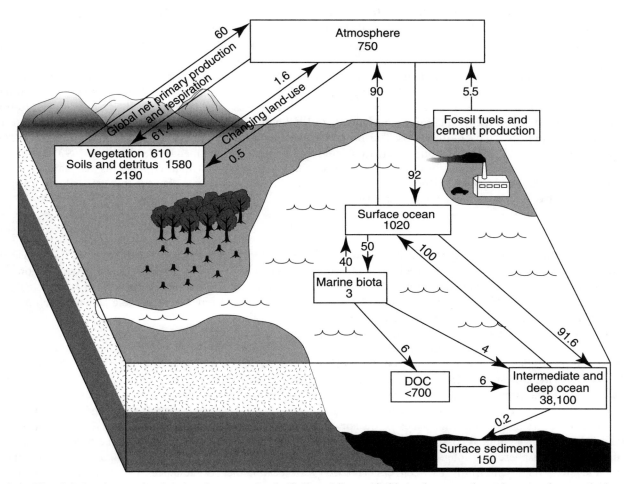

Figure 1. 1: The global carbon cycle, showing the reservoirs (in GtC) and fluxes (GtC/yr) relevant to the anthropogenic perturbation as annual averages over the period 1980 to 1989 (Eswaran et al., 1993; Potter et al., 1993; Siegenthaler and Sarmiento, 1993). The component cycles are simplified and subject to considerable uncertainty. In addition, this figure presents average values. The riverine flux, particularly the anthropogenic portion, is currently very poorly quantified and so is not shown here (see text). While the surface sediment storage is approximately 150 Gt, the amount of sediment in the bioturbated and potentially active layer is of order 400 Gt. Evidence is accumulating that many of the key fluxes can fluctuate significantly from year to year (terrestrial sinks and sources: INPE, 1992; Ciais et al., submitted; export from the marine biota: Wong et al., 1993). In contrast to the static view conveyed by figures such as this one, the carbon system is clearly dynamic and coupled to the climate system on seasonal, interannual and decadal time-scales (Schimel and Sulzman, 1994; Keeling and Whorf, 1994).

RECENT ANOMALIES

The last few decades have been characterised by a number of observed changes in the carbon cycle:

- The early 1980s were characterised by a period of relatively constant or slightly declining fossil carbon emissions. After 1985, emissions again exceeded the 1979 level; each year's release during the latter half of the 1980s was 0. 1 to 0.2 GtC above that for the previous year (Boden *et al.*, 1991).

- Direct measurements and ice core data have revealed a general decrease in atmospheric levels of ^{13}C relative to ^{12}C by about 1‰ over the last century (Friedli *et al.*, 1986; Keeling *et al.*, 1989a; Leuenberger *et al.*, 1992). This decrease is expected from the addition of fossil and/or terrestrial biospheric carbon, both of which are poor in ^{13}C relative to the atmosphere. In contrast, the atmospheric $^{13}C/^{12}C$ ratio remained nearly constant from 1988 to 1993. This constant ratio must reflect changes in the fluxes between the ocean, terrestrial biosphere, and atmosphere which currently are not quantified. The atmospheric record of ^{13}C shows a decrease of ~0. 4‰ between the first measurements in 1978 and the present time. The ice core record provides information on the changing ^{13}C levels over the past few hundred years (Friedli *et al.*, 1986; Leuenberger *et al.*, 1992).

- Relative to the long-term average rate of atmospheric CO_2 concentration increase (~1. 5 ppmv/yr), the years 1988 to 1989 had relatively high CO_2 concentration growth rates (~2. 0 ppmv/yr) while subsequent years (1991 to 1992) had very low growth rates (0. 5 ppmv/yr) (Boden *et al.*, 1991). The magnitude of the 1988 to 1989 anomaly depends on what is defined as "normal" for the long-term trend. At Mauna Loa, Hawaii, the 1988 to 1989 increase was similar to a variation which occurred in 1973 to 1974 (see Figure 1. 2). The subsequent decrease exceeds any previous anomaly since the Mauna Loa record began in 1958. Data for 1993 indicate a higher growth rate than that for 1991 to 1992.

- The CO_2 record exhibits a seasonal cycle, with small peak-to-peak amplitude (about 1 ppmv) in the Southern Hemisphere but increasing northward to about 15 ppmv in the boreal forest zone (55–65°N). This cycle is mainly caused by the seasonal uptake and release of atmospheric CO_2 by terrestrial ecosystems. Part of the seasonal signal is driven by oceanic processes (Heimann *et al.*, 1989). The amplitude of the seasonal atmospheric CO_2 cycle varies with time. For instance, at Mauna Loa (see Figure 1. 3; Keeling *et al.*, 1989a), it was roughly constant at 5. 2 ppmv (peak-to-trough) from the beginning of the Mauna Loa measurements in 1958 until the mid-1970s. It then increased over the late 1970s to reach 5. 8 ppmv for most of the 1980s. The most recent data indicate a further increase. Because the trend is not well-correlated with the CO_2 concentration increase (Thompson *et al.*, 1986; Enting, 1987; Manning, 1993), it provides at best weak evidence for CO_2 fertilisation of terrestrial vegetation, in contrast with interpretations that claimed strong evidence (e. g., Idso and Kimball, 1993). The variation in amplitude does indicate changes in terrestrial metabolism, but not necessarily increased photosynthesis or storage.

riods of millennia in which large carbon exchanges were in near balance, implying nearly constant reservoir contents. Human activities have disturbed this balance through the use of fossil carbon and disruption of terrestrial ecosystems. The consequent accumulation of CO_2 in the atmosphere has caused a number of carbon cycle exchanges to become unbalanced. Fossil fuel burning and cement manufacture, together with forest harvest and other changes of land use, all transfer carbon (mainly as CO_2) to the atmosphere. This anthropogenic carbon then cycles between the atmosphere, oceans, and the terrestrial biosphere. Because the cycling of carbon in the terrestrial and ocean biosphere occurs slowly, on time-scales of decades to millennia, the effect of additional fossil and biomass carbon injected into the atmosphere is a long-lasting disturbance of the carbon cycle. The relationships between concentration changes and emissions of CO_2 are examined through use of models that simulate, in a simplified manner, the major processes of the carbon cycle. The terrestrial and oceanic components of carbon cycle models vary in complexity from a few key equations to spatially explicit, detailed descriptions of ocean and terrestrial biology, chemistry, and transport processes. The simpler models, in general, are designed to reproduce observed behaviour while the more complex models

are aimed at incorporating the processes that cause the observed behaviour. The latter are thus potentially more likely to yield realistic projections of changes in storage under conditions different from the present (e.g., changing climate).

1.2 Time-Scales

Because CO_2 added to the atmosphere by anthropogenic processes is exchanged between reservoirs having a range of turnover times, it is not possible to define a single atmospheric "lifetime." This is in contrast with other anthropogenic compounds such as N_2O and the halogens that are destroyed chemically in the atmosphere. Because the time-scales involved in the CO_2 exchanges range from annual to millennial (thousands of years), the consequences of anthropogenic perturbations will be long-lived. In this regard, the "turnover time" of about 5 years for atmospheric CO_2 deduced from the rate of bomb $^{14}CO_2$ removal is relevant to the initial response of the carbon system, but it does not characterise the much slower, long-term response of atmospheric concentrations to the anthropogenic perturbation.

The remainder of this chapter reviews what is known about the carbon cycle as a basis for understanding past changes and

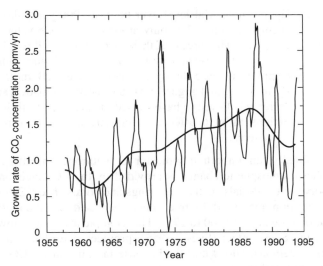

Figure 1. 2: The growth rate of CO_2 concentrations since 1958 (from the Mauna Loa record). The high growth rates of the late 1980s, the extremely low growth rates of the early 1990s, and the recent increase in the growth rate are all evident. The smooth curve shows the same data but filtered to suppress any variations on time-scales less than approximately 10 years.

relationships between future emissions and concentrations. We do not attempt to make specific predictions of likely future changes in CO_2 concentration – rather, we assess the sensitivity of the system to particular scenarios of future emissions and concentrations. We also analyse key areas in which quantitative understanding is deficient. The final section of the chapter presents the results of a set of calculations relating future CO_2 concentrations and future CO_2 emissions. These calculations, produced by modelling groups from many countries following an agreed set of specifications, explored various aspects of uncertainty.

1.3 Past Record of Atmospheric CO_2

1.3.1 Atmospheric Measurements Since 1958

Precise, direct measurements of atmospheric CO_2 started in 1957 at the South Pole, and in 1958 at Mauna Loa, Hawaii. At this time the atmospheric concentration was about 315 ppmv and the rate of increase was ~0.6 ppmv/yr. The growth rate of atmospheric concentrations at Mauna Loa has generally been increasing since 1958. It averaged 0.83 ppmv/yr during the 1960s, 1.28 ppmv/yr during the 1970s, and 1.53 ppmv/yr during the 1980s. In 1992, the atmospheric level of CO_2 at Mauna Loa was 355 ppmv (Figure 1.3) and the growth rate fell to 0.5 ppmv/yr (see "Recent Anomalies" list). Data from the Mauna Loa station are close to, but not the same as, the global mean.

Atmospheric concentrations of CO_2 have been monitored for shorter periods at a large number of atmospheric stations around the world (e.g., Boden *et al.*, 1991). Measurement sites are distributed globally and include sites in Antarctica, Australia, several maritime islands, and high northern latitude sites,

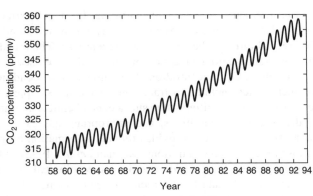

Figure 1. 3: CO_2 concentrations measured at Mauna Loa, Hawaii since 1958 showing trends and seasonal cycle.

but, at present, nowhere on the continents of Africa or South America. The reliability and high precision of the post-1957 record is guaranteed by comparing the measured concentration of CO_2 in air with the concentration of reference gas mixtures calibrated by a constant volume column manometer. The increase shown by the atmospheric record since 1957 can be attributed largely to anthropogenic emissions of CO_2, although considerable uncertainty exists as to the mechanisms involved. The record itself provides important insights that support anthropogenic emissions as a source of the observed increase. For example, when seasonal and short-term interannual variations in concentrations are neglected, the rise in atmospheric CO_2 is about 50% of anthropogenic emissions (Keeling *et al.*, 1989b) with the inter-hemispheric difference growing in parallel to the growth of fossil emissions (Keeling *et al.*, 1989a; Siegenthaler and Sarmiento, 1993; Figure 1.4).

1.3.2 Pre-1958 Atmospheric Measurements and CO_2–Ice Core Record over the Last Millennium

While several sets of relatively precise atmospheric measurements of CO_2 were carried out as early as the 1870s (e.g., Brown and Escombe, 1905), they did not allow assessment of concentration trends, as they were neither adequately calibrated nor temporally continuous.

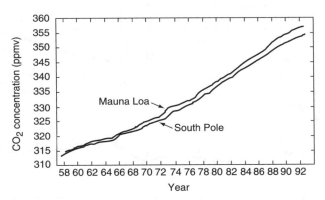

Figure 1. 4: Trends in CO_2 concentration and the growing difference in concentration between the Northern and Southern Hemispheres.

Measurements of CO_2 concentration from air extracted from polar ice cores are presently the best means to extend the CO_2 record through the geologically recent past. The transformation of snow into ice traps air bubbles which are used to determine the CO_2 concentration. Providing certain conditions are met, which include no fracturing of the ice samples, absence of seasonal melting at the surface, no chemical alteration of the initial concentrations, and appropriate gas extraction methods, the ice record provides reliable information on past atmospheric CO_2 concentrations (Raynaud *et al.*, 1993). It has been suggested that, under certain meteorological circumstances, the CO_2 data from Greenland ice cores may be contaminated, apparently influenced by varying levels of carbonate dust interacting with acid (Delmas, 1993) or organic matter deposition onto the ice sheet. Antarctic ice, however, has been uniformly acidic throughout the complete range of climate regimes of the last climatic cycle, and the available evidence suggests that these data are reliable throughout the entire record. Data from appropriate sites with unfractured ice are reliable to within ±3-5 ppmv (Raynaud *et al.*, 1993). The recent ice core record is validated by comparison with direct atmospheric measurements (Neftel *et al.*, 1985; Friedli *et al.*, 1986; Keeling *et al.*, 1989a).

Several high resolution Antarctic ice cores have recently become available in addition to the Siple core (Neftel *et al.*, 1985; Friedli *et al.*, 1986) for documenting both the "industrial era" CO_2 levels and the pre-industrial levels over the last millennium (Figure 1.5). The main results are:

- The ice core record can be used in combination with the direct atmospheric record to estimate, in conjunction with an oceanic model, the net changes in CO_2 flux between terrestrial ecosystems and the atmosphere (Siegenthaler and Oeschger, 1987).

- The pre-industrial level over the last 1000 years shows fluctuations up to 10 ppmv around an average value of 280 ppmv. The largest of these, which occurred roughly between AD 1200 and 1400, was small compared to the 75 ppmv increase during the industrial era (Barnola *et al.*, in press, and Figure 1.5). Short-term climatic variability is believed to have caused the pre-industrial fluctuations through effects on oceanic and/or terrestrial ecosystems.

Finally, an important indicator of anthropogenically induced atmospheric change is provided by the ¹⁴C levels preserved in materials such as tree rings and corals. The ¹⁴C concentration measured in tree rings decreased by about 2% during the period 1800 to 1950. This isotopic decrease, known as the Suess effect (Suess, 1955), provides one of the most clear demonstrations that the increase in atmospheric CO_2 is due to fossil inputs.

1.3.3 The CO_2 Record over the Last Climatic Cycle

Although the magnitude and rate of climate changes observed in the palaeo-record covering the last climatic cycle may differ from those involved in any future greenhouse warming, these records provide an important perspective for recent and potential future changes. The glacial-interglacial amplitudes of temperature change are of similar order to the high estimate of equilibrium temperature shifts predicted for a doubling of CO_2 levels (Mitchell *et al.*, 1990), although the shifts of the past took thousands of years.

The close association between CO_2 and temperature changes during glacial-interglacial transitions was first revealed by data from the ice core record. Samples from Greenland and Antarctica representing the last glacial maximum

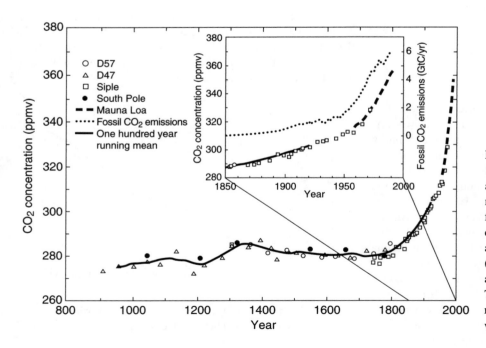

Figure 1.5: CO_2 concentrations over the past 1000 years from the recent ice core record and (since 1958) from the Mauna Loa measurement site. The inset shows the period from 1850 in more detail including CO_2 emissions from fossil fuel. Data sources: D47 and D57 (Barnola et al., in press); Siple (Neftel et al., 1985 and Friedli et al., 1986) and South Pole (Siegenthaler et al., 1988). The smooth curve is based on a 100yr running mean. All ice core measurements were taken in Antarctica.

(about 18,000 years before present) indicate that CO_2 concentrations at that time were 190–200 ppmv, i.e., about 70% of the pre-industrial level (Delmas *et al.*, 1980; Neftel *et al.*, 1982). Thus, an increase of about 80 ppmv occurred in parallel with the warming starting at the end of the glacial period, when the estimated glacial-interglacial rise of the mean surface temperature of the Earth rose by 4°C over about 10,000 years (Crowley and North, 1991). These discoveries have since been confirmed by detailed measurements of the Antarctic Byrd core for the 8,000 to 50,000 year BP period (Neftel *et al.*, 1988). Analyses of ice cores from Vostok, Antarctica, have provided new data on natural variations of CO_2 levels over the last 220,000 years (Barnola *et al.*, 1987, 1991; Jouzel *et al.*, 1993). The record shows a marked correlation between Antarctic temperature, as deduced from the isotopic composition of the ice, and the CO_2 profile (Figure 1.6).

Clear correlations between CO_2 and global mean temperature are evident in much of the glacial-interglacial palaeo-record. This relationship of CO_2 concentration and temperature may carry forward into the future, possibly causing a significant positive climatic feedback on CO_2 fluxes. Information about leads and lags between climatic variations and changes in radiatively active trace gas concentrations is contained in polar ice and deep sea sediment records (Raynaud and Siegenthaler, 1993). There is no evidence that CO_2 changes ever significantly (> 1 kyr) preceded the Antarctic temperature signal. In contrast, CO_2 changes clearly lag behind the Antarctic cooling at the end of the last interglacial. As temperature changes in the South generally preceded temperature changes in the North (CLIMAP Project Members, 1984), it cannot be assumed that CO_2 changes never led northern temperature changes. Comparison of atmospheric CO_2 concentration and continental ice volume suggests that CO_2 started to change ahead of any significant melting of continental ice. It is possible that CO_2 changes may have been caused by changes in climate, and that CO_2 and other trace gases acted to amplify palaeoclimatic changes.

Changes in climate on time-scales of decades to centuries have occurred in the past. The question remains whether these changes have been accompanied by changes in greenhouse trace gas concentrations. The Greenland ice cores (Johnsen *et al.*, 1992; Grootes *et al.*, 1993) show that during the last ice age and the last glacial-interglacial transition, there was a series of rapid (over decades to a century) and apparently large climatic changes in the North Atlantic region (~5 to 7°C in Central Greenland: Johnsen *et al.*, 1992; Dansgaard *et al.*, 1993). These changes may have been global in scale: the methane record suggests the potential for parallel changes in the tropics (Chappellaz *et al.*, 1993). Evidence for rapid climate oscillations during the last interglacial has also recently been reported (GRIP Project Members, 1993). However, because the details were not confirmed by a second core retrieved from the same area (Grootes *et al.*, 1993), the possibility that these features were caused by ice-flow perturbations cannot be discounted. The Dye 3 ice core from Greenland indicates that CO_2 concentration shifts of ~50 ppmv occurred within less than 100 years during the last glacial period (Stauffer *et al.*, 1984). These changes in CO_2 were paralleled by abrupt and drastic climatic events in this region. Such large and rapid CO_2 changes have not been identified in Antarctica, even after accounting for the lower time resolution of the Antarctic core (Neftel *et al.*, 1988). It is possible that impurities have introduced artefacts into the Greenland CO_2 record (Delmas, 1993; Barnola *et al.*, in press), and that these features do not represent real atmospheric CO_2 changes. Nevertheless, taken together with independent support from isotopic studies of mosses (White *et al.*, 1994), rapid CO_2 events recorded during the past cannot be disregarded.

1.4 The Anthropogenic Carbon Budget

1.4.1 Introduction

The phrase "carbon budget" refers to the balance between sources and sinks of CO_2 in the atmosphere, expressed in terms of anthropogenic emissions and fluxes between the main reservoirs – the oceans, the atmosphere, the terrestrial carbon pool – and the build-up of CO_2 in the atmosphere. Because of the relative stability of atmospheric CO_2 concentrations over several thousand years prior to AD 1800, it is assumed that the net fluxes among carbon reservoirs were close to zero prior to anthropogenic disturbance. The data described in Section 1.3

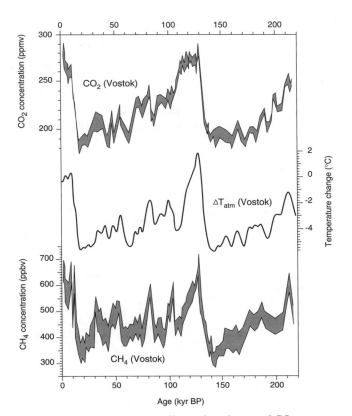

Figure 1.6: Temperature anomalies and methane and CO_2 concentrations over the past 220,000 years as derived from the ice core record at Vostok, Antarctica.

provide the essential background for understanding the carbon budget's changes over time. Several approaches are used to quantify the components of this budget (other than atmospheric mass build-up, which can be measured directly), often in combination.

(a)   direct determination of rates of change of the carbon content in atmospheric, oceanic, and terrestrial carbon pools, either by observations of local inventory changes or by local flux measurements, extrapolated globally;

(b)   indirect assessment of the atmosphere-ocean and atmosphere-terrestrial biosphere fluxes by means of carbon cycle model simulations, either calibrated or partially validated using analogue tracers of CO_2, such as bomb radiocarbon or tritium, or with use of chlorofluorocarbons;

(c)   interpretation of tracers or other substances that are coupled to the carbon cycle (14C/12C and 13C/12C ratios and atmospheric oxygen).

The heterogeneity of some aspects of the oceanic and terrestrial carbon systems makes reliable extrapolation of flux measurements to the entire globe dependent on high resolution geographic information and accurate modelling of processes. Because of this, method (a) is used only for atmospheric carbon and for estimating effects of land use, and estimates of other carbon reservoirs are based primarily on methods (b) and (c).

1.4.2 Methods for Calculating the Carbon Budget

1.4.2.1 Classical Approaches

The calculated ocean uptake rate, together with the estimated fossil emissions and the observed atmospheric inventory change, allows inference of the net terrestrial biospheric balance. Figure 1.7 shows the results of a time-series rate of change calculation for the atmospheric carbon mass, the oceanic component, fossil emissions, and the residual of the first three terms. In this calculation, the time history of fossil plus cement emissions was deduced from statistics (Keeling, 1973; Marland and Rotty, 1984; WEC, 1993; Andres *et al.*, 1994), atmospheric accumulation was determined from the observational and ice core record (e.g., Barnola *et al.*, 1991; Boden *et al.*, 1991), and ocean uptake was modelled with the GFDL ocean general circulation model (Sarmiento *et al.*, 1992). Ocean carbon uptake was determined by forcing the ocean chemistry with the time history of atmospheric concentrations to obtain uptake as a function of the non-linear chemistry of carbon dioxide in the ocean. This calculation illustrates the classic means by which the carbon budget is calculated. Figure 1.8 shows the balance of the terrestrial biosphere, now introducing the time history of land use effects together with a calculation of the inferred terrestrial sink over time. This calculation illustrates the classic estimation of a "missing sink," in which a residual sink arises because the sum of anthropogenic emissions (fossil, cement, changing land use) is greater than the sum of ocean uptake and atmospheric accumulation. Note that this approach is quite different from the estimation of sources and sinks from the spatial distribution of atmospheric concentrations and isotopic composition, discussed in Section 1.4.2.3 (Keeling *et al.*, 1989b; Tans *et al.*, 1990; Enting and Mansbridge, 1991).

1.4.2.2 New Approaches: Budget Assessment Based on Observations of ¹³C/¹²C and O_2/N_2 Ratios

Several promising approaches have recently been proposed to assess the current global carbon budget with less dependence on models. Included are observations of ¹³C/¹²C and O_2/N_2, both of which are strongly influenced by the anthropogenic perturbation of the carbon cycle.

The methods based on ¹³C exploit the fact that the ¹³C/¹²C ratio in fossil fuels and terrestrial biomass is less than that in the atmosphere. There has been a decline in the ¹³C/¹²C ratio of

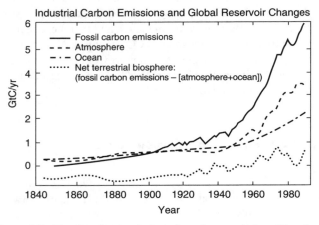

Figure 1.7: Fossil carbon emissions (based on statistics of fossil fuel and cement production), and representative calculations of global reservoir changes: atmosphere (deduced from direct observations and ice core measurements), ocean (calculated with the GFDL ocean carbon model), and net terrestrial biosphere (calculated as remaining imbalance). The calculation implies that the terrestrial biosphere represented a net source to the atmosphere prior to 1940 (negative values) and a net sink since about 1960.

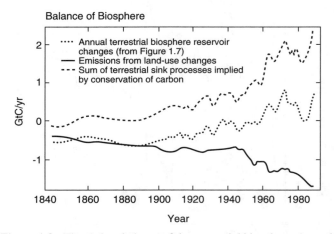

Figure 1.8: The carbon balance of the terrestrial biosphere. Annual terrestrial biosphere reservoir changes (from Figure 1.7), land use flux (plotted negative because it represents a loss of biospheric carbon) and the sum of the terrestrial sink processes (e.g., Northern Hemisphere regrowth, CO_2 and nitrogen fertilization, climate effects) as implied by conservation of carbon mass.

atmospheric CO_2 over the last century (Keeling *et al.*, 1989a). This atmospheric $^{13}C/^{12}C$ ratio change propagates through the global carbon cycle, causing isotopic ratio changes in the ocean and the terrestrial carbon reservoirs.

Quay *et al.* (1992) proposed a method to determine the global budget from repeated measurements of vertical profiles of the $^{13}C/^{12}C$ ratio in oceanic dissolved inorganic carbon. In principle, these observations allow us to determine the rate of change of the oceanic $^{13}C/^{12}C$ ratio which, together with the observed changes in the atmospheric $^{13}C/^{12}C$ ratio, provide another constraint on the global carbon balance. The ratio permits discrimination between oceanic and terrestrial biospheric sinks because terrestrial uptake (photosynthesis) discriminates against the heavy isotope (^{13}C) much more strongly than does ocean uptake. A preliminary analysis based on a data set from seven stations in the Pacific Ocean (sampled in the early 1970s and again in 1990) yielded a mean oceanic sink of 2.1 ± 0.8 GtC/yr (Quay *et al.*, 1992).

Additional information for constraining the oceanic carbon budget may be provided by the $^{13}C/^{12}C$ isotopic disequilibrium between the air and the sea (Tans *et al.*, 1993). The disequilibrium reflects the isotopic adjustment of the ocean to the atmospheric perturbation and can be used to assess the atmospheric balances of ^{13}C and CO_2, thereby discriminating between oceanic and biospheric components. Tans *et al.* (1993) used this method to estimate the net air-sea flux of CO_2 for the period 1970 to 1990, which they found to be less than 0.4 GtC/yr, with unspecified error ranges.

Relative to achievable measurement precision, the anthropogenic perturbation has a greater effect on isotopic composition of dissolved organic carbon than on its concentration, because the former is not affected by chemical buffering reactions. The required analytical quality of the isotopic measurements is still high. The uncertainty range of the current oceanic sink, as estimated by the methods employing isotope measurements, may be reduced substantially if an extended measurement programme is vigorously pursued (Heimann and Maier-Reimer, submitted).

The trend in atmospheric oxygen, as revealed by measurements of the oxygen to nitrogen ratio, can also be used to assess the global carbon budget (Keeling and Shertz, 1992; Bender *et al.*, 1994). Oxygen in many respects is complementary to carbon. It is consumed during combustion and respiration and is released during biotic carbon uptake via photosynthesis. The crucial difference compared to carbon is that, owing to low solubility of O_2 in water, the magnitude of the oceanic oxygen pool is negligible and can be ignored in the global oxygen balance. Therefore, measurements of the temporal trend in atmospheric oxygen together with the known O:C stoichiometric relations during combustion, respiration, and photosynthesis, permit establishment of global oxygen and carbon balances, although interannual variations in marine photosynthesis and respiration may complicate the interpretation of O:C data (Keeling and Severinghaus, 1994). Data are limited, as measurements of the oxygen/nitrogen ratio have been reported for only two years. Within 10 years it is possible that an accurate, model-independent estimate of the oceanic carbon budget will be achieved by this method. The atmospheric oxygen trend during

the last several decades can also be reconstructed from measurements of CO_2 and the O_2/N_2 ratio in glacial surface layers (firn) as a function of depth. Preliminary results indicate past ocean CO_2 uptake rates were large (> 3 GtC/yr), but associated uncertainty ranges are also large (Bender *et al.*, 1994).

Despite high uncertainty, these new approaches provide a different, model-independent means of assessing the global carbon budget. They provide only estimates of the oceanic uptake rate, however. The information gained can help reduce the uncertainty in the net terrestrial sink quantified by differencing, but to partition the uptake into land use fluxes and terrestrial sinks requires other approaches.

1.4.2.3 Constraints from Spatial Distributions

The different spatial patterns of sources and sinks of atmospheric CO_2 create gradients in the concentration that vary inversely with the strength of atmospheric transport. These spatial distributions of concentration can be interpreted quantitatively using models of atmospheric transport to match observed concentration distributions. This process is known as "inversion."

The problem of deducing sources and sinks of trace gases from observations of surface concentrations presents a considerable technical challenge. This is partly because the small-scale features of the spatial distribution of sources and sinks are blurred by atmospheric mixing. The "reconstruction" process amplifies the small-scale details, but the errors and uncertainties introduced along the way are similarly amplified. Hence, there is a trade-off between the spatial resolution sought and the accuracy of the estimates obtained. Consequently, only a small number of independent source/sink components can be reliably determined from the data, and, in general, only the largest spatial scales can be resolved (Enting, 1993). This explains, in part, the range of interpretations obtained in different studies. The uncertainty analysis of Enting *et al.* (1994a) indicates that such analyses cannot, on their own, estimate global totals of net fluxes more accurately than the classical approach to carbon budget analysis. Thus, at present, the constraints from the spatial distribution of CO_2 act mainly as consistency checks with budgets derived from analyses of individual budget components.

Synthesis inversions include additional information beyond atmospheric concentration and transport, and can suppress unwanted amplification of small-scale variations either by restricting the number of source components considered, or through the use of additional constraints such as incorporation of independent estimates of source strengths. Inversions using two-dimensional atmospheric transport models have been presented by Tans *et al.* (1989) and Enting and Mansbridge (1989, 1991). Three-dimensional modelling studies of source and sink distributions have been presented by Keeling *et al.* (1989b), Tans *et al.* (1990), and Enting *et al.* (1994a).

The main results that have been obtained from inverse calculations using atmospheric transport modelling are:

- *Northern Hemisphere source.* Inversion calculations reveal a strong Northern Hemisphere release, as expected from increased use of fossil fuels. This source is so strong that it tends to obscure other details of the source-sink distribution and is additional confirmation of the role of fossil fuel emissions in atmospheric concentration change.

- *Northern Hemisphere sink.* After subtracting the fossil source, there is a strong net sink in the Northern Hemisphere which may involve both marine and terrestrial components (e.g., Tans *et al.*, 1990). The partitioning remains controversial (Keeling *et al.*, 1989b; Ciais *et al.*, submitted), but recent isotopic analyses suggest an appreciable terrestrial component (Ciais *et al.*, submitted). A Northern Hemisphere sink, as estimated by inverse calculations, is consistent with results from terrestrial studies that suggest sinks are the result of changing land use (e.g., Dixon *et al.*, 1994) and nitrogen deposition (largely confined to the industrialised Northern Hemisphere) (e.g., Schindler and Bayley, 1993).

- *Indirect evidence of the existence of a tropical biotic sink.* The net equatorial source from oceanic outgassing and changes in tropical land use is smaller than expected based on other estimates (see Sections 1.4.3.3 and 1.4.3.4), consistent with a role for nutrient and CO_2 fertilisation and forest regrowth.

- *Southern Ocean source.* Most studies indicate a source (probably oceanic) at high southern latitudes.

- *Southern Hemisphere sink.* The net Southern Hemisphere sink (which is presumably oceanic, given the small proportion of Southern Hemisphere land) is weaker than expected in comparison with northern ocean sinks.

1.4.3 Sources and Sinks of Anthropogenic CO_2

1.4.3.1 Fossil Carbon Emissions

The dominant anthropogenic CO_2 source is that generated by the use of fossil fuels (coal, oil, natural gas, etc.) and production of cement. The total emissions of CO_2 from the use of fossil carbon can be estimated based on documented statistics of fossil fuel and cement production (Keeling, 1973; Marland and Rotty, 1984; WEC, 1993; Andres *et al.*, 1994). Average (1980 to 1989) fossil emissions were estimated to be 5.5 GtC/yr (Andres *et al.*, 1994). During 1991, the reported emissions totalled 6.2 GtC (Andres *et al.*, 1994). The cumulative input since the beginning of the industrial revolution (1751 to 1991) is estimated to be approximately 230 GtC (Andres *et al.*, 1994). Uncertainties associated with these estimates are less than 10% for the decade of the 1980s (at the 90% confidence level) based on the methods presented in Marland and Rotty (1984).

1.4.3.2 Atmospheric Increase

The globally averaged CO_2 concentration, as determined through analysis of NOAA/CMDL data (Boden *et al.*, 1991; Conway *et al.*, 1994), increased by 1.53 ± 0.1 ppmv/yr over the period 1980 to 1989. This corresponds to an annual aver-age rate of change in atmospheric carbon of 3.2 ± 0.2 GtC/yr. Other carbon-containing compounds like methane, carbon monoxide, and larger hydrocarbons contain ~1% of the carbon stored in the atmosphere and can therefore be neglected in the atmospheric carbon budget.

1.4.3.3 Ocean Exchanges

The ocean contains more than 50 times as much carbon as the atmosphere. Over 95% of the oceanic carbon is in the form of inorganic dissolved carbon (bicarbonate and carbonate ions); the remainder is composed of various forms of organic carbon (living organic matter, particulate and dissolved organic carbon) (Druffel *et al.*, 1992).

The role of the oceans in the global carbon cycle is twofold: first, it represents a passive reservoir which absorbs excess atmospheric CO_2. It is this role that is discussed in this section. Second, changes in the physical state of the ocean (temperature, circulation) and the marine biota may affect the rate of air-sea exchange, and thus future atmospheric CO_2. This second role, subsumed under "feedbacks," is addressed in Section 1.5.3.

The oceanic uptake of excess CO_2 proceeds by (1) transfer of the CO_2 gas through the air-sea interface, (2) chemical interactions with the oceanic dissolved inorganic carbon, and (3) transport into the thermocline and deep waters by means of water mass transport and mixing processes. Though there are large geographical and seasonal variations of the surface ocean partial pressure of CO_2, averaged globally and annually, the surface water value is close to that for equilibrium with the atmosphere. Therefore, processes (2) and (3) are the main factors limiting the capacity of the ocean to serve as a sink on decadal and centennial time-scales. The chemical buffering reactions between dissolved CO_2 and the HCO_3^- and CO_3^{2-} ions reduce the rate of oceanic CO_2 uptake. At equilibrium, an atmospheric increase in CO_2 concentration of 10% is associated with an oceanic increase of dissolved inorganic carbon of merely 1%. This potential uptake occurs only in those parts of the ocean that are mixed with surface waters on decadal time-scales. Therefore, on time-scales of decades to centuries, the ocean is not as large a sink for excess CO_2 as it might seem from comparison of the relative sizes of the main carbon reservoirs (Figure 1.1). While carbon chemistry is known in sufficient detail to perform accurate calculations, oceanic transport and mixing processes remain the primary uncertainties in the determination of the oceanic uptake of excess CO_2.

The marine biota, if in steady state, are believed to play a minor role, if any at all, in the uptake of excess anthropogenic CO_2. The marine biota, however, play a crucial role in maintaining the steady-state level of atmospheric CO_2. About three-quarters of the vertical gradient in dissolved inorganic carbon is generated by the export of newly produced carbon from the surface ocean and its regeneration at depth (a process referred to as "biological pump"). In the open ocean, however, this process is believed to be limited by the availability of nutrients, light, or by phytoplankton population control via grazing, and not by the abundance of carbon (Falkowski and Wilson, 1992).

Therefore, a direct effect of increased dissolved inorganic carbon (less than 2.5% since pre-industrial times) on carbon fixation and export is unlikely, although a recent study by Riebesell *et al.* (1993) suggested that under particular conditions the rate of photosynthesis and hence phytoplankton growth might indeed be limited by the availability of CO_2 as a dissolved gas. The global significance of this effect, however, remains to be assessed. (See Section 1.5.3 for a discussion of the potential indirect effects of increased dissolved inorganic carbon.)

Flux of carbon from the terrestrial biosphere to the oceans takes place via river transport. Global river discharge of carbon in organic and inorganic forms may be ~1.2–1.4 GtC/yr (Schlesinger and Melack, 1981; Degens *et al.*, 1991; Maybeck, 1993). A substantial fraction of this transport (up to 0.8 GtC/yr), however, reflects the natural geochemical cycling of carbon and thus does not affect the global budget of the anthropogenic CO_2 perturbation (Sarmiento and Sundquist, 1992). Furthermore, the anthropogenically induced river carbon fluxes reflect, to a large extent, increased soil erosion and not a removal of excess atmospheric CO_2.

The role of coastal seas in the global carbon budget is poorly understood. Up to 30% of total ocean productivity is attributed to marine productivity in the coastal seas, which compose only ~8% of the oceanic surface area. Here, discharge of excess nutrients by rivers might have significantly stimulated carbon fixation (up to 0.5–1.0 GtC/yr). At present, however, it is not known how much of this excess organic carbon is simply reoxidised, and how much is permanently sequestered by export to the deep ocean, or in sediments on the shelves and shallow seas. Because of the limited surface area, a burial rate significantly exceeding 0.5 GtC/yr is not very likely, as it would require all coastal seas to be under-saturated in partial pressure of CO_2 by more than 50 μatm on annual average, in order to supply the carbon from the atmosphere. Such under-saturations have been documented, e.g., in the North Sea (Kempe and Pegler, 1991), but these measurements are unlikely to be representative of all coastal oceans. Based on the above considerations, the role of the coastal ocean is judged most likely to be small, but, at present, cannot be accurately assessed and so is neglected in the budget presented in this chapter (Table 1.3).

Exchanges of carbon between the atmosphere and the oceans (net air-sea fluxes) can be deduced from measurements of the partial pressure difference of CO_2 between the air and surface waters. This calculation also requires knowledge of the local gas-exchange coefficient, which is a relatively poorly known quantity (Watson, 1993). Furthermore, while the globally and seasonally averaged partial pressure of surface waters is close to the value for equilibrium with the atmosphere, large geographical and seasonal variations exist, induced both by physical processes (upwelling, vertical mixing, sea surface temperature fluctuations) and by the activity of the marine biota. Representative estimates of the seasonally and regionally averaged net air-sea carbon transfer thus necessitates sampling with high spatial and temporal resolution (Garcon *et al.*, 1992). Estimates of the regional net air-sea carbon fluxes have been obtained, albeit with considerable error margins. However, the global oceanic carbon balance is more difficult to deduce by this method, as it represents a small residual computed from summing up the relatively large emissions from super-saturated and uptake in under-saturated regions (Tans *et al.*, 1990; Takahashi *et al.*, 1993; Wong *et al.*, 1993; Fung and Takahashi, 1994). The World Ocean Circulation Experiment (WOCE) survey has nevertheless demonstrated the feasibility of direct measurement programmes, and the importance of measurements as a cross-check for other approaches should not be underrated.

The oceanic contribution to the global carbon budget can also be assessed by direct observations of changes in the oceanic carbon content. This approach also suffers from the problem of determining a very small signal against large spatial and temporal background variability. Model estimates of the rate of change induced by the anthropogenic perturbation are of the order of 1 part in 2000/yr. Therefore, on a 10-year time-scale, variations in dissolved inorganic carbon would have to be measured with an accuracy of better than 1% in order to determine the carbon balance in a particular oceanic region. While such accuracy can be obtained, a substantial sampling effort is required (Keeling, 1993). Determination of the global oceanic budget by this approach does not appear feasible in the near future. However, repeated observational surveys might reveal regional carbon inventory changes, and thus might provide a cross-check for other approaches.

Model Results

The present-day oceanic uptake of excess CO_2 is estimated using ocean carbon model simulation experiments, with observed atmospheric CO_2 concentration as a prescribed boundary condition. It is necessary to model the transport and mixing of the excess carbon perturbation only if the marine biota are assumed to remain on an annual average in a steady state, as excess CO_2 will disperse within the ocean like a passive tracer, independent of the background natural distribution of dissolved inorganic carbon. In particular, the perturbation excludes a natural cycle of carbon transported by rivers, outgassed by the oceans and then taken up by the terrestrial biota.

Until recently, most models of the oceanic uptake of anthropogenic CO_2 consisted of a series of well-mixed or diffusive reservoirs ("boxes") representing the major oceanic water masses, connected by exchange of water (Oeschger *et al.*, 1975). Global transport characteristics of these models were obtained from simulation studies of the oceanic penetration of bomb radiocarbon (^{14}C) validated by comparison with observations (Broecker *et al.*, 1985). Bomb radiocarbon, however, does not accurately track the oceanic invasion of anthropogenic carbon, because oceanic concentration changes of carbon isotopes depend on the full time history of those inputs and ^{14}C entered the ocean with an atmospheric history different from that of natural CO_2. Furthermore, CO_2 uptake depends on chemical interactions with dissolved inorganic C, while uptake of ^{14}C does not. Nevertheless, bomb radiocarbon provides a powerful constraint on ocean carbon models, a constraint which may be supplemented by analyses of other steady-state and transient tracers, such as halocarbons, tritium from nuclear weapon testing, and possibly [13]C.

Recently, three-dimensional oceanic general circulation models (OGCMs) have been used for modelling CO_2 uptake by the oceans (Maier-Reimer and Hasselmann, 1987; Sarmiento *et al.*, 1992; Orr, 1993). These models calculate oceanic circulation on the basis of the physics of fluid dynamics, and the few adjustable model parameters are primarily tuned to reproduce the relatively well-known large-scale patterns of ocean temperature and salinity. It is known that the OGCMs used in published global carbon cycle studies show significant and similar deficiencies (e.g., too weak a surface circulation, inaccurate deep convection, absence of high resolution features). Simulation of transient tracers, bomb radiocarbon in particular, provides an important validation test.

The average of modelled rates of oceanic carbon uptake is 2.0 GtC/yr over the decade 1980 to 1989 (Orr, 1993; Siegenthaler and Sarmiento, 1993). This value is corroborated by model experiments projecting future concentrations of atmospheric CO_2 (see Section 1.6, and Enting *et al.*, 1994b) and by a simulation with a newly-developed two-dimensional global ocean model (Table 1.1; Stocker *et al.*, 1994). The spread of the modelled uptake rates corresponds to a statistical uncertainty of only about $\pm$ 0.5 GtC/yr (at the 90% confidence level). However, recent assessments of the global radiocarbon balance (Broecker and Peng, 1994; Hesshaimer *et al.*, 1994) suggest that the oceanic carbon uptake estimates of box models tuned by bomb radiocarbon might have to be revised downwards by up to 25%. Furthermore, the spread of the OGCM results listed in Table 1.1 might be fortuitously small in view of the similar deficiencies in these models. Based on these considerations we do not change the estimate of the uncertainty of the ocean uptake from the value of $\pm$ 0.8 GtC/yr as given by IPCC in 1990 (Watson *et al.*, 1990) and in 1992 (Watson *et al.*, 1992).

1.4.3.4 Terrestrial Exchanges

In previous assessments (IPCC, 1990, 1992), the remaining intact terrestrial biosphere has often been assumed to be a significant sink for carbon dioxide, balancing or exceeding emissions derived from changing land use. The calculated imbalance, which has been a persistent feature of global carbon

cycle calculations (Broecker *et al.*, 1979), arises from the difference between measured atmospheric changes, statistically derived fossil fuel and cement emissions, modelled ocean uptake, and estimated emissions from changing land use. In this assessment, part of the previously calculated imbalance is accounted for by effects of changing land use in the middle and high latitudes (0.5 $\pm$ 0.5 GtC/yr: Table 1.3). Several processes may contribute to increased terrestrial carbon storage (Table 1.2), but the problem of detecting these increases is troublesome.

Difficulty in quantifying the role of the terrestrial biosphere in the global carbon cycle arises because of the complex biology underlying carbon storage, the great heterogeneity of vegetation and soils, and the effects of human land use and land management. We consider the following issues:

- *Deforestation and change in land use.* CO_2 is emitted to the atmosphere as a result of land use changes such as biomass burning and forest harvest through the oxidation of vegetation and soil carbon. These "emissions from changing land use" are currently largest in the tropics, but prior to the 1950s, the middle latitudes were a larger source than were the tropics.

- *Forest regrowth.* Carbon is absorbed by regrowing forests following harvest. This absorption is included as a factor in the calculation of net emissions from disturbed regions. In the tropics, emissions are thought to exceed the uptake by secondary growth following forest harvest, but uptake from regrowth in the middle and high latitudes apparently results in a net sink for atmospheric CO_2 (see below).

- *Fertilisation by carbon dioxide.* As atmospheric CO_2 increases, plants increase their uptake of carbon, potentially increasing carbon storage in the terrestrial biosphere. This mechanism has often been hypothesised to account for the calculated imbalance in the global carbon budget (the "missing sink"). There is strong physiological evidence for photosynthetic increase with CO_2 fertilisation (e.g., Woodward, 1992), but interspecific differences in the magnitude of effects on photosynthesis and growth range from marked enhancement to negli-

Table 1.1. *Excess CO_2 uptake rates (average 1980 to 1989) calculated by various ocean carbon cycle models.*

Model	Ocean uptake (GtC/yr)	Reference
Bomb radiocarbon-based box models		
Box-diffusion model	2.32	Siegenthaler & Sarmiento (1993)
HILDA model ("Bern model")	2.15	Siegenthaler & Joos (1992)
Three-dimensional ocean general circulation models		
Hamburg Ocean Carbon Cycle Model (HAMOCC-3)	1.47	Maier-Reimer (1993)
GFDL Ocean General Circulation Model	1.81	Sarmiento *et al.* (1992)
LODYC Ocean General Circulation Model	2.10	Orr (1993)
Two-dimensional ocean circulation models	2.1	Stocker *et al.* (1994)
AVERAGE of all models	**2.0**	

Table 1.2. *Processes leading to increased terrestrial storage and their magnitudes (GtC/yr): average for the 1980s.*

Processes	
Mid-/high latitude forest regrowth	0.5 ± 0.5
CO_2 fertilisation	0.5-2.0
Nitrogen deposition	0.2-1.0
Climatic effects	0-1.0

Table 1.3. *Average annual budget of CO_2 perturbations for 1980 to 1989. Fluxes and reservoir changes of carbon are expressed in GtC/yr, error limits correspond to an estimated 90% confidence interval.*

CO_2 sources	
(1) Emissions from fossil fuel combustion and cement production	5.5 ± 0.5
(2) Net emissions from changes in tropical land use	1.6 ± 1.0
(3) Total anthropogenic emissions (1)+(2)	7.1 ± 1.1
Partitioning among reservoirs	
(4) Storage in the atmosphere	3.2 ± 0.2
(5) Oceanic uptake	2.0 ± 0.8
(6) Uptake by Northern Hemisphere forest regrowth	0.5 ± 0.5*
(7) Additional terrestrial sinks (CO_2 fertilisation, nitrogen fertilisation, climatic effects) $[(1) + (2)] - [(4) + (5) + (6)]$	1.4 ± 1.5

* from Table 1.2

gible and even negative responses. Considerable uncertainty exists as to how to translate laboratory and field results into a global estimate of changing biospheric carbon storage, as ecosystem feedbacks and constraints are numerous (see Section 1.5.2).

- *Nitrogen fertilisation.* Most terrestrial ecosystems are limited by nitrogen, as evidenced by increased carbon storage after nitrogen fertilisation. Additions of nitrogen to the terrestrial biosphere through both intentional fertilisation of agricultural land and deposition of nitrogen arising from fossil fuel combustion and other anthropogenic processes can result in increased terrestrial storage of carbon.

- *Climate.* The processes of terrestrial carbon uptake (photosynthesis) and release (respiration of vegetation and soils) are influenced by climate. Decadal time-scale variations in climate may have caused natural changes in carbon storage by the terrestrial biosphere, acting in conjunction with or even counter to the anthropogenic effects described above. This effect on storage is separate from the potential future changes in carbon storage which might arise from greenhouse gas-induced changes of climate.

- *Interactions.* The processes mentioned above are not independent, and their magnitudes are not necessarily additive. For example, much of the anthropogenic nitrogen has been deposited by the atmosphere on regrowing forests of the middle and high latitudes, thereby potentially enhancing carbon storage in these regions. Recent measurements of mid-latitude and high latitude forest regrowth may therefore reflect the combined effects of nitrogen deposition, elevated CO_2 concentrations, and climate variability. Also, CO_2 fertilisation can affect the nitrogen cycle and vice versa (see Section 1.5.2). It is possible that nitrogen deposition will enhance the effectiveness of carbon storage by offsetting the increased plant demand for nitrogen caused by increased concentrations of CO_2. Interactive effects with air pollutants such as tropospheric ozone may also be important.

1.4.3.4.1 *Emissions from Changing Land Use*
Deforestation and other changes in land use (including land management) cause significant exchanges of CO_2 between the land and atmosphere. Changes in land use from 1850 to 1990

resulted in cumulative emissions of 122 ± 40 GtC (Houghton, 1994a). From the last century through the 1940s, expansion of agriculture and forestry in the middle and high latitudes dominated carbon emissions from the terrestrial biosphere (Houghton and Skole, 1990). Since then, conversion of temperate forests for agricultural use has diminished and forests are regrowing in previously logged forests and on abandoned agricultural lands (Melillo *et al.*, 1988; Birdsey *et al.*, 1993; Dixon *et al.*, 1994). Emissions from the tropics have generally been increasing since the 1950s. Estimates for 1980 range from 0.4 to 2.5 GtC (Houghton *et al.*, 1987; Detwiler and Hall, 1988; Watson *et al.*, 1990).

While uncertainties in the estimated rate of tropical deforestation are large (FAO, 1993), estimation of deforestation rates and area of regrowing forests have improved recently through the use of remote sensing. For example, satellite imagery has reduced estimated rates of deforestation in the Brazilian Amazon (INPE, 1992; Skole and Tucker, 1993). The lack of satellite analyses for other tropical areas may mean that the flux from the tropics as a whole has been overestimated.

While there is considerable uncertainty in estimated emissions from changing land use, the most recent compilations (Dixon *et al.*, 1994; Houghton, 1994b) agree that tropical emissions averaged 1.6 ± 1.0 GtC/yr in the 1980s. These analyses take into account the changing estimates of deforestation rates (INPE, 1992; Skole and Tucker, 1993) as well as other new data. The uncertainty estimate reflects the controversy over quantity and distribution of biomass (e.g., Brown and Lugo, 1992; Fearnside, 1992) as well as the errors associated within and between individual studies. At this writing, considerable work is in progress; it is possible that the estimate of emissions from changing land use in the tropics may be modified in the near future. Of particular concern is the currently poor evaluation of effects of changing land use in the drier, open canopy forests of the seasonal tropics (FAO, 1993), and poor knowledge of carbon accumulation in regrowing forests.

1.4.3.4.2 *Uptake of CO_2 by Changing Land Use*

Several recent analyses suggest a sink of carbon in regrowing Northern Hemisphere forests. Estimates for the magnitude of this term range from essentially zero to 0.74 GtC/yr (Melillo *et al.*, 1988; Houghton, 1993; Dixon *et al.*, 1994). It is difficult to assess the magnitude of this sink for two reasons. The first is the wide disparity in published regional estimates (e.g., a Russian sink of 0.01–0.06 GtC/yr (Melillo *et al.*, 1988; Krankina and Dixon, 1994) versus 0.3–0.5 GtC/yr (Dixon *et al.*, 1994).

A more fundamental problem is the confusion of carbon uptake via forest growth versus forest *regrowth*. As a basis for understanding how fluxes might change in the future, we seek in this report to separate increases in forest carbon storage that result from the regrowth of previously harvested forests versus storage due to processes such as CO_2 fertilisation, nitrogen deposition, and changes in fire frequency, management practices, or climate, all of which can affect forest growth. Inventory-based estimates of forest carbon storage are made by multiplying the age distribution of forest stands (rates of carbon accumulation generally decrease as stands age) by age-specific rates of carbon accumulation, estimated from observations. However, as forest carbon accumulation has been measured during the period of changing CO_2 and nitrogen deposition, and possibly climate, forest inventories cannot distinguish the amount of accumulation attributable to individual processes.

The confusion of mechanisms is not important when inventory data are used to corroborate the results of inverse modelling (e.g., Tans *et al.*, 1990; Enting and Mansbridge, 1991; Ciais *et al.*, in press), but separation of carbon storage mechanisms is essential for projections of future changes in carbon storage. Estimates of carbon storage due to nitrogen deposition or CO_2 fertilisation (Peterson and Melillo, 1985; Schindler and Bayley, 1993; Gifford, 1994) cannot be added to inventory-based land use sinks (Dixon *et al.*, 1994; Houghton, 1994b). Moreover, inventory data are ill-suited for the task of projecting future carbon uptake because the inventory-based method does not distinguish between demographic effects such as forest harvest, agricultural abandonment, and natural disturbance such as fire and wind (Kolchugina and Vinson, 1993). The problem of quantifying the contribution of individual mechanisms to total carbon storage is even greater in the tropics, where few demographic data exist. Thus, it is inappropriate to rely entirely on inventory-based measurements for initialisation of terrestrial carbon models.

In addition to changes on forested lands, the decline in soil carbon storage in agricultural lands may have been reversed in the middle latitudes. Best management practices may even lead to storage increases in the future (Metherell, 1992). While the world-wide extent of management changes and their effect on carbon balance is poorly known, a recent assessment of car-

bon storage in U.S. croplands suggests possible increases over the coming few decades of 0.02–0.05 GtC/yr; the authors note that similar changes are likely to have occurred over the past few decades (Donigian *et al.*, 1994). In terms of the global C budget, increases in soil storage are small (Schlesinger, 1990), although they may be significant at the regional, or even national, scale.

1.4.3.4.3 *Other Terrestrial Sink Processes*

While little has been done to increase understanding of potential sinks in tropical forests, analyses based on models of forest growth, age distribution data, and the age-growth relationship suggest carbon accumulation in Northern Hemisphere forests may be as high as ~1 GtC/yr (Dixon *et al.*, 1994). Analyses based on observed atmospheric CO_2 and $^{13}CO_2$ suggest a strong Northern Hemisphere terrestrial sink. While various pieces of evidence support a substantial terrestrial sink in the Northern Hemisphere, direct observations to confirm the hypothesis and to establish the processes responsible for increasing carbon storage are lacking. For example, tree ring studies of contemporary and pre-industrial forest growth rates are contradictory. These studies have revealed enhanced growth in subalpine conifers (LaMarche *et al.*, 1984), no growth enhancement beyond that explained by climatic variability (Graumlich, 1991; D'Arrigo and Jacoby, 1993), and both positive and negative changes in growth rate dependent upon species and location (Briffa *et al.*, 1990). Models and observational data to confirm tropical sinks are even more deficient (e.g., Brown and Lugo, 1984, 1992; Fearnside, 1992; Skole and Tucker, 1993; Dixon *et al.*, 1994). Improved observations for identifying terrestrial sinks is necessary for the improvement of models capable of projecting future states of the carbon cycle, and also for quantifying regional contributions to terrestrial sources and sinks. Processes that may contribute to a terrestrial sink are explored below (see Table 1.2).

CO₂ Fertilization

Experimental studies of agricultural and wild plant species have shown growth responses of typically 20–40% higher growth under doubled CO_2 conditions (ranging from negligible or negative responses in some wild plants to responses of 100% in some crop species (Körner and Arnone, 1992; Rochefort and Bazzaz, 1992; Coleman *et al.*, 1993; Idso and Kimball, 1993; Owensby *et al.*, 1993; Polley *et al.*, 1993; Idso and Idso, 1994)). The majority of these studies were short-term and were conducted with potted plants (Idso and Idso, 1994). The effect of increased growth under elevated CO_2 conditions, known as the CO_2 fertilisation effect, is often assumed to be the primary mechanism underlying the imbalance in the global carbon budget. This assumption implies a terrestrial sink that increases as CO_2 increases, acting as a strong negative feedback. However, while the potential effect of CO_2 in experimental and some field studies is relatively strong (e.g., Drake, 1992; Idso and Kimball, 1993; see also review by Idso and Idso, 1994), natural ecosystems may be less responsive to increased levels of atmospheric CO_2. Evidence that the effects of CO_2 on long-term carbon storage may be less than is suggested by short-term pot studies of photosynthesis or plant growth includes:

(1) Field studies showing reduced responses over time (Oechel *et al.*, 1993) and zero, small, or statistically insignificant responses (Norby *et al.*, 1992; Jenkinson *et al.*, 1994);

(2) Evidence that plants with low intrinsic growth rates, a common trait in native plants, are less responsive to CO_2 increases than are rapidly growing plants, such as most crop species (Poorter, 1993);

(3) Results of model simulations which suggest that increases in carbon storage eventually become nutrient-limited (Comins and McMurtrie, 1993; Melillo *et al.*, 1994; reviewed in Schimel, 1995). (Long-term nutrient limitation is different from the short-term nutrient limitation of photosynthesis and plant growth observed in experimental studies; it reflects the need for nutrients in addition to carbon to increase organic matter production.)

There is also evidence that while water- or temperature-stressed plants are more responsive to CO_2 increase than are unstressed plants (because CO_2 increases water-use efficiency (Polley *et al.*, 1993)), nitrogen-limited plants are less sensitive to CO_2 level (Bazzaz and Fajer, 1992; Comins and McMurtrie, 1993; Díaz *et al.*, 1993; Melillo *et al.*, 1993; Ojima *et al.*, 1993; Idso and Idso, 1994). The reduction in response of short-term photosynthesis and plant growth to CO_2 caused by nitrogen limitation is highly variable and may on the whole be small (Idso and Idso, 1994), but as all organic matter contains nitrogen, carbon storage increases should eventually become limited by the stoichiometric relationships of carbon and other nutrients in organic matter (Comins and McMurtrie, 1993; Díaz *et al.*, 1993; Schimel, 1995).

While current models assume that nitrogen inputs are not affected by CO_2 fertilisation, this may not be true (Thomas *et al.*, 1991; Gifford, 1993; Idso and Idso, 1994). CO_2 fertilisation could stimulate biological nitrogen fixation, because nitrogen-fixing organisms have high energy (organic carbon) requirements. If nitrogen inputs are stimulated by increasing CO_2, acclimation of carbon storage to higher CO_2 levels could be temporarily offset. However, evidence from native ecosystems suggests that other nutrients, such as phosphorus, may be more limiting to nitrogen fixation than is energy (Cole and Heil, 1981; Eisele *et al.*, 1989; Vitousek and Howarth, 1991).

As a sensitivity analysis, the effects of CO_2 enrichment was varied in one model over the plausible range of values (by the equivalent of 10, 25, and 40% increases in plant growth at doubled CO_2). The results indicate terrestrial carbon storage rates (increased net ecosystem production: NEP) due to CO_2 fertilisation were 0.5, 2.0, and 4.0 GtC/yr during the 1980s (Gifford, 1993). A modelling experiment by Rotmans and den Elzen (1993) indicated the strength of CO_2 fertilisation might be ~1.2 GtC/yr. Overall, it is likely that CO_2 fertilisation plays a role in the current terrestrial carbon budget, and may have amounted to storage of 0.5 to 2.0 GtC/yr during the 1980s.

Nitrogen Fertilization

Many terrestrial ecosystems are nitrogen limited: added fertiliser will produce a growth response and additional carbon storage (e.g., Vitousek and Howarth, 1991; Schimel, 1995). Nitrogen deposition from fertilisers and oxides of nitrogen released from the burning of fossil fuel during the 1980s is estimated to amount to a global total, but spatially concentrated, 0.05–0.06 GtN/yr (Peterson and Melillo, 1985; Duce *et al.*, 1991). The carbon sequestration which results from this added nitrogen is estimated to be of the order 0.2–1.0 GtC/yr (Peterson and Melillo, 1985; Schindler and Bayley, 1993; Schlesinger, 1993), depending on assumptions about the proportion of nitrogen that remains in ecosystems. Estimates significantly higher than 1 GtC/yr are unrealistic because they assume that all of the N would be stored in forms with high carbon to nitrogen ratios, while much atmospheric nitrogen is in reality deposited on grasslands and agricultural lands where storage occurs in soils with low average carbon to nitrogen ratios. Uptake of carbon due to long-term increases in nitrogen deposition could increase as nitrogen pollution increases; but possibly to a threshold, after which additional nitrogen may result in ecosystem degradation (e.g., Aber *et al.*, 1989; Schulze *et al.*, 1989). Because most deposition of anthropogenic nitrogen occurs in the middle latitudes, some of the effect of added nitrogen may already be accounted for in measurement-based estimates of mid-latitude carbon accumulation (discussed above). However, existing analyses do not allow identification of the fraction of measured forest growth that is due to nitrogen addition or effects of CO_2.

Climate Effects

Climate affects carbon storage in terrestrial ecosystems because temperature, moisture, and radiation influence both ecosystem carbon gain (photosynthesis) and loss (respiration) (Houghton and Woodwell, 1989; Schimel *et al.*, 1994). While a number of compensatory processes are possible, warming is thought to reduce carbon storage by increasing respiration, especially of soils (Houghton and Woodwell, 1989; Shaver *et al.*, 1992; Townsend *et al.*, 1992; Oechel *et al.*, 1993; Schimel *et al.*, 1994). Conversely, cooling would be expected to increase carbon storage. In nutrient-limited forests, however, warming may increase carbon storage by "mineralising" soil organic nutrients, which are generally stored in nutrient-rich, or low carbon-to-nutrient ratio forms, thereby allowing increased uptake by trees, which store nutrients in high carbon-to-nutrient forms (Shaver *et al.*, 1992). This is not true in nutrient-limited tundra ecosystems, where recent warming has resulted in a local CO_2 source (Oechel *et al.*, 1993). Increases in precipitation will generally increase carbon storage by increasing plant growth, although in some ecosystems compensatory increases in soil decomposition may reduce or offset this effect (Ojima *et al.*, 1993). While these effects have been discussed as components of future responses of ecosystems to climate change, climate variations during the past century may have influenced the terrestrial carbon budget. In a provocative paper, Dai and

Fung (1993) suggested that climate variations over the past decades could have resulted in a substantial sink. Ciais *et al.* (in press) suggested that cooling arising from the effects of Mt. Pinatubo may have increased terrestrial carbon storage and contributed to the observed reduction in the atmospheric growth rate during the 1991 to 1992 period. Palaeoclimate modelling studies likewise suggest major changes in terrestrial carbon storage with climate (Prentice and Fung, 1990; Friedlingstein *et al.*, 1992). Modelling and observations of ecosystem responses to climate and climate anomalies will be important tools for validating predictions of future changes; meanwhile, investigations of the effects of climate on carbon storage remain suggestive rather than definitive.

1.4.4 Budget Summary

In Table 1.3 we present an estimated budget of carbon perturbations for the 1980s, shown as annual average values.

The budgetary inclusion of a Northern Hemisphere sink in forest regrowth reduces the unaccounted-for sink compared with earlier budgets (IPCC, 1999, 1992) by assigning a portion of this sink to forest regrowth. The remaining imbalance in this budget implies additional net terrestrial sinks of 1.4 ± 1.5 GtC/yr.

1.5 The Influence of Climate and Other Feedbacks on the Carbon Cycle

1.5.1 Introduction

The description of the carbon cycle in the previous sections addressed the budget of anthropogenic CO_2, without emphasising the possibility of more complex interactions within the system. There are, however, a number of processes that can produce feedback loops. One important distinction (see Enting, 1994) is between carbon cycle feedbacks and CO_2-climate feedbacks. The only direct carbon cycle feedback is the CO_2 fertilisation process described in Section 1.4.3.4.3. In contrast, CO_2-climate feedbacks involve the effect of climate change (potentially induced by CO_2 concentration changes) on the components of the carbon cycle. For terrestrial components, the most important effects are likely to be those involving temperature, precipitation, and radiation changes (through changes in cloudiness) on net primary production and decomposition (including effects resulting from changes in species composition). For marine systems, the primary effects to be expected arise through climatic influences on ocean circulation and chemistry. Such changes would affect the physical and biological aspects of carbon distribution in the oceans including physical fluxes of inorganic carbon within the ocean and changes in nutrient cycling (Manabe and Stouffer, 1993). The feedbacks may also include changes in the species composition of the ocean biota, which determines the location and magnitude of oceanic CO_2 uptake.

Factors with a strong influence on the global carbon cycle that are similar to and/or modify feedback loops also exist. Among such factors are the effects of increased UV on terrestrial

and marine ecosystems and the anthropogenic toxification and eutrophication of these ecosystems. While significant effects of possible changes in mid-latitude and high latitude UV radiation, tropospheric ozone and other pollutants have been documented in experimental studies, there is little basis for credible global extrapolation of these studies (Chameides *et al.*, 1994). They will not be discussed further, but should be addressed in subsequent assessments as more data become available.

1.5.2 Feedbacks to Terrestrial Carbon Storage

The responses of terrestrial carbon to climate are complex, with rates of biological activity generally increasing with warmer temperatures and increasing moisture. Because photosynthesis and plant growth increase with warmer temperatures, longer growing seasons and more available water, storage of carbon in living vegetation generally increases as well (Melillo *et al.*, 1993). Storage of carbon in soils generally increases along a gradient from low to high latitudes, reflecting slower decomposition of dead plant material in colder environments (Post *et al.*, 1985; Schimel *et al.*, 1994). Flooded soils, where oxygen becomes depleted, have extremely low rates of decomposition and may accumulate large amounts of organic matter as peat. Global ecosystem models based on an understanding of underlying mechanisms are designed to capture these patterns, and have been used to simulate the responses of terrestrial carbon storage to changing climate. Some models also include the effect of changes in land use. Model results suggest that future effects of changing climate, atmospheric CO_2, and changing land use on carbon storage may be quite large (Vloedbeld and Leemans, 1993).

Effects of Temperature and CO_2 Concentration

Ecosystem models have been used to simulate the response of the terrestrial biosphere to changes in climate, rate of change of climate ("transient" changes), effects of changing CO_2, and the effects of changing land use. Models may consider only the response of plant growth and decomposition, or they may also allow movement of vegetation "types" such as forests and grasslands. The simplest case, involving a general circulation model- (GCM-) simulated climate change and stationary vegetation patterns projected losses of terrestrial carbon of about 200 Gt over an implicit time period of a few hundred years (Melillo *et al.*, 1993). When vegetation types were allowed to migrate, models projected a long-term increase in terrestrial carbon storage: 60–90 GtC over 100–200 years (Cramer and Solomon, 1993; Smith and Shugart, 1993). In Smith and Shugart's (1993) model, climate change and vegetation redistribution led to a transient release of CO_2 from die back before regrowth (~200 GtC over ~100 years), followed by an eventual accumulation of ~90 GtC. This study, like those mentioned above, did not incorporate the effects of CO_2 fertilisation. The TEM model of Melillo *et al.* (1993) was also used to assess the response of ecosystem carbon storage to CO_2 fertilisation under a scenario of climate change resulting from an instantaneous doubling of CO_2. The results of this experiment showed that, over the period of CO_2 doubling, uptake of carbon occurred only when the CO_2 fertilisation effect was included (loss of ~200 GtC without the fertilisation effect, and a gain of ~250 GtC when the effect was included: Melillo *et al.*, 1994). Rotmans and den Elzen (1993) used the IMAGE model to assess the effects of including CO_2 fertilisation and temperature feedbacks on their modelled carbon budget. Estimated biospheric uptake increased by 1.2 GtC/yr over the decade of the 1980s when CO_2 and temperature feedbacks were included. The most complex assessment of the interactive effects on carbon storage are carried out with models of transient changes in climate that include CO_2 fertilisation. Esser (1990) simulated an increase of 170 GtC over 200 years with transient changes of climate and CO_2 fertilisation (this model does not take vegetation redistribution into account). Alcamo *et al.* (1994b) performed a similar experiment, but included redistribution of vegetation. Their results revealed eventual increased carbon storage of ~200–250 Gt, depending on assumptions of future land use. Results from ecosystem models suggest that changes in climate associated with CO_2 doubling are likely to lead to a significant transient release of carbon (1–4 GtC/yr) over a period of decades to more than a century (Smith and Shugart, 1993; Dixon *et al.*, 1994).

Effects of Land Use

Several of these studies also included the effects of changing land use. For example, Esser (1990) simulated a gain of 170 GtC due to the effects of climate and CO_2 change, but losses of 322 GtC when he assumed areal reduction of global forests by 50% by the year 2300, emphasising the importance of maintaining forest ecosystems for terrestrial carbon storage. In the Cramer and Solomon (1993) study, inclusion of dense land clearing also substantially reduced simulated C storage (by up to 152 GtC). In the Alcamo *et al.* (1994b) study, global scenarios that required additional land for agriculture and biomass energy production resulted in lower carbon storage.

While the results from global terrestrial models range widely, all suggest that the terrestrial biosphere could eventually take up 100–300 GtC in response to warming, albeit after a significant transient loss (e.g., Smith and Shugart, 1993). While existing simulations have a number of shortcomings, including the lack of agreed-upon climate and land use scenarios that would allow rigorous comparison of results, they suggest possible exchanges between the atmosphere and the terrestrial biosphere of the order of 100s of GtC over decades to a few centuries (e.g., Alcamo *et al.*, 1994a).

Soil Feedbacks

Because soil carbon is released with increasing temperature, it has been suggested that global warming would result in a large positive feedback (Houghton and Woodwell, 1989; Townsend *et al.*, 1992; Oechel *et al.*, 1993). In order to evaluate this, the sensitivity of carbon storage to temperature was assessed using a number of models (Schimel *et al.*, 1994). Inter-comparison of model results revealed rates of global soil carbon loss of 11–34 GtC per degree warming. Vegetation growth and C storage are stimulated in many of these models because as soil carbon is lost, soil nitrogen is made available to the modelled vegetation. This fertilisation effect can ameliorate or even reverse the overall loss of carbon (Shaver *et al.*, 1992; Gifford, 1994; Schimel *et al.*, 1994). In the Century ecosystem model, the nitrogen feedback reduces the effect of warming on soil carbon loss by 50% (Schimel *et al.*, 1994).

Nutrient Limitation of CO_2 Fertilization

The temperature feedback on nitrogen availability may interact with CO_2 fertilisation. CO_2-fertilised foliage is typically lower in nitrogen than foliage grown under current concentrations of CO_2 (e.g., Coleman and Bazzaz, 1992). As dead foliage from high-CO_2 conditions works its way through the decomposition process, soil decomposer organisms require additional nitrogen (Díaz *et al.*, 1993), and, as a result, vegetation may become more nitrogen limited (attenuating but not eliminating the effects of CO_2 fertilisation (Comins and McMurtrie, 1993; Schimel, 1995)). Additional nitrogen released by warming, as described above, can alleviate the nitrogen stress induced by high-CO_2 foliage: this interaction between warming and CO_2 fertilisation is responsible for the large estimated effect of CO_2 on carbon storage in Melillo *et al.* (1993). Deposition of atmospheric nitrogen could also influence the effectiveness of CO_2 fertilisation (Gifford, 1994), although some modelling studies suggest the effect may be modest (Rastetter *et al.*, 1992; Comins and McMurtrie, 1993). Empirical studies are few, but a recent analysis by Jenkinson *et al.* (1994) revealed no measurable change in hay yield over the past 100 years despite substantial increases in nitrogen inputs via precipitation and a 21% increase in atmospheric CO_2 concentration during the period of study. Nutrient feedbacks clearly influence the response of terrestrial ecosystems to the interactive effects of climate and CO_2.

1.5.3 Feedbacks on Oceanic Carbon Storage

Climate feedbacks influence the storage of carbon in the ocean through physical, chemical, and biological processes. Changes in sea surface temperature affect oceanic CO_2 solubility and carbon chemistry. At equilibrium, a global increase in sea surface temperature of 1°C is associated with an increase of the partial pressure of atmospheric CO_2 of approximately 10 ppmv (Heinze *et al.*, 1991), serving as a weak positive feedback between temperature and atmospheric CO_2. During a transient climate change on time-scales of decades to centuries this temperature feedback would be even weaker (MacIntyre, 1978).

Changes in temperature might also affect the remineralisation of dissolved organic carbon. Climatically driven changes in its turnover time are not expected to result in significant variations in atmospheric CO_2, as the quantity of dissolved organic carbon in the surface ocean is less than 700–750 Gt (Figure 1.1). Earlier suggestions (Sugimura and Suzuki, 1988) that this pool might be much larger than previously thought (i.e., about twice the value shown in Figure 1.1) are now generally discounted (Suzuki, 1993).

Changes in the oceanic circulation and their effects on the oceanic carbon cycle are more difficult to assess. Simulation studies of the transient behaviour of the ocean-atmosphere system using coupled GCMs (Cubasch *et al.*, 1992; Manabe and Stouffer, 1993) indicate a strong reduction of the deep thermohaline circulation, which is driven by cooling and sinking of surface water at high latitudes, especially in the North Atlantic. Smaller effects are expected on the wind-driven features of the oceanic general circulation: the shallow upwelling cells at the equator and the eastern boundaries of the warm sub-tropical gyres and in the cyclonic areas in higher latitudes. A reduction of the vertical water exchange processes would impact the oceanic carbon cycle in several ways:

First, the downward transport of surface waters in contact with the atmosphere and thus enriched with excess CO_2 would be reduced, thereby leading to a smaller oceanic excess CO_2 uptake capacity.

Second, changes in the vertical water mass transports would affect the marine biological pump. This would reduce the flux of nutrients transported to the surface and, in nutrient-limited regions, result in reduced marine production. This effect would constitute a weakening of the marine biological pump (i.e., less export of carbon to depth), and could potentially lead to an increase in dissolved inorganic carbon and partial pressure of CO_2 at the surface. Conversely, smaller vertical water mass transports imply a reduced upward transport of deeper waters enriched in dissolved inorganic carbon. The two effects result from the same process, but affect the surface concentration of dissolved inorganic carbon, and hence the partial pressure and the air-sea flux of CO_2, in opposite directions. The latter of the two effects dominates in models that use constant carbon to nutrient ratios to describe the marine biosphere (Bacastow and Maier-Reimer, 1990; Keir, 1994). These models project a small increase in oceanic carbon storage as a result of the reduction of oceanic circulation and vertical mixing.

Initial model results suggest that the effects of predicted changes in circulation on the ocean carbon cycle are not large (10s rather than 100s of ppmv in the atmosphere). However, exploration of the long-term impacts of warming on circulation patterns has just begun; hence, analyses of impacts on the carbon cycle must be viewed as preliminary. For example, changes in the oceanic environment (temperature and circulation) have the potential to change the composition of marine ecosystems in ways not yet included in global models. This could lead to changes in carbon to nutrient ratios, which are assumed constant in present models, or to changes in the relationship between organic and inorganic carbon fixation, and/or change the efficiency by which marine organisms utilise available nutrients. Furthermore, the remineralisation depths of nutrients and carbon might be affected differently (Evans and Fasham, 1993). An extreme lower bound may be estimated by assuming complete utilisation of the oceanic surface nutrients

which would reduce atmospheric CO_2 by 120 ppmv. Similarly, an upper bound is given by assuming extinction of all marine life, increasing atmospheric CO_2 by almost 170 ppmv (Bacastow and Maier-Reimer, 1990; Shaffer, 1993; Keir, 1994). Such large effects, however, are very unlikely to occur. Indeed, the oceanic carbon system appears to be rather stable as evidenced by the very small fluctuations of the atmospheric CO_2 concentration prior to anthropogenic perturbation. It has also proven very difficult to account for the lower atmospheric CO_2 concentration in glacial times (i.e., by 80 ppmv) merely by changing biospheric parameters within current three-dimensional ocean carbon models (Heinze *et al.*, 1991; Archer and Maier-Reimer, 1994).

In areas of excess nutrients (e.g., in the equatorial and sub-arctic Pacific and in some parts of the Southern Ocean) the micronutrient iron appears to limit marine primary production (Martin, 1990). Changes in atmospheric iron loading thus potentially could affect the biological pump and impact atmospheric carbon dioxide levels. Modelling studies have shown, however, that even excessive "iron fertilisation" of these oceanic areas would have a relatively small impact on atmospheric CO_2 levels (Joos *et al.*, 1991a; Peng and Broecker, 1991; Sarmiento and Orr, 1991; Kurz and Maier-Reimer, 1993). Therefore, at least during the past 200 years, it is very unlikely that iron loading had a significant impact on the present day carbon balance.

External impacts on the oceanic carbon system not directly related to global warming must also be considered. Increased ultraviolet radiation (UV-B, corresponding to light wavelengths of 280–320 nm) resulting from the depletion of stratospheric ozone could affect marine life and thus influence marine carbon storage. However, model simulation studies show that even a complete cessation of marine productivity in high latitudes, where increases in UV-B are expected to occur, would result in an atmospheric CO_2 increase of less than 40 ppmv (Sarmiento and Siegenthaler, 1992). Increased input of anthropogenic nitrogen or other limiting nutrients might also affect the oceanic biota. The global effects on atmospheric CO_2 are most likely much smaller than the extreme bounds discussed above.

1.6 Modelling Future Concentrations of Atmospheric CO_2

1.6.1 Introduction

The clear historical relationship between CO_2 emissions and changing atmospheric concentrations implies that continuing fossil fuel, cement, and land-use-related emissions of CO_2 at or above current rates will result in increasing atmospheric concentrations of this greenhouse gas. Understanding how CO_2 concentrations will change in the future requires quantification of the relationship between CO_2 emissions and atmospheric concentration using models of the carbon cycle. This section presents the results of a set of standardised calculations, carried out by modelling groups from many countries, that analyse the relationships between emissions and concentrations in a number of ways (see Enting *et al.*, 1994b, for full documentation of the modelling exercise).

Two questions are considered:

- For a given CO_2 emission scenario, how might CO_2 concentrations change in the future?
- For a given CO_2 concentration profile leading to stabilisation, what anthropogenic emissions are implied?

As an initial condition, it was required that all models have a balanced carbon budget in which the components matched, within satisfactory limits, a prescribed 1980s-mean budget based on Watson *et al.* (1992). Modelling groups were provided with prescribed future land use fluxes, and a variety of time-series of concentrations and fossil-plus-cement emissions. Because the model intercomparison was conducted simultaneously with the rest of the assessment, the prescribed budget differs from the budget presented in this chapter (Table 1.3) in that:

- (i) the 1980s-mean concentration growth rate used (1.59 ppmv/yr or 3.4 GtC/yr) was higher than the current estimate (1.53 ppmv/yr or 3.2 GtC/yr);
- (ii) the net flux from changing land use was set at 1.6 GtC/yr rather than the current estimate of 1.1 GtC/yr;
- (iii) the only mechanism used in the models to simulate terrestrial uptake was CO_2 fertilisation, whereas we suggest that several other mechanisms may be important (Table 1.2).

Modellers were required to continue whatever processes were used to balance the 1980s budget into the future.

The approach used in balancing the budget probably biases the modelling exercise for this chapter towards lower concentrations when emission profiles were employed and higher anthropogenic emissions when concentrations were prescribed. There are two reasons for this. First, the effect of forest regrowth and nitrogen deposition result in changing terrestrial carbon storage which need not increase with increasing CO_2 concentrations as is required by the CO_2 fertilisation effect. Second, as noted in previous IPCC reports (Watson *et al.*, 1990, 1992) climate-related feedbacks may result in additional transient releases of CO_2 to the atmosphere. In contrast, most of the calculations employed in the 1990 IPCC assessment (Watson *et al.*, 1990) assumed constant terrestrial carbon content after 1990 and probably resulted in biases in the other direction. Revisions to the 1980s budget used in model initialisation for this report, however, are in a direction that would lead to higher concentrations by 5 to 10% (for given emissions) and lower emissions by a similar amount (for given concentrations).

Results from a range of different carbon cycle models are considered in order to assess the sensitivity of calculated emission and concentration profiles to model formulation. The complexity of the models employed varies considerably. The most detailed was a model coupling a three-dimensional ocean circulation and chemistry model to a terrestrial biosphere model incorporating a full geographical representation of ecological processes; but, as with other complex models, only a small set of calculations could be performed. The simplest models were designed to simulate only critical processes and represent terrestrial and oceanic carbon uptake with a minimum number of equations. Because most of the models employed incorporate some representation of terrestrial processes, they tended to have initially higher rates of CO_2 uptake from the atmosphere and produced lower estimates for the effective lifetime of CO_2 than the models used in the 1990 IPCC report (Moore and Braswell, 1994). Thus, the GWPs based on these newer carbon cycle model results are higher for other trace gases than those of earlier IPCC assessments (Shine *et al.*, 1990).

We chose one model, the "Bern model", for a number of important illustrative calculations, because its results were generally near the mid-point of the results obtained with all models, and because complete descriptions exist in the literature (Joos *et al.*, 1991a; Siegenthaler and Joos, 1992). Selected sensitivity analyses from the model of Wigley (1993) were also used as the configuration of that model allowed for ready modification to consider certain issues. Both the Bern and Wigley models have a balanced carbon cycle consisting of a well-mixed atmosphere linked to oceanic and terrestrial biospheric compartments. In the Bern model, the ocean is represented by the HILDA model, which is a box-diffusion model with an additional advective component. It was tuned to observed values of natural and bomb radiocarbon (Joos *et al.*, 1991a, b; Siegenthaler and Joos, 1992) and validated with CFCs and Argon-39 (Joos, 1992). The Wigley model uses a representation of the ocean based on the ocean general circulation carbon cycle model of Maier-Reimer and Hasselmann (1987). Both the models have similar terrestrial components, with representations of ground vegetation, wood, detritus, and soil (Siegenthaler and Oeschger, 1987; Wigley, 1993). A possible enhancement of plant growth due to elevated CO_2 levels is taken into account by a logarithmic dependency between additional photosynthesis and atmospheric CO_2, which, in the Bern model, probably overestimates biological storage at high CO_2 concentrations. The Bern model was chosen for illustrative purposes in discussions where presentation of multiple results would be confusing (and where all models produced similar patterns). This model was also used to define the reference case for the GWPs presented in Albritton *et al.* (1995). Neither model is recommended nor endorsed by the IPCC or the authors of this chapter as having higher credibility than other models.

1.6.2 Calculations of Concentrations for Specified Emissions

Six greenhouse gas emissions scenarios were described in the 1992 IPCC report (Leggett *et al.*, 1992), based on a wide range of assumptions regarding future economic, demographic, and policy factors. The anthropogenic CO_2 emissions for these scenarios are shown in Figure 1.9a. Scenario IS92c, which has the lowest CO_2 emissions, assumes an eventual decrease in population, low economic growth, and severe constraints on the availability of fossil fuel supplies. The highest emission scenario (IS92e) assumes moderate population growth, high economic growth, high fossil fuel availability, and a phase-out of nuclear power. Concentration estimates for these scenarios

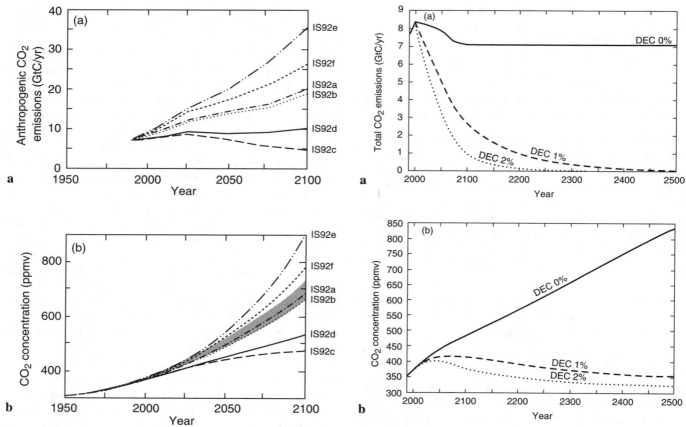

Figure 1.9: (a) Anthropogenic CO₂ emissions for the IS92 Scenarios. (b) Atmospheric CO₂ concentrations calculated from the scenarios IS92a-f (Leggett et al., 1992) using the Bern model (Siegenthaler and Joos; 1992). The typical range of results from different carbon cycle models is indicated by the shaded area.

Figure 1.10: (a) Anthropogenic CO₂ emissions calculated by following the IS92a Scenario to the year 2000 and then either fixed fossil fuel emissions (DEC 0%) or emissions declining at 1%/yr (DEC 1%) or 2%/yr (DEC 2%). Land-use emissions followed the modified IS92a scenario (see Section 1.5.2). (b) Atmospheric CO₂ concentrations resulting from DEC 0%, DEC 1% and DEC 2% emissions. Curves are for the model of Wigley (1993).

have previously been published (e.g., Wigley, 1993). Figure 1.9b shows concentration results from the Bern model that typify the responses of the wide range of models used for the full analysis. This and the other models show strong increases in concentration to well above pre-industrial levels by 2100 (75 to 220% higher). None of the six scenarios leads to stabilisation of concentration before 2100, although IS92c leads to a very slow growth in CO₂ concentration after 2050. IS92a, b, e, and f all produce a doubling of the pre-industrial CO₂ concentration before 2070, with rapid rates of concentration growth. Scenarios developed by the World Energy Council show a similar range of results (Figures 1.11a, b).

In addition to the IS92 emissions cases, three arbitrarily chosen "science" emissions profiles and the newly produced World Energy Council (WEC) Scenarios were also examined. In the former, fossil emissions followed IS92a to the year 2000 and then either stabilised (DEC0%) or decreased at 1% or 2%/yr (see Figure 1.10). For the WEC Scenarios, where only energy-related CO₂ emissions were originally given (WEC, 1993), estimates of gas-flaring and cement production emissions were added (M. Jefferson and G. Marland, personal communications) to ensure consistency with the IS92 Scenarios. Total fos-

sil emissions are given in Figure 1.11a. In all cases net land use emissions were assumed to follow IS92a to 2075. For the WEC cases, IS92a was followed to 2100. For the science scenarios, land use emissions dropped to zero in 2100 and remained zero thereafter. Concentration results are shown in Figures 1.10 and 1.11. Perhaps the most important result (which could be anticipated from the IS92c case) is that stabilisation of emissions at 2000 levels does not lead to stabilisation of CO₂ concentration by 2100; in fact, the calculations show that concentrations continue to increase slowly for at least several hundred years. The lowest of the WEC Scenarios, where emissions were based on policies driven by "ecological" considerations (see WEC, 1993), gives an idea of the sort of emissions profile that could lead to concentration stabilisation.

1.6.3 Stabilization Calculations

The calculations presented in this section illustrate additional aspects of what may be required to achieve stabilisation of atmospheric CO₂ concentrations. The exercise was motivated by

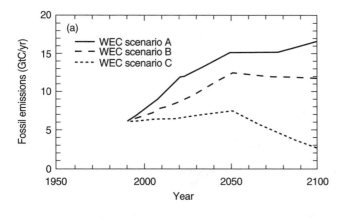

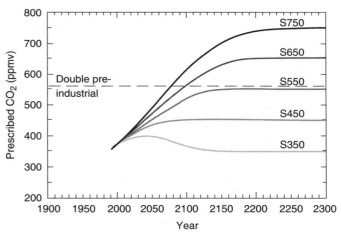

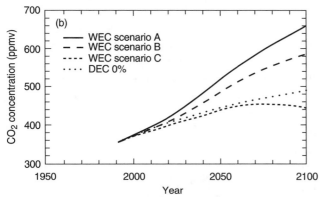

Figure 1.11: (a) Emission scenarios from the World Energy Council (WEC) modified by including gas flaring and cement production. (b) Atmospheric CO_2 concentrations calculated from the WEC Scenarios using the model of Wigley (1993). Concentrations resulting from the fixed emissions case (DEC 0%) (see Figure 1.10) using the same model are included for comparison.

Figure 1.12: CO_2 concentration profiles leading to stabilisation at 350, 450, 550, 650 and 750 ppmv. These are the profiles prescribed for carbon cycle model calculations in which the corresponding emission pathways (shown in Figure 1.13) were determined.

the Framework Convention on Climate Change (United Nations, 1992), which states:

The ultimate objective of this Convention and any related legal instruments that the Conference of the Parties may adopt is to achieve, in accordance with the relevant provisions of the Convention, stabilisation of greenhouse gas concentrations in the atmosphere at a level that would prevent dangerous anthropogenic interference with the climate system. Such a level should be achieved within a time-frame sufficient to allow ecosystems to adapt naturally to climate change, to ensure that food production is not threatened and to enable economic development to proceed in a sustainable manner.

In the context of this objective it is important to investigate a range of emission profiles of greenhouse gases which might lead to atmospheric stabilisation. It is not our purpose here to consider the climate response (this will be done in the 1995 IPCC Scientific Assessment report), nor to define what might constitute "dangerous interference", nor to make any judgement about the rates of change that would meet the criteria of

the objective. In this chapter only CO_2 is considered; the stabilisation of other greenhouse gas concentrations is discussed in Prather *et al.* (1995).

Several carbon cycle models have been used to calculate the emissions of CO_2 that would lead to stabilisation at a range of different concentration levels. These calculations are designed to illustrate the relationship between CO_2 concentration and emissions. Concentration profiles have been devised (Figure 1.12) in which CO_2 concentrations stabilise at levels from 350 to 750 ppmv (for comparison, the pre-industrial CO_2 concentration was close to 280 ppmv and the 1990 concentration was ~355 ppmv). Many different stabilisation levels and routes to stabilisation could have been chosen. Those in Figure 1.12 give a smooth transition from the 1990 rate of CO_2 concentration increase to stabilisation. As a result, the year of stabilisation differs with stabilisation levels from around 2150 for 350 ppmv to 2250 for 750 ppmv. Further details on the concentration profiles, and the implied emissions results are given in Enting *et al.* (1994b).

Figure 1.13 shows the model-derived profiles of total anthropogenic emissions (i.e., the sum of fossil fuel use, changes in land use, and cement production) that lead to stabilisation following the concentration profiles shown in Figure 1.12. Initially emissions rise, followed some decades later by quite rapid and large reductions. Stabilisation at any of the concentration levels studied (350–750 ppmv) is only possible if emissions are eventually reduced well below 1990 levels (Figure 1.13). For comparison, the emissions corresponding to IS92a, c, and e are also shown up to 2100 in Figure 1.13. Emissions for all the stabilisation levels studied are lower than those for IS92a and e, even in the first few decades of the 21st century. For the IS92c Scenario, emissions lie between those which in this study eventually achieve stabilisation between 450 and 550 ppmv.

The concentration profiles here are illustrative. Stabilisation at the same level, via a different route, would produce dif-

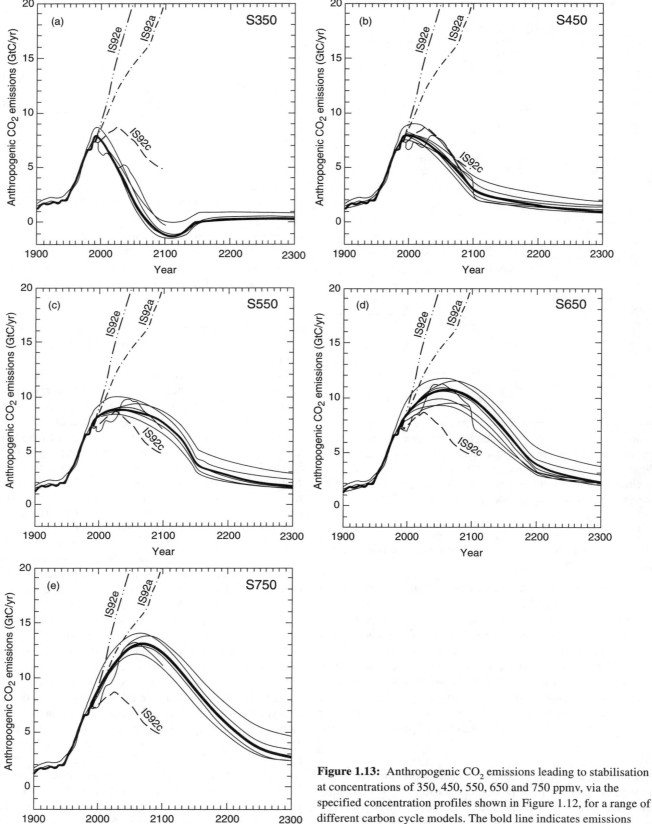

Figure 1.13: Anthropogenic CO₂ emissions leading to stabilisation at concentrations of 350, 450, 550, 650 and 750 ppmv, via the specified concentration profiles shown in Figure 1.12, for a range of different carbon cycle models. The bold line indicates emissions calculated using the Bern model.

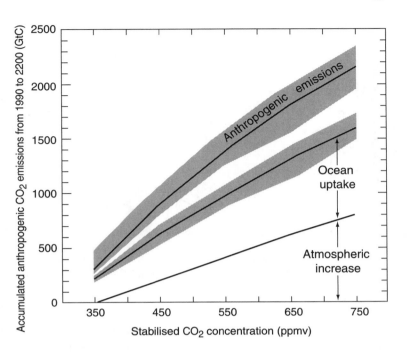

Figure 1.14: Accumulated anthropogenic CO₂ emissions over the period 1990 to 2200 (GtC) plotted against the final stabilised concentration level. Also shown are the accumulated ocean uptake and the increase of CO₂ in the atmosphere. The curves for accumulated anthropogenic emissions and ocean uptake were calculated using the model of Siegenthaler and Joos (1992). The shaded areas show the spread of results from a range of carbon cycle model calculations. The difference (i.e., the accumulated anthropogenic emissions minus the total of the atmospheric increase and the accumulated ocean uptake) gives the cumulative change in terrestrial biomass.

ferent curves from those shown in Figure 1.13. However, the total amount of emitted carbon accumulated over time (the area under the curves in Figure 1.13), is less sensitive to the concentration profile than to the stabilisation level. Cumulative emissions and carbon storage amounts over the period 1990 to 2200 are shown in Figure 1.14. The separate model results are not shown, but the maximum and minimum estimates are indicated, along with the results from the Bern model, identified as a near-average model. Stabilisation at a lower concentration implies lower accumulated emissions (Figure 1.14).

The spread in results is large, in part because the problem of deducing sources given concentrations is inherently less stable than deducing concentrations from sources, analogous to the inverse problems described in Section 1.3.2.3. Any inferences regarding emissions strategies drawn from these results should take into account the inherent uncertainties in the projections and the limitations imposed by assumptions made to derive them (see Section 1.5.4 and Enting *et al.*, 1994b), and allow response to knowledge gained through continuing observation of future trends in emissions and concentrations.

1.6.4 Assessment of Uncertainties

Two types of uncertainty analysis have been performed for the emission-concentration modelling studies. First, because the models used in the intercomparison incorporate model components of differing structure and levels of complexity to calculate terrestrial and ocean uptake, the range of results provides a view of uncertainties arising from different scientific approaches. While this is a useful view, it probably underestimates the overall uncertainty because the experiments were constrained to have a specific value for the flux arising from changing land use during the 1980s, and a terrestrial sink due solely to CO₂ fertilisation. Because ocean sink magnitudes

were not constrained, models with smaller atmosphere-to-ocean fluxes have correspondingly (and possibly unrealistically) larger terrestrial sinks. None of the models used for the terrestrial sink process adequately accounted for the complexities of terrestrial uptake discussed in Section 1.4.3.4. Because the models were constrained to match a particular carbon budget during the 1980s, the model's future projections were closer together than if the experiment had been less constrained. In addition, fossil fuel and cement emissions were prescribed for the 1980s; had the uncertainty in those numbers (about 10%) been allowed for, there would have been a wider range of model parameters for ocean and biospheric processes. Moreover, the results in Figures 1.13 and 1.14 do not account for possible climate feedbacks on the carbon cycle (see Section 1.5).

Although a complete assessment of model sensitivity and uncertainties (e.g., following Gardner and Trabalka, 1985) was outside the scope of the initial exercise carried out for this assessment, several specific model sensitivities have been studied. Figure 1.15 shows the sensitivity of future concentrations under emission scenario IS92a to varied strength of biospheric uptake. In this experiment (Wigley, 1993), the effectiveness of CO₂ fertilisation was varied across the range thought to be reasonable (10–40% enhancements of plant growth for a doubling of CO₂). Significant changes in projected concentrations occur depending upon the CO₂ fertilisation factor used, by approximately ± 14% over 1990 to 2100. Figure 1.16a shows an uncertainty assessment for the emissions corresponding to the S450 and S650 stabilisation cases, using the model of Wigley (1993). The effectiveness of CO₂ fertilisation was varied from 10% to 40% NPP enhancement for a CO₂ doubling (compared with the baseline case for this model of 26%), leading to lower and higher emissions, respectively, by up to ±1.4 GtC/yr for S650 and ±1.1 GtC/yr for S450. Figure 1.16b shows a similar

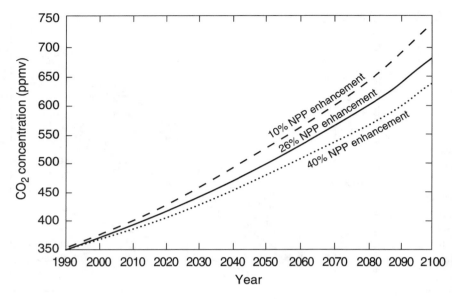

Figure 1.15: Concentration uncertainties associated with the CO_2 fertilisation effect for the IS92a emission scenario. The central curve shows the concentration projection using "best guess" model parameters using the model of Wigley (1993), which is similar to projections using other models. For the upper and lower curves the CO_2 fertilisation effect was decreased to 10% NPP enhancement (from 26% for the "best guess" simulation) and increased to 40% NPP enhancement for a CO_2 doubling from 340 ppmv to 680 ppmv. As explained in Wigley (1993), this assessment also captures some of the uncertainties arising from ocean flux uncertainties, through the need to balance the contemporary carbon budget within realistic limits.

assessment of uncertainty associated with concentrations corresponding to the S450 and S650 stabilisation profiles that result when the strength of terrestrial uptake of CO_2 (via fertilisation of plant growth) is varied.

While the sensitivity of the global carbon budget to the strength of CO_2 fertilisation does not appear great, if biospheric release were to exceed uptake in the future as a result of climate feedbacks (e.g., Smith and Shugart, 1993) or land

use emissions (e.g., Esser, 1990), atmospheric concentrations could be significantly greater than those shown here: effects of changing land use could add 100s of GtC to the atmosphere (for comparison see Figure 1.14), over the next one to several centuries. Different assumptions about land use changes would produce different results. For example, if large areas were deforested, in the absence of substantial regrowth the capacity of the terrestrial biosphere to act as a

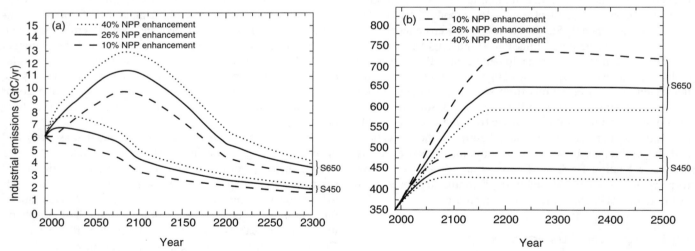

Figure 1.16: (a) Industrial emission uncertainties for concentration stabilisation profiles S450 and S650 associated with the CO_2 fertilisation effect, obtained using the model of Wigley (1993). The baseline emissions results are shown by the bold full lines. These correspond to a CO_2 fertilisation parameter equivalent to an increase in NPP of 26% for a CO_2 doubling. For the upper and lower (dashed) curves, the CO_2 fertilisation effect was increased to 40% NPP enhancement and decreased to 10% NPP enhancement, respectively. Note that it is the relative effects that are important here and that the baseline case is only one of the range of results presented in Figure 1.13. (b) Concentration uncertainties for stabilization profiles S450 and S650 associated with the CO_2 fertilisation effect. The baseline concentration profiles are shown as bold full lines. The dashed curves bounding these represent the range of uncertainty associated with uncertainties in terrestrial sink processes. They were constructed by first estimating the industrial emissions using "best guess" carbon cycle model parameters, which include a 1980s-mean ocean flux of 2.0 Gt C/yr and a CO_2 fertilization parameter equivalent to an increase in NPP of 26% for a CO_2 doubling. The fertilization parameter was then decreased to 10% and increased to 40% to obtain the higher and lower concentration projections. Essentially, this answers the question: what if we devise an industrial emissions policy based on a particular fertilization effect, and this happens to be either too high or too low? Model results were obtained using the model of Wigley (1993)

sink would be reduced; hence, more CO_2 would remain in the atmosphere.

The effects of varying land use practices have not been assessed directly in the stabilisation exercise to date, but the changes in biospheric carbon storage from alternate land use scenarios, mitigation efforts, and climatic effects (e.g., Esser, 1990; Smith and Shugart, 1993; Vloedbeld and Leemans, 1993) discussed in Section 1.4.2 are large enough to influence integrated emissions for a given concentration profile (Figure 1.14) substantially.

References

Aber, J. D., J. K. Nadlehoffer, P. A. Steudler and J. M. Melillo, 1989: Nitrogen saturation in northern forest ecosystems - hypotheses and implications. *BioScience, 39,* 378–386.

Alcamo, J., G. J. J. Kreileman, M. Krol and G. Zuidema, 1994a: Modeling the global society-biosphere-climate system. Part 1: model description and testing. *Water, Air, and Soil Pollution, 76,* 1–35.

Alcamo, J., G. J. Van den Born, A. F. Bouwman, B. de Haan, K. Klein-Goldewijk, O. Klepper, R. Leemans, J. A. Oliver, B. de Vries, H. van der Woerd and R. van den Wijngaard, 1994b: Modelling the global society-biosphere-climate system. Part 2: computed scenarios. *Water, Air, and Soil Pollution, 76,* 37–78.

Albritton, D. L., R. G. Derwent, I. S. A. Isaksen, M. Lal and D. J. Wuebbles, 1995: Trace gas radiative forcing indices. In: *Climate Change 1994: Radiative Forcing of Climate Change and An Evaluation of the IPCC IS92 Emission Scenarios,* J. T. Houghton, L. G. Meira Filho and J. Bruce, Hoesung Lee, B. A. Callander, E. Haites, N. Harris and K. Maskell (eds.), Cambridge University Press, Cambridge, U.K., pp. 205–231.

Andres, R. J., G. Marland, T. Boden and S. Bischoff, 1994: Carbon dioxide emissions from fossil fuel consumption and cement manufacture 1751 to 1991 and an estimate for their isotopic composition and latitudinal distribution. In: *The Carbon Cycle.* T. M. L. Wigley and D. Schimel (eds.), Cambridge University Press, New York, NY (In press)

Archer, D. and E. Maier-Reimer, 1994: Effect of deep-sea sedimentary calcite preservation on atmospheric CO_2 concentration. *Nature, 367,* 260–263.

Bacastow, R. and E. Maier-Reimer, 1990: Ocean-circulation model of the carbon cycle. *Climate Dynamics, 4,* 95–125.

Barnola, J.-M., D. Raynaud, Y. S. Korotkevitch and C. Lorius, 1987: Vostok ice core provides 160,000 year record of atmospheric CO_2. *Nature, 329,* 408–414.

Barnola, J.-M., P. Pimienta, D. Raynaud and T. S. Korotkevich, 1991: CO_2-climate relationship as deduced from the Vostok ice core: a re-examination based on new measurement and on a re-evaluation of the air dating. *Tellus, 43B,* 83–90.

Barnola, J.-M., M. Anklin, J. Porcheron, D. Raynaud, J. Schwander and B. Stauffer, 1994: CO_2 evolution during the last millenium as recorded by Antarctic and Greenland ice. *Tellus.* (In press)

Bazzaz, F. A. and E. D. Fajer, 1992: Plant life in a CO_2-rich world. *Scientific American, 266,* 68–74.

Bender, M. L., T. Sowers, J-M. Barnola and J. Chappellaz, 1994: Changes in the O_2/N_2 ratio of the atmosphere during recent decades reflected in the composition of air in the firn at Vostok Station, Antarctica. *Geophys. Res. Lett., 21,* 189–192.

Birdsey, R. A., A. J. Plantiga and L. S. Heath, 1993: Past and prospective carbon storage in United States forests. *Forest Ecology Management, 58,* 33–40.

Boden, T. A., R. J. Sepanski and F. W. Stoss, 1991: Trends '91: A compendium of data on global change. Oak Ridge National Laboratory, ORNL/CDIAC-46.

Briffa, K. R., T. S. Bartholin, D. Eckstein, P. D. Jones, W. Karlén, F. H. Schweingruber and P. Zetterberg, 1990: A 1,400 tree-ring record of summer temperatures in Fennoscandia. *Nature, 346,* 434–439.

Broecker, W. S. and T.-H. Peng, 1994: Stratospheric contribution to the global bomb radiocarbon inventory: Model versus observation. *Global Biogeochemical Cycles, 8,* 377–384.

Broecker, W. S., T. Takahashi, H. J. Simpson and T-H. Peng, 1979: Fate of fossil fuel carbon dioxide and the global carbon budget. *Science, 206,* 409–418.

Broecker, W. S., T.-H. Peng, G. Ostlund and M. Stuiver, 1985: The distribution of bomb radiocarbon in the ocean. *J. Geophys. Res. 90,* 6953–6970.

Brown, H. T. and F. Escombe, 1905: On the variations in the amount of carbon dioxide in the air of Kew during the years 1893–1901. *Proceedings of the Royal Society of London, Biology, 76,* 118–121.

Brown, S. and A. E. Lugo, 1984: Biomass of tropical forests: a new estimate based on forest volumes. *Science, 223,* 1290–1293.

Brown, S. and A. E. Lugo, 1992: Aboveground biomass estimates for tropical moist forests of the Brazilian Amazon. *Interciencia, 17,* 8–18.

Chameides, W. L., P. S. Kasibhatla, J. Yienger and H. Levy II, 1994: Growth of continental-scale metro-agro-plexes, regional ozone production, and world food production. *Science, 264,* 74–77.

Chappellaz, J., T. Blunier and D. Raynaud, 1993: Synchronous changes in atmospheric CH_4 and Greenland climate between 40 and 8 kyr BP. *Nature, 366,* 443–445.

Ciais, P., P. Tans, J. W. White, M. Trolier, R. Francey, J. Berry, D. Randall, P. Sellers, J. G. Collatz and D. Schimel, 1994: Partitioning of ocean and land uptake of CO_2 as inferred by δ13 measurements from the NOAA/CMDL global air sampling network. *J. Geophys. Res.* (In press)

CLIMAP Project Members, 1984: The last interglacial ocean. *Quaternary Research, 21,* 123–224.

Cole, C. V. and R. D. Heil, 1981: Phosphorus effects on terrestrial nitrogen cycling. In: *Terrestrial Nitrogen Cycles,* F. E. Clark and T. Rosswall (eds.), Ecological Bulletin, Swedish Natural Science Research Council, Stockholm, pp. 363–374.

Coleman, J. S. and F. A. Bazzaz, 1992: Effects of CO_2 and temperature on growth and resource use of co-occurring C_3 and C_4 annuals. *Ecology, 73,* 1244–1259.

Coleman, J. S., K. D. M. McConnaughay and F. A. Bazzaz, 1993: Elevated CO_2 and plant nitrogen use: is reduced tissue nitrogen concentration size-dependent? *Oecologia, 93,* 195–200.

Comins, H. N. and R. E. McMurtrie, 1993: Long-term response of nutrient limited forests to CO_2 enrichment: equilibrium behavior of plant-soil models. *Ecological Applications, 3,* 666–681.

Conway, T. J., P. Tans, L. S. Waterman, K. W. Thoning, D. R. Buanerkitzis, K. A. Maserie and N. Zhang, 1994: Evidence for interannual variability of the carbon cycle from the NOAA/CMDL global air sampling network. *J. Geophys. Res., 99D,* 22, 831–22, 855.

Cramer, W. P. and A. M. Solomon, 1993: Climate classification and future global redistribution of agricultural land. *Climate Research, 3,* 97–110.

Crowley, T. J. and G. R. North, 1991: *Paleoclimatology.* Oxford University Press, Oxford, U.K.

Cubasch, U., K. Hasselmann, H. Hoeck, E. Maier-Reimer, U. Mikolajewicz, B. D. Santer and R. Sausen, 1992: Time-dependent greenhouse warming computations with a coupled ocean-atmosphere model. *Climate Dynamics, 8,* 55–69.

Dai, A. and I. Y. Fung, 1993: Can climate variability contribute to the "missing" CO_2 sink? *Global Biogeochemical Cycles, 7,* 599–609.

Dansgaard, W., S. J. Johnsen, H. B. Clausen, D. Dahl-Jensen, N. S. Gunderstrup, C. U. Hammer, J. P. Steffensen, A. Sveinbjörnsdoltir, J. Jouzel and G. Bond, 1993: Evidence for general instability of past climate from a 250-kyr ice-core record. *Nature, 364,* 218–220.

D'Arrigo, R. D. and G. C. Jacoby, 1993: Tree growth-climate relationships at the northern boreal forest tree line of North America: evaluation of potential response to increasing carbon dioxide. *Global Biogeochemical Cycles, 7,* 525–535.

Degens, E. T., S. Kempe and J. E. Richey, 1991: Summary: biogeochemistry of major world rivers. In: *Biogeochemistry of Major World Rivers,* E. T. Degens, S. Kempe and J. E. Richey (eds.), SCOPE Report 42, John Wiley and Sons, Chichester, pp. 323–347.

Delmas, R. J., 1993: A natural artefact in Greenland ice-core CO_2 measurements. *Tellus, 45B,* 391–396.

Delmas, R. J., J. M. Ascencio and M. Legrand, 1980: Polar ice evidence that atmospheric CO_2 20,000 yr B. P. was 50% of present. *Nature, 284,* 155–157.

Detwiler, R. P. and C. A. S. Hall, 1988: Tropical forests and the global carbon cycle. *Science, 239,* 42–47.

Díaz, S., J. P. Grime, J. Harris and E. McPherson, 1993: Evidence of a feedback mechanism limiting plant response to elevated carbon dioxide. *Nature, 364,* 616–617.

Dixon, R. K., S. A. Brown, R. A. Houghton, A. M. Solomon, M. C. Trexler and J. Wisniewski, 1994: Carbon pools and flux of global forest ecosystems. *Science, 263,* 185–190.

Donigian, A. S., Jr., T. O. Barnwell Jr., R. B. Jackson IV, A. S. Patwardhan, K. B. Weinrich, A. L. Rowell, R. V. Chinnaswamy and C. V. Cole, 1994: Assessment of alternative management practices and policies affecting soil carbon in agroecosystems of the Central United States. Environmental Protection Agency, *EPA/600/R-94/067.*

Drake, B. G., 1992: The impact of rising CO_2 on ecosystem production. *Water, Air, and Soil Pollution, 64,* 25–44.

Druffel, E. R. M., P. M. Williams, J. E. Bauer and J. R. Ertel, 1992: Cycling of dissolved particulate organic matter in the open ocean. *J. Geophys. Res., 97C,* 15,639–15,659.

Duce, R. A., P. S. Liss, J. T. Merrill, E. L. Atlas, P. Buat-Ménard, B. B. Hicks, J. M. Miller, J. M. Prospero, R. Arimoto, T. M. Church, W. Ellis, J. N. Galloway, L. Hansen, T. D. Jickells, A. H. Knap, K. H. Reinhardt, B. Schneider, A. Soudine, J. J. Tokos, S. Tsunogai, R. Wollast and M. Zhou, 1991: The atmospheric input of trace species to the world ocean. *Global Biogeochemical Cycles, 5,* 193–259.

Eisele, K. A., D. S. Schimel, L. A. Kapustka and W. J. Parton, 1989: Effects of available P and N:P ratios on non-symbiotic dinitrogen fixation in tallgrass prairie soils. *Oecologia, 79,* 471–474.

Enting, I. G. 1987: The interannual variation in the seasonal cycle of carbon dioxide concentration at Mauna Loa. *J. Geophys. Res., 92(D),* 5497–5504.

Enting, I. G., 1993: Inverse problems in atmospheric constituent studies. III. Estimating errors in surface sources. *Inverse Problems, 9,* 649–665.

Enting, I. G., 1994: CO_2-climate feedbacks: aspects of detection. In: *Feedbacks in the Global Climate System,* G. M. Woodwell (ed.), Oxford University Press, New York. (In press)

Enting, I. G. and J. V. Mansbridge, 1989: Seasonal sources and sinks of atmospheric CO_2: direct inversion of filtered data. *Tellus, 41,* 111–126.

Enting, I. G. and J. V. Mansbridge, 1991: Latitudinal distribution of sources and sinks of CO_2: results of an inversion study. *Tellus, 43,* 156–170.

Enting, I. G., R. J. Francey and C. M. Trudinger, 1994a: Synthesis study of the atmospheric CO_2 and $^{13}CO_2$ budgets. *Tellus* (In press)

Enting, I. G., T. M. L. Wigley and M. Heimann, 1994b: Future emissions and concentrations of carbon dioxide: key ocean/atmosphere/land analyses. CSIRO Division of Atmospheric Research Technical Paper No. 31.

Esser, G., 1990: Modeling global terrestrial sources and sinks of CO_2 with special reference to soil organic matter. In: *Soils and the Greenhouse Effect,* A. F. Bouwman (ed.), John Wiley and Sons, New York, pp. 247–262.

Eswaran, H., E. Van den Berg and P. Reich, 1993: Organic carbon in soils of the world. *Soil Science Society of America Journal, 57,* 192–194.

Evans, G. T. and M. J. R. Fasham (eds.), 1993: *Towards a Model of Ocean Biogeochemical Processes.* NATO Workshop, Springer-Verlag, New York, 350 pp.

Falkowski, P. G. and C. Wilson, 1992: Phytoplankton productivity in the North Pacific ocean since 1990 and implications for absorption of anthropogenic CO_2. *Nature, 358,* 741–743.

FAO (United Nations Food and Agriculture Organisation), 1993: Forest Resources Assessment 1990: Tropical Countries. FAO Forestry Paper No. 112, Rome, Italy.

Fearnside, P. M., 1992: Forest biomass in Brazilian Amazonia: comments on the estimate by Brown and Lugo. *Interciencia, 17,* 19–27.

Friedli, H., H. Loetscher, H. Oeschger, U. Siegenthaler and B. Stauffer, 1986: Ice core record of the $^{13}C/^{12}C$ ratio of atmospheric CO_2 in the past two centuries. *Nature, 324,* 237–238.

Friedlingstein, P., C. Delire, J-F. Müller and J-C. Gérard, 1992: The climate induced variation of the continental biosphere: a model simulation of the Last Glacial Maximum. *Geophys. Res. Lett., 19,* 897–900.

Fung, I. Y. and T. Takahashi, 1994: A re-evaluation of empirical estimates of the oceanic sink of anthropogenic CO_2. In: *The Carbon Cycle,* T. M. L. Wigley and D. Schimel (eds.), Cambridge University Press, New York, NY (In press)

Garcon, V., F. Thomas, C. S. Wong and J. F. Minster, 1992: Gaining insight into the seasonal variability of CO_2 at O. W. S. P. using an upper ocean model. *Deep Sea Research, 30,* 921–938.

Gardner, R. H. and J. R. Trabalka, 1985: *Methods of Uncertainty Analysis for a Global Carbon Dioxide Model,* U.S. Department of Energy, Carbon Dioxide Research Division, Technical Report TR024, 41 pp.

Gifford, R. M., 1993: Implications of CO_2 effects on vegetation for the global carbon budget. In: *The Global Carbon Cycle,* M. Heimann (ed.), Proceedings of the NATO Advanced Study Institute, Il Ciocco, Italy, September 8–20, 1991, pp. 165–205.

Gifford, R. M., 1994: The global carbon cycle: a viewpoint on the missing sink. *Australian Journal of Plant Physiology, 21,* 1–15.

Graumlich, L. J., 1991: Subalpine tree growth, climate, and increasing CO_2: an assessment of recent growth trends. *Ecology, 72,* 1–11.

GRIP Project Members, 1993: Climatic instability during the last interglacial period revealed in the Greenland summit ice-core. *Nature, 364,* 203–207.

Grootes, P. M., M. Stuvier, J. W. C. White, S. Johnsen and J. Jouzel, 1993: Comparison of oxygen isotope records from GISP2 and GRIP Greenland ice cores. *Nature, 366,* 552–554.

Heimann, M. and E. Maier-Reimer, 1994: On the relations between the uptake of CO_2 and its isotopes by the ocean. *Global Biogeochemical Cycles.* (Submitted)

Heimann, M., C. D. Keeling and C. J. Tucker, 1989: A three dimensional model of atmospheric CO_2 transport based on observed winds: 3. Seasonal cycle and synoptic time-scale variations. In:

Aspects of Climate Variability in the Pacific and the Western Americas, D. H. Peterson (ed.), American Geophysical Union, Washington, DC, pp. 277–303.

Heinze, C., E. Maier-Reimer and K. Winn, 1991: Glacial pCO$_2$ reduction by the world ocean: experiments with the Hamburg carbon cycle model. *Paleoceanography, 6,* 395–430.

Hesshaimer, V., M. Heimann and I. Levin, 1994: Radiocarbon evidence suggesting a smaller oceanic CO$_2$ sink than hitherto assumed. *Nature, 370,* 201–203.

Houghton, R. A., 1993: Is carbon accumulating in the northern temperate zone. *Global Biogeochemical Cycles, 7,* 611–617.

Houghton, R. A., 1994a: Emissions of carbon from land-use change. In: *The Carbon Cycle*, T. M. L. Wigley and D. Schimel (eds.), Cambridge University Press, Stanford, CA. (In press)

Houghton, R. A., 1994b: Effects of land-use change, surface temperature and CO$_2$ concentration on terrestrial stores of carbon. In: *Biotic Feedbacks in the Global Climate System: Will the Warming Speed the Warming?*, G. M. Woodwell and F. T. Mackenzie (eds.), Oxford University Press, Oxford. (In press)

Houghton, R. A. and G. M. Woodwell, 1989: Global climatic change. *Scientific American, 260,* 36–47.

Houghton, R. A. and D. L. Skole, 1990: Carbon. In: *The Earth as Transformed by Human Action*, B. L. Turner II, W. C. Clark, R. W. Kates, J. F. Richards, J. T. Mathews and W. B. Meyer (eds.), Cambridge University Press, New York, pp. 393–408.

Houghton, R. A., R. D. Boone, J. R. Fruci, J. E. Hobbie, J. M. Melillo, C. A. Palm, B. J. Peterson, G. R. Shaver, G. M. Woodwell, B. Moore, D. L. Skole and N. Myers, 1987: The flux of carbon from terrestrial ecosystems to the atmosphere in 1980 due to changes in land use: geographic distribution of the global flux. *Tellus, 39B,* 122–139.

Idso, K. E. and S. B. Idso, 1994: Plant responses to atmospheric CO$_2$ enrichment in the face of environmental constraints: a review of the last 10 years' research. *Agricultural and Forest Meteorology, 69,* 153–203.

Idso, S. B. and B. A. Kimball, 1993: Tree growth in carbon dioxide enriched air and its implications for global carbon cycling and maximum levels of atmospheric CO$_2$. *Global Biogeochemical Cycles, 7,* 537–555.

INPE, 1992: *Deforestation in Brazilian Amazonia.* Instituto Nacional de Pesquisas Espaciais, São Paulo, Brazil.

IPCC (Intergovernmental Panel on Climate Change), 1990: *Climate Change: the IPCC Scientific Assessment*, J. T. Houghton, G. J. Jenkins and J. J. Ephraums (eds.). Cambridge University Press, Cambridge, U.K. 365 pp.

IPCC, 1992: *Climate Change 1992: The Supplementary Report to the IPCC Scientific Assessment*, J. T. Houghton, B. A. Callander and S. K. Varney (eds.). Cambridge University Press, Cambridge, U.K. 200 pp.

Jenkinson, D. S., J. M. Potts, J. N. Perry, V. Barnett, K. Coleman and A. E. Johnston, 1994: Trends in herbage yields over the last century on the Rothamsted long-term continuous hay experiment. *Journal of Agricultural Science, 122,* 365–374.

Johnsen, S. J., H. B. Clausen, W. Dansgaard, K. Furher, N. Gundestrup, C. U. Hammer, P. Iverson, J. Jouzel, B. Stauffer and J. P. Steffensen, 1992: Irregular glacial interstadials recorded in a new Greenland ice core. *Nature, 359,* 311–313.

Joos, F., 1992: Modellierung und verteilung von spurenstoffen im ozean und des globalen kohlenstoffkreislaufes. Ph.D. Thesis, University of Bern, Bern.

Joos, F., J. L. Sarmiento and U. Siegenthaler, 1991a: Estimates of the effect of Southern Ocean iron fertilization on atmospheric CO$_2$ concentrations. *Nature, 349,* 772–774.

Joos, F., U. Siegenthaler and J. L. Sarmiento, 1991b: Possible effects of iron fertilization in the Southern Ocean on atmospheric CO$_2$ concentration. *Global Biogeochemical Cycles, 5,* 135–150.

Jouzel, J., N. I. Barkov, J-M. Barnola, M. Bender, J. Chappellaz, C. Genthon, V. M. Kotlyakov, V. Lipenkov, C. Lorius, J. R. Petit, D. Raynaud, G. Raisbeck, C. Ritz, T. Sowers, M. Stievenard, F. Yiou and P. Yiou, 1993: Extending the Vostok ice-core record of paleoclimate to the penultimate glacial period. *Nature, 364,* 407–412.

Keeling, C. D., 1973: Industrial production of carbon dioxide from fossil fuels and limestone. *Tellus, 25,* 174–198.

Keeling, C. D., 1993: Surface ocean CO$_2$. In: *The Global Carbon Cycle*, M. Heimann (ed.), Springer-Verlag, Heidelberg, pp. 413–429.

Keeling, C. D. and T. P. Whorf, 1994: Decadal oscillations in global temperature and atmospheric carbon dioxide. In: *Natural Variability of Climate on Decade-to-Century Time Scales*, W. A. Sprigg (ed.), National Academy of Sciences, Washington, DC. (In press)

Keeling, C. D., R. B. Bacastow, A. F. Carter, S. C. Piper, T. P. Whorf, M. Heimann, W. G. Mook and H. Roeloffzen, 1989a: A three-dimensional model of atmospheric CO$_2$ transport based on observed winds: 1. Analysis and observational data. In: *Aspects of Climate Variability in the Pacific and Western Americas. Geophysical Monograph 55*, D. H. Peterson (ed.), American Geophysical Union, Washington, DC, pp. 165–236.

Keeling, C. D., S. C. Piper and M. Heimann, 1989b: A three-dimensional model of atmospheric CO$_2$ transport based on observed winds: 4. Mean annual gradients and interannual variations. In: *Aspects of Climate Variability in the Pacific and Western Americas. Geophysical Monograph 55*, D. H. Peterson (ed.), American Geophysical Union, Washington, DC, pp. 305–363.

Keeling, R. F. and R. Shertz, 1992: Seasonal and interannual variations in atmospheric oxygen and implications for the global carbon cycle. *Nature, 358,* 723–727.

Keeling, R. F. and J. Severinghaus, 1994: Atmospheric oxygen measurements and the carbon cycle. In: *The Carbon Cycle*, T. M. L. Wigley and D. Schimel (eds.), Cambridge University Press, Stanford, CA. (In press)

Keir, R. S., 1994: Effects of ocean circulation changes and their effects on CO$_2$. In: *The Carbon Cycle*. T. M. L. Wigley and D. Schimel (eds.), Cambridge University Press, Stanford, CA. (In press)

Kempe, S. and K. Pegler, 1991: Sinks and sources of CO$_2$ in coastal seas: the North Sea. *Tellus, 43,* 224–235.

Kolchugina, T. P. and T. S. Vinson, 1993: Carbon sources and sinks in forest biomes of the former Soviet Union. *Global Biogeochemical Cycles, 7,* 291–304.

Körner, C. and J. A. Arnone III, 1992: Responses to elevated carbon dioxide in artificial tropical ecosystems. *Science, 257,* 1672–1675.

Krankina, O. N. and R. K. Dixon, 1994: Forest management options to conserve and sequester terrestrial carbon in the Russian Federation. *World Resources Review, 6,* 88–101.

Kurz, K. D. and E. Maier-Reimer, 1993: Iron fertilization of the austral ocean – a Hamburg model assessment. *Global Biogeochemical Cycles, 7,* 229–244.

LaMarche, V. C. J., D. A. Graybill and M. R. Rose, 1984: Increasing atmospheric carbon dioxide: tree ring evidence for growth enhancement in natural vegetation. *Science, 225,* 1019–1021.

Leggett, J., W. J. Pepper and R. J. Swart, 1992: Emissions scenarios for IPCC: an update. In: *Climate Change 1992: The Supplementary Report to the IPCC Scientific Assessment*, J. T. Houghton, B. A. Callander and S. K. Varney (eds.), Cambridge University Press, Cambridge, U.K. pp. 69–95.

Leuenberger, M., U. Siegenthaler and C. C. Langway, 1992: Carbon isotope composition of atmospheric CO_2 during the last ice age from an Antarctic ice core. *Nature, 357,* 488–490.

MacIntyre, F., 1978: On the temperature coefficient of pCO_2 in seawater. *Climatic Change, 1,* 349–354.

Maier-Reimer, E., 1993: The biological pump in the greenhouse. *Global and Planetary Change, 8,* 13–15.

Maier-Reimer, E. and K. Hasselmann, 1987: Transport and storage in the ocean - an inorganic ocean-circulation carbon cycle model. *Climate Dynamics, 2,* 63–90.

Manabe, S. and R. J. Stouffer, 1993: Century-scale effects of increased atmospheric CO_2 on the ocean-atmosphere system. *Nature, 364,* 215–218.

Manning, M. R., 1993: Seasonal cycles in atmospheric CO_2 concentrations. In: *The Global Carbon Cycle,* M. Heimann (ed.), Springer-Verlag, Heidelberg, pp. 65–94.

Marland, G. and R. M. Rotty, 1984: Carbon dioxide emissions from fossil fuels: a procedure for estimation and results for 1950–1982. *Tellus, 36B,* 232–261.

Martin, J. H., 1990: Glacial-interglacial CO_2 change: the iron hypothesis. *Paleoceanography, 5,* 1–13.

Maybeck, M., 1993: Natural sources of C, N, P, and S. In: *Interactions of C, N, P, and S Biogeochemical Cycles and Global Change,* R. Wollast (ed.), Springer-Verlag, Berlin, pp. 163–193.

Melillo, J. M., J. R. Fruci, R. A. Houghton, B. Moore III and D. L. Skole, 1988: Land-use change in the Soviet Union between 1850 and 1980: causes of a net release of CO_2 to the atmosphere. *Tellus, 40B,* 166–128.

Melillo, J. M., A. D. McGuire, D. W. Kicklighter, B. Moore III, C. J. Vorosmarty and A. L. Schloss, 1993: Global climate change and terrestrial net primary production. *Nature, 363,* 234–240.

Melillo, J. M., D. W. Kicklighter, A. D. McGuire, W. T. Peterjohn and K. Newkirk, 1994: Global change and its effects on soil organic carbon stocks. In: *Dahlem Conference Proceedings,* John Wiley and Sons, New York. (In press)

Metherell, A. K., 1992: Simulation of soil organic matter dynamics and nutrient cycling in agroecosystems. Ph.D. Dissertation, Colorado State University, Fort Collins.

Mitchell, J. F. B., S. Manabe, V. Meleshko and T. Tokioka, 1990: Equilibrium climate change – and its implications for the future. In: *Climate Change: the 1990 Scientific Assessment,* J. T. Houghton, G. J. Jenkins and J. J. Ephraums (eds.), Cambridge University Press, Cambridge, U.K., pp. 131–172.

Moore, B. III, and B. H. Braswell Jr., 1994: The lifetime of excess atmospheric carbon dioxide. *Global Biogeochemical Cycles, 8,* 23–38.

Neftel, A., H. Oeschger, J. Schwander, B. Stauffer and R. Zumbrunn, 1982: Ice core sample measurements give atmospheric CO_2 content during the past 40,000 years. *Nature, 295,* 220–223.

Neftel, A., E. Moor, H. Oeschger and B. Stauffer, 1985: Evidence from polar ice cores for the increase in atmospheric CO_2 in the past two centuries. *Nature, 315,* 45–47.

Neftel, A., H. Oeschger, T. Staffelbach and B. Stauffer, 1988: CO_2 record in the Byrd ice core 50,000–5,000 years BP. *Nature, 331,* 609–611.

Norby, R. J., C. A. Gunderson, S. D. Wullschleger, E. G. O'Neill and M. K. McCracken, 1992: Productivity and compensatory response of yellow-poplar trees in elevated CO_2. *Nature, 357,* 322–324.

Oechel, W. C., S. J. Hastings, G. Vourlitis, M. Jenkins, G. Riechers and N. Grulke, 1993: Recent change of Arctic tundra ecosystems from a net carbon dioxide sink to a source. *Nature, 361,* 520–523.

Oeschger, H., U. Siegenthaler and A. Guglemann, 1975: A box-diffusion model to study the carbon dioxide exchange in nature. *Tellus, 27,* 168–192.

Ojima, D. S., W. J. Parton, D. S. Schimel, J. M. O. Scurlock and T. G. F. Kittel, 1993: Modelling the effects of climatic and CO_2 changes on grassland storage of soil carbon. *Water, Air, and Soil Pollution, 70,* 643–657.

Orr, J. C., 1993: Accord between ocean models predicting uptake of anthropogenic CO_2. *Water, Air, and Soil Pollution, 70,* 465–481.

Owensby, C. E., P. I. Coyne, J. M. Ham, L. M. Auen and A. K. Knapp, 1993: Biomass production in a tallgrass prairie ecosystem exposed to ambient and elevated CO_2. *Ecological Applications, 3,* 644–653.

Peng, T.-H. and W. S. Broecker, 1991: Dynamic limitations on the Antarctic iron fertilization strategy. *Nature, 349,* 227–229.

Peterson, B. J. and J. M. Melillo, 1985: The potential storage of carbon caused by eutrophication of the biosphere. *Tellus, 37B,* 117–127.

Polley, H. W., H. B. Johnson, B. D. Marino and H. S. Mayeux, 1993: Increase in C_3 plant water-use efficiency and biomass over Glacial to present CO_2 concentrations. *Nature, 361,* 61–63.

Poorter, H., 1993: Interspecific variation in the growth response of plants to an elevated ambient CO_2 concentration. *Vegetatio, 104/105,* 77–97.

Post, W. M., J. Pastor, P. J. Zinke and A. G. Stangenberger, 1985: Global patterns of soil nitrogen storage. *Nature, 317,* 613–616.

Potter, C. S., J. T. Randerson, C. B. Field, P. A. Matson, P. M. Vitousek, H. A. Mooney and S. A. Klooster, 1993: Terrestrial ecosystem production: a process model based on global satellite and surface data. *Global Biogeochemical Cycles, 7,* 811–841.

Prather, M., R. G. Derwent, D. Ehhalt, P. Fraser, Sanhueza and X. Zhou, 1995: Other trace gases and atomspheric chemistry. In: *Climate Change 1994: Radiative Forcing of Climate Change and An Evaluation of the IPCC IS92 Emission Scenarios,* J. T. Houghton, L. G. Meira Filho and J. Bruce, Hoesung Lee, B. A. Callander E. Haites, N. Harris and K. Maskell (eds.), Cambridge University Press, Cambridge, U.K., pp. 205–231.

Prentice, K. C. and I. Y. Fung, 1990: The sensitivity of terrestrial carbon storage to climate change. *Nature, 346,* 48–51.

Quay, P. D., B. Tilbrook and C. S. Wong, 1992: Oceanic uptake of fossil fuel CO_2: carbon-13 evidence. *Science, 256,* 74–79.

Rastetter, E. B., R. B. McKane, G. R. Shaver and J. M. Melillo, 1992: Changes in C storage by terrestrial ecosystems: how C-N interactions restrict responses to CO_2 and temperature. *Water, Air, and Soil Pollution, 64,* 327–344.

Raynaud, D. and U. Siegenthaler, 1993: Role of trace gases: the problem of lead and lag. In: *Global Changes in the Perspective of the Past,* J. A. Eddy and H. Oeschger (eds.), John Wiley and Sons, Chichester, pp. 173–188 .

Raynaud, D., J. Jouzel, J.-M. Barnola, J. Chappellaz, R. J. Delmas and C. Lorius, 1993: The ice record of greenhouse gases. *Science, 259,* 926–934.

Riebesell, U., D. A. Wolf-Gladrow and V. Smetacek, 1993: Carbon dioxide limitation of marine phytoplankton growth rates. *Nature, 361,* 249–251.

Rochefort, L. and F. A. Bazzaz, 1992: Growth response to elevated CO_2 in seedlings of four co-occurring birch species. *Canadian Journal of Forestry Research, 22,* 1583–1587.

Rotmans, J. and M. G. J. den Elzen, 1993: Modelling feedback mechanisms in the carbon cycle: balancing the carbon budget. *Tellus, 45B,* 301–320.

Sarmiento, J. L. and J. C. Orr, 1991: Three dimensional ocean model simulations of the impact of Southern Ocean nutrient depletion on atmospheric CO$_2$ and ocean chemistry. *Limnology and Oceanography, 36,* 1928–1950.

Sarmiento, J. L. and U. Siegenthaler, 1992: New production and the global carbon cycle. In: *Primary Productivity and Biogeochemical Cycles in the Sea,* P. G. Falkowski and A. D. Woodhead (eds.), Plenum Press, New York, pp. 317–332.

Sarmiento, J. L. and E. T. Sundquist, 1992: Revised budget for the oceanic uptake of anthropogenic carbon dioxide. *Nature, 356,* 589–593.

Sarmiento, J. L., J. C. Orr and U. Siegenthaler, 1992: A perturbation simulation of CO$_2$ uptake in an ocean general circulation model. *J. Geophys. Res., 97,* 3621–3645.

Schimel, D. S., 1995: Terrestrial biogeochemical cycles: global estimates with remote sensing. *Remote Sensing of Environment.* (In press)

Schimel, D. S., B. H. Braswell Jr., E. A. Holland, R. McKeown, D. S. Ojima, T. H. Painter, W. J. Parton and A. R. Townsend, 1994: Climatic, edaphic and biotic controls over storage and turnover of carbon in soils. *Global Biogeochemical Cycles , 8,* 279–293.

Schimel, D. S. and E. W. Sulzman, 1994: Variability in the Earth-climate system: decadal and longer timescales. In: *The U. S. National Report (1991–1994) to the International Union of Geophysics and Geodessey,* S. P. Nelson (ed.), American Geophysical Union, Washington, DC. (In press)

Schindler, D. W. and S. E. Bayley, 1993: The biosphere as an increasing sink for atmospheric carbon: estimates from increased nitrogen deposition. *Global Biogeochemical Cycles, 7,* 717–734.

Schlesinger, W. H., 1990: Evidence from chronosequence studies for a low carbon-storage potential of soils. *Nature, 348,* 232–234.

Schlesinger, W. H., 1993: Response of the terrestrial biosphere to global climate change and human perturbation. *Vegetatio, 104/105,* 295–305.

Schlesinger, W. H. and J. M. Melack, 1981: Transport of organic carbon in the world's rivers. *Tellus, 33,* 172–187.

Schulze, E. D., W. De Vries, M. Hauhs, K. Rosén, L. Rasmussen, O-C. Tann and J. Nilsson, 1989: Critical loads for nitrogen deposition in forest ecosystems. *Water, Air, and Soil Pollution, 48,* 451–456.

Shaffer, G., 1993: Effects of the marine biota on global carbon cycling. In: *The Global Carbon Cycle,* M. Heimann (ed.), Springer-Verlag, Heidelberg, pp 431–456.

Shaver, G. R., W. D. Billings, F. S. Chapin III, A. E. Giblin, K. J. Nadelhoffer, W. C. Oechel and E. B. Rastetter, 1992: Global change and the carbon balance of Arctic ecosystems. *BioScience, 42,* 433–441.

Shine, K., R. G. Derwent, D. J. Wuebbles, and J.-J. Morcrette, 1990: Radiative forcing of climate. In: *Climate Change: The IPCC Scientific Assessment,* J. T. Houghton, G. J. Jenkins and J. J. Ephraums (eds.), Cambridge University Press, Cambridge, U.K., pp. 41–68.

Siegenthaler, U. and H. Oeschger, 1987: Biospheric CO$_2$ emissions during the past 200 years reconstructed by deconcolution of ice core data. *Tellus, 39B,* 140–154.

Siegenthaler, U. and F. Joos, 1992: Use of a simple model for studying oceanic tracer distributions and the global carbon cycle. *Tellus, 44B,* 186–207.

Siegenthaler, U. and J. L. Sarmiento, 1993: Atmospheric carbon dioxide and the ocean. *Nature, 365,* 119–125.

Siegenthaler, U., H. Friedli, H. Loeeetscher, E. Moor, A. Neftel, H. Oeschger and B. Stauffer, 1988: Stable-isotope ratios and concentration of CO$_2$ in air from polar ice cores. *Ann. Glaciol., 10,* 151–156.

Skole, D. and C. Tucker, 1993: Tropical deforestation and habitat fragmentation in the Amazon: satellite data from 1978 to 1988. *Science, 260,* 1905–1910.

Smith, T. M. and H. H. Shugart, 1993: The transient response of terrestrial carbon storage to a perturbed climate. *Nature, 361,* 523–526.

Stauffer, B., H. Hofer, H. Oeschger, J. Schwander and U. Siegenthaler, 1984: Atmospheric CO$_2$ concentration during the last glaciation. *Ann. Glaciol., 5,* 160–164.

Stocker, T. F., W. S. Broecker and D. G. Wright, 1994: Carbon uptake experiments with a zonally-averaged global ocean circulation model. *Tellus, 46B,* 103–122.

Suess, H. E., 1955: Radiocarbon concentration in modern wood. *Science, 122,* 415–417.

Sugimura, Y. and Y. Suzuki, 1988: A high temperature catalytic oxidation method for non-volatile dissolved organic carbon in seawater by direct injection of a liquid sample. *Marine Chemistry, 24,* 105–131.

Suzuki, Y., 1993: On the measurement of DOC and DON in seawater. *Marine Chemistry, 41,* 287–288.

Takahashi, T., J. Olafsson, J. G. Goddard, D. W. Chipman and S. C. Sutherland, 1993: Seasonal variation of CO$_2$ and nutrients in the high-latitude surface oceans: a comparative study. *Global Biogeochemical Cycles, 7,* 843–878.

Tans, P. P., T. J. Conway and T. Nakazawa, 1989: Latitudinal distribution of the sources and sinks of atmospheric carbon dioxide derived from surface observations and an atmospheric transport model. *J. Geophys. Res., 94(D),* 5151–5172.

Tans, P. P., I. Y. Fung and T. Takahashi, 1990: Observational constraints on the global atmospheric CO$_2$ budget. *Science, 247,* 1431–1438.

Tans, P. P., J. A. Berry and R. F. Keeling, 1993: Oceanic ^{13}C/^{12}C observations: a new window on ocean CO$_2$ uptake. *Global Biogeochemical Cycles, 7,* 353–368.

Thomas, R. B., D. D. Richter, H. Ye, P. R. Heine and B. R. Strain, 1991: Nitrogen dynamics and growth of seedlings of an N-fixing tree (*Gliricidia sepium* (Jacq.) Walp.) exposed to elevated atmospheric carbon dioxide. *Oecologia, 88,* 415–421.

Thompson, M. L., I. G. Enting, G. I. Pearman and P. Hyson, 1986: Internal variations of atmospheric CO$_2$ concentration. *J. Atmos. Chem., 4,* 125–155.

Townsend, A. R., P. M. Vitousek and E. A. Holland, 1992: Tropical soils could dominate the short-term carbon cycle feedbacks to increased global temperatures. *Climatic Change, 22,* 293–303.

United Nations, 1992: *Earth Summit Convention on Climate Change, 3–14 June 1992,* United Nations Conference on Environment and Development, Rio de Janeiro, Brazil.

Vitousek, P. M. and R. W. Howarth, 1991: Nitrogen limitation on land and in the sea: how can it occur? *Biogeochemistry, 13,* 87–115.

Vloedbeld, M. and R. Leemans, 1993: Quantifying feedback processes in the response of the terrestrial carbon cycle to global change: the modeling approach of IMAGE-2. *Water, Air, and Soil Pollution, 70,* 615–628.

Watson, A. J., 1993: Air-sea gas exchange and carbon dioxide. In: *The Global Carbon Cycle,* M. Heimann (ed.), Springer-Verlag, Heidelberg, pp. 397–411.

Watson, R. T., H. Rhode, H. Oeschger and U. Siegenthaler, 1990: Greenhouse gases and aerosols. In: *Climate Change: the IPCC Scientific Assessment,* J. T. Houghton, G. J. Jenkins and J. J. Ephraums (eds.), Cambridge University Press, Cambridge, U.K., pp. 1–40.

Watson, R. T., L. G. Meira Filho, E. Sanhueza and A. Janetos, 1992: Sources and sinks. In: *Climate Change 1992: The Supplementary*

Report to the IPCC Scientific Assessment, J. T. Houghton, B. A. Callander and S. K. Varney (eds.), Cambridge University Press, Cambridge, U.K., pp. 25–46.

WEC (World Energy Council), 1993: *Energy for Tomorrow's World - the Realities, the Real Options and the Agenda for Achievement*, St. Martin's Press, New York, 320 pp.

White, J. W. C., P. Ciais, R. A. Figge, R. Kenny and V. Markgraf, 1994: A high resolution atmospheric pCO$_2$ record from carbon isotopes in peat. *Nature, 367,* 153–156.

Wigley, T. M. L., 1993: Balancing the global carbon budget. Implications for projections of future carbon dioxide concentration changes. *Tellus, 45B,* 409–425.

Wong, C. S., Y-H. Chan, J. S. Page, G. E. Smith and R. D. Bellegay, 1993: Changes in equatorial CO$_2$ flux and new production estimated from CO$_2$ and nutrient levels in Pacific surface waters during the 1986/87 El Niño. *Tellus, 45B,* 64–79.

Woodward, F. I., 1992: Predicting plant responses to global environmental change. *New Phytologist, 122,* 239–251.

2

CO$_2$ and the Carbon Cycle

(Extract from the 1995 Intergovernmental Panel on Climate Change (IPCC) "Second Assessment Report," *Climate Change 1995: The Science of Climate Change*

D. SCHIMEL, D. ALVES, I. ENTING, M. HEIMANN,
F. JOOS, D. RAYNAUD, T. WIGLEY (2.1)

Summary

The following summary contains some material more fully discussed in Chapter 1 (which was extracted from the 1994 IPCC Report): Bullets containing significant new information are marked "***"; those containing updated information are marked "**"; and those which contain information which is essentially unchanged are marked "*".

Climate change can be driven by changes in the atmospheric concentrations of a number of radiatively active gases and aerosols. We have clear evidence that human activities have affected concentrations, distributions, and life cycles of these gases. These matters, discussed in this chapter, were assessed at greater length in IPCC WGI report "Radiative Forcing of Climate Change" (IPCC 1994).

* Carbon dioxide concentrations have increased by almost 30% from about 280 ppmv in the late 18th century to 358 ppmv in 1994. This increase is primarily due to combustion of fossil fuel and cement production, and to land-use change. During the last millennium, a period of relatively stable climate, concentrations varied by about ±10 ppmv around the pre-industrial value of 280 ppmv. On the century time-scale these fluctuations were far less rapid than the change observed over the 20th century.

*** The growth rate of atmospheric CO_2 concentrations over the last few years is comparable to, or slightly above, the average of the 1980s (~1.5 ppmv/yr). On shorter (interannual) time-scales, after a period of slow growth (0.6 ppmv/yr) spanning 1991 to 1992, the growth rate in 1994 was higher (~2 ppmv/yr). This change in growth rate is similar to earlier short time-scale fluctuations, which reflect large but transitory perturbations of the carbon system. Isotope data suggest that the 1991 to 1994 fluctuations resulted from natural variations in the exchange fluxes between the atmosphere and both the land biota and the ocean, possibly partly induced by interannual variations in climate.

*** As well as the issue of natural fluctuations discussed above, other issues raised since IPCC (1994) have been addressed. There are some unresolved concerns about the ¹⁴C budget which may imply that previous estimates of the atmosphere-to-ocean flux were slightly too high. However, the carbon budget remains within our previously quoted uncertainties and the implications for future projections are minimal. Suggestions that the observed decay of bomb-¹⁴C implies a very short atmospheric lifetime for CO_2 result from a misunderstanding of reservoir lifetimes. Current carbon cycle modelling is based on principles that have been well-understood since the 1950s and correctly accounts for the wide range of reservoir time-scales that affect atmospheric concentration changes.

** The major components of the anthropogenic perturbation to the atmospheric carbon budget, with estimates of their magnitudes over the 1980s, are: (a) emissions from fossil fuel combustion and cement production (5.5 ± 0.5 GtC/yr); (b) atmospheric increase (3.3 ± 0.2 GtC/yr); (c) ocean uptake (2.0 ± 0.8 GtC/yr); (d) tropical land-use changes (1.6 ± 1.0 GtC/yr); and (e) Northern Hemisphere forest regrowth (0.5 ± 0.5 GtC/yr).

Other potential terrestrial sinks include enhanced terrestrial carbon storage due to CO_2 fertilisation (0.5–2.0 GtC/yr) and nitrogen deposition (0.2–1.0 GtC/yr), and possibly response to climatic anomalies. The latter is estimated to be a sink of 0–1.0 GtC/yr over the 1980s, but this term could be either a sink or a source over other periods. This budget is changed from IPCC (1994) by a small adjustment (from 3.2 to 3.3 GtC/yr) to the atmospheric rate of increase and a corresponding decrease in "other terrestrial sinks" from 1.4 to 1.3 GtC/yr.

* In IPCC (1994), calculations of future CO_2 concentrations and emissions from 18 different carbon cycle models were presented based on the IPCC (1992) carbon budget. Concentrations were derived for the IS92 emission scenarios. Future CO_2 emissions were derived leading to stable CO_2 concentration levels at 350, 450, 550, 650, and 750 ppmv. Inter-model differences varied with time and were up to ±15% about the median value. Biogeochemical (apart from CO_2 fertilisation) and climate feedbacks were not included in these calculations. The results showed that in order for atmospheric concentrations to stabilise at 750 ppmv or below, anthropogenic emissions must eventually fall well below today's levels. Stabilisation of emissions at 1990 levels is not sufficient to stabilise atmospheric CO_2: If anthropogenic emissions are held constant at 1990 levels, modelled atmospheric concentrations of CO_2 continue to increase throughout the next century and beyond.

*** These calculations have been re-run based on the revised IPCC (1994) carbon budget discussed above. These changes result in higher concentration projections by 15 ppmv (IS92c) to 40 ppmv (IS92e) in the year 2100. In terms of radiative forcing, the additional amounts in the year 2100 are 0.2–0.3 Wm⁻². Emissions requirements to achieve stabilisation are correspondingly reduced by up to about 10%. These changes are within the overall uncertainties in the calculations.

*** The implied future CO_2 emissions were calculated for additional cases to investigate the effect of lower reductions in CO_2 emissions in the early years by using CO_2 concentration profiles which closely followed the IS92a emission scenario for 10–30 years after 1990. Concentrations stabilised at the same date and levels as in the earlier scenarios, *viz* 350, 450, 550, 650, and 750 ppmv. Stabilisation at 1000 ppmv, above the range previously examined, was also investigated. These new calculations, while not necessarily spanning the full range of future emission options, still show the same characteristic long-term emissions behaviour as found previously; namely an eventual decline to emissions well below present levels.

2.1 Introduction

In 1994, IPCC WGI produced a report entitled "Radiative Forcing of Climate Change". Two main topics were addressed:

(a)   the relative climatic importance of anthropogenically induced changes in the atmospheric concentrations and distribution of different greenhouse gases and aerosols;

(b)   possible routes to stabilisation of greenhouse gas concentrations in the atmosphere.

This chapter contains a summary and update of the material presented in chapter 2 of the 1994 report, to which the reader is referred for a fuller discussion.

Clear evidence has been presented in past IPCC WGI reports (1990, 1992, 1994) that the atmospheric concentrations of a number of radiatively active gases have increased over the past century as a result of human activity. Most of these trace gases possess strong absorption bands in the infrared region of the spectrum (where energy is emitted and absorbed by the Earth's surface and atmosphere) and they thus act to increase the heat trapping ability of the atmosphere and so drive the climate change considered in this report. The other atmospheric constituents which are important in climate change are aerosols (suspensions of particles in the atmosphere), which tend to exert a cooling effect on the atmosphere.

The relative importance of the various constituents is assessed using the concept of radiative forcing. A change in the concentration of an atmospheric constituent can cause a radiative forcing by perturbing the balance between the net incoming radiation and the outgoing terrestrial radiation. A radiative forcing is defined to be a change in average net radiation (either solar or terrestrial in origin) at the top of the troposphere (the tropopause). As defined here, the incoming solar radiation is not considered a radiative forcing, although a change in the amount of incoming solar radiation would be a radiative forcing. Similarly changes in clouds and water vapour resulting from alterations to the general circulation of the atmosphere are considered to be climate feedbacks rather than radiative forcings. However, changes caused, for example, in clouds through the indirect aerosol effect and in water vapour through the oxidation of methane in the stratosphere, are counted as indirect radiative forcings. Radiative forcing was discussed in detail in IPCC (1994).

The concentration of an atmospheric constituent depends on the size of its sources (emissions into and production within the atmosphere) and sinks (chemical loss in the atmosphere and removal at the Earth's surface). These processes act on different time-scales which, singly and on aggregate, define the

---

DEFINITION OF TIME-SCALES

Throughout this report different time-scales are used to characterise processes affecting trace gases and aerosols. The following terminology is used in this chapter.

Turnover time (T) is the ratio of the mass (M) of a reservoir – e.g., a gaseous compound in the atmosphere – and the total rate of removal (S) from the reservoir: $T = M/S$

In cases where there are several removal processes (S_i), separate turnover times (T_i) can be defined with respect to each removal process: $T_i = M/S_i$

Adjustment time or response time (T_a) is the time-scale characterising the decay of an instantaneous pulse input into the reservoir. Adjustment time is also used to characterise the adjustment of the mass of a reservoir following a step change in the source strength.

Lifetime is a more general term often used without a single definition. In the Policymakers Summary and elsewhere in this report, lifetime is sometimes used, for simplicity, as a surrogate for adjustment time. In atmospheric chemistry, however, lifetime is often used to denote the turnover time.

In simple cases, where the global removal of the compound in question is directly proportional to the global reservoir content ($S = kM$, with k, the removal frequency, being a constant), the adjustment time almost equals the turnover time ($T_a = T$). An example is CFC-11 in the atmosphere, which is removed only by photochemical processes in the stratosphere. In this case $T = T_a = 50$ years.

In other situations, where the removal frequency is not constant or there are several reservoirs that exchange with each other, the equality between T and T_a no longer holds. An extreme example is that of CO_2. Because of the rapid exchange of CO_2 between the atmosphere and the oceans and the terrestrial biota, the turnover time of CO_2 in the atmosphere (T) is only about 4 years. However, a large part of the CO_2 that leaves the atmosphere each year is returned to the atmosphere from these reservoirs within a few years. Thus, the adjustment time of CO_2 in the atmosphere (T_a) is actually determined by the rate of removal of carbon from the surface layer of the oceans into the deeper layers of the oceans. Although an approximate value of about 100 years may be given for the adjustment time of CO_2 in the atmosphere, the actual adjustment is faster in the beginning and slower later on.

Methane is another gas for which the adjustment time is different from the turnover time. In the case of methane the difference arises because the removal (S) – which is mainly through chemical reaction with the hydroxyl radical in the troposphere – is related to the amount of methane (M) in a non-linear fashion, i.e., $S = kM$, with k decreasing as M increases.

various lifetimes of the atmospheric constituent (see box "Definition of Time-scales"). The past changes in the concentrations of the stable gases are relatively well known as they can be found by measuring the concentrations in air bubbles trapped in ice cores in Greenland and Antarctica. The past changes in the concentrations of the less stable gases and of aerosols are harder to quantify.

Estimation of possible future changes in the concentration of an atmospheric constituent requires a quantitative understanding of the processes that remove the constituent, and estimates of its future emissions (and/or atmospheric production). The former is discussed in this chapter. Estimating the emissions, however, is outside the scope of this report, and the so-called IS92 scenarios (IPCC, 1992) are used here because they cover a wide range of possible future emissions and because both their strengths and weaknesses are relatively well-known (Alcamo et al., 1995).

The present chapter discusses the factors that affect the atmospheric abundance of carbon dioxide. An intriguing, but still unresolved issue is whether there is an underlying cause for the variations in the trends of a number of trace gases including carbon dioxide (CO_2), methane (CH_4), and nitrous oxide (N_2O). In the early 1990s, particularly between 1991 and 1993, the rates of increases of these gases became smaller and, at certain times and places, negative. A number of mechanisms have been proposed for each gas, but no quantitative resolution has yet been made. One common factor is the eruption of Mt. Pinatubo in June 1991, which may have affected the sources and sinks of CO_2, CH_4, and N_2O through changes in meteorology, atmospheric chemistry, and/or biogeochemical exchange at the Earth's surface (see Section 2.2.2). However, no consensus has been reached as to whether the changes in the growth rates are linked or not.

2.2 CO₂ and the Carbon Cycle

2.2.1 Introduction

Two factors have increased attention on the carbon cycle: the observed increase in levels of atmospheric CO_2 (~280 ppmv in 1800; ~315 ppmv in 1957; ~358 ppmv in 1994); and the Framework Convention on Climate Change (FCCC), under which nations have to assess their contributions to sources and sinks of CO_2 and to evaluate the processes that control CO_2 accumulation in the atmosphere. Over the last several years, our understanding of the carbon cycle has improved, particularly in the quantification and identification of mechanisms for terrestrial exchanges, and in the preliminary quantification of feedbacks. An overview of the carbon cycle is presented in Figure 2.1.

IPCC (1994) (Schimel et al., 1995) specifically addressed four areas:

(a) the past and present atmospheric CO_2 levels;
(b) the atmospheric, oceanic, and terrestrial components of the global carbon budget;
(c) feedbacks on the carbon cycle;
(d) the results of a model-based examination of the relationship between future emissions and atmospheric concen-

trations (addressing, in particular, the requirements for achieving stabilisation of atmospheric CO_2 concentrations).

In this introduction we summarise the earlier review, updating it where appropriate, before discussing four specific issues.

In Section 2.2.2 we re-address the issue of the early 1990s slow-down in the atmospheric growth rate of CO_2, as recent evidence shows that growth rates are once again rising. We also present modelled emissions and concentrations calculated on the basis of the carbon budget presented in 1994, and compare the results with those presented earlier based on the 1992 budget. In addition, we address some concerns of the reviewers of the 1994 report. For example, an expanded analysis of the sensitivity of the model calculations to the extremely uncertain estimates of global net deforestation is presented in Section 2.2.3.2. Further, we discuss the published suggestion that the carbon cycle calculations carried out to date could be in error (Starr, 1993), and we explain why these suggestions are wrong. Lastly, we present recent evidence that suggests a slightly smaller magnitude for the oceanic CO_2 sink, and discuss the implications of this in terms of concentration and emissions projections.

Atmospheric CO₂ Levels

Precise, direct measurements of atmospheric CO_2 started in 1957 at the South Pole, and in 1958 at Mauna Loa, Hawaii. At this time the atmospheric concentration was about 315 ppmv and the rate of increase was ~0.6 ppmv/yr. The growth rate of atmospheric concentrations at Mauna Loa has generally been increasing since 1958. It averaged 0.83 ppmv/yr during the 1960s, 1.28 ppmv/yr during the 1970s, and 1.53 ppmv/yr during the 1980s. In 1994, the atmospheric level of CO_2 at Mauna Loa was 358 ppmv. Data from the Mauna Loa station are close to, but not the same as, the global mean.

Atmospheric concentrations of CO_2 have been monitored for shorter periods at a large number of atmospheric stations around the world (e.g., Boden et al., 1991). Measurement sites are distributed globally and include sites in Antarctica, Australia, Asia, Europe, North America, and several maritime islands, but, at present, nowhere on the continents of Africa or South America. The globally averaged CO_2 concentration, as determined through analysis of NOAA/CMDL data (Boden et al., 1991; Conway et al., 1994), increased by 1.53 ± 0.1 ppmv/yr over the period 1980 to 1989. This corresponds to an annual average rate of change in atmospheric carbon of 3.3 ± 0.2 GtC/yr. Other carbon-containing compounds like methane, carbon monoxide and larger hydrocarbons contain ~1% of the carbon stored in the atmosphere (with even smaller percentage changes) and can be neglected in the atmospheric carbon budget. There is no doubt that the increase shown by the atmospheric record since 1957 is due largely to anthropogenic emissions of CO_2. The record itself provides important insights that support anthropogenic emissions as a source of the observed increase. For example, when seasonal and short-term interannual variations in concentrations are neglected, the rise in atmospheric CO_2 is about 50% of anthropogenic emissions (Keeling et al., 1989a, 1995) with the inter-hemispheric differ-

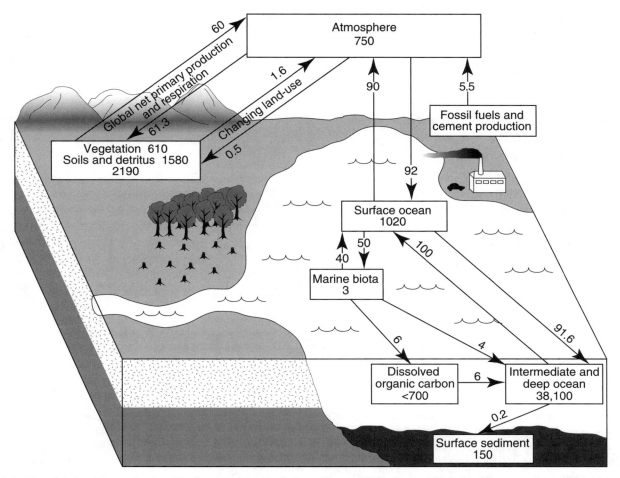

Figure 2.1: The global carbon cycle, showing the reservoirs (in GtC) and fluxes (GtC/yr) relevant to the anthropogenic perturbation as annual averages over the period 1980 to 1989 (Eswaran *et al.*, 1993; Potter *et al.*, 1993, Siegenthaler and Sarmiento, 1993). The component cycles are simplified and subject to considerable uncertainty. In addition, this figure presents average values. The riverine flux, particularly the anthropogenic portion, is currently very poorly quantified and so is not shown here. Evidence is accumulating that many of the key fluxes can fluctuate significantly from year to year (terrestrial sinks and sources: INPE, 1992; Ciais *et al.*, 1995a; export from the marine biota: Wong *et al.*, 1993). In contrast to the static view conveyed by figures such as this one, the carbon system is clearly dynamic and coupled to the climate system on seasonal, interannual, and decadal time-scales (e.g., Schimel and Sulzman, 1995).

ence growing in parallel to the growth of fossil emissions (Keeling *et al.*, 1989b; Siegenthaler and Sarmiento, 1993).

The atmospheric CO₂ concentration records prior to 1957 mainly come from air bubbles in ice cores, although some values have been inferred indirectly from isotopic data. Ice cores provide a direct record of past atmospheric composition back to well before the industrial revolution. Information between 1000 and 9000 years ago is less certain because of ice defects. Prior to that, the data show natural variations, the most noticeable of which is an increase in CO₂ level of about 80 ppmv that paralleled the last interglacial warming. Throughout this record, there is no direct evidence that past changes in CO₂ levels were as rapid as those of the 20th century; early indications that such rapid changes may have occurred during the last glacial period have not been confirmed (Neftel *et al.*, 1988). Over the last 1000 years, CO₂ concentrations in the atmosphere have fluctuated ±10 ppmv around 280 ppmv, until the recent increase to a concentration of ~358 ppmv, with a current rate of increase of ~1.6 ppmv/yr (1994).

The Anthropogenic Carbon Budget

The major components of the atmospheric carbon budget are anthropogenic emissions, the atmospheric increase, exchanges between the ocean and the atmosphere, and exchanges between the terrestrial biosphere and the atmosphere (Table 2.1). Emissions from fossil fuels and cement production averaged 5.5 ± 0.5 GtC/yr over the decade of the 1980s. In 1990 the emissions were 6.1 ±0.6 GtC. The measured average annual rate of atmospheric increase in the 1980s was 3.3 ± 0.2 GtC/yr. Average ocean uptake during the decade has been estimated by a combination of modelling and measurements of carbon isotopes and atmospheric oxygen/nitrogen ratios to be 2.0 ± 0.8 GtC/yr.

Averaged over the 1980s, terrestrial exchanges include a tropical source of 1.6 ± 1.0 GtC/yr from ongoing changes in land-use, based on land clearing rates, biomass inventories and modelled forest regrowth. Recent satellite data have reduced uncertainties in the rate of deforestation for the Amazon, but

Table 2.1: *Average annual budget of CO2 perturbations for 1980 to 1989. Fluxes and reservoir changes of carbon are expressed in GtC/yr, error limits correspond to an estimated 90% confidence interval.*

	IPCC 1992[†]	IPCC 1994*	IPCC 1995
		Estimates for 1980s budget	
CO_2 sources			
(1) Emissions from fossil fuel combustion and cement production	$5.5 \pm 0.5^\Delta$	5.5 ± 0.5	$5.5 \pm 0.5^\S$
(2) Net emissions from changes in tropical land-use	$1.6 \pm 1.0^\Delta$	1.6 ± 1.0	$1.6 \pm 1.0^\S$
(3) Total anthropogenic emissions = (1) + (2)	7.1 ± 1.1	7.1 ± 1.1	7.1 ± 1.1
Partitioning amongst reservoirs			
(4) Storage in the atmosphere	$3.4 \pm 0.2^\Delta$	3.2 ± 0.2	$3.3 \pm 0.2^\S$
(5) Ocean uptake	$2.0 \pm 0.8^\Delta$	2.0 ± 0.8	$2.0 \pm 0.8^\S$
(6) Uptake by Northern Hemisphere forest regrowth	not accounted for	0.5 ± 0.5	$0.5 \pm 0.5^\S$
(7) Other terrestrial sinks = (3) – ((4) + (5) + (6))			
(CO_2 fertilisation, nitrogen fertilisation, climatic effects)	1.7 ± 1.4	1.4 ± 1.5	1.3 ± 1.5

† Values given in IPCC (1990, 1992).
* Values given in IPCC (1994).
Δ Values used in the carbon cycle models for the calculations presented in IPCC (1994).
§ Values used in the carbon cycle models for the calculations presented here.

rates for the rest of the tropics remain poorly quantified. For the tropics as a whole, there is incomplete information on initial biomass and rates of regrowth. There is currently no estimate available for the years 1990 to 1995. The tropical analyses do not account for a number of potential terrestrial sinks. These include the regrowth of mid- and high latitude Northern Hemisphere forests (1980s mean value of 0.5 ± 0.5 GtC/yr), enhanced forest growth due to CO_2 fertilisation (0.5–2.0 GtC/yr), nitrogen deposition (0.2–1.0 GtC/yr), and, possibly, response to climatic anomalies (0–1.0 GtC/yr). The latter term, although thought to be positive over the 1980s, and in 1992 to 1993 in response to the Mt. Pinatubo eruption of 1991 (Keeling *et al.*, 1995), could be either positive or negative during other periods. Partitioning the sink among these processes is difficult, but it is likely that all components are significant. While the CO_2 fertilisation effect is the most commonly cited terrestrial uptake mechanism, existing model studies indicate that the magnitudes of the contributions from each process are comparable, within large ranges of uncertainty. For example, some model-based evidence suggests that the magnitude of the CO_2 fertilisation effect is limited by interactions with nutrients and other ecological processes. Experimental confirmation from ecosystem-level studies, however, is lacking. As a result, the role of the terrestrial biosphere in controlling past atmospheric CO_2 concentrations is uncertain, and its future role is difficult to predict.

The Influence of Climate and Other Feedbacks on the Carbon Cycle

The responses of terrestrial carbon to climate are complex, with rates of biological activity generally increasing with warmer temperatures and increasing moisture. Storage of carbon in soils generally increases along a gradient from low to high latitudes, reflecting slower decomposition of dead plant material in colder environments (Post *et al.*, 1985; Schimel *et al.*, 1994). Global ecosystem models based on an understanding of underlying mechanisms are designed to capture these patterns, and have been used to simulate the responses of terrestrial carbon storage to changing climate. Models used to assess the effects of warming on the carbon budget point to the possibility of large losses of terrestrial carbon (~200 GtC) over the next few hundred years, offset by enhanced uptake in response to elevated CO_2 (that could eventually amount to 100–300 GtC). Effects of changing land-use on carbon storage also may be quite large (Vloedbeld and Leemans, 1993). Experiments that incorporate the record of past changes of climate (e.g., the climate of 18,000 years ago) suggest major changes in terrestrial carbon storage with climate.

Storage of carbon in the ocean may also be influenced by climate feedbacks through physical, chemical, and biological processes. Initial model results suggest that the effects of predicted changes in circulation on the ocean carbon cycle are not large (10s rather than 100s of ppmv in the atmosphere). However, exploration of the long-term impacts of warming on ocean circulation patterns has just begun; hence, analyses of climate impacts on the oceanic carbon cycle must be viewed as preliminary.

Modelling Future Concentrations of Atmospheric CO_2

For IPCC (1994), modelling groups from many countries were asked to use published carbon cycle models to evaluate the degree to which CO_2 concentrations in the atmosphere might be expected to change over the next several centuries, given a standard set of emission scenarios (including changes in land-use) and levels for stabilisation of CO_2 concentrations (350, 450, 550, 650, and 750 ppmv (S350–S750)). Models were constrained to balance the 1992 IPCC version of the 1980s mean carbon budget and to match the atmospheric record of past CO_2 variations using CO_2 fertilisation as the sole sink for the terrestrial biosphere. These analyses were re-done for this report, using the Bern (Siegenthaler and Joos, 1992; Joos *et al.*, 1996), Wigley (Wigley, 1993), and Jain (Jain *et al.*, 1995) models as representatives of the whole model set. The new calculations incorporated more recent information about the atmospheric CO_2 increase and a revised net land-use flux. As complete data are not yet available for a 1990s budget, the 1980s mean values were again used as a reference calibration period to account for the influence of natural variability. This choice does not affect the results in any significant way.

The stabilisation analyses explored the relationships between anthropogenic emissions and atmospheric concentrations. The analyses were based on a specific set of concentration profiles constrained to match present-day (1990) conditions and to achieve stabilisation at different levels and different future dates. The levels chosen (350–750 ppmv) spanned a realistic range (which has been extended in this report). Stabilisation dates were chosen so that the emissions changes were not unrealistically rapid. Precise pathways were somewhat arbitrary, loosely constrained by the need for smooth and not-too-rapid emissions changes (see Enting *et al.* (1994) for detailed documentation). Alternative pathways are considered in this report. The models were used to perform a series of inverse calculations to determine fossil emissions (land-use emissions were prescribed). These calculations:

(1)   determined the time course of carbon emissions from fossil fuel combustion required to arrive at the selected CO_2 concentration stabilisation profiles while matching the past atmospheric record and the 1980s mean budget; and

(2)   assessed (by integration) the total amount of fossil carbon released.

For stabilisation at 450 ppmv in the 1994 calculations, fossil emissions had to be reduced to about a third of today's levels (i.e., to about 2 GtC/yr) by the year 2200. For stabilisation at 650 ppmv, reductions by 2200 had to be about two-thirds of current levels (i.e., 4 GtC/yr). This clearly indicates that stabilisation of emissions at 1990 levels is not sufficient to stabilise atmospheric CO_2. These results are independent of the assumed pathway to concentration stabilisation (as shown later in this report), although the detailed changes in future anthropogenic emissions do depend on the pathway selected.

For the 18 models used for IPCC (1994), the implied fossil emissions differed. Initial differences were small, increasing with time to span a range up to ±15% about the median. In addition, the maximum range of uncertainty in fossil emissions associated with the parametrization of CO_2 fertilisation (evaluated with one of the models) varied ±10% about the median for low stabilisation values and ±15% for higher stabilisation

values. For these 1994 calculations, the use of CO_2 fertilisation alone to control terrestrial carbon storage, when in fact other ecological mechanisms are likely to be involved, probably results in an underestimate of future concentrations (for given emissions) or an overestimate of emissions (for the stabilisation profiles). The revised calculations presented here use a lower fertilisation factor and so the associated bias is likely to be smaller. IPCC (1994) reported these results in terms of total anthropogenic emissions, directly addressing the requirements of the FCCC and coincidentally removing the loss of generality arising from choosing a specific land-use flux.

2.2.2 Atmospheric CO_2 Concentrations and the Status of the CO_2 Growth Rate Anomaly

The decline in the growth rate of CO_2 in 1992 is one of the most noticeable changes in the carbon cycle in the recent record of observation (Conway *et al.*, 1994). The decline in the short time-scale growth rate (based on interannual data filtered to remove annual cycle variations and considering the mean growth rates for overlapping 12-month periods) to 0.6 ppmv/yr in 1992 is considerable when contrasted with the mean rate over 1987 to 1988 of 2.5 ppmv/yr, until recently the highest one-year mean growth rate ever recorded. Examination of the CO_2 growth rate record reveals, in fact, considerable variability over time (Figure 2.2, which is the updated version of Figure 1.2 in Schimel *et al.*, 1995). The magnitude of individual anomalies in the CO_2 growth rate depends on what is defined as "normal" for the long-term trend. At Mauna Loa, for example, the 1987 to 1988 increase was similar to a variation which occurred in 1972 to 1973 (see Figure 2.2).

How are these changes explained, and do they matter? Analyses of the recent changes suggest that both the relative and absolute magnitudes of ocean and terrestrial processes vary substantially from year to year and that these changes cause marked annual time-scale changes in the CO_2 growth rate. Although the oceanic and biospheric sources of CO_2 cannot be distinguished by examining the concentration data alone, they can be distinguished by looking at observations of the stable carbon isotope ratio (^{13}C to ^{12}C) of atmospheric CO_2. Carbon dioxide release from the ocean has nearly the same ^{13}C to ^{12}C ratio as atmospheric CO_2, whereas carbon of biospheric origin is substantially depleted in ^{13}C. As such, it is clearly distinguishable from carbon originating from the ocean. The use of atmospheric transport modelling to analyse spatial distributions of CO_2 and $^{13}CO_2$ provides additional information (Ciais *et al.*, 1995a, b). To study interannual CO_2 fluctuations, Keeling *et al.* (1989a, 1995) removed the seasonal cycle and the long-term fossil fuel-induced trend from the direct atmospheric record to obtain anomalies in CO_2 concentration at Mauna Loa and the South Pole. Anomalies of the order of 1–2 ppmv are found: a small signal compared to the long-term increase (about 80 ppmv). The interannual CO_2 variations must reflect imbalances in the exchange fluxes between the atmosphere and the terrestrial biosphere and/or the ocean.

The isotopic data and modelling studies suggest that earlier short-term CO_2 anomalies on the El Niño-Southern Oscilla-

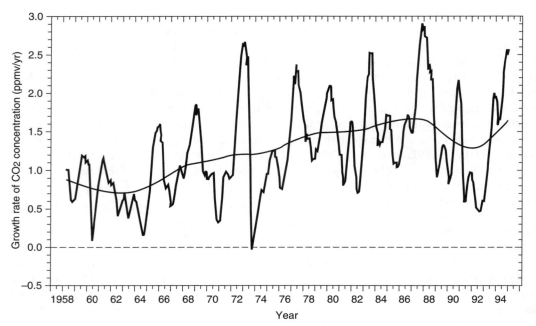

Figure 2.2: Growth rate of CO_2 concentrations since 1958 in ppmv/year at the Mauna Loa, Hawaii, station. The high growth rates of the late 1980s, the low growth rates of the early 1990s, and the recent upturn in the growth rate are all apparent. The smoothed curve shows the same data but filtered to suppress variations on time-scales less than approximately 10 years. (Sources: C. D. Keeling and T. P. Whorf, Scripps Institute of Oceanography, and P. Tans, NOAA CMDL. The Keeling and NOAA results are in close agreement. The Mauna Loa Observatory is operated by the NOAA.)

tion (ENSO) time-scale reflected two opposing effects: reduced net primary production of the terrestrial biosphere possibly due to reduced precipitation in monsoon regions, and a concomitant, temporary increase in oceanic CO_2 uptake due to a reduction of the CO_2 outgassing in the Pacific equatorial ocean (Keeling *et al.*, 1989b, 1995; Volk, 1989; Siegenthaler, 1990; Winguth *et al.*, 1994). The most recent anomaly seems to be unusual as it cannot be directly related to an ENSO event. Keeling and co-workers suggest that it may have been induced by a global anomaly in air temperature. While the evidence for pronounced interannual variability in both marine and terrestrial exchange fluxes is strong (Bacastow, 1976; Francey *et al.*, 1995), the exact magnitude and timing of the variations in these exchange fluxes remains controversial (Keeling *et al.*, 1989a, 1995; Feely *et al.*, 1995; Francey *et al.*, 1995).

Most of the global carbon cycle models used in IPCC (1994) (Schimel *et al.*, 1995) address only the longer term (10 year time-scale) direct perturbation of the global carbon cycle due to anthropogenic emissions. These models assume that the physical, chemical, and biological processes that control the exchange fluxes of carbon between the different reservoirs change only as a function of the carbon contents. Hence these models ignore any perturbations due to fluctuations in climate and are thus not able to reproduce these shorter-term atmospheric CO_2 variations. Nevertheless, some of the more complex models are founded on physico-chemical and biological principles that also operate on shorter time-scales, and first attempts to model these interannual variations have been attempted (Kaduk and Heimann, 1994; Winguth *et al.*, 1994; Sarmiento *et al.*, 1995). An improved understanding of the rapid fluctuations of the global carbon cycle would help to further develop and validate these models.

We now know that short-term growth rates increased markedly through 1993 and 1994, and are currently at levels

above the long-term (decadal time-scale) mean (Figure 2.2). As best as can be established, the anomaly of the early 1990s represented a large but transient perturbation of the carbon system.

2.2.3 Concentration Projections and Stabilization Calculations

Fossil fuel burning and cement manufacture, together with forest harvest and other changes of land-use, all transfer carbon (mainly as CO_2) to the atmosphere. This anthropogenic carbon then cycles between the atmosphere, oceans, and the terrestrial biosphere. Because an important component of the cycling of carbon in the ocean and terrestrial biosphere occurs slowly, on time-scales of decades to millennia, the effect of additional fossil and biomass carbon injected into the atmosphere is a long-lasting disturbance of the carbon cycle. The record itself provides important insights that support anthropogenic emissions as a source of the observed increase. For example, when seasonal and short-term interannual variations in concentrations are neglected, the rise in atmospheric CO_2 is about 50% of anthropogenic emissions (Keeling *et al.*, 1989a) with the inter-hemispheric difference growing in parallel to the growth of fossil emissions (Keeling *et al.*, 1989a; Siegenthaler and Sarmiento, 1993). These aspects are in accord with our understanding of the carbon cycle, and agree with model simulations of it. An additional important indicator of anthropogenically induced atmospheric change is provided by the ¹⁴C levels preserved in materials such as tree rings and corals. The ¹⁴C concentration measured in tree rings decreased by about 2% during the period 1800 to 1950. This isotopic decrease, known as the Suess effect (Suess, 1955), provides one of the most clear demonstrations that the increase in atmospheric CO_2 is due largely to fossil fuel inputs.

In this section, the relationships between future concentration changes and emissions of CO_2 are examined through use of models that simulate the major processes of the carbon cycle. In the context of future climate change, carbon cycle model calculations play a central role because the bulk of projected radiative forcing changes comes from CO_2. In IPCC (1994), two types of carbon cycle calculations relating future CO_2 concentration and emissions were presented: concentration projections for the IPCC 1992 (IS92) emission scenarios; and emissions estimates for a range of concentration stabilisation profiles directly addressing the stabilisation goal in Article 2 of the FCCC.

As noted above, these emissions-concentrations relationships have been re-calculated because the previous results were based on a 1980s mean budget from IPCC (1992) that has been revised in three ways:

(1) In the 1992 budget the net land-use flux for the decade of the 1980s was 1.6 GtC/yr and the atmospheric increase was 3.4 GtC/yr. The more recent estimate of the net land-use flux, however, is significantly lower (1.1 GtC/yr).

(2) IPCC (1994) presented a change in atmospheric CO_2 over the 1980s of 3.2 GtC/yr, a value that has been further revised to 3.3 GtC/yr (Komhyr *et al.*, 1985; Conway *et al.*, 1994; Tans, pers. comm.).

(3) Minor changes have been made to the industrial emissions data for the 1980s, although these do not noticeably change the 1980s mean. These changes mean that the additional sinks required to balance the budget (which were and still are assigned to the CO_2 fertilisation effect in the model calculations) are smaller than assumed in calculations presented in IPCC (1994). Consequently, when the new budget is used, future concentration projections for any given emission scenario are larger relative to those previously presented, and the emissions consistent with achieving stabilisation of concentrations are lower.

Three different models were used to carry out these calculations: the Bern model (Siegenthaler and Joos, 1992; Joos *et al.*, 1996) and the models of Wigley (Wigley, 1993) and Jain (Jain *et al.*, 1995). Changes relative to the previous calculations are virtually the same for each model. The methodology used was the same as in IPCC (1994), as documented in Enting *et al.* (1994). New results were produced for concentrations associated with emission scenarios IS92a–f and emissions associated with stabilisation profiles S350–S750. Impulse response functions were also revised for input into GWP calculations.

In addition, stabilisation calculations were performed for a profile with a 1000 ppmv stabilisation level (S1000) and for variations on S350–S750 in which the pathway to stabilisation was changed markedly. These latter profiles (WRE350–WRE750), from Wigley *et al.* (1996), were designed specifically to follow closely the IS92a "existing policies" emission scenario for 10–30 years after 1990. Coupled with S350–S750 results, these new profiles give insights into the range of emissions options available for any given stabilisation level, although they still do not necessarily span the full option range. It should be noted that, even though concentration sta-

bilisation is still possible if emissions initially follow an existing policies scenario, this cannot be interpreted as endorsing a policy to delay action to reduce emissions. Rather, the WRE scenarios were designed to account for inertia in the global energy system and the potential difficulty of departing rapidly from the present level of dependence on fossil fuels. Wigley *et al.* (1996) stress the need for a full economic and environmental assessment in the choice of pathway to stabilisation. Both the S350–S750 and the WRE350–WRE750 profile sets require substantial reductions in emissions at some future time and an eventual emissions level well below that of today.

The S1000 profile was designed to explore the emissions consequences of an even higher stabilisation level; the emissions implied by this case follow IS92a closely out to 2050. The results show that, even with such a high stabilisation level, the same characteristic emissions curve arises as found for S350–S750. This scenario is by no means derived to advocate such a high stabilisation level. The environmental consequences of such a level have not been assessed, but they are certain to be very large. It should be noted that recent emissions are low compared to IS92a-projected emissions (emissions were essentially the same in 1990 and in 1992, at 6.1 GtC/yr), although the slow-down may be temporary.

2.2.3.1 Effects of Carbon Cycle Model Recalibration

The effects of using the 1992 versus the 1994 budget with IS92 Scenarios a, c, and e are shown in Figure 2.3. The results show that the impact of the new budget on projected future concentrations is noticeable but relatively minor. In the year 2100, the concentrations increase by about 15 ppmv (IS92c), 25 ppmv (IS92a), and 40 ppmv (IS92e). All three models gave similar results. In terms of radiative forcing increases, the 2100 values are 0.20 Wm⁻² (IS92c), 0.25 Wm⁻² (IS92a), and 0.26 Wm⁻² (IS92e). These changes are all small compared

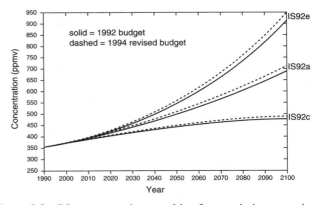

Figure 2.3: CO_2 concentrations resulting from emission scenarios IS92a, c, and e plotted using both the 1992 budget and the 1994 budget. Solid lines represent model results based on the 1992 budget calibration; dashed lines, calibration with the 1994 budget. The differences between results is small (of order 20 ppmv change in projected concentration by the year 2100). The changes from revising the budget result in changes in radiative forcing of approximately 0.2 Wm⁻².

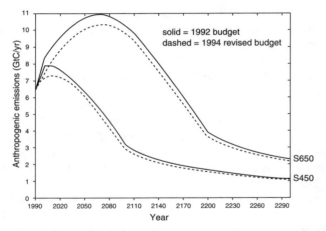

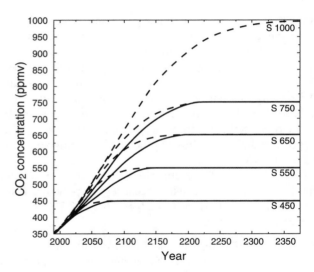

Figure 2.4: Anthropogenic CO₂ emissions for revised concentration stabilisation profiles S450 and S650 resulting from model projections initialised with the 1992 budget (solid curve: net land-use flux = 1.6 GtC/yr) and the 1994 budget (dashed curve: net land-use flux = 1.1 GtC/yr).

Figure 2.5: Stabilisation pathways used to illustrate sensitivity of allowed emissions to choice of pathway. The solid curves are based on the revised versions of the original S450 to S750 stabilisation profiles, incorporating minor changes to account for the revised concentration history. The dashed curves show slower approaches to stabilisation and are from Wigley *et al.* (1995). Stabilisation at 1000 ppmv is plotted for comparison (S1000).

to the overall forcing changes over 1990 to 2100 for these scenarios.

Figure 2.4 shows changes in the emissions requirements for the S450 and S650 profiles due to changes between the 1992 and 1994 budgets. The net effect of these changes is quite complex as it involves compensating effects due to different budget factors. Changing the 1980s mean net land-use flux from 1.6 GtC/yr to 1.1 GtC/yr (by reducing the fertilisation factor required to give a balanced budget) leads to emissions that are about 10% lower than given previously (Schimel *et al.*, 1995). This reduction is modified slightly by the changes in concentration history and profiles, and by an even smaller amount due to the change in the fossil fuel emissions history, leading to the results presented in Figure 2.4. The changes shown in Figure 2.4 arise from the complex interplay of a number of factors; but the overall conclusion (that emissions must eventually decline to substantially below current levels) remains unchanged.

We have carried out several new calculations to illustrate the effects of the choice of pathway leading to stabilisation of modelled emissions and uptake. In Figures 2.5 and 2.6 we show the effect of pathway in more detail by employing the profiles of Wigley *et al.* (1995) (WRE below). Figure 2.5 shows the different pathways, reaching stabilisation at the same time in each case, for target levels of 450 ppmv to 750 ppmv. The corresponding anthropogenic emissions are shown in Figure 2.6, which shows two things: first, that the emissions required to achieve any given stabilisation target are quite sensitive to the pathway taken to reach that target; and second, that it is possible to achieve stabilisation even if the initial emissions pathway follows the IS92a Scenario for some period of time. There is, of course, a penalty for this: having higher emissions initially requires a larger and possibly sharper drop in emissions later, to lower levels than otherwise (also see Enting, 1995). This is a direct consequence of the fact that cumulative emissions tend to become (at the stabilisation point or

beyond) similar no matter what the pathway. If cumulative emissions are constrained to be nearly constant, then what is gained early must be lost later.

Figures 2.5 and 2.6 also show the case for stabilisation at 1000 ppmv (in the year 2375). The profile here was constructed to follow the IS92a Scenario out to 2050 (using the model of Wigley (1993)) and is just one of a number of possible pathways. Even with this high target, to achieve it requires

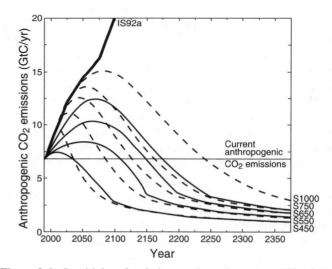

Figure 2.6: Sensitivity of emissions to the pathway to stabilisation. The pathways, which stabilise at 450 ppmv to 750 ppmv, are shown in Figure 2.5. Results for stabilisation at 1000 ppmv (also from Figure 2.5) are shown for comparison. (The results shown are calculated using the Bern model.)

an immediate and substantial reduction below IS92a after 2050, with emissions peaking around 2080 and then undergoing a long and steady decline to, eventually, values well below the current level. Although offset substantially, the general shape of the emissions curve for S1000 is similar to all other curves.

2.2.3.2 *Effects of Uncertainties in Deforestation and CO₂ Fertilisation*

The future uptake of CO_2 by the terrestrial biosphere is critical to the global carbon balance. In the above concentration projections and emissions results, the terrestrial biosphere uptake was determined by the CO_2 fertilisation factor, which in turn is determined mainly by the value assumed for the 1980s mean land-use emissions (if the ocean uptake is taken as a given). Thus, as in Wigley (1993) and Enting *et al.* (1994), we can assess the effects of fertilisation and future terrestrial uptake uncertainties by changing the 1980s mean land-use term used in initialising the carbon cycles models. Because the ocean uptake is held constant, this results in changing the modelled CO_2 effect. The sensitivity of the calculated emissions for CO_2 stabilisation at 450 and 650 ppmv to these uncertainties is shown in Figure 2.7 (cf. Figure 1.16 in Schimel *et al.*, 1995).

The results from all three models used in the present analysis show that uncertainty in the strength of the CO_2 fertilisation effect leads to significant differences in emissions deduced for any specified concentration profile. Reducing the uncertainty in emissions due to past land-use change is therefore very important in order to constrain the overall range of projections from carbon cycle models. At the same time, independent estimates of the global fertilisation effect may be used as a direct test of the value used in model calculations and so can also help to reduce uncertainties in projections. Estimates of the CO_2 fertilisation effect from terrestrial ecosystem models are now becoming available.

2.2.4 Bomb Lifetime vs. Perturbation Lifetime

Some recently published work (Starr, 1993) has argued that the post-industrial increase of atmospheric CO_2 may be due largely to natural variation rather than to human causes. This argument hinges on a presumed low value for the "lifetime" of CO_2 in the atmosphere. The argument is as follows: if the atmospheric lifetime were short, then large net CO_2 fluxes into the atmosphere would have been required to produce the observed concentration increases; if it were long, smaller fluxes would yield the observed increases. Proponents of a natural cause assert that the CO_2 lifetime is (and has remained) only a decade or less. The "evidence" for such a short lifetime includes the rapid decline of bomb ^{14}C in the atmosphere following its 1964 peak. A lifetime this short would require fluxes much larger than those estimated from fossil fuel burning in order to produce the observed atmospheric increase; therefore, the reasoning goes, the build-up can only be partly human-induced.

Atmospheric CO_2 is taken up by the ocean and the biospheric carbon pools if a disequilibrium exists between atmosphere and the ocean (i.e., if the partial pressure of CO_2 in the atmosphere is higher than in the surface ocean) and if the uptake by plant growth is larger than the CO_2 released by the biosphere by respiration and by decay of organic carbon (heterotrophic respiration). For terrestrial ecosystems, the removal of radiocarbon and that of anthropogenic CO_2 from the atmosphere are different. Carbon storage corresponds to the difference in net primary production (plant growth, NPP) and decay of organic carbon plus plant respiration. Radiocarbon, however, is assimilated in proportion to the product of NPP and the atmospheric ^{14}C/^{12}C ratio, and is released in proportion to the decay rate of organic matter and its ^{14}C/^{12}C ratio.

For oceans, one needs to differentiate between the amount or concentration of radiocarbon in the atmosphere and the isotopic ratio ^{14}C/[12]C. The ratio, a non-linear function of the CO_2 and ^{14}C concentration, is often confused with concentration, giving rise to misunderstanding. The removal of bomb-radiocarbon atoms is governed by the same mechanisms as the removal of anthropogenic CO_2. The removal of small pulse inputs of CO_2 and ^{14}C into the atmosphere by air-sea exchange follows the same pathways for both model tracers. However, an atmospheric perturbation in the isotopic ratio disappears *much faster* than the perturbation in the number of ^{14}C atoms as shown in Figure 2.8 (Siegenthaler and Oeschger, 1987; Joos *et al.*, 1996).

Thus a ^{14}C perturbation, such as the bomb ^{14}C input into the atmosphere, induces a net transfer of ^{14}C into oceans and biosphere. Because the [14]CO₂ molecules generated from the bomb tests represent a negligible CO_2 excess and, more importantly, because the atmospheric ^{14}C/^{12}C ratio is changed considerably by the ^{14}C input, this results in a large isotopic disequilibrium

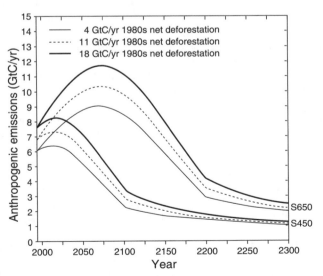

Figure 2.7: Sensitivity of the CO_2 emissions for the S450 and S650 concentration stabilisation profiles to the magnitude of the net land-use flux. The 1994 "best guess" for the magnitude of this flux is 1.1 GtC/yr (dashed curve). The range shown is for net land-use fluxes of 0.4–1.8 GtC/yr. The 1992 "best guess" was 1.6 GtC/yr. This sensitivity is equivalent to assessing the sensitivity to the magnitude of the CO_2 fertilisation effect. (The results shown are calculated using the Bern model.)

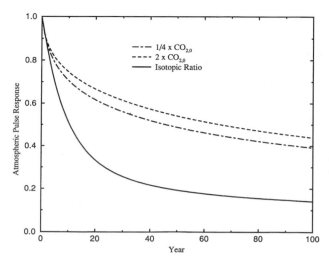

Figure 2.8: Atmospheric response to a pulse input of carbon at time t = 0 for CO_2 and for a perturbation of the isotopic ratio ($^{13}C/^{12}C$ or $^{14}C/^{12}C$), obtained by using the ocean-atmosphere compartments of the Bern model with no biosphere. The response of the atmospheric CO_2 concentration to a pulse input depends on the pulse size and the CO_2 background concentration. The dashed line is for a doubling of pre-industrial CO_2 concentration (280 ppmv) at t = 0; the dashed-dotted line is for an increase of pre-industrial CO_2 concentration by one quarter (70 ppmv) at t = 0; the solid line shows the decrease of an isotopic perturbation. Note the more rapid decline in the isotopic ratio, in accord with observations.

between the atmosphere and the oceanic and terrestrial biospheric reservoirs.

Because of the different dynamic behaviour, ^{14}C can not be taken as a simple analogue tracer for the excess anthropogenic CO_2, a fact that has long been known to the carbon cycle modelling community (e.g., Revelle and Suess, 1957; Oeschger *et al.*, 1975; Broecker *et al.*, 1980).

2.2.5 Recent Bomb Radiocarbon Results and Their Implication for Oceanic CO_2 Uptake

Two recent assessments of bomb radiocarbon in the entire global carbon system (i.e., the troposphere, stratosphere, terrestrial biosphere and the ocean (Broecker and Peng, 1994; Hesshaimer *et al.*, 1994)) imply a surprising global imbalance during the decade after the bomb tests, when, apart from the relatively very small natural cosmic production, no major ^{14}C source existed.

This imbalance may be explained either by a hidden high (above 30 km height) stratospheric bomb ^{14}C inventory not detected in the existing observations, which is unlikely, or by an overestimated bomb radiocarbon inventory in the terrestrial biosphere or the ocean or both. The calculated biospheric reservoir uptake would have to be reduced by about 80% in order to achieve the global bomb radiocarbon balance. Such a reduction is very unlikely, as it strongly contradicts observations of bomb ^{14}C in wood (e.g., in tree rings: Levin *et al.*, 1985) and soils (Harrison *et al.*, 1993). A more likely, albeit tentative, explanation is that previous estimates of the oceanic bomb ^{14}C inventory compiled from the observations of the GEOSECS program

(1973–78) (Broecker *et al.*, 1985) are too high by approximately 25%. This is slightly larger than the generally accepted uncertainty of this quantity. Because this explanation is inconsistent with a new assessment of the oceanic observations (Broecker *et al.*, 1995) the issue is still not fully resolved.

Observations of the distribution and inventories of bomb-produced radiocarbon constitute the key tracer either to calibrate simple ocean carbon models or to validate the upper ocean transport calculated by three-dimensional ocean carbon models. Therefore a downward revision of the oceanic ^{14}C inventory during the GEOSECS time period would have implications for our quantitative view of the global carbon cycle, requiring a smaller role for the oceans, and hence larger terrestrial uptake (or lower releases from land-use change). The implied changes in the carbon budget, although large (~0.5 GtC/yr), are within existing uncertainties and thus do not require any fundamental changes in understanding, but rather indicate the importance of adequate global sampling of critical variables.

References

Alcamo, J., A. Bouwman, J. Edmonds, A. Grübler, T. Morita and A. Sugandhy, 1995: An Evaluation of the IPCC IS92 Emission Scenarios. In: *Climate Change 1994: Radiative Forcing of Climate Change and An Evaluation of the IPCC IS92 Emission Scenarios.*, J. T. Houghton, L. G. Meira Filho, J. Bruce, Hoesung Lee, B. A. Callander, E. Haites, N. Harris and K. Maskell (eds.), Cambridge University Press, Cambridge, UK.

Bacastow, R. B., 1976: Modulation of atmospheric carbon dioxide by the Southern Oscillation. *Nature, 261,* 116–118.

Boden, T. A., R. J. Sepanski and F. W. Stoss, 1991: Trends '91: A compendium of data on global change. Oak Ridge National Laboratory, *ORNL/CDIAC-46.*

Broecker, W. S. and T.-H. Peng, 1994: Stratospheric contribution to the global bomb radiocarbon inventory: Model vs. observation. *Global Biogeochem. Cycles, 8,* 377–384.

Broecker, W. S., T.-H. Peng and R. Engh, 1980: Modelling the carbon system. *Radiocarbon, 22,* 565–598.

Broecker, W. S., T.-H. Peng, G. Ostlund and M. Stuiver, 1985: The distribution of bomb radiocarbon in the ocean. *J. Geophys. Res., 90,* 6953–6970.

Broecker, W. S., S. Sutherland, W. Smethie, T.-H. Peng and G. Oestlund, 1995: Oceanic radiocarbon: Separation of the natural and bomb components. *Global Biogeochem. Cycles, 9,* 263–288.

Ciais, P., P. P. Tans, J. W. C. White, M. Trolier, R. J. Francey, J. A. Berry, D. R. Randall, P. J. Sellers, J. G. Collatz and D. S. Schimel, 1995a: Partitioning of ocean and land uptake of CO_2 as inferred by $\delta^{13}C$ measurements from the NOAA Climate Monitoring and Diagnostics Laboratory Global Air Sampling Network. *J. Geophys. Res., 100D,* 5051–5070.

Ciais, P., P. P. Tans, M. Trolier, J. W. C. White and R. J. Francey, 1995b: A large northern hemisphere terrestrial CO_2 sink indicated by $^{13}C/^{12}C$ of atmospheric CO_2. *Science, 269,* 1098–1102.

Conway, T. J., P. P. Tans, L. S. Waterman, K. W. Thoning, D. R. Buanerkitzis, K. A. Maserie and N. Zhang, 1994: Evidence for interannual variability of the carbon cycle from the NOAA/CMDL global air sampling network. *J. Geophys. Res., 99D,* 22831–22855.

Enting, I. G., 1995: Analysing the conflicting requirements of the framework convention on climate change. *Climatic Change, 31,* 5–18.

Enting, I. G., T. M. L. Wigley and M. Heimann, 1994: Future emissions and concentrations of carbon dioxide: Key ocean/atmosphere/land analyses. *CSIRO Division of Atmospheric Research Technical Paper No. 31,* 120 pp.

Eswaran, H., E. Van den Berg and P. Reich, 1993: Organic carbon in soils of the world. *Soil Sci. Soc. America J., 57,* 192–194.

Feely, R. F., R. Wanninkhof, C. E. Cosca, P. P. Murphy, M. F. Lamb and M. D. Steckley, 1995: CO_2 distributions in the equatorial Pacific during the 1991–92 ENSO event. *Deep-Sea Research* II, *42,* 365–386.

Francey, R. J., P. P. Tans, C. E. Allison, I. G. Enting, J. W. C. White and M. Trolier, 1995: Changes in oceanic and terrestrial carbon uptake since 1982. *Nature, 373,* 326–330.

Harrison, K., W. Broecker and G. Bonani, 1993: A strategy for estimating the impact of CO_2 fertilisation on soil carbon storage. *Global Biogeochem. Cycles, 7,* 69–80.

Hesshaimer, V., M. Heimann and I. Levin, 1994: Radiocarbon evidence suggesting a smaller oceanic CO_2 sink than hitherto assumed. *Nature, 370,* 201–203.

INPE, 1992: *Deforestation in Brazilian Amazonia.* Instituto Nacional de Pesquisas Especiais, Sao Paulo, Brazil.

IPCC, 1990: *Climate Change: The IPCC Scientific Assessment,* J. T. Houghton, G. J. Jenkins and J. J. Ephraums (eds.). Cambridge University Press, Cambridge, UK.

IPCC, 1992: *Climate Change 1992: The Supplementary Report to the IPCC Scientific Assessment,* J. T. Houghton, B. A. Callander and S. K. Varney (eds.). Cambridge University Press, Cambridge, UK.

IPCC, 1994: *Climate Change 1994: Radiative Forcing of Climate Change and an Evaluation of the IPCC IS92 Emission Scenarios,* J. T. Houghton, L. G. Meira Filho, J. Bruce, Hoesung Lee, B. A. Callander, E. F. Haites, N. Harris and K. Maskell (eds.). Cambridge University Press, Cambridge, UK.

Jain, A. K., H. S. Kheshgi, M. I. Hoffert and D. J. Wuebbles, 1995: Distribution of radiocarbon as a test of global carbon cycle models. *Global Biogeochem. Cycles, 9,* 153–166.

Joos, F., M. Bruno, R. Fink, U. Siegenthaler, T. Stocker, C. Le Quéré and J. L. Sarmiento, 1996: An efficient and accurate representation of complex oceanic and biospheric models of anthropogenic carbon uptake. *Tellus* (In press).

Kaduk, J. and M. Heimann, 1994: The climate sensitivity of the Osnabrück Biosphere Model on the ENSO timescale. *Ecological Modelling, 75/76,* 239–256.

Keeling, C. D., S. C. Piper and M. Heimann, 1989a: A three-dimensional model of atmospheric CO_2 transport based on observed winds: 4. Mean annual gradients and interannual variations. In: *Aspects of Climate Variability in the Pacific and Western Americas, Geophysical Monograph 55,* D. H. Peterson (ed.), American Geophysical Union, Washington, DC, pp. 305–363.

Keeling, C. D., R. B. Bacastow, A. F. Carter, S. C. Piper, T. P. Whorf, M. Heimann, W. G. Mook and H. Roeloffzen, 1989b: A three-dimensional model of atmospheric CO_2 transport based on observed winds: 1. Analysis and observational data. In: *Aspects of Climate Variability in the Pacific and Western Americas. Geophysical Monograph 55,* D. H. Peterson (ed.), American Geophysical Union, Washington, DC, pp. 165–236.

Keeling, C. D., T. P. Whorf, M. Wahlen and J. van der Plicht, 1995: Interannual extremes in the rate of rise of atmospheric carbon dioxide since 1980. *Nature, 375,* 666–670.

Komhyr, W. D., R. H. Gammon, T. B. Harris, L. S. Waterman, T. J. Conway, W. R. Taylor and K. W. Thoning, 1985: Global atmospheric CO_2 distribution and variations from 1968–1982 NOAA/ GMCC CO_2 flask sample data. *J. Geophys. Res., 90,* 5567–5596.

Levin, I., B. Kromer, H. Schoch-Fischer, M. Bruns, M. Münnich, D. Berdau, J. C. Vogel and K. O. Münnich, 1985: 25 years of tropospheric ^{14}C observations in Central Europe. *Radiocarbon, 27,* 1–19.

Neftel, A., H. Oeschger, T. Staffelbach and B. Stauffer, 1988: CO_2 record in the Byrd ice core 50,000–5,000 years BP. *Nature, 331,* 609–611.

Oeschger, H., U. Siegenthaler and A. Guglemann, 1975: A box-diffusion model to study the carbon dioxide exchange in nature. *Tellus, 27,* 168–192.

Post, W. M., J. Pastor, P. J. Zinke and A. G. Stangenberger, 1985: Global patterns of soil nitrogen storage. *Nature, 317,* 613–616.

Potter, C. S., J. T. Randerson, C. B. Field, P. A. Matson, P. M. Vitousek, H. A. Mooney and S. A. Klooster, 1993: Terrestrial ecosystem production: A process model based on global satellite and surface data. *Global Biogeochem. Cycles, 7,* 811–841.

Revelle, R. and H. E. Suess, 1957: Carbon dioxide exchange between atmosphere and ocean and the question of an increase of atmospheric CO_2 during the past decades. *Tellus, 9,* 18–27.

Sarmiento, J. L., C. Le Quéré and S. W. Pacala, 1995: Limiting future atmospheric carbon dioxide. *Global Biogeochem. Cycles, 9,* 121–137.

Schimel, D. S. and E. Sulzman, 1995: Variability in the earth climate system: Decadal and longer timescales. *Reviews of Geophysics,* Supplement July 1995, 873–882.

Schimel, D. S., B. H. Braswell Jr., E. A. Holland, R. McKeown, D. S. Ojima, T. H. Painter, W. J. Parton and A. R. Townsend, 1994: Climatic, edaphic and biotic controls over carbon and turnover of carbon in soils. *Global Biogeochem. Cycles, 8,* 279–293.

Schimel, D. I., Enting, M. Heimann, T. M. L. Wigley, D. Raynaud, D. Alves and U. Siegenthaler, 1995: CO_2 and the carbon cycle. In: *Climate Change 1994: Radiative Forcing of Climate Change and an Evaluation of the IS92 Emission Scenarios,* J. T. Houghton, L. G. Meira Filho, J. Bruce, Hoesung Lee, B. A. Callander, E. Haites, N. Harris and K. Maskell (eds.). Cambridge University Press, Cambridge, UK.

Siegenthaler, U. 1990: El Niño and atmospheric CO_2. *Nature, 345,* 295–296.

Siegenthaler, U. and F. Joos, 1992: Use of a simple model for studying oceanic tracer distributions and the global carbon cycle. *Tellus, 44B,* 186–207.

Siegenthaler, U. and H. Oeschger, 1987: Biospheric CO_2 emissions during the past 200 years reconstructed by deconvolution of ice core data. *Tellus, 39B,* 140–154.

Siegenthaler, U. and J. L. Sarmiento, 1993: Atmospheric carbon dioxide and the ocean. *Nature, 365,* 119–125.

Starr, C., 1993: Atmospheric CO_2 residence time and the carbon cycle. *Energy, 18,* 1297–1310.

Suess, H. E., 1955: Radiocarbon concentration in modern wood. *Science, 122,* 415–417.

Vloedbeld, M. and R. Leemans, 1993: Quantifying feedback processes in the response of the terrestrial carbon cycle to global change: The modelling approach of IMAGE-2. *Water, Air and Soil Pollution, 70,* 615–628.

Volk, T., 1989: Effect of the equatorial Pacific upwelling on atmospheric CO_2 during the 1982–83 El Niño. *Global Biogeochem. Cycles, 3,* 267–279.

Wigley, T. M. L., 1993: Balancing the global carbon budget. Implications for projections of future carbon dioxide concentration changes. *Tellus, 45B,* 409–425.

Wigley, T. M. L., R. Richels and J. A. Edmonds, 1996: Economic and environmental choices in the stabilization of CO_2 concentrations: choosing the "right" emissions pathway. *Nature, 379,* 240–243.

Winguth, A. M. E., M. Heimann, K. D. Kurz, E. Maier-Reimer, U. Mikolajewicz and J. Segeschneider, 1994: El Niño Southern Oscillation related fluctuations of the marine carbon cycle. *Global Biogeochem. Cycles, 8,* 39–63.

Wong, C. S., Y. -H. Chan, J. S. Page, G. E. Smith and R. D. Bellegay, 1993: Changes in equatorial CO_2 flux and new production estimated from CO_2 and nutrient levels in Pacific surface waters during the 1986/87 El Niño. *Tellus, 45B,* 64–79.

Part II

The Missing Carbon Sink

3

Carbon Dioxide Emissions from Fossil Fuel Consumption and Cement Manufacture, 1751–1991, and an Estimate of Their Isotopic Composition and Latitudinal Distribution

ROBERT J. ANDRES, GREGG MARLAND, TOM BODEN, AND STEVE BISCHOF

Abstract

This work briefly discusses four of the current research emphases at Oak Ridge National Laboratory regarding the emission of CO_2 from fossil fuel consumption, natural gas flaring, and cement manufacture. These emphases include: (1) updating the 1950 to present time series of CO_2 emissions from fossil fuel consumption and cement manufacture, (2) extending this time series back to 1751, (3) gridding the data at $1° \times 1°$ resolution, and (4) estimating the isotopic signature of these emissions.

A latitudinal distribution of carbon emissions is being completed. A southward shift in the major mass of CO_2 emissions is occurring from European–North American latitudes toward Central–Southeast Asian latitudes, reflecting the growth of population and industrialization at these lower latitudes.

The carbon isotopic signature of these CO2 emissions has been reexamined. The emissions of the past two decades were approximately 1% lighter than previously estimated.

3.1 Introduction

Emissions of CO_2 from the consumption of fossil fuels have resulted in an increasing concentration of CO_2 in the atmosphere of the Earth. Combined with CO_2 releases from changes in land use, these emissions have perturbed the natural cycling of carbon, resulting in the accumulation of CO_2 in the atmosphere and concern that this may significantly change the climate of the Earth (Houghton et al., 1990, 1996).

Understanding the changes currently being observed and changes likely to occur in the future requires the best possible information on the flows of carbon in the Earth system. Analyses summarized here attempt to inventory the global magnitude and distribution of CO_2 emissions from fossil fuel consumption. The recent interest in limiting the growth of atmospheric CO_2 dictates an understanding of the sources of CO_2 as a first step in trying to limit emissions. The 1992 Framework Convention on Climate Change, signed by 154 nations, requires that each nation conduct an inventory of the sources and sinks of greenhouse gases. This increasing concern and interest has motivated updates and improvements of earlier CO_2 emission inventories (e.g., Keeling, 1973; Marland and Rotty, 1984; Marland et al., 1989; Marland and Boden, 1991; Marland and Boden, 1993).

International and national organizations have been systematically compiling statistics on world energy production and consumption since the early 1970s. These statistics are reported for individual countries and can be used with data on fuel chemistry and combustion characteristics to determine CO_2 emissions from fossil fuel consumption on national, regional, and global scales. CO_2 emissions from cement manufacture contribute approximately 1–2% of that from fossil fuel production globally and are included here.

3.2 The 1950–91 Annual Time Series

The 1950–91 time series of CO_2 emissions from fossil fuel consumption and cement manufacture was calculated follow-

ing the procedures in Marland and Rotty (1984) and Marland and colleagues (1989). Briefly, CO_2 emissions from fossil fuels are calculated from:

$$CO_{2,i} = FP_i \times FO_i \times C_i \qquad (3.1)$$

where the FP_i represents the net fuel production, FO_i is the fraction of the fuel oxidized, and C_i is the carbon content of the fuel, respectively; i represents solid, liquid, or gaseous fuels. Global totals of CO_2 emissions are calculated from national production data. National totals of CO_2 emissions are calculated from estimates of apparent consumption, with FC_i replacing FP_i in Equation (3.1), where FC_i represents the national production data modified by national data on global trade and changes in stocks. The sum of FC_i for all countries is less than the sum of FP_i for all countries, mainly because of (1) the exclusion in national consumption data of bunker fuels, that is, those fuels used for conducting international commerce; (2) inaccuracies in reporting of imports and exports; and (3) difficulty in accounting for nonfuel uses of fossil fuels, particularly the use of petroleum products in the chemical industry. Data on fuel production, trade, and changes in stocks are obtained on magnetic media from the United Nations (UN, 1993). Fuel oxidation and carbon content data come from a variety of sources, as reported in Marland and Rotty (1984).

For cement manufacture, data are obtained from the U.S. Bureau of Mines (Solomon, 1992). An equation analogous to Equation (3.1) is used to calculate CO_2 emissions from calcining $CaCO_3$, with the cement chemistry data assembled by Griffin (1987).

Energy production data published by the UN Statistics Office in 1993 led to a global CO_2 emission estimate of $6,188 \times 10^6$ tonnes C for 1991 (see Figure 3.1 and Table 3.1). This is approximately 1.5% more than the revised 1990 CO_2 emission estimate. The total includes 123×10^6 metric tonnes C for oil combusted in the Kuwaiti oil-field fires (estimation procedure described below). From 1990 to 1991, there was an increase in the apparent consumption of natural gas and in cement manufacture, but the global total of CO_2 would have shown a slight decline had it not been for the Kuwaiti oil-field fires.

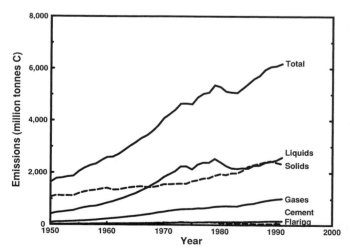

Figure 3.1: Global CO_2 emissions from fossil fuel production and cement manufacture during 1950–91.

Table 3.1. *Global CO$_2$ emissions from fossil fuel production and cement manufacture during 1950–91*

Year	Total	Gases	Liquids	Solids	Flaring	Cement	δ^{13}C
1950	1,638	97	423	1,077	93	18	−25.86
1951	1,775	115	479	1,137	24	20	−26.00
1952	1,803	124	504	1,127	26	22	−26.11
1953	1,848	131	533	1,132	27	24	−26.16
1954	1,871	138	557	1,123	27	27	−26.21
1955	2,050	150	625	1,215	31	30	−26.24
1956	2,185	161	679	1,281	32	32	−26.29
1957	2,278	178	714	1,317	35	34	−26.37
1958	2,338	192	732	1,344	35	36	−26.42
1959	2,471	214	790	1,390	36	40	−26.51
1960	2,586	235	850	1,419	39	43	−26.61
1961	2,602	254	905	1,356	42	45	−26.79
1962	2,708	277	981	1,358	44	49	−26.91
1963	2,855	300	1,053	1,404	47	51	−27.00
1964	3,016	328	1,138	1,442	51	57	−27.12
1965	3,154	351	1,221	1,468	55	59	−27.23
1966	3,314	380	1,325	1,485	60	63	−27.35
1967	3,420	410	1,424	1,455	66	65	−27.55
1968	3,596	445	1,552	1,456	73	70	−27.70
1969	3,809	487	1,674	1,494	80	74	−27.85
1970	4,084	516	1,838	1,564	87	78	−27.89
1971	4,235	554	1,946	1,564	88	84	−28.01
1972	4,403	583	2,055	1,580	94	89	−28.07
1973	4,641	608	2,240	1,588	110	95	−28.13
1974	4,649	618	2,244	1,585	107	96	−28.16
1975	4,622	623	2,131	1,679	93	95	−28.12
1976	4,889	647	2,313	1,717	109	103	−28.16
1977	5,028	646	2,389	1,780	104	108	−28.09
1978	5,076	674	2,383	1,796	107	116	−28.17
1979	5,358	714	2,534	1,892	100	119	−28.19
1980	5,292	726	2,407	1,950	89	120	−28.20
1981	5,121	736	2,270	1,922	72	121	−28.24
1982	5,081	731	2,176	1,985	69	121	−28.21
1983	5,072	733	2,161	1,991	63	125	−28.20
1984	5,237	785	2,185	2,083	57	128	−28.26
1985	5,413	818	2,170	2,239	55	131	−28.23
1986	5,601	835	2,276	2,300	53	137	−28.17
1987	5,727	897	2,287	2,351	48	143	−28.28
1988	5,953	940	2,394	2,413	55	152	−28.29
1989	6,068	976	2,429	2,454	53	156	−28.31
1990	6,098	1,004	2,482	2,393	621	57	−28.42
1991	6,188	1,024	2,591	2,342	701	62	−28.45

Note: Total is the sum of CO$_2$ emissions from the production of gas, liquid, and solid fuels; gas flaring; and cement manufacture in million metric tonnes C. The 1991 emissions include 123 and 7.3 million metric tonnes C from oil consumed and gas flared during the Kuwaiti oil-field fires.

Emissions data reported here are updated annually. Updates can be obtained by Internet using the file transfer protocol from cdiac.esd.ornl.gov or via the worldwide web from http://cdiac.esd.ornl.gov or by contacting the Carbon Dioxide Information Analysis Center, Oak Ridge National Laboratory, Oak Ridge, Tennessee, 37831–6335, USA (423–574–3645 or cdp@ornl.gov).

With each annual release of the UN energy data, past data are also revised. The largest number of revisions occur in the first year after the initial release, and the number of revisions drops considerably in subsequent years. Based on past years' experience (Table 3.2), the 1991 number is likely within 2% of its final value.

An interesting facet of the 1991 results is the estimation of CO_2 emissions from Kuwait (Figure 3.2). The emissions from the oil fields that were set ablaze were considerable, which raises the issue of their proper allocation. The United Nations included in their energy statistics an estimate of the total gas flared from Kuwait in 1991 (Figure 3.2B). The total CO_2 emissions showed a decline in 1990, reflecting a decline in Kuwaiti energy production. However, in 1991, total emissions returned to a level apparently indicative of the longer-term, steady growth in CO_2 emissions from Kuwait. However, this total emission curve is strongly affected by the gas flaring that resulted from the oil fields being set ablaze.

Figure 3.2A is a more realistic picture of Kuwaiti CO_2 emissions resulting from their energy production. It differs from Figure 3.2B in that the 1991 emissions from gas flaring (8.1 million metric tonnes C) have been reduced by 90%. This reduction is based on a UN estimate that only 10% of the gas flaring resulted from normal oil-field production procedures and the other 90% from the oil-field fires (S. Hussein, UN Energy Statistics Office, personal communication, 1993). Ferek and colleagues (1992) estimated that 7% of the carbon discharged from the oil-field fires during May and June 1991 was from the burning of natural gas rather than oil. Combined with our estimate of oil burning, this implies that 8.6 million metric tonnes C were discharged from natural gas burning, which is consistent with the estimate based on UN (1993) energy data. The differences between Figures 3.2A and 3.2B if limited to

the year 1991 are not major. They suggest that CO_2 emissions caused by the gas flaring from oil-field fires could be allocated to Kuwait without significantly changing the long-term emissions pattern of Kuwait.

However, if CO_2 emissions from the gas flaring from oil-field fires are allocated to Kuwait, should the CO_2 emissions from the burning oil also be allocated to Kuwait (Figure 3.2C)? The UN (1993) energy data do not include oil lost during the fires. Based on the detailed data from tables of average flow rate per week for the interval from March 16 until the last fires were extinguished on November 6 (the first fires were ignited in late January, and the first one was extinguished on March 23) (NOAA, 1991), the amount of oil released in the fires is estimated at 1,060 million barrels combusted with another 60 million barrels spilled but not combusted. Of the spilled oil, some small fraction was recovered, and the rest either seeped into the ground or was covered with sand. Based on measurements made in the plume during May and June 1991, 96% of the fuel carbon emitted was as CO_2, while only 0.53% was as soot (Ferek et al., 1992; Laursen et al., 1992). Carbon was also discharged as carbon monoxide, methane, and other organic volatiles and particulates. Because much of this carbon is soon converted to CO_2 in the atmosphere, approximately 99% of the fuel carbon ended up in the atmosphere as CO_2. Given an oil density of 0.86 kg/l and a carbon content of 85% (Laursen et al., 1992), the total CO_2 discharged contained approximately 123 million tonnes C.

Figure 3.2C differs from Figure 3.2B, in that it includes the CO_2 emissions from the oil combusted during the 1991 oil-field fires. The inclusion of this source dwarfs the other emissions from Kuwait. In a global context, this source is greater than the increase in total CO_2 emissions from 1990 to 1991. Without including the CO_2 emissions released from the oil-field fires (liquids and gas flaring) in Kuwait, the 1991 CO_2 emissions estimate for the global total is only 99.3% of the revised 1990 estimate.

Because of unique situations such as the oil-field fires, and a basic data set that is annually updated, the global CO_2 calculations are periodically revised. Ongoing work with global CO_2 calculations currently includes refining the supporting data used to calculate emissions from the UN energy data. This includes the acquisition of new data on the carbon content of fossil fuels, especially solids, and efforts to refine estimates of the geographic distribution of emissions on a 1° × 1° global grid. The gridded data are part of the Global Emissions Inventory Activity, an activity of the International Global Atmospheric Chemistry Project of the International Geosphere-Biosphere Programme. Also included in the refinement is a reexamination of the higher and lower heating values of fossil fuels and the role they have played in past calculations. Finally, the reevaluation of the slow oxidation of fossil fuels used in nonfuel uses will be included in future work.

Marland and Rotty (1984) estimated that the uncertainty in the global total CO_2 emissions estimates was 6–10%; because of the subjectivity of those estimates, they have not been reevaluated. Ongoing work can provide refinements in the chemistry and the fate of fuels and fuel products; however, the major source of uncertainty is the national statistics on fuel

Table 3.2. *Changing estimates of CO_2 emissions (in million metric tonnes C) from fossil fuel and cement production based on successive annual releases of UN energy data*

Emission Year	Energy Data Release Date				
	1993[a]	1992[b]	1991[c]	1990[d]	1989[e]
1991	6,188	—	—	—	—
1990	6,098	6,097	—	—	—
1989	6,068	6,024	5,967	—	—
1988	5,953	5,912	5,897	5,893	—
1987	5,727	5,698	5,661	5,680	5,650

Note: There is about an 18-month lag between the end of a year and the first release of the UN energy data for that year. References are for publication of the CO_2 emissions estimates based on annual UN energy data. All emissions are given in millions of metric tonnes carbon.
[a]This study.
[b]Previously unpublished data.
[c]Marland and Boden (1991).
[d]Marland (1990).
[e]Marland and Boden (1989).

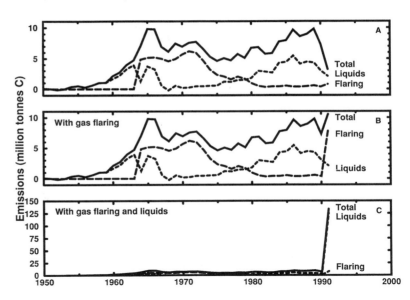

Figure 3.2: Three CO_2 emissions scenarios for Kuwait. (A) Includes no emissions resulting from the oil field fires. (B) Contains gas flaring emissions from the oil-field fires and is the set of curves that would be calculated from the UN (1993) data. (C) Contains emissions from both gas flaring and oil consumed during the 1991 oil-field fires. Individual curves for emissions from natural gas and coal consumption and from cement manufacture are not shown, for clarity. However, these emissions estimates are included in the total emissions curve. The three figures are identical, except for 1991. Based on Etemad et al. (1991) and UN (1993) data.

production and trade. These statistics vary in quality among countries, and international statistical offices, such as that of the United Nations, are obligated to rely on data they can obtain from national sources or energy industries.

An assessment conducted by the Energy Information Administration (EIA) of the U.S. Department of Energy provides an insight into the uncertainty of national data by looking at some of the supporting data used to generate its own energy data sets (EIA, 1989). One would expect data on the quantities of fuel received and consumed by electric utilities in the United States to be among the best energy data available. The EIA compared data on fuel receipts at electric utilities, compiled from one questionnaire, against data on fuel consumption plus changes in stocks, compiled from another questionnaire. Both questionnaires were completed by the utilities themselves. Receipts should equal consumption plus the change in stocks. With focused inquiry, the EIA was able to offer a variety of reasons why the equality might not hold (including different sampling strategies between the two forms, reporting lags between the two forms, loss of coal from stockpiles by wind and oxidation, etc.). It is interesting, however, to see the magnitude of the difference between two numbers that one expects to be equal. For coal, receipts and consumption plus changes in stock were within 3% for all 10 U.S. census regions, and the national totals differed by less than 0.8%. For petroleum and natural gas, the difference exceeded 5% in several census regions, and the national totals differed by approximately 1.3% for petroleum and approximately 6.3% for natural gas.

With increasing interest in greenhouse gas emissions and increasing numbers of people making greenhouse gas emissions estimates, the data presented here provide opportunity for comparisons with other estimates. Some of these estimates rely on the same UN energy data, and others are based on national energy data that may not be broadly available. The most comprehensive alternate set of CO_2 emissions calculations is that compiled by S. Subak and her colleagues at the Stockholm Environment Institute (von Hippel et al., 1993). They calculated CO_2 emissions for 1988 based on energy data from the

International Energy Agency/Organization of Economic Cooperation and Development (IEA/OECD) when possible and the United Nations when necessary. Their global total differs from that presented here by less than 2.1%, although there are some significant differences in some of the national totals. Because their treatment of the national energy data is slightly different, direct comparison between the two studies is difficult and requires caution.

3.3 A 1751–1991 Time Series

Etemad and colleagues (1991) published a volume on world energy production at varying temporal resolutions from 1800 to 1985. Footnotes in the text extend production data back to 1751. These data can be treated in the same manner as the UN energy data to obtain a time series of CO_2 emissions from fossil fuel production.

Preliminary analysis shows good agreement between this time series and others currently available (Table 3.3). In comparison to the estimates of CO_2 emissions from 1860 to 1953 compiled by Keeling (1973), calculations based on the data from Etemad and colleagues (1991) are always within 5%, usually within 3%. Similar agreement is found between CO_2 estimates based on the data from Etemad and colleagues (1991) data and on the 1950–91 UN (1993) data. For the four years of overlap between Keeling (1973) and the calculations based on UN (1993) energy data, the agreement is within 1%.

A preliminary time series of cumulative CO_2 emissions since 1751, whose values may change slightly as the full energy set becomes available, is presented in Figure 3.3 and Table 3.4. For 1751–1949, the time series is based on calculations using summary data for a limited number of years (Etemad et al., 1991) with extrapolation between those years. For 1950–91, the time series is based on calculations using the UN (1993) energy data.

The time series shows over 2 million metric tonnes of carbon had been released to the atmosphere by 1751. Cumulative

Table 3.3. *Comparison of selected annual CO_2 emissions estimates based on three different time series of energy data*

Year	Etemad et al. (1991)[a]	Keeling (1973)[b]	U.N. (1993)[c]
1751	0.10		
1761	0.12		
1771	0.15		
1781	0.19		
1791	1.05		
1800	8		
1810	11		
1820	14		
1830	25		
1840	34		
1850	55		
1860	93	93	
1870	151	145	
1880	238	227	
1890	365	350	
1900	548	525	
1910	837	805	
1913	967	929	
1928	1,093	1,091	
1938	1,173	1,161	
1950	1,635	1,613	1,620
1951		1,744	1,755
1952		1,767	1,781
1953		1,809	1,824
1955	2,045		2,020
1960	2,623		2,543
1965	3,234		3,095
1970	4,194		4,006
1975	4,720		4,527
1980	5,408		5,172
1985	5,449		5,282

[a]CO_2 estimates based on the Etemad et al. (1991) energy data.
[b]CO_2 estimates of Keeling (1973), based on energy data largely from the UN, circa 1973.
[c]CO_2 estimates based on the 1993 UN energy data.
Note: This table does not include carbon emissions from cement manufacture. All emissions in millions of metric tonnes C.

emissions topped 100 million metric tonnes C by 1781, and over 1 billion metric tonnes C had been released to the atmosphere by the mid-1840s.

Further refinement of the pre-1950 part of the time series depends on a fuller analysis of the data from Etemad and colleagues (1991). This requires access to supporting data not published in the book. These supporting data are currently being sought. Completion of the ongoing analysis of this data set

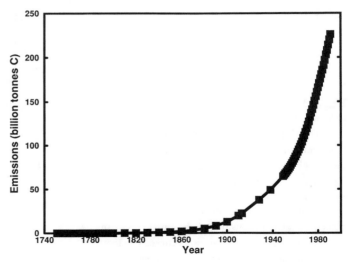

Figure 3.3: Preliminary cumulative emissions of carbon to the atmosphere based on fossil fuel production. Based on energy data from Etemad et al. (1991) for 1751–1949 and from the United Nations (1993) for 1950–91. Note that carbon emissions from cement manufacture are not included. The boxes indicate years with calculated annual CO_2 emissions (Table 3.4). Between boxes, emissions are linearly extrapolated.

will supply a better time series of fossil fuel CO_2 emissions to the atmosphere back to 1751. This will complement carbon cycle modeling studies, which typically begin their analyses in the early 1700s.

3.4 The Latitudinal Distribution

A data set of CO_2 emissions on a uniform geographic basis can be constructed from the country CO_2 emissions data. The challenge is to allocate emissions within countries, based on either subcountry data on energy use or proxy data such as population density. The current effort to distribute CO_2 emissions from fossil fuels on a 1° latitude × 1° longitude grid updates the 5° × 5° grid used by Marland and colleagues (1985). The 360 longitudinal grid spaces in each latitudinal band can be summed to determine the latitudinal emissions.

A sense of the latitudinal changes that are occurring can be gained from the within-country energy consumption distributions estimated for 1980 (Marland et al., 1985) and national fossil fuel CO_2 emissions calculated with 1989 energy use data (Marland and Boden, 1989) (Figure 3.4). The bulk of fossil fuel CO_2 emissions is shifting southward with increasing global industrialization.

3.5 The Isotopic Signature

The carbon isotopic signature ($\delta^{13}C$ measured against the PeeDee belemnite standard) of fossil fuel emissions has decreased during the past century, reflecting the changing mix of fossil fuels produced. Previous estimates of the $\delta^{13}C$ of anthro-

Table 3.4. *Preliminary cumulative emissions of carbon to the atmosphere based on fossil fuel production*

Year	Annual Emissions	Cumulative Emissions
1751	2.6	0.0026
1760	2.9	0.027
1770	3.6	0.059
1780	4.2	0.10
1790	5.6	0.15
1800	7.8	0.22
1810	11	0.31
1820	14	0.44
1830	25	0.63
1840	34	0.92
1850	55	1.4
1860	93	2.1
1870	151	3.3
1880	238	5.3
1890	365	8.3
1900	548	13
1910	837	20
1913	967	22
1928	1,093	38
1938	1,173	49
1949	1,635	65
1950	1,620	66
1960	2,543	87
1970	4,006	119
1980	5,172	166
1990	5,941	220
1991	6,026	226

Note: Based on energy data from Etemad et al. (1991) for 1751–1949 and from UN (1993) for 1950–91. Annual emissions in million metric tonnes C and cumulative emissions in billion metric tonnes C. Only selected years used to create Figure 3.3 are shown here. Note that carbon emissions from cement manufacture are not included.

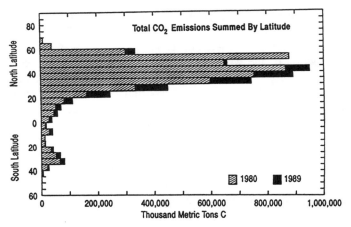

Figure 3.4: The latitudinal change in CO_2 emissions from 1980 to 1989 as seen in 5° bands. There is a small decrease in 1989 emissions from 1980 emissions in the 50–55°N latitude band. Energy consumption data from 1980 and 1989 (Marland and Boden, 1989) were distributed within countries with data on population density in 1980 (Marland et al., 1985).

pogenic CO_2 emissions have assumed an average worldwide value for each fuel type that is invariant with time and area of production (Tans, 1981). Combining updated $\delta^{13}C$ signatures with the anthropogenic CO_2 emissions estimates allows a new estimate of global, average $\delta^{13}C$ signature that considers emissions type and geographic origin. The improved fossil fuel $\delta^{13}C$ signature should provide an additional constraint for balancing the sources and sinks of the global carbon cycle.

The variability in the $\delta^{13}C$ values of coals is small, and thus the average isotopic ratio of -24.1% adopted by Tans (1981) is within $\pm0.3\%$, regardless of source. Over 80% of coal is composed of humic materials or type III kerogen, which shows little variability in isotopic composition (Redding et al., 1980). Even after significant thermal alteration, coal retains the isotopic signature of its source material to within $\pm0.3\%$ (Craig, 1953; Jeffrey et al., 1955). It is the relatively unchanged isotopic composition of the humic materials and kerogen that determines the isotopic signature of coals.

Oil shows a significant spread of $\delta^{13}C$ isotopic values, from -19% to -35%. The modal value of $\delta^{13}C$ of oil is near -26.5% (Degens, 1967). However, this modal value does not represent the average value of oil produced for energy. The average value for the oil consumed must be weighted by the production from various oil-producing regions since different regions, often have characteristic $\delta^{13}C$ values.

Several important oil-producing regions during the last two decades have $\delta^{13}C$ values near -30% or lighter. These areas include the North Slope of Alaska, the North Sea, the Western Siberian Basin, and the large nonmarine basins in China. The individual isotopic values for oil production in the top 17 oil-producing countries have been taken from the published literature (Galimov, 1973; Yeh and Epstein, 1981; Wanli, 1985; Golyshev et al., 1991; Chung et al., 1992; Philp et al., 1992; and references contained therein). For minor petroleum-producing countries the modal value of -26.5% is used (Degens, 1967; Tans, 1981). In the United States, the shift in production from isotopically heavy California oil to isotopically light Alaskan oil has shifted the oil production–weighted $\delta^{13}C$ value from approximately -26.5% in 1950 to -27.5% in 1991. A similar shift, from approximately -28% in 1965 to -30% in 1980, has occurred in the former Soviet Union, as production moved during the late 1960s and 1970s from the Caspian and Volga–Ural regions to the fields in western Siberia (Riva, 1991). Between 1950 and 1990 the global oil production–weighted $\delta^{13}C$ signature decreased by 1.3%, from -26.5% to -27.8%.

Natural gas shows the greatest variability in $\delta^{13}C$ of any natural substance (Schoell, 1988). Methane associated with coal may be as heavy as -20%, while methane from marine sediments may be lighter than -100% (Schoell, 1988). Natural gas produced for fuel can be divided into separate genetic groups, thermogenic and biogenic gases, with different $\delta^{13}C$ isotopic signatures (Schoell, 1984). Since 80% of the gas production of the world is thermogenic (Rice and Claypool, 1981), a weighted average of 193 thermogenic (-40%) and 55 biogenic (-65%) methane gases would have a mean of -45%. However, natural gas also contains C2 and C3 gases, which are often 10% to 20% heavier than methane but are rarely more than 10% of natural gas. This results in an average $\delta^{13}C$ of approximately -44% for natural gases.

It would be appropriate to weight the values for natural gas production by source within each region of production or country; however, the variability of $\delta^{13}C$ values of natural gases, even within a single field, makes it difficult to estimate a weighted average value. An example is the Anadarko Basin, a major natural gas production zone in the United States, which has $\delta^{13}C$ ranging from -33.2% to -49.8%, with a mean value near -43.5% (Rice et al., 1988). The world average of -44.0% may be a heavy estimate when natural gas production in the former Soviet Union is considered. The UN (1993) energy data show that the former Soviet Union produced 32% of the world's total natural gas in 1990. Much of this production was concentrated in the gas fields of western Siberia, which is characterized by $\delta^{13}C$ of -46% to -50% and hydrocarbon composition of greater than 99% methane (Galimov, 1988). Because of the variability in the $\delta^{13}C$ isotopic signature and the lack of sufficient field production data, the average $\delta^{13}C$ of -44% calculated above is used for all natural gas production.

Methane associated with liquid petroleum has an average value near -40% (this study, the average for thermogenic gases). Therefore, this is the $\delta^{13}C$ isotopic signature used for gas flaring.

As is the case for coal, the variability in the $\delta^{13}C$ values of cement is small, and the 0% average isotopic ratio adopted by Tans (1981) is within ±0.3%. Limestones, the source material for cement, have a mean isotopic value near 0%, which reflects the equilibrium with the ocean at the time of formation (Garrels and Mackenzie, 1971). The calcite in limestones may undergo diagenesis, which can change the $\delta^{13}C$ value toward lighter values. However, most limestones used for cement manufacture are considered unaltered and retain a value near 0%.

The $\delta^{13}C$ value of total CO_2 emitted from anthropogenic sources was calculated by using the production of carbon for each fuel type by country and multiplying it by the appropriate $\delta^{13}C$ value. Values of -24.1% for coal, -44% for natural gas, -40% for gas flaring, and 0% for cement production were used for all countries and all years. Fifteen of the top oil-producing countries were assigned unique isotopic values for oil production that did not vary with time. In addition, for the United States and the former Soviet Union, isotopic values for oil production changed with time as the location of production fields changed. The $\delta^{13}C$ used for U.S. oil decreased linearly between 1950 and 1991, and the $\delta^{13}C$ used for the former Soviet Union oil decreased in two linear steps: 1965-75 and 1975-80. For all remaining countries, the modal value for

oil, -26.5%, was used. Based on these $\delta^{13}C$ isotopic signatures (Table 3.5) and the national CO_2 emissions data calculated from Equation (3.1), the $\delta^{13}C$ value of CO_2 produced for each year from fossil fuels and cement was calculated (Figure 3.5 and Table 3.1). The absolute accuracy of the method for calculating the $\delta^{13}C$ of CO_2 emissions from fossil fuel production and cement manufacture is believed to be approximately ±0.5%. Table 3.1 shows $\delta^{13}C$ with two decimal places so that the relative structure of the time series can be seen.

3.6 Epilogue

The estimation of CO_2 emissions from fossil fuel consumption and cement manufacture is fundamental to understanding the human perturbation of the global carbon cycle. As the magni-

Table 3.5. *d13C isotopic values used for carbon isotope calculations*

	$\delta^{13}C$ (PDB)
Fuel or cement	
Coal	-24.1
Natural gas	-44.0
Gas flaring	-40.0
Cement	0.0
Petroleum (mode)	-26.5
Top petroleum producers	
Canada	-29.9
China	-29.3
Egypt	-28.5
Indonesia	-27.7
Iran	-26.8
Iraq	-27.1
Kuwait	-27.3
Libya	-28.2
Mexico	-26.5
Nigeria	-26.6
Norway	-28.9
Saudi Arabia	-26.4
USSR, former, 1950-65	-28.0
1966-75	-28.1 to -29.0
1976-79	-29.2 to -29.8
1980-91	-30.0
United Arab Emirates	-27.0
United Kingdom	-29.6
United States of America	-26.5 to -27.525
Venezuela	-26.1

Note: Data sources cited in text.

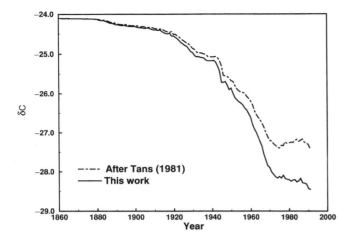

Figure 3.5: $\delta^{13}C$ of annual CO_2 emissions from fossil fuel production and cement manufacture. The upper curve was calculated with the $\delta^{13}C$ values suggested by Tans (1981), and the lower curve was calculated with the $\delta^{13}C$ values described herein. For years prior to 1950, the lower curve was calculated with a $\delta^{13}C = -26.52$ for oil, which was the 1950 production-weighted mean. Both curves use the same fuel production data (Keeling, 1973, for pre-1950 and UN, 1993, for post-1949). Cement manufacturing data (Solomon, 1992) are included for post-1949 years only, and their inclusion causes the slight increase seen in both curves for 1950.

tude, distribution, and $\delta^{13}C$ isotopic signature become better known, our chance to understand the current functioning of the global carbon cycle and the changes that are anticipated to occur in the future will improve.

As efforts to control global CO_2 emissions increase, national and regional political bodies are searching for means to reduce or offset the impact of the emissions originating from within their jurisdictions. Calculations of greenhouse gas emissions are currently taking place in many national and international groups. The annual data set on CO_2 emissions from fossil fuel consumption and cement manufacture is intended to be a well-documented, technically sound data set of CO_2 emissions for use by the scientific community and a benchmark for the policy community. It is periodically revised as new data become available. These estimates should improve over time as increased international interest improve the accuracy of the national data sets of energy production and consumption on which these CO_2 emission estimates ultimately rely.

Acknowledgments

Robert J. Andres was supported in part by an appointment to the Oak Ridge National Laboratory Postdoctoral Research Associates Program, administered jointly by the Oak Ridge National Laboratory and the Oak Ridge Institute for Science and Education (ORISE). Gregg Marland and Tom Boden were supported by the Global Change Research Program, Environmental Sciences Division, Office of Health and Environmental Research, U.S. Department of Energy (DOE), under contract DE-AC05–84OR21400 with Lockheed Martin Energy Research, Inc. Steve Bischof was supported by a Graduate Fellowship for Global Change, sponsored by DOE, administered by ORISE.

References

Chung, H. M., M. A. Rooney, M. B. Toon, and G. E. Claypool. 1992. Carbon isotope composition of marine crude oils. *Bulletin of the American Association of Petroleum Geologists 76*, 1000–1007.

Craig, H. 1953. The geochemistry of stable carbon isotopes. *Geochimica et Cosmochimica Acta 3*, 53–92.

Degens, E. T. 1967. Biogeochemistry of stable carbon isotopes. In *Organic Geochemistry: Methods and Results,* G. Elginton and M. T. J. Murphy, eds. Springer-Verlag, New York, 304–329.

EIA (Energy Information Administration). 1989. *An Assessment of the Quality of Selected EIA Data Series.* DOE/EIA-0292(87), Energy Information Administration, U.S. Department of Energy, Washington, D.C.

Etemad, B., J. Luciani, P. Bairoch, and J.-C. Toutain. 1991. *World Energy Production: 1800–1985.* Librarie DROZ, Geneva, Switzerland.

Ferek, R. J., P. V. Hobbs, J. A. Herring, K. K. Laursen, R. E. Weiss, and R. A. Rasmussen. 1992. Chemical composition of emissions from the Kuwait oil fires. *Journal of Geophysical Research 97*, 14483–14489.

Galimov, E. M. 1973. *Isotopy ugleroda v neftegazovoy geologii.* Nedra, Moscow, Russia. [English translation: Galimov, E. M. 1975. *Carbon Isotopes in Oil and Gas Geology.* TT 682, National Aeronautics and Space Administration, Washington D.C.]

Galimov, E. M. 1988. Sources and mechanisms of formation of gaseous hydrocarbons in sedimentary rocks. *Chemical Geology 71*, 77–95.

Garrels, R. M., and F. T. Mackenzie. 1971. *Evolution of Sedimentary Rocks.* Norton and Co., New York.

Golyshev, S. I., N. A. Verkhovskaya, and V. A. Burkova. 1991. Stable carbon isotopes in source-bed organic matter of West and East Siberia. *Organic Geochemistry 17*, 277–291.

Griffin, R. C. 1987. *CO2 Release from Cement Production 1950–1985.* Institute for Energy Analysis, Oak Ridge Associated Universities, Oak Ridge, Tenn.

Houghton, J. T., G. J. Jenkins, and J. J. Ephraums (eds.). 1990. *Climate Change: The IPCC Scientific Assessment.* Cambridge University Press, Cambridge, U.K.

Houghton, J. T., L. G. Meira Filho, B. A. Callander, N. Harris, A. Kattenberg, and K. Maskell (eds.). 1996. *Climate Change 1995: The Science of Climate Change, Contribution of Working Group I to the Second Assessment Report of the Intergovernmental Panel on Climate Change.* Cambridge University Press, New York.

Jeffrey, P. M., W. Compston, D. Greenhalgh, and B. C. Heezen. 1955. On the carbon-13 abundance of limestones and coals. *Geochimica et Cosmochimica Acta 7*, 255–286.

Keeling, C. D. 1973. Industrial production of carbon dioxide from fossil fuels and limestone. *Tellus 2*, 174–198.

Laursen, K. K., R. J. Ferek, P. V. Hobbs, and R. A. Rasmussen. 1992. Emission factors for particles, elemental carbon, and trace gases from the Kuwait oil fires. *Journal of Geophysical Research 97*, 14491–14497.

Marland, G. 1990. CO_2 emissions–Modern record. In *TRENDS '90: A Compendium of Data on Global Change,* T. A. Boden, P. Kanciruk, and M. P. Farrell, eds. ORNL/CDIAC-36, Carbon Dioxide

Information Analysis Center, Oak Ridge National Laboratory, Oak Ridge, Tenn., pp. 92–133.

Marland, G., and T. A. Boden. 1989. Carbon dioxide releases from fossil-fuel burning, testimony before the Senate Committee on energy and natural resources, 26 July 1989. *Senate Hearing 101–235.* DOE's National Energy Plan and Global Warming, U.S. Senate, U.S. Government Printing Office, Washington, D.C., pp. 62–84.

Marland, G., and T. A. Boden. 1991. CO_2 emissions – Modern record. In *Trends '91: A Compendium of Data on Global Change,* T. A. Boden, R. J. Sepanski, and F. W. Stoss, eds. ORNL/CDIAC-46, Carbon Dioxide Information Analysis Center, Oak Ridge National Laboratory, Oak Ridge, Tenn., pp. 386–507.

Marland, G., and T. A. Boden. 1993. The magnitude and distribution of fossil-fuel–related carbon releases. In *The Global Carbon Cycle,* M. Heimann, ed. NATO Advanced Study Institute, il Ciocco, Italy, Springer-Verlag, New York, pp. 117–139.

Marland, G., T. A. Boden, R. C. Griffin, S. F. Huang, P. Kanciruk, and T. R. Nelson. 1989. *Estimates of CO_2 Emissions from Fossil Fuel Burning and Cement Manufacturing, Based on the United Nations Energy Statistics and the U.S. Bureau of Mines Cement Manufacturing Data.* NDP-030, Oak Ridge National Laboratory, Oak Ridge, Tenn.

Marland, G., and R. M. Rotty. 1984. Carbon dioxide emissions from fossil fuels: A procedure for estimation and results for 1950–1982. *Tellus 36B,* 232–261.

Marland, G., R. M. Rotty, and N. L. Treat. 1985. CO_2 from fossil fuel burning: Global distribution of emissions. *Tellus 37B,* 243–258.

NOAA (National Oceanic and Atmospheric Administration). 1991. *Kuwait Oil Fire Extinguishing Chronology.* Office of the Chief Scientist, Gulf Program Office, U.S. Department of Commerce, Washington, D.C.

Philp, R. P., J. H. Chen, J. M. Fu, and G. Y. Sheng. 1992. A geochemical investigation of crude oils and source rocks from Biyang Basin, China. *Organic Geochemistry 18,* 933–945.

Redding, C. E., M. Schoell, J. C. Monin, and B. Durand. 1980. Hydrogen and carbon isotopic composition of coals and kerogen. In *Advances in Organic Geochemistry,* A. G. Douglas and J. R. Maxwell, eds. Pergamon Press, Oxford, U.K., pp. 711–723.

Rice, D. D., and G. E. Claypool. 1981. Generation, accumulation and resource potential of biogenic gas. *Bulletin of the American Association of Petroleum Geologists 65,* 5–25.

Rice, D. D., C. N. Threlkeld, and A. K. Vuletich. 1988. Character, origin and occurrence of natural gases in the Anadarko Basin, southwestern Kansas, western Oklahoma and Texas Panhandle, United States. *Chemical Geology 71,* 149–157.

Riva, Jr., J. P. 1991. Soviets may halt production drop with outside funds, technologies. *Oil and Gas Journal 89,* 110–112.

Schoell, M. 1984. Wasserstoff- und Kohlenstoffisotope in organische Substanzen, Erdolen, und Erdgasen. *Geologisches Jahrbuch, Riehe D 67,* 1–164.

Schoell, M. 1988. Multiple origins of methane in the earth. *Chemical Geology 71,* 1–10.

Solomon, C. 1993. *Cement.* U.S. Department of Interior, Bureau of Mines, Washington, D.C.

Tans, P. 1981. $^{13}C/^{12}C$ of industrial CO_2. In *Carbon Cycle Modelling,* B. Bolin, ed. SCOPE 16, John Wiley and Sons, Chichester, U.K., pp. 127–129.

UN (United Nations). 1993. *The United Nations Energy Statistics Database.* United Nations Statistical Division, New York.

von Hippel, D., P. Raskin, S. Subak, and D. Stavisky. 1993. Estimating greenhouse gas emissions from fossil fuel consumption: Two approaches compared. *Energy Policy 21,* 691–702.

Wanli, Y. 1985. Daqing oil field, People's Republic of China: A giant field with oil of nonmarine origin. *Bulletin of the American Association of Petroleum Geologists 69,* 1101–1111.

Yeh, H., and S. Epstein. 1981. Hydrogen and carbon isotopes of petroleum and related organic matter. *Geochimica et Cosmochimica Acta 45,* 753–762.

4

Emissions of Carbon from Land-Use Change

R. A. HOUGHTON

Abstract

This chapter reviews the approach and data used to determine the flux of carbon from changes in land use. Both net and gross fluxes are presented, and uncertainties in the data and in the calculated fluxes are discussed. Analyses of land-use change – based on rates of agricultural expansion, logging, and regrowth and their accompanying changes in C/ha – show that terrestrial ecosystems were a net source of approximately 120 Pg C to the atmosphere between 1850 and 1990. In 1990, the net release from changes in land use was approximately 1.7 Pg C, essentially all of it from the tropics.

These estimates of flux are higher than estimates obtained from analyses of data from forest inventories and from inverse calculations with geochemical data. Data from forest inventories show a net accumulation of approximately 0.8 Pg C/yr in northern midlatitude forests, as opposed to a flux of nearly 0 Pg C from changes in land use. Apparently these northern forests are either recovering from harvests more rapidly than they did in the past or accumulating carbon in areas not directly affected by human management. If the analyses of forest inventories are correct, the imbalance in the global carbon equation is approximately half (0.7–0.8 Pg C/yr) what it was when defined on the basis of land-use change. The smaller estimate of missing carbon is unlikely to be found in the trees of midlatitude forests. The additional sink would require a systematic error of about 100% in the observed rates of growth.

4.1 Introduction

One of the four terms in the global carbon balance is change in terrestrial carbon storage. Part of this change results from human activities, such as logging, reforestation, and the clearing of forests for agriculture. Analyses of terrestrial carbon storage based on these changes in land use have consistently shown a net loss of carbon to the atmosphere globally, although some regions have accumulated carbon. The current imbalance in the global carbon equation suggests, however, that the net emissions of carbon from changes in land use may not account for all the changes in terrestrial carbon storage. Indirect geochemical evidence suggests that terrestrial ecosystems are approximately balanced with respect to carbon. Therefore, if changes in land use are responsible for a release of carbon to the atmosphere, undisturbed ecosystems (or unaccounted responses of disturbed ecosystems) must be accumulating an almost equivalent amount of carbon. Such an accumulation has never been observed, and may not be observable against the large and variable stocks and flows of carbon in natural systems. Indeed, the lack of observation of any such change has led ecologists to consider land-use change as the major terrestrial contributor to atmospheric carbon. Changes in the storage of terrestrial carbon as a result of changes in land use are certainly more easily determined than are changes in carbon outside of land-use change. Changes in land use generally occur in well-defined areas (agricultural fields), and the attendant changes in carbon per unit area are large and well documented.

This chapter reviews the approach and data used to determine the flux of carbon from changes in land use. Both net and gross fluxes are presented. Uncertainties in the data and in the calculated flux are also discussed. The approach is based on the rate of loss of carbon from ecosystems following disturbance, including the fate of wood products removed from forests, and the rate of accumulation of carbon in vegetation and soil in recovering ecosystems. Three types of data are used to calculate the emissions and accumulations of carbon: rates and types of land-use change, stocks of carbon per unit area in vegetation and soil, and changes in these stocks as a result of land-use change.

4.2 Methods

The annual flux of carbon is calculated with a bookkeeping model that tracks the areas and ages of different land uses and the fate of carbon initially held in the ecosystem (that is, the burning and decay of wood and wood products as well as the oxidation of soil carbon). The calculation also includes the reaccumulation of carbon in vegetation and soil of regrowing forests following logging, abandonment of agriculture, and reforestation. For global calculations, the approach requires enormous amounts of data, and the analysis is data-limited. This review briefly discusses the types of data used in the analyses. The reader is referred to previously published analyses for specific data and sources and to Houghton and Hackler (1995) for the most recent compilation.

4.2.1 Data

The global analysis considered nine geopolitical regions. Initial analyses (Houghton et al., 1983; Woodwell et al., 1983) showed that the largest fluxes of carbon were from tropical regions, and subsequent analyses have emphasized Latin America (Houghton et al., 1991c,b) and South and Southeast Asia (Palm et al., 1986; Houghton, 1991b; Houghton and Hackler, 1994). As new data on rates of tropical deforestation have become available, they have been incorporated into analyses to update and, in some cases, revise estimates of flux for the entire tropics (Houghton et al., 1985, 1987; Houghton, 1991c) and the globe (Houghton et al., 1987; Houghton and Skole, 1990). Revisions have also been made for temperate and boreal regions (Melillo et al., 1988).

4.2.1.1. Rates and Types of Land-Use Change

Because forests contain so much more carbon than the lands that replace them, analyses of terrestrial carbon have emphasized forests and rates of deforestation and reforestation. Prior to 1980, however, there were no estimates of deforestation for much of the Earth, and analyses of long-term trends in carbon storage relied on changes in land use (e.g., croplands) that were better documented. Six major types of land use were included in these analyses: cultivated lands, pastures, shifting cultivation, land degradation, logging and degradation of forests (one category), and reforestation.

Changes in the area of cultivated lands and pastures, together with maps or other descriptions of these lands, were used to infer the rates at which different types of ecosystems were converted to agricultural use. According to these analyses, the global area of croplands increased from approximately 500×10^6 ha in 1850 to somewhat more than $1,400 \times 10^6$ ha in 1990 (FAO, 1991). About half of the 900×10^6 ha increase is estimated to have occurred in tropical regions. The area in pastures, globally, is more than twice the area in croplands (FAO, 1991); however, most pastures were once natural grasslands, and they are assumed not to have lost carbon as a result of the conversion. Historical changes in the tropical areas used for shifting cultivation are difficult to document. In this analysis the areas were assumed not to have changed until the 1940s, after which they expanded exponentially (FAO/UNEP, 1981).

Logging and regrowth are important to include in these analyses because they affect not only the carbon held in forests, but also the carbon accumulated in wood products of long-term storage. Rates of logging (ha/yr) were calculated from forestry data on wood production (m³/yr) and harvesting efficiencies (m³/ha). Unless harvested forests were converted to other uses, forests were assumed to regrow on logged lands, thus accumulating carbon.

4.2.1.2 Stocks of Carbon per Unit Area in Vegetation and Soil

The average biomass, t C/ha, in different types of ecosystems was obtained from ecological reviews (e.g., Ajtay et al., 1979; Olson et al., 1983; Brown et al., 1989) and from individual studies cited in the original analyses. About 50 types of ecosystems were included in these analyses, 3–8 per geopolitical region.

Estimates of the organic carbon in the top meter of soil were obtained from reviews by Brown and Lugo (1982), Post and colleagues (1982), Sanchez and colleagues (1982), Schlesinger (1984), and Zinke and colleagues (1986). On average, the soils of the world contain 1,400–1,500 Pg C, approximately 3 times more organic carbon than is contained in vegetation, 500–600 Pg C. When inventories of soil carbon are not limited to the top 1 m, the total organic carbon of soils, including peats, may be considerably greater than usually thought.

4.2.1.3 Changes in Carbon Stocks as a Result of Land-Use Change

The net flux of carbon to the atmosphere from deforestation depends not only on rates of deforestation and stocks of carbon in forests but on the fate of biomass and soil carbon affected by land-use change. Rates of decay and growth and the residence times for wood products vary with land use, ecosystem, and geopolitical region. In this analysis, 6 types of land use in approximately 50 types of ecosystems (in effect, 75, because primary and secondary forests required different response curves) were distributed among 9 regions. Because not all land uses apply in all ecosystems or regions, the total number of land-use/ecosystem interactions was between 300 and 400. Changes in carbon for each interaction were defined by a set of

response curves based on rates of decay and regrowth obtained from the ecological literature and land-use statistics.

Figure 4.1 shows an example of a group of response curves for the clearing of a temperate evergreen forest for agriculture with subsequent abandonment of agriculture and regrowth of the forest. The response curves specify the changes in carbon that result from disturbance. One curve describes the reduction (and accumulation) of carbon in live vegetation in response to clearing (and regrowth). A second curve specifies the exponential decay of dead plant material left on site after logging (slash). Another curve describes the fate of plant material removed from the site and assigned to as many as four pools with different decay constants: 1-yr (burned), 10-yr (short-lived products), 100-yr (long-lived products), and 1,000-yr

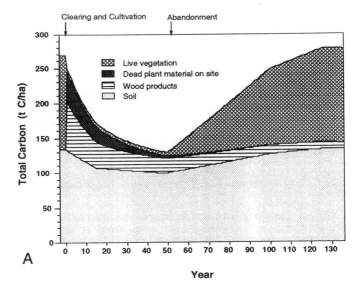

A

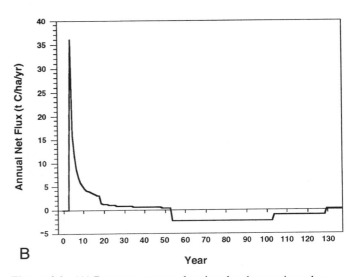

B

Figure 4.1: (A) Response curves showing the changes in carbon (t C/ha) that follow clearing of a temperate evergreen forest for cultivation and subsequent abandonment of cultivation. (B) Annual net flux of carbon to or from the atmosphere as a result of these changes in carbon stocks.

(charcoal or elemental carbon formed as a result of fire). Products were apportioned among these decay pools according to forestry statistics that describe the amount of wood in different end products and according to anthropological data that describe the uses of natural products by those practicing shifting cultivation. A final curve describes the reduction (and accumulation) of soil carbon following cultivation (and abandonment).

Disturbance of soil, and particularly cultivation, generally results in a loss of organic carbon. The variability is large, but, on average, 25–30% of the carbon in the surface meter is lost to the atmosphere when soils are cleared of natural vegetation and cultivated (Schlesinger, 1984, 1986; Detwiler, 1986; Davidson and Ackerman, 1993). Most of the loss occurs rapidly within the first few years of clearing. Conversion of forests to pastures, which are not cultivated, generally results in a smaller loss of carbon or no loss at all.

The release of carbon from soils following tropical deforestation may be greater than previously thought. Work in the northeast region of the Brazilian Amazon has shown that roots and soil organic matter extend to 8 m or more (Nepstad et al., 1991). Although the concentration of carbon in these deep soils is low, the volume of soil, and hence total carbon content, is large. More important, some of the deep carbon is not refractory but actively cycling. Recent work with isotopes in Amazonian soils shows that approximately 20% of the carbon released with cultivation was from depths greater than 1 m (Trumbore et al., 1995). More research is needed to determine the importance of this loss globally.

Although logging is not generally considered deforestation, it causes a release of carbon to the atmosphere from the mortality and decay of trees damaged in harvest operations, from the decay of logging debris, and from the oxidation of wood products. It also causes a net withdrawal of carbon from the atmosphere if logged forests are allowed to regrow. If there is no other use of land following harvest, the net flux of carbon from logging and regrowth is probably close to zero in the long term. In the short term, however, logging may cause either release or accumulation of carbon, depending on whether rates of logging are increasing or decreasing, respectively, and depending on the relative rates of decay and regrowth.

4.2.2 A Model

The response curves, described above, form the structure of a bookkeeping model. They define the annual amount of carbon lost or gained in each age class of a particular land use and ecosystem. Annual rates of land-use change start an age class along its appropriate set of response curves. The model keeps track of the area in each ecosystem and land use. It also keeps track of the area in each age class until that age class has reached steady state with respect to carbon (i.e., when forests have regrown to their initial carbon level or when agricultural soils have reached their new equilibrium value). For each age, the response curves define the exchange of carbon with the atmosphere. For live vegetation and soils, the model tracks both area and carbon. For the exponential decay of material left on site, the model tracks the carbon remaining in each age class.

Products removed from the forest and assigned to decay pools are not assigned an age class; they are simply apportioned among the pools, from which a fraction of the carbon is released to the atmosphere each year. Similar bookkeeping models have been used in other analyses (Moore et al., 1981; Houghton et al., 1983, 1985, 1987; Detwiler and Hall, 1988).

Although the net flux of carbon is calculated annually in these analyses, the data on land-use change are generally interpolations within 5–10-year intervals for recent decades, and 25-year intervals for older periods. Therefore, short-term variations in the calculated net flux are not necessarily accurate.

4.3 Results

4.3.1 1980s

Globally, the annual flux of carbon from changes in land use is calculated to have been approximately 1.4 Pg C in 1980 (1.3 Pg from the tropics and 0.1 Pg from outside the tropics) and 1.7 Pg C in 1990 (essentially all of it from the tropics). Thus for the decade of the 1980s the average net flux was approximately 1.6 Pg C/yr. This estimate is intermediate between previous estimates, which, for the tropics, varied between 0.6 and 2.5 for 1980 (Houghton et al., 1987; Hall and Uhlig, 1991; Houghton, 1991c) and between 1.1 and 3.6 Pg C for 1990 (Houghton, 1991c). The high end of the earlier ranges now seems too high.

The high estimates were based on a Brazilian rate of deforestation of 5.0×10^6 ha/yr in 1989 (Myers, 1991). More recent estimates based on data from satellites include those by the Brazilian Space Agency (average 2.2×10^6 ha/yr between 1978 and 1989) and those by Skole and Tucker (1993) (average 1.52×10^6 ha/yr between 1978 and 1988). Use of the Brazilian estimate of deforestation reduced the calculated flux for all of Latin America from approximately 0.7 Pg C/yr (Houghton et al., 1991b) to approximately 0.5 Pg C/yr in 1980, and from approximately 0.9 to 0.7 Pg C/yr in 1990. Use of the Skole and Tucker (1993) estimate reduced the estimate of flux by another 0.1 Pg C/yr for the decade of the 1980s.

Recent estimates of biomass and deforestation in tropical Asia and Africa have modified estimates of carbon flux for those regions as well. Rates of deforestation in Southeast Asia were revised upward recently by the UN Food and Agriculture Organization (FAO) (1993), and historical rates of degradation in the region were reassessed by Flint and Richards (1994). These revisions gave an estimate of flux for South and Southeast Asia that was approximately 0.7 Pg C in 1990 (Houghton and Hackler, 1994). A reanalysis of land-use change in Africa, using improved estimates of biomass (Brown et al., 1989), gave an estimate of flux there of 0.35 Pg C/yr in 1990.

The estimate of carbon flux for regions outside the tropics was 0.1 Pg C/yr according to this analysis. For the world's temperate and boreal zones, the accumulation of carbon in forests recovering from harvest was approximately balanced by releases of carbon from the decay of logging debris and wood products (Houghton et al., 1987; Melillo et al., 1988; Houghton, 1993).

4.3.2 1850–1990

The long-term (1850–1990) flux of carbon to the atmosphere from global changes in land use is estimated to have been 120 Pg C (Figure 4.2). The annual flux increased from approximately 0.4 Pg C/yr in 1850 to approximately 1.7 Pg C/yr in 1990. It took 100 years for the first increase of 0.6 Pg C/yr (from 0.4 to 1.0 Pg C/yr) and only 35 years (1950 to 1985) for the next increase of 0.6 Pg C/yr. Until about 1940, the region with the greatest flux was the temperate zone. Since 1950, the tropics have been increasingly important.

About two-thirds of the total long-term flux, or 80 Pg C, was from oxidation of plant material, either burned or decayed. About one-third, or 40 Pg C, was from oxidation of soil carbon, largely as a result of cultivation. Relative to the stocks of carbon in 1850, carbon in vegetation was reduced by approximately 12% over this 140-year interval, and organic carbon in soil was reduced by approximately 4%, worldwide. The estimated emissions for the tropics may be underestimated if the mobilization of carbon deeper than 1 m is widespread (Trumbore et al., 1995).

4.3.3 Extrapolation to 1765

The net annual flux of carbon was extrapolated back to 1765 based on the relationship between the average annual rate of deforestation and the average net flux of carbon from total changes in land use, as determined above. Rates of deforestation before 1850 were obtained from interpolation between 1650 and 1850 (Williams, 1990). The net emission of carbon in 1765, according to this relationship, was 0.38 Pg C/yr (range 0.32–0.44 Pg C/yr), increasing to approximately 0.44 Pg C/yr by 1850. There is considerable uncertainty in estimates of deforestation prior to 1850 (indeed, even at present),

particularly for the three regions where human habitation has had a long history: Europe, Asia, and Africa. In other regions, a greater proportion of the clearing has occurred in more recent decades and can be better documented. In an earlier analysis, Houghton and Skole (1990) estimated the average net flux from changes in land use to have been approximately 0.33 Pg C/yr between 1700 and 1850. Thus, uncertainty for estimates of flux prior to 1850 may be ±20%.

4.4 Uncertainties

4.4.1 Comparison with Other Analyses

The estimate presented here of 120 Pg C released between 1850 and 1990 is lower than previous estimates (Figure 4.2). It is almost 50% lower than the intermediate estimate from Houghton and colleagues (1983), for three reasons. First, estimates of the amount of soil carbon oxidized per hectare as a result of cultivation were reduced from 50% to 25%. Second, estimates of forest biomass in Latin America and Africa were lowered (Brown et al., 1989). And finally, the demand for tropical fuel wood was assumed to be satisfied when the wood was made available through agricultural expansion, including shifting cultivation, rather than through additional logging. Rates of logging were too high in the earlier analysis.

The estimate of 120 Pg C is lower than the estimate of 136 Pg C (Houghton and Skole, 1990) because of revisions in the estimated rate of Brazilian deforestation (see Section 4.4.2), because an earlier analysis of Latin America (Houghton et al., 1991a) overestimated the rate of land degradation, and because results from a more comprehensive analysis of land-use change in South and Southeast Asia were incorporated (Flint and Richards, 1994; Houghton and Hackler, 1994).

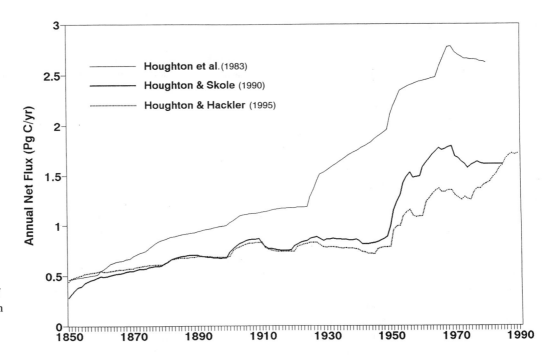

Figure 4.2: Three estimates of the long-term net flux of carbon to the atmosphere from global changes in land use.

Two observations can be made about these estimates of flux: First, estimates of emissions have become lower; second, successive revisions have resulted in smaller changes. The smaller revisions result, in part, from an accumulation of better data and, in part, from the fact that new findings have worked in opposite directions. For example, estimates of biomass were revised downward (Brown and Lugo, 1984), hence lowering estimates of flux. However, the recognition that part of the reduction in biomass represented real changes in biomass, and not simply in estimates, led to a consideration of degradation, which, in turn, increased the estimate of flux (see Section 4.4.3).

For South and Southeast Asia alone, the long-term estimate of flux given here is approximately 30% lower than the one determined independently by Flint and Richards (1994). The major difference is in the amount of carbon estimated to have been lost from forests as a result of degradation. Based on the difference between their analysis and the one presented here, the overall uncertainty of the calculated flux is estimated to be within ±30%.

A similar upper limit to uncertainty probably applies to the contemporary flux. The estimate reported here for 1980 is approximately 45% higher than the recent estimate by Hall and Uhlig (1991). However, Hall and Uhlig did not explicitly consider reduction of biomass within forests. Because such reductions in tropical Asia accounted for 50% or more of the net flux (Flint and Richards, 1994; Houghton and Hackler, 1994), the estimate provided here and the estimate by Hall and Uhlig (1991) agree within ±30%. They are not independent, however. Both studies relied on rates of deforestation provided by the FAO/United Nations Environment Programme (FAO/UNEP, 1981).

Outside the tropics, recent analyses suggest that the net accumulation of carbon from land-use change may be greater than estimated here. Citing several recent studies, Dixon and colleagues (1994) reported a net accumulation of 0.7 ± 0.2 Pg C/yr for temperate and boreal forests. If the analyses are correct, the net global flux was a release of 0.9 ± 0.4 Pg C/yr rather than the 1.7 ± 0.5 Pg C/yr reported here. The data and analyses bear further investigation, however. More than one-half of the net flux to temperate and boreal forests was attributed to growth of Russian forests where data are uncertain.

4.4.2 Rates of Land-Use Change

One of the only ways to estimate the errors associated with rates of land-use change is by comparing different estimates. The approach is clearly flawed because the actual rate may lie outside the range of available estimates. Nevertheless, a comparison of recent estimates of tropical deforestation provides an example of the uncertainties. In 1980 and again in 1990, Myers (1980, 1991) and the FAO (FAO/UNEP, 1981; FAO, 1993) reported rates of deforestation for most of the tropics. For the years around 1980, Myers' estimates were 1% higher in Africa, 28% higher in Asia, and 11% lower in Latin America than the FAO/UNEP estimate. For the tropics as a whole, the rates were within 5% of each other (Houghton et al., 1985).

A major difference between these early estimates of tropical deforestation involved shifting cultivation. According to the FAO study (FAO/UNEP, 1981), the area of fallow increased between 1980 and 1985 because shifting cultivation was responsible for much of the deforestation in the tropics. Myers (1980), on the other hand, argued that shifting cultivation was largely being replaced by permanently cleared land, and that the area of fallow was not increasing but decreasing. He estimated that fallow forests were being cleared at a rate of approximately 10×10^6 ha/yr. The discrepancy concerning the fate of forest fallow has not been resolved, although satellite data have the potential to resolve it.

By 1989, Myers (1991) estimated that the annual loss of tropical forests had increased by 90%, from 7.34×10^6 ha/yr in 1979 to 13.86×10^6 ha/yr in 1989. The FAO's recent estimate (FAO, 1993) was also higher than its earlier one (approximately a 40% increase); however, the FAO maintains that some of the apparent increase resulted from its underestimate of deforestation in the earlier period. The largest difference between these recent estimates of deforestation was in the rate for Brazil. Myers (1991) based his estimate for Brazil on a study by Setzer and Pereira (1991), who used Advanced Very High Resolution Radiometer (AVHRR) data from the NOAA-7 satellite to determine the number of fires burning during the dry season (mid-July through September). They accounted for the facts that some fires burned for more than one day (and should not be counted twice) and that a small, hot fire would saturate the entire pixel (1 km square) and cause overestimation of the area actually burned. With these adjustments, Setzer and Pereira (1991) estimated that fires burned approximately 20×10^6 ha in the Brazilian Amazon in 1987, approximately 60% of which, or 12×10^6 ha, was land that had already been deforested. Their estimate of deforestation was 8×10^6 ha/yr.

Myers (1991) reduced their estimate to 5×10^6 ha to account for other factors. Nevertheless, even this reduced rate now seems high, according to other recent studies. Using data from Landsat (80-m resolution rather than 1-km resolution), the National Institute for Space Research (INPE) in Brazil found the rate of deforestation of closed forests in Brazil's Legal Amazonia to have averaged approximately 2.2×10^6 ha/yr between 1978 and 1989 (Fearnside, 1993), about one-fourth the rate initially determined by Setzer and Pereira (1991). Recent work by Skole and Tucker (1993) found the average rate of deforestation to have been even lower than that reported by the INPE: 1.52×10^6 ha/yr between 1978 and 1988. The actual rate probably increased between 1978 and 1987, but fell substantially after 1987 to 1.8×10^6 ha in 1988–89, to 1.38 in 1989–90, and to 1.11 in 1991 (Fearnside, 1993).

The highest estimate (8.0×10^6 ha/yr; Setzer and Pereira, 1991) is 5 times higher than the lowest (1.5×10^6 ha; Skole and Tucker, 1993). The estimates however, are not considered equally credible. The coarser resolution of the AVHRR data used by Setzer and Pereira (1991) has been shown to overestimate deforested areas (Skole, 1992), although not by as much as a factor of 5. In contrast, the Landsat data used by INPE and Skole and Tucker (1993) have a resolution of 80 m (multispectral scanner system, or MSS, data) or 30 m (thematic mapper, or TM, data). These high-resolution data have the potential to determine rates of deforestation to within 10%, perhaps less. Such data are expensive, however, and only recently has there been support, through the Landsat Pathfinder program of NASA, the U.S. Environmental Protection Agency, and the

U.S. Geological Survey, to determine rates of deforestation in major tropical regions. If, as a result of modifications in the Climate Convention, emissions of carbon need to be determined with high precision, the funds available to measure rates of deforestation and reforestation may permit routine use of Landsat or other satellite data. Under such circumstances, annual emissions would best be determined with an approach combining annual satellite data with a geographic information system (GIS), such that the location, age, and fate of deforested lands could be monitored. Such an approach would not only determine rates of deforestation accurately. It would help constrain estimates of carbon stocks by determining age and former use of forests. It would also resolve the uncertainty surrounding recent changes in the area of shifting cultivation.

Before the availability of Landsat data in 1972, estimates of land-use change were obtained from statistics reported by different countries or international organizations. One measure of uncertainty is apparent from comparisons of land areas within different yearbooks of the FAO (1949–91) (Figure 4.3). In a worst case, revisions between the 1975 yearbook and the 1985 yearbook show a reduction of approximately 200×10^6 ha in "other land" in Latin America. "Other land" is a catchall term used to account for total land area. The "change" in other land and most of the other changes shown in Figure 4.3 represent not actual changes in area but revisions based on better data or reclassification. Estimates within any one yearbook are probably internally consistent and reasonably accurate, however – at least after 1970 (note how similar the slopes are for the estimates from the 1981 and 1985 yearbooks). For data before 1970, other sources must be obtained and evaluated for internal consistency. Figure 4.3 shows the danger inherent in using different sources of data through time. Major discontinuities are possible unless great care is taken.

One of the ironies of the data on land use is that there appears to be more certainty in the data for 100 years ago than in more recent data (Figure 4.2). Obviously, such is not the case; the greater apparent uncertainty at present results from the fact that there are many more estimates available for recent changes than for changes occurring a century ago. However, although the relative uncertainty of historic data may be greater than that of current data, the absolute uncertainty is greater at present. In the tropics, at least, rates of change are much greater at present than they ever were in the past.

4.4.3 Stocks of Carbon in Vegetation and Soil

Deforestation of forests with high biomass releases more carbon to the atmosphere than deforestation of forests with low biomass. The biomass of tropical forests is variable, however, and not well known. Estimates of tropical forest biomass based on surveys of wood volumes are substantially lower than estimates based on direct measurements of biomass (Brown and Lugo, 1984; Brown et al., 1989, 1991). Because the lower estimates are based on a much larger sampling of area, they are often assumed to be more accurate than the higher estimates and more appropriate for calculation of carbon flux (Brown and Lugo, 1984; Brown et al., 1989, 1991; Detwiler and Hall, 1988). Recent evidence suggests, however, that use of low estimates of biomass leads to an underestimate

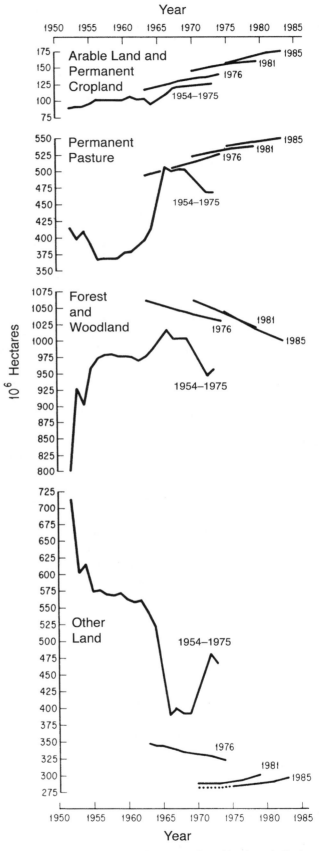

Figure 4.3: Changes in the major categories of land use in Latin America according to successive *Production Yearbooks* of the FAO (1949–91). (Data from Houghton et al., 1991c.) Reprinted with permission from Elsevier Science.

of both the current and the long-term (>10 yr) emissions of carbon. The evidence comes from the observed, widespread reduction in carbon density (t C/ha) of tropical forests over time. This reduction of biomass within forests implies a loss of carbon to the atmosphere. The loss does not appear in the calculated flux, however, if low estimates of biomass are used throughout the period of calculation. For the same area deforested, then and now, more carbon will have been released in the past.

Use of current (low) estimates of biomass also underestimates the current release of carbon if the calculations are based on deforestation alone and do not include the carbon lost as a result of reduction of biomass within forests (degradation). The actual loss of carbon includes losses resulting from deforestation (change in area) as well as losses resulting from degradation within forests. Most analyses have not considered the loss of carbon from within forests, although some have approached the issue of degradation by considering logging (Houghton et al., 1983; Palm et al., 1986; Detwiler and Hall, 1988). Logging reduces biomass if the rates of logging remove biomass more rapidly than it regrows. The net flux of carbon calculated from logging has generally been small, however (Houghton et al., 1985; Palm et al., 1986), and the flux from degradation has been large (Brown et al., 1991; Flint and Richards, 1994; Houghton, 1991b; Houghton and Hackler, 1994). The difference is thought to result from the fact that much logging is illicit and not reported in official statistics.

Uncertainty in biomass has at least three different components. First, there is uncertainty in average aboveground biomass, as discussed above. Second, total biomass includes not only live aboveground biomass but also dead aboveground biomass and belowground biomass. In the Brazilian Amazon, preliminary estimates of the carbon held in these components add another 30% to aboveground live estimates (Brown et al., 1992). Estimates of flux may be low to the extent these other components of biomass have been ignored in the analyses.

Third, the biomass of any particular type of ecosystem is spatially heterogeneous and varies widely about the mean. Use of a mean value, therefore, assumes that deforestation is randomly distributed among stands of different biomass. Clearing may be systematically biased, however, such that stands of lower or higher biomass are preferentially cleared or logged. Recent attempts have been made to assign geographically specific estimates of biomass to areas deforested in tropical Asia and in the Brazilian Amazon. Iverson and colleagues (1994) attempted to model biomass in tropical Asia on the basis of environmental variables and human activity. Skole (personal communication) used geographically specific data from a survey of forest biomass in the Brazilian Amazon with more than 2,000 measured stands. Systematic coverage over such a large area is unusual, however. Determination of aboveground biomass with data from satellite would be ideal, but has received little attention. There has been some success with using radar to determine aboveground biomass, but radar saturates at approximately 100 t C/ha and needs considerable development if it is to be useful. Ultimately, spatially specific estimates of aboveground biomass with remotely sensed data will be required if estimates of carbon emissions are to be precise.

4.4.4 Changes in Carbon Stocks as a Result of Human Disturbance

The rates at which carbon is lost to the atmosphere or accumulated again on land depend on ecological factors, such as rates of decay and regrowth that are environmentally determined, and on human factors, such as the end products of harvest (short or long residence times) and efficiency of harvest (the fraction of original biomass damaged and killed during harvest).

As long as these rates and proportions have been constant over the last 150 years or so, estimates of flux are not sensitive to uncertainties in them. For example, if the rate of growth of temperate evergreen forests were 3.7 t C/ha/yr (Wofsy et al., 1993) instead of 2.4 t C/ha/yr (as assumed in the analyses summarized here), the total net flux for North America would change by less than 10%. On the other hand, if the rate of regrowth has increased over time, both the timing and magnitude of the net flux may be modified significantly. In general, however, the net flux of carbon is not as sensitive to uncertainty in these parameters as it is to uncertainties in rates of land-use change and initial stocks of carbon.

4.5 Gross versus Net Emissions

There are at least four different levels of gross versus net fluxes of carbon. The first level of gross flux is the annual flux associated with the processes of gross primary production (GPP), or photosynthesis, and ecosystem respiration (including plant and microbial, or soil, respiration). Globally, these gross fluxes are thought to be between 100 and 120 Pg C/yr.

Second, there is the potential difference between the net flux of carbon from changes in land use and the total net flux of carbon from terrestrial ecosystems. Indirect evidence from geochemical analyses suggests that the two are not identical (see Section 4.6.1). They might be identical if undisturbed ecosystems were indeed in steady state, neither accumulating nor losing carbon. However, other processes that are thought to have changed the storage of terrestrial carbon are not included in analyses of land-use change (see Section 4.6, below). Regional or global changes in terrestrial carbon resulting from eutrophication, toxification, changes in climate, or other inadvertent effects were ignored in these analyses of land use.

A third level of gross versus net flux relates to land use alone. The net flux given here, 120 Pg C over the period 1850–1990, is the result of gross emissions and accumulations. The gross emissions are calculated to have been 490 Pg C (Figure 4.4). The gross accumulations in regrowing vegetation and redeveloping soils were 370 Pg C. The difference between gross and net emissions results from rotational or cyclic uses of land, for example, logging with regrowth or shifting cultivation with fallow. Globally, the ratio of gross to net flux decreased from approximately 5:1 in 1850 to 2.5:1 in 1990, indicating the relative decline in the importance of shifting cultivation and logging (Figure 4.4). Outside the tropics, the ratio increased as logging replaced clearing of forests for agriculture as the major contributor to flux. In 1990, the global net

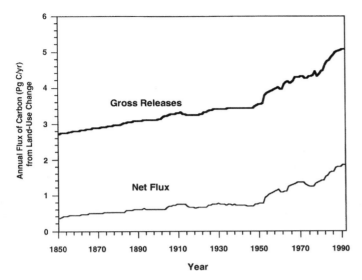

Figure 4.4: Annual gross and net emissions of carbon to the atmosphere from changes in land use.

flux of carbon from changes in land use was the difference between gross emissions of 5 Pg C and gross accumulations of 3.3 Pg C.

A major confusion about terrestrial emissions of carbon centers about this third level of distinction between gross and net emissions. Attempts to estimate the emissions from use of wood fuels (Rotty, 1986) or the emissions of carbon from biomass burning (Crutzen and Andreae, 1990) are gross estimates. They do not account for regrowth of harvested or burned ecosystems. Emissions of carbon from biomass burning are not comparable to the emissions from land-use change. Different processes are included in each estimate (Houghton, 1991a).

The gross fluxes associated with land-use change are somewhat underestimated here (see Figure 4.4) because rates of land-use change are generally determined from net changes in area. For example, although the net loss of forest area may have been 10×10^6 ha in 1980, a gross deforestation of 12×10^6 ha may have been partially offset by a reforestation of 2×10^6 ha. In the United States during the early part of the twentieth century, increased agricultural lands in the Midwest were partially offset by decreasing agricultural areas in the East and Midwest (Hart, 1968). Use of net rather than gross changes in land use is thought to underestimate gross emissions here by less than 20%. It affects estimates of the net flux by much less.

4.6 The Unidentified Sink

4.6.1 The Historical Pattern of the Unidentified Sink

The long-term flux of carbon from changes in land use (120 Pg C between 1850 and 1990) is considerably larger than the terrestrial source calculated indirectly by inverse methods using models of oceanic uptake and historical records of atmospheric CO_2 concentrations. Estimates of the biotic flux based

on inverse calculations vary between 50 and 25 Pg C over this period (Siegenthaler and Oeschger, 1987; Keeling et al., 1989; Sarmiento et al., 1992). The disagreement between estimates based on land-use change and those based on inverse calculations, however, may not be the result of errors in one or the other of the approaches. Instead, it may represent a flux of carbon between terrestrial ecosystems and the atmosphere that is not related to changes in land use but to changes in the global environment. Analyses of flux based on changes in land use assume that ecosystems not directly affected by land-use change are in steady state with respect to carbon. This assumption is probably not valid. The amount of carbon stored in vegetation and soils is changed not only by deliberate human activity but also by inadvertent changes in climate, CO_2 concentrations, nutrient deposition, and pollution. The effects of these environmental factors were ignored in this analysis, as were changes in the frequency of wildfires and changes in the area of arid lands (desertification). The area unaffected by land-use change and assumed to be in steady state is large. In the analysis reported here, almost 80% of the forest area in 1980 was undisturbed, that is, either recovered or never disturbed in the first place (Figure 4.5).

It is useful to distinguish between these two causes or types of change: deliberate and inadvertent. Deliberate changes result directly from management, for example, cultivation. Inadvertent changes occur through changes in the metabolism of ecosystems, more specifically through a change in the ratio of the rates of primary production and respiration (respiration includes autotrophic and heterotrophic respiration, or plant and microbial respiration). Clearly, management practices affect metabolism, and the distinction between deliberate and inadvertent effects is blurred. Nevertheless, the distinction is a useful one because changes in carbon stocks that result from deliberate human activity are more easily documented than changes resulting from metabolic changes alone. Deliberate changes usually involve a well-defined area, for example, the area in agriculture or the area reforested. Further, changes in the density of carbon (t C/ha) associated with land-use change are generally large: Forests hold 20–50 times more carbon in their vegetation than the ecosystems that replace them following deforestation. In contrast to these large and well-defined changes, those that occur as a result of metabolic changes are generally subtle. They occur slowly over poorly defined areas. They are difficult to measure against the large background levels of carbon in vegetation and soils and against the natural variability of diurnal, seasonal, and annual metabolic rates.

Assuming that the estimate obtained from inversion of ocean models is an estimate of the total net terrestrial flux, it includes both the flux from land-use change and a flux resulting from other changes in terrestrial ecosystems. The flux of carbon from land-use change, on the other hand, includes only those lands directly and deliberately managed. Thus, the difference between the two estimates of flux may define the flux of carbon to or from terrestrial ecosystems not directly modified by humans.

The difference is shown in Figure 4.6 (see the thin, solid line). The flux has always been negative and is interpreted to mean that some terrestrial ecosystems were accumulating car-

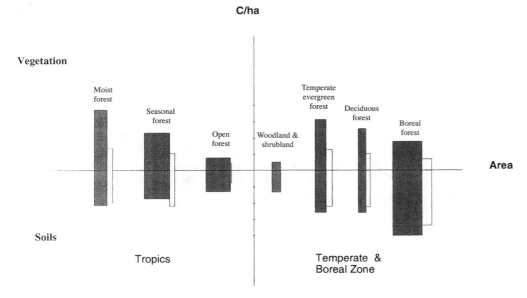

Figure 4.5: Areas and average carbon stocks of primary (undisturbed or recovered) and secondary (recovering from logging or agricultural abandonment) forests in 1980.

bon independent of land-use change. Positive differences, if they occurred, would indicate that some terrestrial ecosystems were releasing carbon to the atmosphere in addition to that released from changes in land use. The short-term variation in Figure 4.6 is an artifact of the smoothed oceanic and atmospheric data used in the inversion (Sarmiento et al., 1992). The short-term variation is entirely the result of year-to-year variation in fossil fuel use. Only the longer-term patterns are of interest here. The difference between the results of the inversion

and the land-use analyses exhibits three broad features over time: first, a period before 1920 showing a small terrestrial sink (not different from zero) with little variation; second, a period between about 1920 and 1975 or so, when some of the world's terrestrial ecosystems were apparently accumulating carbon at an increasing rate; and third, a period since the mid-1970s during which the rate of accumulation has decreased.

Over the period 1920–90, the generally increasing sink is proportional to increasing industrial activity. This suggests

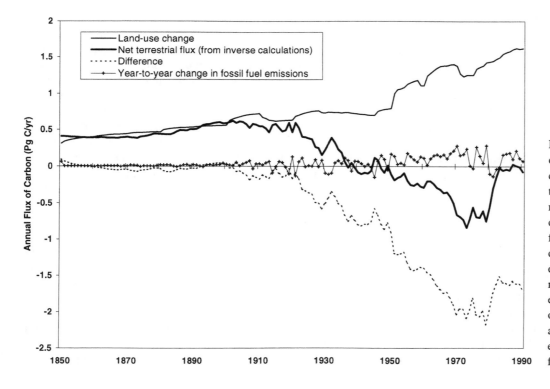

Figure 4.6: The annual net flux of carbon to the atmosphere from changes in land use; the total net terrestrial flux of carbon (the residual flux of carbon determined by inverse methods; from Sarmiento et al., 1992); the difference between these two estimates of flux (assumed to represent changes in terrestrial ecosystems caused by factors other than changes in land use); and year-to-year changes in the emissions of carbon from fossil fuels.

that CO_2 fertilization or increased nitrogen deposition may have been responsible for an accumulation of carbon in undisturbed ecosystems. Since 1940, the difference is correlated more with temperature than with atmospheric concentrations of CO_2 (Houghton, 1995). Larger sinks were associated with cooler years; reduced sinks were associated with warmer years. The fact that a similar relationship was not apparent before 1940 suggests either that the relationship is accidental or that different mechanisms were dominating in the two periods. Perhaps before 1940, ecosystems were largely rebounding from the Little Ice Age, while after 1940 they were responding to a more rapid warming. Analyses by Keeling and colleagues (1989 and Chapter 10, this volume) have also pointed to a positive correlation between terrestrial sources of carbon and temperature.

4.6.2 A Northern Midlatitude Sink?

Just as estimates of flux based on inverse calculations with atmospheric data show greater terrestrial sinks of carbon than do estimates based on land-use change, recent analyses of forest growth based on data from forest inventories show greater accumulations of carbon than do analyses based on land-use change (in this case, previous harvests and agricultural abandonment) (Houghton, in press). Forest inventories for the northern midlatitudes show a combined net accumulation of 0.7–0.8 Pg C/yr (Dixon et al., 1994; Houghton, 1995). The net accumulation contrasts with a source of 0–0.1 Pg C/yr estimated from past rates of harvest in these regions (Houghton et al., 1987; Melillo et al., 1988; Houghton, 1993). The difference between the two types of analyses (forest inventories and land-use change) suggests either that forests recovering from harvests are recovering more rapidly than they did in the past, or that forests unaffected by direct human activity (and hence not considered in analyses of land-use change) are accumulating carbon. Possible explanations for the latter accumulations include environmental changes (CO_2, nitrogen deposition, or climatic fluctuations) or historic variations in natural disturbances of forests (storms, insects, disease) that may have created young, vigorously growing forests currently accumulating carbon. Fire is another agent of change, the result of either natural or human processes. Suppression of wildfires has clearly been a deliberate management strategy in midlatitude forests, yet it was not considered as a land-use change in the analyses reported here.

If the analyses based on data from forest inventories are correct, half the global carbon imbalance for the 1980s (defined by the terrestrial flux from land-use change) is explained. Although the mechanisms for the sink are not known, the observed sink accounts for some of the global carbon imbalance. The forest inventory data for the 1980s suggest that another curve could be added to Figure 4.6, identifying a portion of the missing sink defined by the difference between a net global flux (from inverse methods) and the flux from land-use change. The lesson from the analyses of forest inventory data is that the global carbon (im)balance need no longer be defined by the flux of carbon from changes in land use. In many ways the inventory data provide a more comprehensive analysis of net flux. Unfortunately, analyses of forest inventory data were not carried out prior to 1980, and the data may not permit historic analyses.

Although an accumulation of 0.8 Pg C/yr in the forests of the northern temperate and boreal zones is not large enough to account for the global imbalance of 1–2 Pg C/yr (defined by the terrestrial flux from land-use change), the global carbon

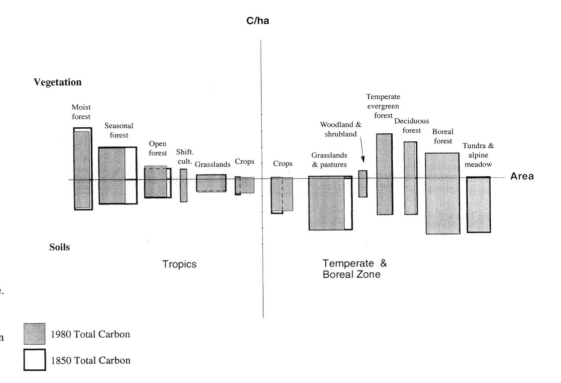

Figure 4.7: Areas and average carbon stocks of different ecosystems in 1850 and 1980. Changes in total stocks of carbon are the result of land use. Ecosystems with small areas (such as marshes and swamps) or with small amounts of carbon (such as deserts, rock, and ice) are not shown.

budget might be balanced if the same processes responsible for the sink in temperate zone and boreal forests were also active in tropical regions. Unfortunately, forest inventories are much less common in the tropics, so it will be difficult to evaluate the total net flux from such data. Furthermore, the boreal and temperate zone sink of 0.8 Pg C/yr is considerably less than the sink of 3.5 Pg C/yr (for 1988) calculated for northern midlatitude regions on the basis of spatial variations in atmospheric CO_2 (Ciais et al., 1995). Application of the method to a longer series of years may reveal a more moderate sink, but whether it will average as low as 0.8 Pg C/yr is unclear.

4.6.3 The Capacity for a Terrestrial Sink to Continue

If the unidentified sink has been spread throughout the world's vegetation and soils, it may be largely undetectable. Assuming that the sink has, nevertheless, been terrestrial, one can ask whether the magnitude of the sink will continue to expand in the future as it seems to have done over the past century (Figure 4.6). If the cause of the sink is CO_2 fertilization, one might expect it to increase. However, the global area of forests, where the sink seems most likely, has been shrinking (Figure 4.7). Approximately 15% of the forests in existence in 1850 have already been lost, and rates of tropical deforestation appear to be increasing. Furthermore, the trend in carbon storage within many of the remaining forests has been negative (see Section 4.4.3). According to the analysis presented here, the net effect of changes in land use has been to reduce not only the area of forests but also the average amount of C/ha held in their vegetation and soil. Although the potential exists for growth and recovery, and hence increased accumulation, trends in land use have been in the opposite direction and appear likely to continue in that direction for the foreseeable future.

4.7 Summary and Conclusions

Terrestrial ecosystems may either accumulate or lose carbon in response to deliberate changes in land use and in response to changes in the environment. Deliberate changes are more easily documented. Analyses of land-use change – based on rates of agricultural expansion, logging, and regrowth and their accompanying changes in C/ha – show that terrestrial ecosystems were a net source of carbon to the atmosphere over the entire period of 1850–1990. The analysis provided here shows a net release of 120 Pg C over this period. Approximately one-third of the long-term net release was from soils and the rest from burning or decay of plant material, including wood products. The net release was the difference between gross emissions of approximately 490 Pg C and gross accumulations in growing forests of approximately 370 Pg C.

In 1990, the net release was approximately 1.7 Pg C, essentially all from the tropics. Gross sources and sinks were 5 Pg C and 3.3 Pg C, respectively. Tropical sinks accounted for approximately 1.8 Pg C, temperate and boreal sinks for approximately 1.3 Pg C. The magnitude of the gross sink in the northern temperate zone and boreal forests of Canada, the United States, Europe, and the former U.S.S.R. (0.8 Pg C/yr) is approximately half the sink calculated for those forests on the basis of data from forest inventories. The forest inventories show a net accumulation of 0.8 Pg C/yr, a value equal to approximately one-half of the global carbon imbalance when the imbalance is defined by the terrestrial flux of carbon from land-use change. Apparently these northern forests are either recovering from earlier harvests more rapidly than they did in the past or accumulating carbon in areas not directly affected by human activities of land management. If the analyses based on data from forest inventories are correct, the residual imbalance in the global carbon equation (0.7–0.8 Pg C/yr) is unlikely to be found in forests of the northern midlatitudes. The additional sink in these forests would require a systematic error of about 100% in the observed growth.

If the results of the forest inventories are assumed to apply to the decades prior to 1980 in the same proportion to the apparent missing sink as observed for the 1980s (50%), the imbalance in the carbon cycle is reduced to about half of what is shown in Figure 4.6, or to between 35 and 50 Pg C over the period 1920–90. If this entire sink were to have accumulated in trees of the northern temperate zone, it represents a 25–40% increase in biomass. Tree rings should show a response of this order over the past 100 years or less. The most remarkable finding from studies of tree rings, however, is the lack of consistent growth trends. Either the analysis described here has overestimated the net release of carbon from land-use change, or analyses of tree rings have not considered representative trees, or the terrestrial sink is distributed evenly over the surface of the Earth in both trees and soils and hence is undetectable, or only a portion of the missing sink is actually on land. The imbalance in the global carbon budget remains.

References

Ajtay, G. L., P. Ketner, and P. Duvigneaud. 1979. Terrestrial primary production and phytomass. In *The Global Carbon Cycle*, B. Bolin, E. T. Degens, S. Kempe, and P. Ketner, eds. John Wiley and Sons, New York, pp. 129–182.

Brown, I. F., D. C. Nepstad, I. O. Pires, L. M. Luz, and A. S. Alechandre. 1992. Carbon storage and land-use in extractive reserves, Acre, Brazil. *Environmental Conservation* 19, 307–315.

Brown, S., A. J. R. Gillespie, and A. E. Lugo. 1989. Biomass estimation methods for tropical forests with applications to forest inventory data. *Forest Science* 35, 881–902.

Brown, S., A. J. R. Gillespie, and A. E. Lugo. 1991. Biomass of tropical forests of South and Southeast Asia. *Canadian Journal of Forest Research* 21, 111–117.

Brown, S., and A. E. Lugo. 1982. The storage and production of organic matter in tropical forests and their role in the global carbon cycle. *Biotropica* 14, 161–187.

Brown, S., and A. E. Lugo. 1984. Biomass of tropical forests: A new estimate based on volumes. *Science* 223, 1290–1293.

Ciais, P., P. P. Tans, J. W. C. White, M. Trolier, R. J. Francey, J. A. Berry, D. R. Randall, P. J. Sellers, J. G. Collatz, and D. S. Schimel. 1995. Partitioning of ocean and land uptake of CO_2 as inferred by $\delta^{13}C$ measurements from the NOAA Climate Monitoring and Diagnostics Laboratory global air sampling network. *Journal of Geophysical Research* 100, 5051–5070.

Crutzen, P. J., and M. O. Andreae. 1990. Biomass burning in the tropics: Impact on atmospheric chemistry and biogeochemical cycles. *Science 250*, 1669–1678.

Davidson, E. A., and I. L. Ackerman. 1993. Changes in soil carbon inventories following cultivation of previously untilled soils. *Biogeochemistry 20*, 161–193.

Detwiler, R. P. 1986. Land use change and the global carbon cycle: The role of tropical soils. *Biogeochemistry 2*, 67–93.

Detwiler, R. P., and C. A. S. Hall. 1988. Tropical forests and the global carbon cycle. *Science 239*, 42–47.

Dixon, R. K., S. Brown, R. A. Houghton, A. M. Solomon, M. C. Trexler, and J. Wisniewski. 1994. Carbon pools and flux of global forest ecosystems. *Science 263*, 185–190.

FAO (UN Food and Agriculture Organization). 1949–91. *Production Yearbooks*. FAO, Rome, Italy.

FAO (UN Food and Agriculture Organization). 1991. *1990 Production Yearbook*. FAO, Rome, Italy.

FAO (UN Food and Agriculture Organization). 1993. *Forest Resources Assessment 1990: Tropical Countries*. FAO Forestry Paper No. 112, FAO, Rome, Italy.

FAO/UNEP (UN Food and Agriculture Organization/UN Environment Program). 1981. *Tropical Forest Resources Assessment Project*. FAO, Rome, Italy.

Fearnside, P. M. 1993. Deforestation in Brazilian Amazonia: The effect of population and land tenure. *Ambio 22*, 537–545.

Flint, E. P., and J. F. Richards. 1994. Trends in carbon content of vegetation in south and southeast Asia associated with changes in land use. In *Effects of Land Use Change on Atmospheric CO$_2$ Concentrations: South and Southeast Asia as a Case Study*, V. H. Dale, ed. Springer-Verlag, New York, pp. 201–299.

Hall, C. A. S., and J. Uhlig. 1991. Refining estimates of carbon released from tropical land-use change. *Canadian Journal of Forest Research 21*, 118–131.

Hart, J. F. 1968. Loss and abandonment of cleared farmland in the eastern United States. *Annals of the Association of American Geographers 58*, 417–440.

Houghton, R. A. 1991a. Biomass burning from the perspective of the global carbon cycle. In *Global Biomass Burning*, J. S. Levin, ed. MIT Press, Cambridge, Mass., pp. 321–325.

Houghton, R. A. 1991b. Releases of carbon to the atmosphere from degradation of forests in tropical Asia. *Canadian Journal of Forestry Research 21*, 132–142.

Houghton, R. A. 1991c. Tropical deforestation and atmospheric carbon dioxide. *Climatic Change 19*, 99–118.

Houghton, R. A. 1993. Is carbon accumulating in the northern temperate zone? *Global Biogeochemical Cycles 7*, 611–617.

Houghton, R. A. 1995. Effects of land-use change, surface temperature, and CO$_2$ concentration on terrestrial stores of carbon. In *Biotic Feedbacks in the Global Climate System: Will the Warming Speed the Warming?* G. M. Woodwell and F. T. Mackenzie, eds. Oxford University Press, New York, pp. 333–350.

Houghton, R. A., R. D. Boone, J. R. Fruci, J. E. Hobbie, J. M. Melillo, C. A. Palm, B. J. Peterson, G. R. Shaver, G. M. Woodwell, B. Moore, D. L. Skole, and N. Myers. 1987. The flux of carbon from terrestrial ecosystems to the atmosphere in 1980 due to changes in land use: Geographic distribution of the global flux. *Tellus 39B*, 122–139.

Houghton, R. A., R. D. Boone, J. M. Melillo, C. A. Palm, G. M. Woodwell, N. Myers, B. Moore, and D. L. Skole. 1985. Net flux of CO$_2$ from tropical forests in 1980. *Nature 316*, 617–620.

Houghton, R. A., and J. L. Hackler. 1994. The net flux of carbon from deforestation and degradation in South and Southeast Asia. In *Effects of Land Use Change on Atmospheric CO$_2$ Concentra-*

tions: South and Southeast Asia as a Case Study, V. H. Dale, ed. Springer-Verlag, New York, pp. 301–327.

Houghton, R. A., and J. L. Hackler. 1995. *Continental Scale Estimates of the Biotic Carbon Flux from Land Cover Change: 1850–1980*. ORNL/CDIAC-79, NDP-050, Oak Ridge National Laboratory, Oak Ridge, Tenn.

Houghton, R. A., J. E. Hobbie, J. M. Melillo, B. Moore, B. J. Peterson, G. R. Shaver, and G. M. Woodwell. 1983. Changes in the carbon content of terrestrial biota and soils between 1860 and 1980: A net release of CO$_2$ to the atmosphere. *Ecological Monographs 53*, 235–262.

Houghton, R. A., D. S. Lefkowitz, and D. L. Skole. 1991a. Changes in the landscape of Latin America between 1850 and 1980. I. A progressive loss of forests. *Forest Ecology and Management 38*, 143–172.

Houghton, R. A., and D. L. Skole. 1990. Carbon. In *The Earth as Transformed by Human Action*, B. L. Turner II, W. C. Clark, R. W. Kates, J. F. Richards, J. T. Mathews, and W. B. Meyer, eds. Cambridge University Press, Cambridge, U. K., pp. 393–408.

Houghton, R. A., D. L. Skole, and D. S. Lefkowitz. 1991b. Changes in the landscape of Latin America between 1850 and 1980. II. A net release of CO$_2$ to the atmosphere. *Forest Ecology and Management 38*, 173–199.

Iverson, L. R., S. Brown, A. Prasad, H. Mitasova, A. J. R. Gillespie, and A. E. Lugo. 1994. Use of GIS for estimating potential and actual forest biomass for continental South and Southeast Asia. In *Effects of Land Use Change on Atmospheric CO$_2$ Concentrations: South and Southeast Asia as a Case Study*, V. H. Dale, ed. Springer-Verlag, New York, pp. 67–116.

Keeling, C. D., R. B. Bacastow, A. F. Carter, S. C. Piper, T. P. Whorf, M. Heimann, W. G. Mook, and H. Roeloffzen. 1989. A three-dimensional model of atmospheric CO$_2$ transport based on observed winds: 1. Analysis of observational data. In *Aspects of Climate Variability in the Pacific and the Western Americas*, D. H. Peterson, ed. Geophysical Monograph 55, American Geophysical Union, Washington, D.C., pp. 165–236.

Melillo, J. M., J. R. Fruci, R. A. Houghton, B. Moore, and D. L. Skole. 1988. Land-use change in the Soviet Union between 1850 and 1980: Causes of a net release of CO$_2$ to the atmosphere. *Tellus 40B*, 116–128.

Moore, B., R. D. Boone, J. E. Hobbie, R. A. Houghton, J. M. Melillo, B. J. Peterson, G. R. Shaver, C. J. Vorosmarty, and G. M. Woodwell. 1981. A simple model for analysis of the role of terrestrial ecosystems in the global carbon budget. In *Carbon Cycle Modelling*, B. Bolin, ed. John Wiley and Sons, New York, pp. 365–385.

Myers, N. 1980. *Conversion of Tropical Moist Forests*. National Academy of Sciences Press, Washington, D.C.

Myers, N. 1991. Tropical forests: Present status and future outlook. *Climatic Change 19*, 3–32.

Nepstad, D. C., C. Uhl, and E. A. S. Serrao. 1991. Recuperation of a degraded Amazonian landscape: Forest recovery and agricultural restoration. *Ambio 20*, 248–255.

Olson, J. S., J. A. Watts, and L. J. Allison. 1983. *Carbon in Live Vegetation of Major World Ecosystems*. TR004, U. S. Department of Energy, Washington, D.C.

Palm, C. A., R. A. Houghton, J. M. Melillo, and D. L. Skole. 1986. Atmospheric carbon dioxide from deforestation in Southeast Asia. *Biotropica 18*, 177–188.

Post, W. M., W. R. Emanuel, P. J. Zinke, and A. G. Stangenberger. 1982. Soil carbon pools and world life zones. *Nature 298*, 156–159.

Rotty, R. 1986. Estimates of CO$_2$ from wood fuel based on forest harvest data. *Climatic Change 9*, 311–325.

Sanchez, P. A., M. P. Gichuru, and L. B. Katz. 1982. Organic matter in major soils of the tropics and temperate regions. In *Non-Symbiotic Nitrogen Fixation and Organic Matter in the Tropics*, 12th International Congress of Soil Science, New Delhi, India, pp. 99–114.

Sarmiento, J. L, J. C. Orr, and U. Siegenthaler. 1992. A perturbation simulation of CO_2 uptake in an ocean general circulation model. *Journal of Geophysical Research 97*, 3621–3645.

Schlesinger, W. H. 1984. The world carbon pool in soil organic matter: A source of atmospheric CO_2. In *The Role of Terrestrial Vegetation in the Global Carbon Cycle: Measurement by Remote Sensing*, G. M. Woodwell, ed. John Wiley and Sons, New York, pp. 111–124.

Schlesinger, W. H. 1986. Changes in soil carbon storage and associated properties with disturbance and recovery. In *The Changing Carbon Cycle: A Global Analysis*, J. R. Trabalka and D. E. Reichle, eds. Springer-Verlag, New York, pp. 194–220.

Setzer, A. W., and M. C. Pereira. 1991. Amazonia biomass burnings in 1987 and an estimate of their tropospheric emissions. *Ambio 20*, 19–22.

Siegenthaler, U., and H. Oeschger. 1987. Biospheric CO_2 emissions during the past 200 years reconstructed by deconvolution of ice core data. *Tellus 39B*, 140–154.

Skole, D. L. 1992. *Measurement of Deforestation in the Brazilian Amazon Using Satellite Remote Sensing*. Ph.D. dissertation, University of New Hampshire, Durham, New Hampshire.

Skole, D. L., and C. Tucker. 1993. Tropical deforestation and habitat fragmentation in the Amazon: Satellite data from 1978 to 1988. *Science 260*, 1905–1910.

Trumbore, S. E., E. A. Davidson, P. B. de Camargo, D. C. Nepstad, and L. A. Martinelli. 1995. Belowground cycling of carbon in forests and pastures of Eastern Amazonia. *Global Biogeochemical Cycles 9*, 515–528.

Williams, M. 1990. Forests. In *The Earth as Transformed by Human Action*, B. L. Turner, W. C. Clark, R. W. Kates, J. F. Richards, J. T. Mathews, and W. B. Meyer, eds. Cambridge University Press, Cambridge, U.K., pp. 179–201.

Wofsy, S. C., M. L. Goulden, J. W. Munger, S.-M. Fan, P. S. Bakwin, B. C. Daube, S. L. Bassow, and F. A. Bazzaz. 1993. Net exchange of CO_2 in a mid-latitude forest. *Science 260*, 1314–1317.

Woodwell, G. M., J. E. Hobbie, R. A. Houghton, J. M. Melillo, B. Moore, B. J. Peterson, and G. R. Shaver. 1983. Global deforestation and the atmospheric carbon dioxide problem. *Science 222*, 1081–1086.

Zinke, P. J., A. G. Stangenberger, W. M. Post, W. R. Emanuel, and J. S. Olson. 1986. *Worldwide Organic Soil Carbon and Nitrogen Data*. ORNL/CDIC-18, Oak Ridge National Laboratory, Oak Ridge, Tenn.

5

The CO_2 Fertilizing Effect: Relevance to the Global Carbon Cycle

ROGER M. GIFFORD, DAMIAN J. BARRETT, JASON L. LUTZE, AND
ANANDA B. SAMARAKOON

Abstract

The CO$_2$ fertilizing effect on vegetation growth arises from primary effects of CO$_2$ concentration on photosynthetic CO$_2$ fixation, suppression of photorespiration (and possibly of dark respiration), and reduction in stomatal conductance. These mechanisms increase the efficiency of use of growth-restricting inputs of light, water, and nitrogen in the formation of dry matter. It is of critical significance that the C:N ratio of plant tissues varies considerably when CO$_2$ is varied. The relative response of plant stand seasonal growth to high CO$_2$ is typically similar to that which would be calculated on basic photosynthetic biochemical and stomatal diffusion grounds. Researchers are still determining the full extent of various negative and positive feedbacks and other factors that accentuate or attenuate the propagation of this primary response into the size of live and dead C pools. However, on the basis of present evidence it seems unjustified to assume that all such modifiers act to annul completely the primary stimulus of high CO$_2$ in terms of increase in C pool sizes. Indeed, the likely magnitude of the CO$_2$ fertilizing effect is such that it can comfortably account for the "missing carbon sink" of approximately 1–2 Gt C/yr, according to several independent terrestrial C cycle models.

For modeling the response of net primary production (NPP) to CO$_2$, there are five approaches. The once common approach of assuming a flat response (i.e., nonresponse) above the preindustrial CO$_2$ concentration of 280 μmol/mol is highly unlikely to be correct. A single-parameter logarithmic function is popular but does not allow satisfactory representation of the relationship close to the CO$_2$ compensation point and near CO$_2$ saturation. Biochemically based models are too inflexible in simple form and perhaps too complex in full form for the global C cycle, considering other major C cycle uncertainties. Under conditions where productivity is proportional to rainfall, productivity is also approximately proportional to CO$_2$ concentration, depending on the closeness of coupling between transpiration and the atmosphere. We suggest that a two-parameter rectangular hyperbola provides the requisite user flexibility to allow for the fact that the function under water-limited conditions is different from that under adequate water conditions, and that the CO$_2$ compensation point must be independent of the CO$_2$ sensitivity that is assigned.

On the scale of decades to centuries, it is plausible that CO$_2$-enhanced productivity fosters N retention and fixation in ecosystems to match their increased C content, but this remains to be proved. Anthropogenic N deposition (as fertilizer and from atmospheric pollutants) is of such a scale that, if entirely taken up into all terrestrial ecosystems and stored as organic matter with typical C:N ratios, it could alone account for more than the missing sink. Although that assumption is obviously unjustified, the calculation shows that the concurrent global eutrophication with CO$_2$ and N has a powerful potential to lead to net C storage in the terrestrial biosphere.

The failure of tree ring records unequivocally to express by their width an increase in C storage is not surprising. It could require an increase in ring width of only 0.07%/yr or less to account for the missing C sink. This would be difficult to detect against typical interannual variability in ring width. Analysis of current basal area increment rather than ring width is a more promising approach for detecting increased C storage.

5.1 Introduction

Gross fixation of CO$_2$ by terrestrial vegetation of approximately 120 Gt C/yr (Gifford, 1982) is approximately 20 times the rate of emission of CO$_2$ by fossil fuel burning and approximately 40 times the annual rate of CO$_2$ increase in the atmosphere. However, this uptake is approximately balanced by CO$_2$ emissions from photorespiration; plant, animal, and microbial respiration; and vegetation burning. Buffering this carbon turnover between atmosphere and vegetation are substantial C pools of standing vegetation (approximately 600 Gt C) and soil organic matter (approximately 1,200–1,500 Gt C). Given that processes of C turnover are sensitive to atmospheric composition, climate, nutrient deposition, and human land use, it is unlikely that they are currently in equilibrium. It has been normal to include tropical deforestation, to the extent of approximately 0.5–3.5 Gt C/yr (Watson et al., 1992), into atmospheric C budgets: Deforestation is very evident from satellite imagery and to the casual observer. Less evident, however, are the possible effects of changing atmospheric composition on C storage. The long-term effects on the amount of C stored in vegetation and soil of the changing atmospheric composition (e.g., CO$_2$, NO$_x$, SO$_2$), radiation (cloudiness, turbidity, ultraviolet B), nutrient deposition (fertilizer, atmospheric deposition) and gradual climate change (temperature, humidity, rainfall change) are neither visible to the casual observer nor easily quantified even with instrumentation or satellite imagery. Of these effects, fertilization of vegetation growth by elevated CO$_2$ concentration and by N deposition are, at present, probably the major ones of global significance.

There is a marked division of opinion, however, as to whether the terrestrial biosphere is a net source or a net sink for C. One view is that "a vast number of natural and managed ecosystems are currently accreting carbon, and the quantity may be large enough to account for the so-called 'missing carbon'" and that "rising CO$_2$ in the atmosphere . . . probably increases carbon storage in terrestrial ecosystems" (Lugo and Wisniewski, 1992, p. 456). In contrast, other analysts conclude that the terrestrial biosphere is not now a large net global C sink and is unlikely to act as such in the foreseeable future (Schlesinger, 1993), and that C supply does not appear to provide a major limit to growth at the ecosystem level (Körner, 1993).

5.2 Mechanism of the CO$_2$ Fertilizing Effect

The CO$_2$ fertilizing effect is the increased rate of plant growth, and potentially of C storage, shown by plants that are exposed to CO$_2$ concentrations above the preindustrial atmospheric level of 280 μmol/mol. The mechanism has several interrelated components, involving photosynthetic, photorespiratory, stomatal, and respiratory responses to CO$_2$.

5.2.1 Photosynthesis and Photorespiration

The large investment in leaf nitrogen (about 25% of total leaf N in C$_3$ species) that plants make in the main carbon-fixing enzyme, ribulose bisphosphate 1,5-carboxylase/oxygenase (or rubisco), attests to the evolutionary difficulty that they have faced in capturing CO$_2$ from the air. Catalytically, rubisco is an extremely slow enzyme, the active site of which discriminates poorly between CO$_2$ and O$_2$ as competitive reactants (Morell et al., 1992). The products of the oxygenase reaction would be toxic if they accumulated. Two mechanisms cope with this. In the C$_3$ species, which dominate the terrestrial biosphere except in tropical grasslands, the photorespiratory pathway removes the products of oxygenation, converting half of their C into compounds useful to metabolism while releasing the other half as photorespiratory CO$_2$. In succulents having crassulacean acid metabolism, or CAM (under certain environmental conditions), and in the C$_4$ species, there are CO$_2$ "pumps" that increase the CO$_2$ concentration around the rubisco enzyme to such an extent that the O$_2$ is outcompeted by CO$_2$. so that there is virtually no photorespiration. At present-day atmospheric CO$_2$ concentrations, the CO$_2$ pumping mechanisms of C$_4$ and CAM species are nearly CO$_2$ saturated, while C$_3$ rubisco is only about half CO$_2$-saturated.

For C$_3$ species under high radiation, the rate of carboxylation by rubisco increases as CO$_2$ concentration increases. Also, at all incident radiation levels, the rate of oxygenation decreases with increasing CO$_2$ concentration even at the light compensation point where respiration balances photosynthesis. Thus, the net fixation of CO$_2$ by leaves increases with CO$_2$ concentration at all light levels. Stated another way, the quantum yield of light-limited C$_3$-species' CO$_2$ fixation is CO$_2$-dependent. The theoretical relationship, based on knowledge of the enzyme-kinetic properties of rubisco (Figure 5.1), is supported by field measurements (Long et al., 1993).

5.2.2 Stomatal Conductance

CO$_2$ passes the impermeable waxy surface of leaves through apertures – the stomata. Most water loss through plants also occurs through stomata. Thus, regulation of water loss and carbon gain are closely interrelated (Cowan and Farquhar, 1977). Stomatal conductance is under physiological control and decreases with decreasing irradiance, soil water content, and atmospheric humidity, and with increasing CO$_2$ concentration. Summarizing responses of numerous diverse species, Morison (1985) found that, for a CO$_2$ doubling from 340 to 680 μmol/mol, stomatal conductance declines by approximately 40%. The sensitivity of stomatal conductance relative to the sensitivity of photosynthesis to CO$_2$ is such that the ratio of the CO$_2$ concentration in the intercellular air spaces of the leaf to that outside (c_i/c_a) is often approximately independent of at-

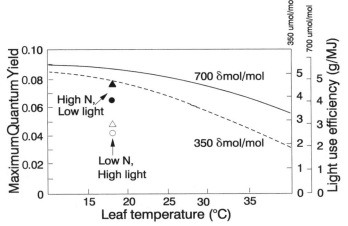

Figure 5.1: The curves are the maximum quantum yield (mol CO$_2$ per mol absorbed photons, 340–700-nm waveband) of light-limited C$_3$ photosynthesis calculated from a biochemical model at 350 and at 700 μmol CO$_2$/mol air (Long, 1991). The right-hand axes are the equivalent of the left axis but scaled to the canopy level, with assumptions indicated in the text. Note that the scaling is slightly different for canopies grown at 350 μmol/mol and at 700 μmol/mol, owing to respiratory effects. The points (right-hand axes) are for experimental wheat crops grown at 350 (circles) and 700 (triangles) μmol/mol CO$_2$ with unrestricted N nutrient supply (high N) and low irradiance, and growth-limiting N (low N) but high irradiance.

mospheric CO2 concentration (Wong et al., 1979; Morison, 1987). Near-constant c_i/c_a has also been observed for several crop species grown between 120 and 500 μmol/mol CO$_2$ (Masle et al., 1990; Polley et al., 1993). The significance of this observation is explained in Section 5.3.3.

Depending on windiness, which determines the aerodynamic conductance of the boundary layer to CO$_2$ transport, and the proportion of regional evaporation from the soil via plants, the stomatal response to CO$_2$ may lead to a net reduction in water loss from vegetated land despite compensatory adjustments that must occur involving leaf temperature and atmospheric humidity (Jarvis and McNaughton, 1986; and see Section 5.3.3).

5.2.3 Dark Respiration

Dark respiration is often reported to decrease at elevated CO$_2$ concentration. The mechanism(s), however, are unclear. Some individuals have seen instantaneous effects of CO$_2$ on plant respiration (e.g., Reuveni and Gale, 1985; Mousseau, 1993). Others have noted long-term effects of the CO$_2$ environment during growth on the ratio of respiration to photosynthesis at the whole-plant level for herbaceous and tree species (Gifford et al., 1985; Bunce and Caulfield, 1991; Bunce, 1992; Gifford, 1991). A literature survey on respiration per unit tissue mass or leaf respiration per unit leaf area (involving diverse methodologies) gave no clear trend; respiration was observed both to decrease and to increase (Poorter et al., 1992). Decline of respiration at high CO$_2$ concentration may be related to a lower protein content of the tissues; however, this does not explain the whole effect. In general, the respiratory effect is probably a relatively minor part of the CO$_2$ fertilizing effect.

5.3 Interactions of CO$_2$ with Other Growth-Determining Environmental Factors

Seasonal vegetation growth is almost never limited by a single environmental factor. It is usually co-limited by several factors (Bloom et al., 1985; Warrick et al., 1985; Gifford, 1974, 1992a; Rastetter and Shaver, 1992). The character of these limitations, and the interactions between them, may differ according to the timescale, from days to millennia (Gifford et al., 1996).

5.3.1 Light

Highly light-limited plant growth is CO$_2$-dependent (Gifford, 1979) because the oxygenase function of rubisco leads to the dissipation of CO$_2$ through photorespiration even at low light (see Section 5.2.1). The CO$_2$ effect on the light-limited quantum yield of photosynthesis is well established (e.g., Ehleringer and Bjorkman, 1977; Long, 1991). The theoretical quantum yield for C$_3$ photosynthesis can be calculated (Long, 1991) from the kinetic parameters of rubisco determined in vitro (Jordan and Ogren, 1984). Quantum yield decreases with increasing temperature (Figure 5.1), owing to the increased oxygenase relative to carboxylase activity of rubisco with increasing temperature and, to a lesser degree, because of differential responses of the solubilities of CO$_2$ and O$_2$ to temperature.

The quantum yield of leaf photosynthesis (mol CO$_2$/mol photons absorbed) can be scaled to field productivity under light-limiting conditions (Figure 5.1, right-hand axis) by converting it to the light use efficiency (LUE) in plant dry matter production. The scaling assumes that:

- There is 85% absorption of the photosynthetically active radiation that is intercepted by green leaves.
- Solar radiation has 4.57 mol photons/MJ (in the 400–700-nm waveband) (Meek et al., 1984).
- Plant dry matter contains 45% C.
- Carbon use efficiency (CUE) in the conversion of gross photosynthetic assimilate to plant dry matter at the community level is 58% at ambient CO$_2$ and 61% at 700 µmol/mol owing to diminished respiration (Gifford, 1995).

The vertical axes on the right-hand side of Figure 5.1 show the LUE (g dry matter/MJ of intercepted photosynthetic irradiance) equivalent to the quantum yields on the left-hand axis. The data points are experimentally measured full-season LUE for wheat crops grown either at high N nutrition under low winter irradiances (Gifford and Morison, 1993) or at low N nutrition under high summer irradiances (as determined by Morison and Gifford), with or without CO$_2$ enrichment. The points are plotted at daytime mean air temperature. The results show

that at low irradiance the crops responded only slightly below that expected on the basis of light-limited leaf quantum yield. High CO$_2$ gave a relative response similar to that calculated from rubisco kinetics alone. However, with high irradiance the LUE was only about half that expected from consideration of light-limited rubisco alone. The lower performance could also be partly an artifact of the experimental difficulty of measuring radiation intercepted by green leaf that is intermingled with dead or partially dead leaves under the poor N nutrition, and partly also because N stress might cause departures from the leaf-level assumptions. Mostly, however, the departures from the curves are probably simply because many of the leaves were often light-saturated, not light-limited as assumed for the theoretical curves. Nevertheless, the main point is that light-limited as well as light-saturated growth is sensitive to CO$_2$ concentration.

5.3.2 Temperature

Theory proposes (Pearcy et al., 1983; Long, 1991; Gifford, 1992b) that the sensitivity of photosynthesis to CO$_2$ concentration increases as temperature increases under both light-limiting and light-saturating conditions (Figure 5.2). However, the experimental evidence on plant growth is not systematically consistent with that. Growth response to CO$_2$ is often, but not always, less temperature-dependent than expected (Kimball, 1986; Rawson, 1992). The reasons are not clear. Temperature also affects the growth potential of carbohydrate sinks and dark respiration; however, both these effects would also be expected to increase growth responses to elevated CO$_2$ at high

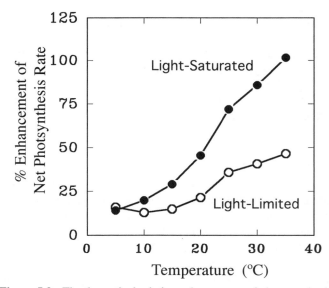

Figure 5.2: The theoretical relative enhancement of photosynthesis at 700 µmol/mol CO$_2$ compared with 350 µmol/mol based on the kinetics of rubisco alone under light-saturated and light-limited conditions. (Data from Gifford, 1992b.) Reprinted with permission from R. M. Gifford, Interaction of carbon dioxide and growth-limiting environmental factors in vegetation productivity: Implications for the global carbon cycle. In *Advances in BioClimatology* 1:23–57. © 1992, Springer-Verlag.

temperature (Gifford, 1992b). Given that global warming is expected to accompany the CO_2 increase, this uncertainty is particularly important to resolve.

5.3.3 Water

The net primary production (NPP) of most ecosystems is closely related to the water budget (Uchijima and Seino, 1985; Woodward, 1987; Stephenson, 1990). Thus reduction of leaf conductance (g) and, hence, potentially of transpiration, by elevated CO_2 is probably significant to most ecosystems. Assuming that g varies with water status and CO_2 in such a manner that the ratio of leaf intercellular CO_2 to atmospheric CO_2 (c_i/c_a) is nearly constant (see Section 5.2.2), the following relationships show that where net photosynthesis (P) is directly proportional to transpiration (T), it is also directly proportional to atmospheric CO_2 (Gifford, 1981):

$$P = (c_a - c_i)g \tag{5.1}$$
$$T = (e_i - e_a)1.6\,g \tag{5.2}$$

where e_i is the water vapor mole fraction at leaf temperature and e_a is the atmospheric water vapor mole fraction, and the factor 1.6 accounts for the ratio of molecular diffusion coefficients of CO_2 and water vapor.

Thus:

$$P = [T\,c_a(1 - (c_i/c_a))]/[1.6e_a((e_i/e_a - 1)] \tag{5.3}$$

If we regard c_i/c_a as nearly independent of c_a and consider leaf temperature to be close enough to air temperature that e_i/e_a is close to the reciprocal of relative humidity of the air, then Equation (5.3) approximates to

$$P = kTc_a \tag{5.4}$$

where $k = (1 - (c_i/c_a))\,/\,1.6\,(e_i - e_a)$ is relatively conservative (see also Morison, 1993).

Equation (5.4) suggests that leaf water use efficiency (P/T) has close to a linear dependence on atmospheric CO_2 concentration for both C$_3$ and C$_4$ species. A survey of leaf-level studies covering 34 species (Morison, 1993) indeed found water use efficiency (WUE) to be close to directly proportional to CO_2 concentration for both C$_3$ and C$_4$ species, WUE being increased by about the same proportion as CO_2 was increased.

Translating this leaf-level argument to community-level growth, when transpiration consumes all the rainfall and hence productivity tends toward linear variation with annual rainfall, productivity tends toward direct proportionality with CO_2 concentration. This is borne out by experiments in controlled environments. Whereas under well-watered conditions a doubling of CO_2 typically increases growth rate of C$_3$ species by approximately 30–40% (Kimball, 1983; Poorter, 1993), when water is highly growth-limiting, growth of both C$_3$ and C$_4$ species (Morison, 1993) approaches direct proportionality with the CO_2 concentration. However, the leaf-level argument translates directly to the natural ecosystem level only where there is a minimal negative feedback effect of surface temperature and atmospheric humidity on the sensitivity of transpiration to a change in leaf conductance. Strong negative feedback effects on transpiration occur when aerodynamic conductance

is small relative to canopy conductance (McNaughton and Jarvis, 1991). In this case, the saturation deficit near the leaf surface is uncoupled from that in the mixed air above the vegetation. However, many native plant communities having aerodynamically rough surfaces tend to be more coupled to the atmosphere than typical uniform short stature field crops (McNaughton and Jarvis, 1983). Thus we may expect some reduction in canopy transpiration with a CO_2-induced reduction in leaf conductance, although some increase in evaporation directly from the soil surface may attenuate the effect. Alternatively, in dry environments CO_2 fertilization may increase canopy leaf area. Then we may expect little effect of elevated CO_2 on overall canopy conductance, since any reduction in leaf conductance is counterbalanced by an increase in the leaf area index. Such a condition is likely to yield similar rates of water use in future compared with today but greater WUE for growth (Morison and Gifford, 1984; Gifford, 1988; Samarakoon et al., 1995), at least for some species (Samarakoon and Gifford, 1995). For wet ecosystems at typical wind speeds, when growth is unrestricted by water and photosynthetically active radiation is fully intercepted by leaves, elevated CO_2 conserves water (Drake, 1992); hence, waterlogging may become more frequent in some places. This could be detrimental to primary productivity in those areas.

5.3.4 Nutrients

The interaction of CO_2 and nutrients is more poorly characterized than are the light and water interactions. Some experiments have involved nitrogen and phosphorus as variables (Conroy, 1992). However, insufficient attention has been paid to the way that nutrient supply to plants is experimentally restricted. Typically, the experimental distinction between nutrient *concentration* limitation and nutrient *replenishment* limitation (Rastetter and Shaver, 1992) has not been clearly drawn. Commonly, N or P supply is restricted experimentally by varying the concentration of the nutrient without restricting the volume of nutrient solution supplied. In such experiments, the plant might use CO_2-stimulated carbohydrate production as an energy source to foster the acquisition of more of the available nutrients, for example, via a larger root system. In the field, plants usually take up the net amount of nutrient mineralized in any one season. Consequently, the roots may not have an opportunity to acquire more nutrients in the short term. Thus, since annual productivity of most vegetation is nutrient- (especially N-) "limited," productivity might not be responsive to CO_2 concentration. This view does not allow for the possibility that elevated CO_2 might promote higher nutrient use efficiency within the plant (Luxmoore et al., 1986) or greater mobilization of "unavailable" nutrients from the environment (Luxmoore, 1981; Gifford, 1992b), whether this is achieved by accelerated mineralization from organic matter via increased soil or root hydrolase activity (e.g., Barrett et al., 1998), by mineral solubilization from rock (Gifford Lutze and Barrett, 1996), or, for nitrogen, by enhanced fixation from the atmosphere (e.g., Arnone and Gordan, 1990).

Because a large fraction of plant N is involved in acquiring C, increased C availability at elevated levels of CO_2 might induce a redistribution of N from rubisco to other functions so that the plant can make optimum use of environmental resources. A review of experiments (Conroy, 1992) shows that plant leaves and other tissues typically do exhibit decreased N content when grown at elevated CO_2 concentration, even when the N supply is abundant. If this is generally true, then inevitably C storage increases for the same amount of N in the vegetation simply because of the higher C:N ratios.

Although that first-order effect would probably persist, in the long term the second-order or higher-order effects may be important moderators. Plant tissues with a high C:N ratio are of lower quality for herbivory, decomposition, and mineralization. Opinion varies (Lincoln, 1993) as to whether this would lead to more plant tissue loss through herbivory because individual insects forage more in order to obtain scarce N (Johnson and Lincoln, 1990), or whether there would be less herbivory entirely. The latter condition might exist because lower-quality diets support lower populations (Scriber and Slansky, 1981) and because high plant C content (low N) can lead to more investment in C-based defensive chemical production (Jonasson et al., 1986). One view (Mattson, 1980) is that there is an intermediate plant N concentration at which attack is maximal, with herbivory declining at both higher and lower N contents. Herbivory typically consumes 5–10% of primary production; however, at times it can be much higher (Fox and Morrow, 1983) and can therefore be quantitatively significant to the terrestrial carbon cycle.

If a higher plant C:N ratio leads to a higher litter C:N ratio, as Gifford and colleagues (1996) report, then there are repercussions for C storage and N cycling. First, there will be more carbon stored in litter per unit of litter N. Second, there could be slower litter decomposition, again causing more storage of C in litter. However, concomitantly slower decomposition may mean slower N mineralization, which would reduce primary production and litterfall. In the long term, this negative feedback can be compensated for if the ecosystem fixes or retains more N (see Section 5.7). A further complication is the effect that CO_2 might have on the content of lignin, tannin, and other secondary plant products that also determine the quality of tissues for herbivores and decomposers (Norby et al., 1986).

In summary, nutrient feedbacks are extremely complex and will require much work to comprehend. It is clearly too simplistic to argue that ecosystems that are in part nutrient-limited are not responding to the increasing atmospheric CO_2 concentration with changes in productivity or C storage.

5.4 Applicability of Controlled Environment Results to Natural Ecosystems

5.4.1 The Issues

The following reasons have been given, not always justifiably, as to why controlled environments may yield results that differ from findings in natural ecosystems:

- Plants in pots may respond differently from vegetation in natural conditions.

- Nutrient supplies may be too high in controlled environments.

- Controlled environment studies may be too short-term.

- Native plants may not be as CO_2-responsive as agricultural plants.

- Responses may differ between communities of plants and single plants.

- Over time, physiological adaptation and community dynamics might change vegetation responses.

- Increased NPP might not necessarily lead to increased C storage.

Unequivocal answers are not yet available for all these opinions; however, some things are known. For agricultural species Lawlor and Mitchell (1991) concluded that the field CO_2 enrichment studies conducted thus far confirm results from controlled environment studies. Nevertheless, the design of many controlled environment studies demands that their results be applied with caution. Some of the problems are discussed below.

5.4.2 Leaf Expansion

Spaced plants grown at high CO_2 concentration can experience a strong positive feedback by investing incremental growth into leaf. More leaf area absorbs more light, leading to greater C gain. During exponential growth of a plant, even a small fractional CO_2 effect on photosynthesis per unit leaf area can be amplified by "compound interest" to a severalfold effect on growth rate and C storage per plant (Gifford and Morison, 1993). This, together with the use of high temperature and abundant nutrient supply, is a probable explanation for cases in which more than a direct proportionality was observed between increases in CO_2 and growth rate (e.g., for sour orange trees in open-topped chambers in Arizona; Idso et al., 1991). Results from experiments with single spaced plants or with communities where population density is reduced as plant size increases – owing, for example, to serial harvesting–require cautious interpretation.

Currently, in many nonarid ecosystems, radiation is fully intercepted. There is therefore little prospect for a strong positive feedback through leaf area increase. One could not expect to observe growth rate increases in doubled CO_2 conditions much greater than the 10–50% per unit leaf area that is appropriate for the largely light-limited leaf photosynthetic effect at typical average temperatures (Figure 5.2).

5.4.3 Root Restriction

Plant pots can physically restrict root growth and, thus, shoot growth. It has been suggested that such a photoassimilate-sink restriction feeds back to photosynthesis such that growth enhancement by CO_2 is not always fully expressed owing to photosynthetic "down-regulation" (Arp, 1991). Thus, pot experiments might understate the CO_2 responsiveness of plant growth. However, in the field the density of plants is often such that the available soil per plant is lower than is often used in pot experiments, particularly in shallow or rocky soils. Photosynthetic down-regulation might occur under such field conditions. Root restriction is also confounded with questions of nutrient and water availability per plant. Total nutrient and water supply are greater in larger pots. Experiments on cotton (Barrett and Gifford, 1995) in which soil phosphate concentration has been used as a covariate to eliminate confounding of pot size with nutrient supply have indicated that acclimation of photosynthesis at elevated CO_2 is independent of root growth restriction or phosphate limitation. This is an area of active research that will not be clarified quickly.

5.4.4 Photosynthetic Acclimation

Sometimes the response of leaf photosynthesis rate to light changes – usually lessening, but not always – when plants are grown continuously under elevated CO_2 concentration. The initial slope of the photosynthetic response to light (apparent quantum yield) is unaltered (Xu et al., 1994); however, the light-saturating rate can decrease owing to reduced rates of carboxylation. Described variously as photosynthetic down-regulation, homeostatic adjustment, or photosynthetic adjustment to CO_2, the mechanism of this decrease and its significance to the response of productivity to CO_2 are not completely understood. It is seen as a whole-plant property rather than something that a single CO_2-enriched leaf (Xu et al., 1994) would show; however, that perception has not been tested thoroughly. There is a widespread view that acclimation implies that growth enhancement under high CO_2 will not be sustained. This is not necessarily so, however: We have observed that photosynthetic acclimation in individual plants occurs in association with a larger leaf area, which sustains an enhanced growth rate (Gifford, 1992a). These plants have an initial, short-lived, higher relative growth rate, which results in larger plants with larger leaf areas than those of control plants in ambient CO_2. After this initial period, the growth rates become similar to those of the controls (Badger, 1992). This convergence in relative growth rate, however, does not necessarily imply photosynthetic acclimation, because relative growth rate inevitably declines with increasing plant size. The comparisons should therefore be made at equal plant size (Gifford, Lutze, and Barrett, 1996).

If this typical mode of CO_2 response were expressed at the canopy level, one might argue that since a closed canopy that intercepts all the incident light cannot take advantage of a higher leaf area index, photosynthetic acclimation would not be compensated for by greater light interception, and the growth enhancement would disappear. However, this is not what occurs. Instead, the canopy seems to act more like a big leaf; the efficiency of conversion of intercepted radiation increases by elevated CO_2 (Gifford and Morison, 1993; see Figure 5.1). Presumably, this is because a canopy is, for most of the time, on or near the initial slope of the canopy light response curve, which is uninfluenced by acclimation.

There are two popular hypotheses about the origin of photosynthetic acclimation to elevated CO_2 concentration. One is that it is a sink feedback phenomenon. The other is that it is an expression of the capability of whole plants to optimize the

distribution of scarce N throughout their tissues to maximize whole-plant carbon gain. The latter could be, in part, an expression of the former via a regulatory effect of carbohydrate accumulation in the leaf on synthesis of photosynthetic enzymes (Stitt, 1991) and hence the N investment in those enzymes. For example, accumulation of hexoses has been implicated in suppressing the expression of photosynthetic genes (Jang and Sheen, 1994).

It is well established that when growth sinks cannot grow fast enough to utilize all the photoassimilate, photosynthetic capacity per unit leaf area may decline (Neales and Incoll, 1968). The biochemical regulatory mechanism is beginning to be understood (Stitt, 1991). Such inadequacy of sinks might be caused by inherent properties of the species, such as being of determinate habit (Kramer, 1981) or being genetically adapted for slow growth in resource-impoverished environments. Alternatively, this inadequacy could be caused by low temperatures or other stresses such as root restriction or limited nutrient supply that might inhibit "sink-pull" more than "source-push." The extent to which this is a factor in the field is not known.

Acclimation that is an expression of a plant reducing its leaf N content is paradoxical: It leads to reduced photosynthesis per unit leaf area while actually maximizing the growth response to CO_2 at the whole-plant and community levels. The lower leaf N concentration at high CO_2 involves many photosynthetic components, not solely rubisco (Makino, 1994). This appears to contradict the concept (Field and Mooney, 1986) that plants optimize the distribution of N within them, because one might think that N released from carbon fixation machinery would be best invested in light-harvesting machinery instead (Makino, 1994). However, perhaps the N released from leaves is more effective if reinvested in the N acquisition machinery, such as in the roots (Xu et al., 1994). Wherever N goes, such redistribution away from leaves (and the concomitant reduced photosynthesis rate) is seen as the plant's way of readjusting its resource deployment to take advantage of higher CO_2, thereby facilitating biomass gain.

5.4.5 Nutrient Availability

Controlled environments do not preclude experimental use of low nutrient availability. Indeed, if the experiment is properly designed, it can arrange much lower nutrient availability than would occur in nature. For example, for Australian native grass swards growing in a controlled environment in soil with severely restricted N supply rate, the CO_2 effect on increasing the C:N ratio and C accumulation rate by vegetation plus soil was found for the first year of the experiment (Lutze and Gifford, 1995) and has persisted unabated for an additional three years (Lutze, 1996).

Two studies reported that communities of native plants grown with restricted nutrients had no significant shoot biomass growth increase despite, in one study, a sustained increase in leaf photosynthesis. This study was conducted in open-topped chambers (Norby et al., 1992), and the other was conducted in a closed microcosm within a greenhouse (Körner and Arnone, 1992). In both studies the measured biomass increase was numerically 15–20% above ambient CO_2 controls. With the coefficient of variation one expects from plant growth studies, however, there was insufficient replication in each of these experiments to statistically distinguish a 15–20% effect. Therefore, the results cannot be used to conclude that, in the long term, natural communities do not respond to high CO_2. In addition, under nutrient-limited conditions, the CO_2 response can be allocated substantially belowground.

5.4.6 Native Plants versus Agricultural Cultivars

It is a common view that species having a high growth potential, as agricultural species often do, will be more CO_2-responsive than slow-growing species that occur, for example, in nutrient-stressed environments. Poorter (1993) collated data from 89 reports that documented the CO_2 effect on growth of 156 species, covering different functional groups. In each experiment the plant dry weight was determined after individual plants had been grown for a certain number of days in elevated CO_2. The ratio of plant weight at an approximately doubled CO_2 concentration to that at ambient CO_2 was calculated. The results are summarized in Table 5.1. This table supports the prediction that C_4 species and CAM species will be relatively insensitive to CO_2. In fact, from our own experience (Samarakoon and Gifford, 1996), we believe that when C_4 plants responded it was because they had experienced some (perhaps unrecognized) water deficits. The table gives only weak support to the theory that crops are more responsive than wild species. With such wide ranges between maximal and minimal responses, the means are not significantly different. However, on the assumption that the extreme values may be spurious, Poorter (1993) analyzed the data set again, this time ignoring the lowest and highest 10% of weight ratios. Then the C_3 herbaceous crops responded to CO_2 significantly more strongly (enriched:control weight ratio for crops, 1.58; for wild species, 1.35). When Poorter classified plants as fast, medium, or slow growers, he also found a relationship between the plant's growth potential and CO_2 responsiveness (fast growers weight ratio, 1.54; medium, 1.38; slow, 1.23).

However, the coefficient of determination (R^2) was weak (0.21). A subsequent study of diverse pasture species did not support the hypothesis that CO_2 responsiveness of growth is related to the inherent relative growth rate of a species (Bolger et al., 1997).

In conclusion, although crops and fast-growing plants may be slightly more responsive to elevated CO_2 concentration, the analysis does not indicate that wild species are not responsive.

5.4.7 Competition and Community Dynamics

Little work has been done on CO_2 effects on species in competition. Competition studies in controlled environments indicate that the responses of individuals become highly modified in competition (Bazzaz, 1992). However, it seems plausible to propose that any species that grows better than others in high CO_2 may become relatively more abundant in time, and hence that the productivity of the whole community might increase as long as restraints by nutrient cycles do not override in the long term.

5.4.8 C Storage in Relation to NPP

Even if NPP is responding to the increasing CO_2 level, does such an increase in NPP translate into more C storage? Inevitably NPP appears in the short term either as storage in live biomass or as litter or as both. The fate of incremental litter is discussed in Section 5.7. One approach to the above question is to examine whether the standing live phytomass of existing ecosystems is related to their NPP. To that end, estimates of live phytomass of major ecosystem complexes of the world have been plotted (Figure 5.3) against their NPP. In Figure 5.3, ecosystem complexes are grouped into three types: forests and woodlands, grasslands/nonwoodlands, and deserts. Linear regressions through the origin are fitted for each type. Clearly, with only three groups there is much scatter within each group. However, there is no suggestion in these plots that live phytomass tends toward a plateau as NPP increases to the maximal

Table 5.1. *The ratio of plant weight grown in high CO_2 concentration (600–720 μmol/mol) to that for the control treatment (300–360 umol/mol) for diverse species*

Functional Group	No. of Species	Average Age at Harvest (days)	Enriched/Control Ratio		
			Maximum	Average	Minimum
C_3 crops	19	30	2.10	1.58	1.08
C_3 wild (herb.)	62	43	3.66	1.35	0.92
C_3 wild (woody)	49	136	4.13	1.41	0.92
C_4 wild	17	38	1.73	1.21	0.63
C_4 crops	2	31	1.52	1.31	1.09
CAM wild	62	214	1.36	1.15	0.90

Note: The information is extracted from a compilation by Poorter (1993), in which the 10% largest and smallest observations were removed from each category to avoid possible experimental artifacts.

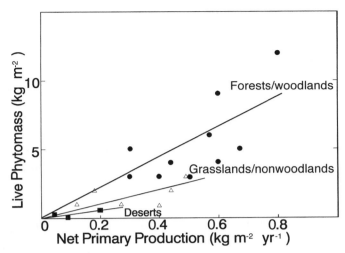

Figure 5.3: Standing live phytomass of natural ecosystem complexes plotted against annual NPP, both expressed in terms of carbon. Each point represents a different ecosystem complex, grouped into forests/woodlands, grasslands/nonwoodlands, and deserts. The linear regressions are forced through the origin. (Data from Olson et al., 1985.)

values. This suggests that live phytomass may increase approximately linearly with NPP.

5.5 What Is the CO₂ Response Function of Primary Productivity?

Numerous climate change analysts have explicitly or implicitly assumed that productivity of natural vegetation is saturated above the preindustrial concentration of 280 μmol/mol (i.e., the response function is flat); however, in the light of material reviewed here, it is unlikely to be so for most ecosystems. Well-watered vegetation will have a response different from water-limited vegetation. Furthermore, given several feedbacks between plant and soil processes that have different relaxation times up to centuries, including species composition change and nutrient cycling, the precise response curve of primary productivity to atmospheric CO₂ concentration probably depends on how long the ecosystem has been exposed to the step change in CO₂ concentration conceptually imposed. Since we can never conduct the appropriate CO₂ experiment on an ecosystem over several centuries while holding all else constant, we will never know the actual realized equilibrium response function of productivity. It would not be of much use if we did, because we will always be dealing with transients. Consequently we must accept that whatever function we use will involve substantial approximation of a complex, dynamically interactive system.

Two inevitable characteristics of the response function for well-watered vegetation are that, at equilibrium at least, it must pass through the CO₂ compensation point at which oxygenation equals carboxylation (Gifford, 1993), and it must saturate at some concentration. The compensation concentration is dependent on the characteristics of rubisco, atmospheric O₂,

and temperature (T°C): At 20.9%, O₂ it is given by the relation (McMurtrie et al., 1992):

$$42\exp[9.46(T - 25)/(T + 273.2)]\ \mu\text{mol/mol}$$

The saturating characteristic depends not only on the biochemistry of photosynthesis but also on all the negative and positive feedbacks involved in acclimation and in scaling from mesophyll biochemistry to annual field productivity, some of which were discussed above. In fact, there is the possibility that above 1,200–1,500 μmol/mol, growth could decline slightly from the peak.

For strictly water-limited vegetation well coupled to the atmosphere, theory and experiment indicate that the growth function is close to directly proportional for CO₂ between present and doubled concentrations. Extrapolation of the data of Polley and colleagues (1993) over the CO₂ range 160–330 μmol/mol suggests that linearity holds from the CO₂ compensation point up to 330 μmol/mol – in controlled environments, at least. Experiments of Morison and Gifford (1983) and Gifford and Morison (1985) suggest that near-linearity holds up to 600–700 μmol/mol (Morison, 1993).

For most real ecosystems for which the annual productivity is partially-water limited, but which nevertheless probably accrue a substantial portion of their annual growth during periods of no water stress, the annual CO₂ response curve following a notional step change in CO₂ concentration would be some form of saturating relationship, with the steepest part of the function being essentially linear.

Given the above considerations, we now consider three functions that have been used in the global C cycle literature. Discounting the existence of a postindustrial CO₂ effect on terrestrial NPP implies a remarkably strong response between 50 and 280 μmol/mol, followed by a sharp, biologically implausible transition to a horizontal plateau. This is most unlikely. For β-function, a logarithmic relationship is the most used:

$$P = P_0\left[1 + \beta \ln(c/c_0)\right] \tag{5.5}$$

where P is annual NPP, c is CO₂ concentration, and subscript 0 refers to a reference condition. This was actually applied originally (Bacastow and Keeling, 1973) to the C flux into long-lived biomass; however, it has the same problems whether it refers to that or to NPP. These problems are that if the biotic growth factor (β) is considered independent of CO₂, the functional form has the wrong properties (Gifford, 1980). If β is allowed to be a function of CO₂, the issue becomes one of finding the CO₂ response function of β, and the analyst is no further ahead than if he had started with an appropriate function to relate P to CO₂. However, Goudriaan (Goudriaan et al., 1985; Goudriaan, 1993) suggests that Equation (5.5) is adequate between 200 and 1,000 μmol/mol with a constant β.

Instead of using an empirical function for NPP versus CO₂ parameterized against growth experiments, one can model it from lower-level biochemical and ecophysiological information and concepts. The biochemical model of Farquhar and colleagues (1980) has been one approach used either directly (e.g., Long, 1991) or to derive a CO₂ response of β (Polglase and Wang, 1992). That model, based on rubisco kinetics, is for C₃ leaf photosynthesis as a function of CO₂ and light. The CO₂ response function differs according to photon irradiance. It has been assumed that only the light-limited part of the model is applicable to whole ecosystems (e.g., Polglase and Wang, 1992). This is not strictly true. Moreover, C$_4$ species require a different model.

Another drawback of applying a simple light-limited leaf photosynthesis model to whole-ecosystem annual productivity is that the model takes no account of the feedbacks discussed above, involving respiration, leaf area modulation, sinks, and – above all – interactions with the hydrologic cycle via the CO$_2$ effect on stomatal conductance and hence water use and water use efficiency. Fully modeling all these phenomena is a major but necessary objective; however, the results need to be validated against stand or community growth experiments. For computational efficiency in global C cycle models, such complex NPP models may need to be used to parameterize much simpler functional representations of NPP versus CO$_2$, such as the rectangular hyperbola.

Given these considerations, it may be preferable to use a rectangular hyperbola (Gifford, 1980) passing through the CO$_2$ compensation point to express the productivity/CO$_2$ relationship, scaling the coefficients according to the effect of a CO$_2$ doubling above the recent ambient atmospheric concentration, using as a guide a combination of theory, experiment, whole-ecosystem models, and informed judgment about longer-term feedbacks (Gifford, 1993). This approach gives no misleading impressions about the specific mechanistic comprehensiveness of the function and forces presentation of a range of estimates based on uncertainty in the assumptions. The equation, $P = k_1 (c - c_b)/[1 + k_2 (c - c_b)]$, includes the equilibrium CO$_2$ compensation point, c_b. Such a function has been shown to fit canopy photosynthesis and yield excellently (Allen et al., 1987) under nonstressed conditions. In practice, Farquhar and colleagues' (1980) model of photosynthesis under light-limiting conditions gives a CO$_2$ response that is close to a rectangular hyperbola in form (Gifford et al., 1996).

5.6 The Potential Magnitude of the CO$_2$ Fertilizing Effect on the Global C Cycle

Several recent assessments of the current rate of global C sequestration, which would follow from CO$_2$ fertilizing effects, are of magnitudes similar to those discussed here (Table 5.2). The works listed have modeled the fate in phytomass, litter, and soil of assumed or modeled incremental increases in NPP in response to the increasing atmospheric CO$_2$ concentration. (Other estimates of terrestrial C sequestration of comparable magnitude could not be presented in simple tabular form, e.g., Esser, 1987; Kohlmaier et al., 1989.) All these estimates give or imply recent global net sequestration values deriving from the CO$_2$ fertilizing effect alone that fall within the range of 0.5–4.1 Gt C/yr. This is the amount necessary to balance the global C budget, allowing for various estimates of net deforestation in the range of 0.5–4.2 Gt C/yr (Houghton et al., 1985; Houghton, 1993). Thus, in order to predict future atmospheric CO$_2$ levels it is of quantitative significance to establish whether or not feedbacks at the ecosystem level over the long

Table 5.2. *Some photosynthetically based estimates of the annual rate of C sequestration by the global terrestrial vegetation/soil system owing to the CO$_2$ fertilizing effect*

Author	Sequestration Rate (Gt C/yr)	Applicable Year
Gifford (1980)[a]	1.6	1979
Gifford (1993)[b]	2.3±1.8	1991
Taylor and Lloyd (1992)[c]	2.4	1985
Goudriaan (1992)[d]	1.2	1980
Goudriaan (1993)[e]	1.3	1980
Polglase and Wang (1992)[f]	1.26	1990
Gifford et al. (1996)	2.6	1995

[a]Based on CO$_2$ increase since 1958; assumes biotic growth factor for NPP = 0.6.
[b]Globally aggregated model; uses hyperbolic NPP vs. CO$_2$ function; assumes that NPP at 680 μmol/mol/NPP at 340 μmol/mol = 1.25 + 0.15; includes effects of global temperature increase since 1750 AD.
[c]Uses the Gifford (1993) model, but on biome-disaggregated basis.
[d]Biosphere disaggregated into six vegetation types; use logarithmic NPP vs. CO$_2$ function; biotic growth factor = 0.5.
[e]An extension of Goudriaan (1992); biotic growth factor = 0.5.
[f]Biosphere disaggregated into 10 biomes; biotic growth factor calculated as function of CO$_2$ from biochemistry of light-limited C$_3$ photosynthesis; C$_4$ assumed half as CO$_2$ responsive as C$_3$.

term do in fact eliminate the assumed CO$_2$ fertilizing effects. One of the major questions in determining whether short-term CO$_2$ fertilizing effects would be expressed in accumulated live and dead organic C is whether the N cycle would feed back to constrain the CO$_2$ fertilizing effect.

5.7 The Role of the N Cycle and N Deposition

The commonly observed increase of C:N ratio under elevated CO$_2$ would alone tend to increase C storage in ecosystems even if the total pool of N in vegetation and soil were to remain fixed with elevated CO$_2$. We do not yet have much information on the flexibility of the C:N ratios of different pools (leaf, stem, litter, root, and soil organic matter fractions) with varying CO$_2$ concentrations. In ongoing long-term work on C storage by N-limited native grass swards (Lutze and Gifford, 1995), the C:N ratio of the total organic matter (dead, alive, and soil organic matter) was, after three years of exposure to 700 μmol/mol CO$_2$, 17.2 – compared with 15.3 in the ambient CO$_2$ control (Gifford, Lutze, and Barrett, 1996).

There are three routes through which such longer-term attenuation of the C:N ratio effect might come about during decomposition:

1. The excess carbon could be oxidized to CO$_2$, leaving the residue with a "normal" C:N ratio. As a consequence, extra C would not be sequestered into long-term soil pools; however, the N contained would be remineralized

and potentially taken up into new growth and hence into deposited litter.

2. Decomposer organisms could compete more strongly with roots for the available soil N such that a "normal" C:N ratio would be restored in the soil organic matter and less N would be available for further primary production and litter formation (leading to a negative feedback).

3. More N could be acquired from the environment to match the CO$_2$-driven C sequestration (leading to a positive feedback).

All three of these routes could contribute concurrently, although the third possibility is conventionally disregarded because it is thought that the N cycle drives the C cycle in natural ecosystems. However, there are grounds to doubt that in the long term, N is a primary limiting factor to forest productivity (Attiwell and Adams, 1993). Gifford (1992a) hypothesized that on the longer time scales of decades to centuries that are relevant to global atmospheric change, the relationship might be the other way around, with the ecosystem C cycle providing the energy to drive N acquisition and retention in the N cycle.

Hypothetically, there are three routes whereby more N could be acquired from the environment: by more biological N fixation, by less N loss from ecosystems via leaching and gaseous emissions, and by entrainment into the ecosystem of anthropogenic N deposition. Whether by symbiotic or by free-living microbes, the energy-costly process of N fixation is critically dependent on light- or plant-derived organic C as an energy source (Gutschick, 1978; Sprent, 1993). It is likely that plants growing in elevated CO$_2$ exude more of the easily decomposed, soluble, energy-rich organic compounds (sugars, organic acids, amino acids, enzymes) from their roots (Oechel and Strain, 1985) and/or inject more organic matter into the soil by fine root turnover (Norby et al., 1987). This material has a relatively high C:N ratio. In the short run, microbial attack of this material could cause increased rates of decomposition of the old soil organic matter owing to the increased demand for N mobilization for microbial growth. However, in the longer run, as the easier organic N sources in the soil are depleted, it seems likely that omnipresent free-living N-fixing organisms would fix their own N, using rhizosphere organic C as an energy source, and that they would have a competitive advantage over the other microbes. Indeed, increased input of high C:N straw to soil fosters heterotrophic N fixation (Gibson et al., 1988). Thus, N could enter the ecosystem to match the increased C inputs from CO$_2$ fertilization.

Less speculative is that symbiotic N fixation is fostered by elevated CO$_2$ concentration. Both rhizobial and actinorrhizal symbiotic N fixation in legumes and trees has been shown to increase under elevated CO$_2$ when soil N levels are growth-limiting to plants (e.g., Masterton and Sherwood, 1978, for legumes; Norby, 1987, and Arnone and Gordan, 1990, for trees). This increase occurs primarily because high CO$_2$ fosters bigger plants with bigger root systems that have a larger N-fixing capacity, and perhaps also because of an increase, in some cases, in the specific activity (intensity) of the N-fixing nodules (Arnone and Gordan, 1990).

From the available evidence, it seems likely that in any ecosystem where symbiotic N fixation plays a role, it will be fostered under elevated CO$_2$ concentration. One uncertainty is for those ecosystems where the amount of symbiotic N fixation is limited by availability of other minerals, such as phosphate. It is unclear whether elevated CO$_2$ can increase the efficiency of use of phosphate for N fixation, perhaps via stimulation of mycorrhizal activity (Norby, 1987).

The global organic N pool experiences natural gaseous losses, largely by denitrification to N$_2$ or N$_2$O and volatilization as NH$_3$, as well as gains, largely by microbial fixation. Global litterfall of 60 Gt C/yr must deposit approximately 1 Gt N/yr on and in the soil (Gifford, 1994). By comparison, global biological N fixation is variously estimated at 0.044–0.2 Gt N/yr (Smil, 1992). At equilibrium, this fixation is matched primarily by gaseous losses. Thus gaseous N turnover via fixation/denitrification amounts to 4–20% of N uptake into vegetation for primary production. Given C turnover rates in soil and vegetation, the current annual increase in NPP (owing to the annual increment of CO$_2$) needed to account for terrestrial C sequestration of approximately 2 Gt C/yr would be 0.18 Gt C/yr (Gifford, 1994), involving 0.003 Gt N/yr. Since this annual increase in N requirement is between 1 and 2 orders of magnitude smaller than the fixation/loss turnover, it is well within the bounds of possibility, quantitatively, for the N needed to match the C to be diverted from the turnover. Whether this is in fact happening is not known. To be sustained over decades, it would have to be accompanied ultimately by increased fixation and deposition or decreased losses.

Whether or not biological N fixation is stimulated under elevated CO$_2$, human-induced N deposition may be globally significant to the potential of CO$_2$-induced C sequestration (Gifford, 1987). Nitrogen fertilizer manufacture is now 80–90 Mt N/yr globally, and wet and dry deposition from atmospheric pollutants is estimated to be 15–45 Mt N/yr (Smil, 1992). As an upper limit, one can calculate what this anthropogenic input could mean for C sequestration were it all to be absorbed and retained in vegetation and soil organic matter, assuming average C:N ratios for soil and vegetation to be 12 and 60, respectively. If it were distributed 50:50 into soil organic matter and standing biomass, then the associated annual C sequestration would be 0.6–0.84 Gt C/yr into soil and 3–4.2 Gt C/yr into vegetation, making a total of 3.6–5 Gt C/yr. This is about twice the size of the missing C sink. In reality, because N fertilization and N deposition are neither uniform nor only on the land, actual N-driven C sequestration would be lower than the potential. Some ecosystems in industrialized regions are experiencing N saturation (Aber, 1992). Furthermore, much of the anthropogenic N (especially from fertilizer) is converted to gaseous NO, N$_2$O, or N$_2$ (McLaughlin et al., 1992). However, in concert with annual increases in CO$_2$ concentration, there is clearly substantial scope for C and N "eutrophication" to be reinforcing, allowing more organic C to accumulate without large shifts in the C:N ratio. If the terrestrial biosphere has in fact been sequestering C, are there ways of detecting it directly? One possibility is the tree ring record.

5.8 Should the CO$_2$ Fertilizing Effect Be Detectable in Tree Rings?

Evidence suggesting a historic CO$_2$ and/or N fertilizing effect on tree ring width is equivocal (Hari et al., 1984; La Marche et al., 1984; Gates, 1985; Kienast and Luxmoore, 1988; Briffa, 1992; Graumlich, 1991; Luxmoore et al., 1993). Some studies show an increasing trend not explicable by local climatic variations; others do not. According to the CQUESTN model (Gifford, Lutze, and Barrett, 1996) with standard parameterization of interacting global C and N cycles, global NPP increased by 12.8% between 1750 AD and 1995 AD as CO$_2$ concentration increased from 28020µmol/mol to 363 µmol/mol. By 1995, model terrestrial C sequestration from CO$_2$ fertilization was 2.6 Gt C/yr and NPP was increasing by 0.27%/yr. As a first approximation, let us assume that, in forests, woody storage and litterfall would be increasing at the same rate. Should this be detectable in tree rings?

Tree ring width is a component of wood volume. The woody component of NPP (NPP$_w$, g/m^2/yr) can be divided into five components:

$$NPP_w = n \times 1 \times g \times w \times d$$

where n is stand density (trees per m^{-2}); 1 is cumulative trunk, branch (i.e., stem), and woody root length per tree; g is a weighted average circumference of stems and woody roots; w is average ring width; and d is average C density of the ring.

Any or all of these five multiplicative components could account for an annual increment in NPP arising from a gradual increase in atmospheric CO$_2$ concentration. There is no reason to expect ring width to be the only component affected by CO$_2$. If it happened that four of the five subcomponents had been affected equally by CO$_2$, over the 243 years, each subcomponent would have increased by 3.2% to accommodate the 12.8% increase in annual tree growth during that period. Also, the current annual increase in ring width deriving from the annual increase in CO$_2$ would be 0.07%. Such small effects would be extremely difficult to detect against the enormous normal interannual variability of tree ring width, some of which is associated with variations in site microenvironment and some of which cannot be ascribed. Under that circumstance, the fact that there were preindustrial years when ring widths in an area exceeded recent ring widths (Graumlich, 1991) is hardly a basis for discarding the hypothesis of a CO$_2$ fertilizing effect. We may be seeking a tiny signal against massive variability. However, would all components of the above equation be equally affected by CO$_2$?

It is not inevitable that tree ring width must increase with a CO$_2$-induced increase in productivity. Another possibility is that incremental productivity is partitioned into root rather than stem growth (Luxmoore et al., 1993). Yet another possibility is that more understory tree seedlings could establish successfully with higher CO$_2$, thereby increasing stand density. If this were so, the consequential increased shading of any one tree could tend to decrease ring width. Isolated trees could more visibly show ring width effects of CO$_2$; however, a growth response may still be "shared" with increased stem length, stem diameter, ring density, and root growth. Although

use of isolated high-altitude trees has been favored on the grounds that the reduced partial pressure of CO$_2$ at altitude may make the growth more CO$_2$-sensitive, the fact that much of the CO$_2$ fertilizing effect derives from suppression of photorespiration may invalidate this hypothesis. Oxygen partial pressure is also lower at high altitudes; consequently, the competitive inhibition from O$_2$ is also reduced.

C$_3$ plants often become taller and branch more under elevated CO$_2$ (Acock and Allen, 1985; Oechel and Strain, 1985). This group includes young trees (Kramer and Sionit, 1987). Total stem length has not often been determined on trees from high-CO$_2$ experiments. Branching increased in some species and conditions, for example, *Liquidambar styraciflua,* but not in others, for example, *Pinus taeda* (Sionit et al., 1985); it decreased in another (*Castanea sativa*) under some conditions but not under others (Mousseau, 1993). There is even less information on wood density effects of CO$_2$; however, Rogers and colleagues (1983) found that the volume of *Liquidambar styracifluva* seedlings did not increase although dry weight did, suggesting that tissue density increased.

Thus it is not necessarily incompatible for global phytomass to increase without a statistically significant increase in tree ring width. To maximize the chance of picking up CO$_2$-related effects, future tree ring work should emphasize ring area (basal area increment or BAI; Briffa, 1992) and C density in rings, rather than ring width alone. Ring width tends to decline with individual tree age; BAI does not. Using such an approach to analyze numerous records from diverse conifer species, Briffa (1992) found that European unmanaged conifer growth has been increasing during the past century relative to that in the previous century.

5.9 Conclusions

Examination of the complexity of the interactions of CO$_2$ concentration and other co-limiting factors that influence natural vegetation productivity and the size of the living and dead organic C pools indicates that it is much too simplistic to assume that those pools currently cannot be sequestering C in response to the atmospheric CO$_2$ increase. It is inevitable that there be a saturating CO$_2$ response function for C pool size, starting from zero at the CO$_2$ compensation point of photosynthesis. Atmospheric CO$_2$ concentration has ranged extremely widely during the evolution of land plants. Present CO$_2$ concentration is toward the lower end of that range. The biochemical mechanism of CO$_2$ capture by C$_3$ plants is now well below saturation. This chapter has discussed various ways in which feedbacks that mitigate against ecosystem responses to CO$_2$ in the short term may relax in the long term. The modeled magnitude of the CO$_2$ fertilizing effect on net C storage by the biosphere (assuming that the negative feedbacks are not strong in the long term) is approximately the amount needed to account for the missing sink in the global C budget. Given this fact, the option must remain open that the terrestrial biosphere is responding to the increasing global atmospheric CO$_2$ concentration, with support in some areas from the deposition of anthropogenic N.

Acknowledgments

We are grateful to J. I. L. Morison, R. J. Luxmoore, and F. I. Woodward for helpful comments on a draft.

References

Aber, J. D. 1992. Nitrogen cycling and nitrogen saturation in temperate forest ecosystems. *Tree 7*, 220–224.

Acock, B., and L. H. Allen, Jr. 1985. Crop responses to elevated carbon dioxide concentrations. In *Direct Effects of Increasing Carbon Dioxide on Vegetation*, B. R. Strain and J. D. Cure, eds. United States Department of Energy, DOE/ER-0238, Washington D.C., pp. 53–97.

Allen, L. H. Jr., K. J. Boote, D. W. Jones, P. H. Jones , R. R. Valle, B. Acock, H. Rogers, and R. C. Dahlman. 1987. Responses of vegetation to rising carbon dioxide: Photosynthesis, biomass, and seed yield of soybean. *Global Biogeochemical Cycles 1*, 1–14.

Arnone, J. A., III, and J. C. Gordan. 1990. Effect of nodulation, nitrogen fixation and CO_2 enrichment on the physiology, growth and dry mass allocation of seedlings of *Alnus rubra* Bong. *New Phytologist 116*, 55–66.

Arp, W. J. 1991. Effects of source-sink relations on photosynthetic acclimation to elevated CO_2. *Plant Cell and Environment 14*, 869–875.

Attiwell, P. M., and M. A. Adams. 1993. Nutrient cycling in forests. *New Phytologist 124*, 561–582.

Bacastow, R., and C. D. Keeling. 1973. Atmospheric carbon dioxide and radiocarbon in the natural carbon cycle. II. Changes from A. D. 1700 to 2070 as deduced from a geochemical model. In *Carbon and the Biosphere*, G. M. Woodwell and E. V. Pecan, eds. U.S. Atomic Energy Commission, Washington, D.C., pp. 86–135.

Badger, M. 1992. Manipulating agricultural plants for a future high CO_2 environment. *Australian Journal of Botany 40*, 421–429.

Barrett, D. J., and Gifford, R. M. 1995. Acclimation of photosynthesis and growth by Cotton to elevated CO_2: Interactions with severe deficiency and restricted rooting volume. *Australian Journal of Plant Physiology 22*, 955–963.

Barrett, D. J., Richardson, A. E., and Gifford, R. M. 1998. Elevated atmospheric CO_2 concentrations increase wheat boot phosphatase activity when growth is limited by phosphorus. *Australian Journal of Plant Physiology 25:* in press.

Bazzaz, F. A. 1992. The response of natural ecosystems to the rising global CO_2 levels. *Annual Review of Ecology and Systematics 21*, 167–196.

Bloom, A. J., F. S. Chapin III, and H. A. Mooney. 1985. Resource limitation in plants – An economic analogy. *Annual Reviews of Ecology and Systematics 16*, 363–392.

Bolger, T. P., J. M. Lilley, R. M. Gifford, and J. R. Donnelly. 1997. Growth response of Australian temperate pasture species to CO_2 enrichments. *Proceedings of the International Grassland Congress*, Canada, June 1997.

Briffa, K. R. 1992. Increasing productivity of 'natural growth' conifers in Europe over the last century. In Proceedings of the International Symposium on Tree Rings and the Environment, T. S. Bartholin, B. E. Berglund, D. Eckstein, and F. H. Schweingruber, eds. Department of Quaternary Geology, Lund University, Sweden, pp. 64–71.

Bunce, J. A. 1992. Stomatal conductance, photosynthesis and respiration of temperate deciduous tree seedlings grown outdoors at an elevated concentration of carbon dioxide. *Plant Cell Environment 15*, 541–549.

Bunce, J. A., and F. Caulfield. 1991. Reduced respiratory carbon dioxide efflux during growth at elevated carbon dioxide in three herbaceous perennial species. *Annals of Botany 67*, 325–330.

Conroy, J. P. 1992. Influence of elevated atmospheric CO_2 concentration on plant nutrition. *Australian Journal of Botany 40*, 445–456.

Cowan, I. R., and G. D. Farquhar. 1977. Stomatal function in relation to leaf metabolism and environment. *Symposium of the Society of Experimental Biology 31*, 471–505.

Drake, B. G. 1992. A field study of the effects of elevated CO_2 on ecosystem processes in a Chesapeake Bay wetland. *Australian Journal of Botany 40*, 579–595.

Ehleringer, J., and O. Bjorkman. 1977. Quantum yields for CO_2 uptake in C_3 and C_4 plants: Dependence on temperature, CO_2 and O_2 concentration. *Plant Physiology 59*, 86–90.

Esser, G. 1987. Sensitivity of global carbon pools and fluxes to human and potential climatic impacts. *Tellus 39B*, 245–260.

Farquhar, G. D., S. von Caemmerer, and J. A. Berry. 1980. A biochemical model of photosynthetic CO_2 assimilation in leaves of C_3 species. *Planta 149*, 8–90.

Field, C., and H. A. Mooney. 1986. The photosynthesis-nitrogen relationship in wild plants. In *On the Economy of Plant Form and Function*, T. J. Givnish, ed. Cambridge University Press, Cambridge, U.K., pp. 25–55.

Fox, L. R., and P. A. Morrow. 1983. Estimates of damage by herbivorous insects on Eucalyptus trees. *Australian Journal of Ecology 8*, 139–147.

Gates, D. M. 1985. Global biospheric responses to increasing atmospheric carbon dioxide concentration. In *Direct Effects of Increasing Carbon Dioxide on Vegetation*, B. R. Strain and J. D. Cure, eds. DOE/ER-0238, U.S. Department of Energy, Washington, D.C., pp. 172–184.

Gibson, A. H., M. M. Roper, and D. M. Halsall. 1988. Nitrogen fixation not associated with legumes. In *Advances in Nitrogen Cycling in Agricultural Ecosystems*, J. R. Wilson, ed. CAB International, Wallingford, Oxford, U.K., pp. 66–88.

Gifford, R. M. 1974. Photosynthetic limitations to cereal yield. *Royal Society of New Zealand Bulletin 12*, 887–893.

Gifford, R. M. 1979. CO_2 and plant growth under water and light stress: Implications for balancing the global carbon budget. *Search 10*, 316–318.

Gifford, R. M. 1980. Carbon storage by the biosphere. In *Carbon Dioxide and Climate: Australian Research*, G. I. Pearman, ed. Australian Academy of Science, Canberra, Australia, pp. 167–181.

Gifford, R. M. 1981. Increasing productivity through photosynthesis by breeding and by modification of the global carbon cycle. In *Integration of Photosynthetic Carbon Metabolism as a Basis of Plant Productivity*, T. Akazawa and C. B. Osmond, eds. The First Binational Joint Seminar under the Japan–Australia Cooperative Science Programme, Nagoya University, Nagoya, Japan, pp. 67–68.

Gifford, R. M. 1982. Global photosynthesis in relation to our food and energy needs. In *Photosynthesis: II. Development, Carbon Metabolism and Plant Productivity*, Govindjee, ed. Academic Press, New York, pp. 460–495.

Gifford, R. M. 1987. Global photosynthesis, atmospheric carbon dioxide and man's requirements. In *Current Perspectives in Environmental Biogeochemistry*, G. Giovannozzi-Sermanni and P. Nannipieri, eds. C.N.R.-I.P.R.A., Rome, Italy, pp. 413–443.

Gifford, R. M. 1988. Direct effects of higher carbon dioxide concentrations on vegetation. In *Greenhouse: Planning for Climate Change*, G. I. Pearman, ed. E. J. Brill, Leiden, Germany, pp. 506–519.

Gifford, R. M. 1991. *Impact of the Increasing Atmospheric Carbon Dioxide Concentration on the Carbon Balance of Vegetation.* Project Report No. 37, Energy Research and Development Corporation, Canberra, Australia.

Gifford, R. M. 1992a. Implications of the globally increasing atmospheric CO_2 concentration and temperature for the Australian terrestrial carbon budget: Integration using a simple model. *Australian Journal of Botany 40,* 527–543.

Gifford, R. M. 1992b. Interaction of carbon dioxide and growth-limiting environmental factors in vegetation productivity: Implications for the global carbon cycle. *Advances in Bioclimatology 1,* 24–57.

Gifford, R. M. 1993. Implications of CO_2 effects on vegetation for the global carbon budget. In *The Global Carbon Cycle,* Martin Heimann, ed. Proceedings of the NATO ASI at Il Ciocco, Italy, September 8–20, 1991. Springer-Verlag, Berlin, Germany, pp. 165–205.

Gifford, R. M. 1994. The global carbon cycle: A viewpoint on the missing carbon sink. *Australian Journal of Plant Physiology 21,* 1–15.

Gifford, R. M. 1995. Whole plant respiration and photosynthesis of wheat under increased CO_2 concentration and temperature: Long term vs. short term distinctions for modelling. *Global Change Biology 1,* 385–396.

Gifford, R. M., D. J. Barrett, J. L. Lutze, and A. B. Samarakoon. 1996. Agriculture and global change: Scaling direct carbon dioxide impacts and feedbacks through time. In *Global Change and Terrestrial Ecosystems,* B. H. Walker and W. L. Steffen, eds. Cambridge University Press, Cambridge, U.K., pp. 229–259.

Gifford, R. M., H. Lambers, and J. I. L. Morison. 1985. Respiration of crop species under CO_2 enrichment. *Physiologia Plantarum 63,* 51–356.

Gifford, R. M., J. L. Lutze, and D. Barrett. 1996. Global atmospheric change effects on terrestrial carbon sequestration: Exploration with a global C- and N-cycle model (CQUESTN). *Plant and Soil 187,* 369–387.

Gifford, R. M., and J. I. L. Morison. 1985. Photosynthesis, growth and water use of a C_4 stand at high CO_2. *Photosynthesis Research 7,* 69–76.

Gifford, R. M., and J. I. L. Morison. 1993. Crop responses to the global increase in atmospheric CO_2 concentration. In *Proceedings of the First International Crop Science Congress,* Ames, Iowa, July 1992, Agronomy Society of America, Madison, Wisconsin.

Goudriaan, J. 1992. Biosphere structure, carbon sequestering potential and the atmospheric ^{14}C record. *Journal of Experimental Botany 43,* 1111–1119.

Goudriaan, J. 1993. Interaction of ocean and biosphere in their transient responses to increasing atmospheric CO_2. *Vegetatio 104/105,* 329–337.

Goudriaan, J., H. H. van Laar, H. van Keulen, and W. Louwerse. 1985. Photosynthesis, CO_2 and plant production. In *Wheat Growth and Modeling,* W. Day and H. Atkins, eds. NATO ASI Series, Series A: Life Sciences, Vol. 86, Plenum, New York, pp. 107–122.

Graumlich, L. J. 1991. Subalpine tree growth, climate, and increasing CO_2: An assessment of recent growth trends. *Ecology 72,* 1–11.

Gutschick, V. P. 1978. Energy and nitrogen fixation. *BioScience 28,* 571–575.

Hari, P., H. Arovaara, T. Raunemaa, and A. Hautojarvi. 1984. Forest growth and the effects of energy production: A method for detecting trends in the growth potential of trees. *Canadian Journal of Forest Research 14,* 437–440.

Houghton, R. A. 1993. Changes in terrestrial carbon over the last 135 years. In *The Global Carbon Cycle,* M. Heimann, ed. Springer-Verlag, Berlin, Germany, pp. 145–163.

Houghton, R. A., R. D. Boone, J. M. Melillo, C. A. Palm, G. M. Woodwell, N. Myers, B. Moore, and D. L. Skole. 1985. Net flux of CO_2 from tropical forests. *Nature 316,* 617–620.

Idso, S. B., B. A. Kimball, and S. G. Allen. 1991. CO_2 enrichment of sour orange trees: 2. 5 years into a long term experiment. *Plant Cell and Environment 14,* 351–353.

Jang, J. C., and J. Sheen. 1994. Sugar sensing in higher plants. *The Plant Cell 6,* 1665–1679.

Jarvis, P. G., and K. G. McNaughton. 1986. Stomatal control of transpiration. *Advance in Ecological Research 15,* 1–49.

Johnson, R. H., and D. E. Lincoln. 1990. Sagebrush and grass hopper responses to atmospheric carbon dioxide concentration. *Oecologia 84,* 103–110.

Jonasson, S., J. P. Bryant, F. S. Chapin, and M. Andersson. 1986. Plant phenols and nutrients in relation to variations in climate and rodent grazing. *American Naturalist 128,* 394–408.

Jordan, D. B., and W. L. Ogren. 1984. The CO_2/O_2 specificity of ribulose 1, 5-bisphosphate carboxylase/oxygenase. *Planta 161,* 308–313.

Kienast, F., and R. J. Luxmoore. 1988. Tree-ring analysis and conifer growth responses to increased atmospheric CO_2 levels. *Oecologia 76,* 487–495.

Kimball, B. A. 1983. *Carbon Dioxide and Agricultural Yield: An Assemblage and Analysis of 770 Prior Observations.* Water Conservation Laboratory Report 14, U.S. Department of Agriculture, ARS, Phoenix, Arizona.

Kimball, B. A. 1986. CO_2 stimulation of growth and yield under environmental restraints. In *Carbon Dioxide Enrichment of Greenhouse Crops: II. Physiology, Yield and Economics,* H. Z. Enoch and B. A. Kimball, eds. CRC Press, Baton Rouge, La.

Kohlmaier, G. H., E-O. Sire, A. Janacek, C. D. Keeling, S. C. Piper, and R. Revelle. 1989. The seasonal contribution of a CO_2 fertilizing effect of the terrestrial vegetation to the amplitude increase in atmospheric CO_2 at Mauna Loa observatory. *Tellus 41B,* 487–510.

Körner, C. 1993. CO_2 fertilization: The great uncertainty in future vegetation development. In *Vegetation Dynamics and Global Change,* A. M. Solomon and H. H. Shugart, eds. Chapman and Hall, New York, pp. 53–70.

Körner, C., and J. A. Arnone, III. 1992. Responses to elevated carbon dioxide in artificial tropical ecosystems. *Science 257,* 1672–1675.

Kramer, P. J. 1981. Carbon dioxide concentration, photosynthesis, and dry matter production. *BioScience 31,* 29–33.

Kramer, P. J., and N. Sionit. 1987. Effects of increasing carbon dioxide concentration on the physiology and growth of forest trees. In *The Greenhouse Effect, Climate Change, and U.S. Forests,* W. E. Shands and J. S. Hoffman, eds. Conservation Foundation, Washington, D.C., pp. 219–246.

La Marche, V. C. Jr., D. A. Graybill, H. C. Fritts, and M. R. Rose. 1984. Increasing atmospheric carbon dioxide: Tree ring evidence for growth enhancement in natural vegetation. *Science 225,* 1019–1021.

Lawlor, D. W., and A. C. Mitchell. 1991. The effects of increasing CO_2 on crop photosynthesis and productivity: A review of field studies. *Plant Cell and Environment 14,* 807–818.

Lincoln, D. E. 1993. The influence of plant carbon dioxide and nutrient supply on susceptibility to insect herbivores. *Vegetatio 104/105,* 273–280.

Long, S. P. 1991. Modification of the response of photosynthetic productivity to rising temperature by atmospheric CO_2 concentrations: Has its importance been underestimated? *Plant Cell and Environment 14,* 729–739.

Long, S. P., N. R. Baker, and C. A. Raines. 1993. Analyzing the responses of photosynthetic CO_2 assimilation to long-term elevation of atmospheric CO_2 concentration. *Vegetatio 104/105,* 33–45.

Lugo, A. E., and J. Wisniewski. 1992. Natural sinks of CO$_2$: Conclusions, key findings and research recommendations from the Palmas Del Mar workshop. *Water, Air and Soil Pollution 64,* 455–459.

Lutze, J. L. 1996. Carbon and nitrogen relationships in swards of *Danthonia richardsonii* in response to carbon dioxide enrichment and nitrogen supply. Ph.D. dissertation. Australian National University, Canberra, Australia.

Lutze, J. L., and R. M. Gifford. 1995. Carbon storage and productivity of a carbon dioxide enriched nitrogen limited grass sward after one year's growth. *Journal of Biogeography 22,* 227–233.

Luxmoore, R. J. 1981. CO$_2$ and phytomass. *BioScience 31,* 626.

Luxmoore, R. J., R. J. Norby, and E. G. O'Neill. 1986. Seedling tree responses to nutrient stress under atmospheric CO$_2$ enrichment. In *Proceedings of the 18th IUFRO World Congress,* Division 2, Vol. 1, IUFRO Secretariat, Vienna, Austria, pp. 178–183.

Luxmoore, R. J., S. D. Wullschleger, and P. J. Hanson. 1993. Forest responses to CO$_2$ enrichment and climate warming. *Water, Air, and Soil Pollution 70,* 309–323.

Makino, A. 1994. Biochemistry of C$_3$-photosynthesis in high CO$_2$. *Journal of Plant Research 107,* 79–84.

Masle, J., G. D. Farquhar, and R. M. Gifford. 1990. Growth and carbon economy of wheat seedlings as affected by soil resistance to penetration and ambient partial pressure of CO$_2$. *Australian Journal of Plant Physiology 17,* 465–487.

Masterton, C. L., and M. T. Sherwood. 1978. Some effects of increased atmospheric carbon dioxide on white clover (*Trifolium repens*) and pea (*Pisum sativum*). *Plant and Soil 49,* 421–426.

Mattson, W. J., Jr. 1980. Herbivory in relation to plant nitrogen content. *Annual Review of Ecology and Systematics 11,* 119–161.

McLaughlin, M. J., I. R. Fillery, and A. R. Till. 1992. Operation of the phosphorus, sulphur and nitrogen cycles. In *Australia's Renewable Resources, Sustainability and Global Change: IGBP-Australia Planning Workshop, October 3–4, 1990,* R. M. Gifford and M. M. Barson, eds. pp. 67–116.

McMurtrie, R. E., H. N. Comins, M. U. F. Kirschbaum, and Y-P. Wang. 1992. Modifying existing forest growth models to take account of effects of elevated CO$_2$. *Australian Journal of Botany 40,* 657–677.

McNaughton, K. G., and P. G. Jarvis. 1983. Predicting the effects of vegetation changes on transpiration and evaporation. In *Water Deficits and Plant Growth,* T. T. Kozlowski, ed. Vol. VII, Academic Press, New York, pp. 1–47.

McNaughton, K. G., and P. G. Jarvis. 1991. Effects of spatial scale on stomatal control of transpiration. *Agricultural and Forest Meteorology 54,* 279–301.

Meek, D. W., J. L. Hatfield, T. A. Howell, S. B. Idso, and R. J. Reginato. 1984. A generalized relationship between photosynthetically active radiation and solar radiation. *Agronomy Journal 76,* 939–945.

Morell, M. K., K. Paul, H. J. Kane, and T. J. Andrews. 1992. Rubisco: Maladapted or misunderstood? *Australian Journal of Botany 40,* 431–441.

Morison, J. I. L. 1985. Sensitivity of stomata and water use efficiency to high CO$_2$. *Plant Cell and Environment 8,* 467–474.

Morison, J. I. L. 1987. Intercellular CO$_2$ concentration and stomatal response to CO$_2$. In *Stomatal Function,* E. Zeiger, I. Cowan, and G. D. Farquhar, eds. Stanford University Press, Stanford, Calif., pp. 229–251.

Morison, J. I. L. 1993. Response of plants to CO$_2$ under water limited conditions. *Vegetatio 104/105,* 193–209.

Morison, J. I. L., and R. M. Gifford. 1983. Stomatal sensitivity to carbon dioxide and humidity. *Plant Physiology 71,* 789–796.

Morison, J. I. L., and R. M. Gifford. 1984. Plant growth and water use with limited water supply in high CO$_2$ concentrations: I. Leaf area, water use and transpiration. *Australian Journal of Plant Physiology 11,* 361–374.

Mousseau, M. 1993. Effects of elevated CO$_2$ on growth, photosynthesis and respiration of sweet chestnut (*Castanea sativa Mill.*) *Vegetatio 104/105,* 413–419.

Neales, T. F., and L. D. Incoll. 1968. The control of leaf photosynthesis by the level of assimilate concentration in the leaf: A review of the hypothesis. *Botanical Reviews 34,* 107–125.

Norby, R. J. 1987. Nodulation and nitrogenase activity in the nitrogen-fixing woody plants stimulated by CO$_2$ enrichment of the atmosphere. *Physiologia Plantarum 71,* 77–82.

Norby, R. J., C. A. Gunderson, S. D. Wullschleger, E. G. O'Neill, and M. K. McCracken. 1992. Productivity and compensatory responses of yellow-poplar trees in elevated CO$_2$. *Nature 357,* 322–324.

Norby, R. J., E. G. O'Neill, W. G. Hood, and R. J. Luxmoore. 1987. Carbon allocation, root exudation and mycorrhizal colonization of *Pinus echinata* seedlings grown under CO$_2$ enrichment. *Tree Physiology 3,* 203–210.

Norby, R. J., J. Pastor, and J. M. Melillo. 1986. Carbon-nitrogen interactions in CO$_2$-enriched white oak: Physiological and long term perspectives. *Tree Physiology 2,* 33–241.

Oechel, W., and B. R. Strain. 1985. Native species responses to increased carbon dioxide. In *Direct Effects of Increasing Carbon Dioxide on Vegetation,* B. R. Strain and J. D. Cure, eds. DOE/ER-0238, U.S. Department of Energy, Washington, D.C., pp. 117–154.

Olson, J. S., J. A. Watts, and L. J. Allison. 1985. *Major World Ecosystem Complexes Ranked by Carbon in Live Vegetation: A Database.* Carbon Dioxide Information Center Numeric Data Collection, Oak Ridge National Laboratory, NDP-017, Oak Ridge, Tenn.

Pearcy, R. W., O. Bjorkman, R. W. Pearey, O. Bjorkman, L. H. Allen, Jr., E. W. R. Barlow, J. R. Ehleringer, G. D. Farquar, D. R. Geiger, R. T. Gjaquinta, R. M. Gifford, T. C. Hsiao, T. Jurik, D. A. Phillips, K. Raschke, and S. C. Wong. 1983. Physiological effects. In *CO$_2$ and Plants: The Response of Plants to Rising Levels of Atmospheric Carbon Dioxide,* E. R. Lemon, ed. AAAS Symposium 84, Westview Press, Boulder, Colo., pp. 65–105.

Polglase, P. J., and Y.-P. Wang. 1992. Potential CO$_2$-enhanced carbon storage by the terrestrial biosphere. *Australian Journal of Plant Physiology 40,* 641–656.

Polley, H. W., H. B. Johnson, B. D. Marino, and H. S. Mayeux. 1993. Increase in C$_3$ plant water-use efficiency and biomass over glacial to present CO$_2$ concentrations. *Nature 361,* 61–64.

Poorter, H. 1993. Interspecific variation in the growth response of plants to an elevated ambient CO$_2$ concentration. *Vegetatio 104/105,* 77–97.

Poorter, H., R. M. Gifford, P. E. Kriedemann, and S. C. Wong. 1992. A quantitative analysis of dark respiration and carbon content as factors in the growth response of plants to elevated CO$_2$. *Australian Journal of Botany 40,* 501–513.

Rastetter, E. B., and G. R. Shaver. 1992. A model of multiple-element limitation for acclimating vegetation. *Ecology 73,* 1157–1174.

Rawson, H. M. 1992. Plant responses to temperature under conditions of elevated CO$_2$. *Australian Journal of Botany 40,* 473–490.

Reuveni, J., and J. Gale. 1985. The effect of high levels of carbon dioxide on dark respiration and growth of plants. *Plant Cell and Environment 8,* 623–628.

Rogers, H. H., G. E. Bingham, J. D. Cure, J. M. Smith, and K. A. Surano. 1983. Responses of selected plant species to elevated carbon dioxide in the field. *Journal of Environmental Quality 12,* 569–574.

Samarakoon, A. B., and R. M. Gifford. 1995. Soil water content of plants at high CO_2 concentration and interaction with the direct CO_2 effects: A species comparison. *Journal of Biogeography 22,* 193–202.

Samarakoon, A. B., and R. M. Gifford. 1996. Elevated CO_2 effects on water use and growth of maize in wet and drying soil. *Australian Journal of Plant Physiology 23,* 53–62.

Samarakoon, A. B., W. J. Müller, and R. M. Gifford. 1995. Transpiration and leaf area under elevated CO_2: Effects of soil water status and genotype in wheat. *Australian Journal of Plant Physiology 22,* 33–44.

Schlesinger, W. H. 1993. Response of the terrestrial biosphere to global climate change and human perturbation. *Vegetatio 104/105,* 295–305.

Scriber, J. M., and F. Slansky. 1981. The nutritional ecology of immature insects. *Annual Review of Entomology 26,* 183–211.

Sionit, N., B. R. Strain, H. Hellmers, G. H. Riechers, and C. H. Jaeger. 1985. Atmospheric CO_2 enrichment affects the growth and development of *Liquidambar styraciflua* L. and *Pinus taeda* L. seedlings. *Canadian Journal of Forest Research 15,* 468–471.

Smil, V. 1992. Nitrogen and phosphorus. In *The Earth as Transformed by Human Action: Global and Regional Changes in the Biosphere Over the Past 300 Years,* B. L. Turner II, W. C. Clark, R. W. Kates, J. F. Richards, J. T. Matthews, and W. B. Meyer, eds. Cambridge University Press, Cambridge, U.K., pp. 423–436.

Sprent, J. I. 1993. The role of nitrogen fixation in primary succession on land. In *Primary Succession on Land,* J. Miles and D. W. H. Walton, eds. Blackwell Scientific, Oxford, U.K., pp. 209–219.

Stephenson, N. L. 1990. Climatic control of vegetation distribution: The role of water balance. *American Naturalist 135,* 649–670.

Stitt, M. 1991. Rising CO_2 levels and their potential significance for carbon flow in photosynthetic cells. *Plant Cell and Environment 14,* 741–762.

Taylor, J. A., and J. Lloyd. 1992. Sources and sinks of atmospheric CO_2. *Australian Journal of Botany 40,* 407–418.

Uchijima, Z., and H. Seino. 1985. Agroclimatic evaluation of net primary productivity of natural vegetations. I. Chikugo model evaluating net primary productivity. *Journal of Agricultural Meteorology 40,* 43–352.

Warrick, R. A., R. M. Gifford, and M. L. Parry. 1985. CO_2 climatic change and agriculture: Assessing the response of food crops to the direct effects of increased CO_2 and climate change. In *The Greenhouse Effect, Climatic Change, and Ecosystems,* B. Bolin, B. R. Doos, J. Jäger, and R. A. Warrick, eds. SCOPE 29, John Wiley and Sons, Chichester, U.K., pp. 393–473.

Watson, R. T., L. G. Meiro Filho, E. Sanhueza, and A. Janetos. 1992. Sources and sinks. In *Climate Change: The Supplementary Report to the IPCC Scientific Assessment,* J. T. Houghton, B. A. Callander, and S. K. Varney, eds. Cambridge University Press, Cambridge, U.K., pp. 29–46.

Wong, S. C., I. R. Cowan, and G. D. Farquhar. 1979. Stomatal conductance correlates with photosynthetic capacity. *Nature 282,* 424–426.

Woodward, F. I. 1987. *Climate and Plant Distribution.* Cambridge University Press, Cambridge, U.K.

Xu, D-Q., R. M. Gifford, and W. S. Chow. 1994. Photosynthetic acclimation in pea and soybean to high atmospheric CO_2 partial pressure. *Plant Physiology 106,* 661–671.

6

Soils and the Global Carbon Cycle

WILLIAM H. SCHLESINGER, JULIA PALMER WINKLER, AND
J. PATRICK MEGONIGAL

Abstract

Soils hold one of the largest near-surface pools in the global carbon cycle, containing at least 1,500 Pg C in organic forms, with a large proportion of this amount lying near the surface. Largely as a result of the human disturbance of soils, especially in cultivation, 36 Pg C was lost from soils between 1860 and 1960, with a current rate of loss of approximately 0.8 Pg C/yr. Thus, the loss of carbon from soils is a significant component of the biotic flux of CO_2 to the atmosphere The soil carbon pool does not appear likely to house the missing sink. In fact, as a result of global warming, substantial amounts of CO_2 are likely to be lost from soils.

6.1 Introduction

Soils hold one of the largest near-surface pools in the global carbon cycle, containing at least 1,500 Pg C in organic forms. Although some fractions of soil organic matter are very old, the global mean residence time for organic carbon in soils is approximately 30 years. The soil carbon pool is large and dynamic; increases or decreases in the amount of carbon in soils could have significant effects on the concentration of CO_2 in the atmosphere (Trumbore et al., 1996). A large literature shows that human activities – especially cultivation – reduce the pool of carbon in soils and that most of this carbon is transferred to the atmosphere. We know much less about the potential for soils to serve as a sink for CO_2 under various scenarios of global change. Higher global temperatures may stimulate the rate of decomposition in soils, but the input of organic residues to soils may also increase as net primary production responds to the higher temperatures, rainfall, and atmospheric CO_2 concentrations that characterize most global models of future climate (Melillo et al., 1993; Oechel et al., 1994).

6.2 Pool of Carbon in Soil

Various workers have estimated the pool of organic carbon in soils. Most of the values fall in the range of 1,400–1,600 Pg C, regardless of whether the global pool is estimated from an aggregation of vegetation types (Schlesinger, 1977), climatic life zones (Post et al., 1982), or soil orders (Eswaran et al., 1993). Of this global total, approximately 55 Pg C resides in fresh litter, or detritus, on the surface of the soil, known informally as the forest floor. Because this value is similar to global terrestrial net primary production, the mean residence time of surface litter globally is approximately 1 year. Within the mineral soil, the concentration of soil organic matter nearly always declines exponentially with depth, reflecting its origin from aboveground plant debris (litterfall) and root turnover (Nakane, 1976). Thus, a large proportion of the soil carbon pool lies near the surface of the soil where it is subject to microbial decay, erosion, and disruption by human activities.

In the mineral soil, organic matter consists of a variety of fractions; typically, these are separated by laboratory methods based on differential chemical reactivity or on different physical properties, especially density. A portion of the soil organic matter consists of undecomposed cellular residues of plant materials that have been mixed into the soil; approximately 5% of the organic carbon in soils is living microbial biomass (Wardle, 1992). By far the largest fraction of the soil organic matter is classified as humic material, which appears to be a metabolic by-product of microbial activity. Humus is typically separated into humic and fulvic acids on the basis of the different solubilities of these fractions in acidic and alkaline solutions. The molecular structure of humic and fulvic materials is poorly understood. Schulten and Schnitzer (1993) have recently proposed a generic structure for humic acid that encompasses many of its bulk properties; however, it is likely that the soil humic fraction comprises a wide variety of forms.

Humic materials are widely regarded as refractory, and the radiocarbon age of specific fractions is commonly measured in hundreds to thousands of years (Campbell et al., 1967). Therefore, the pool of organic carbon in soils is best viewed as a multicompartment system, consisting of a small pool of fresh debris that decomposes relatively rapidly near the surface, and a larger pool of humic materials, dispersed through the soil profile, with relatively slow turnover (Figure 6.1). The flux of CO_2 from soils to the atmosphere is derived primarily from the decomposition of fresh residues in the litter layer (Edwards and Sollins, 1973; Bowden et al., 1993).

Some of the organic matter in soils is found as charcoal, especially in ecosystems in which fire is a frequent natural phenomenon. We know relatively little about the proportion of the global pool of soil carbon that is held as charcoal. Pine forests on the Virginia Piedmont in the United States contain 0.46 kg C/m^2 charcoal in the forest floor (Schiffman and Johnson, 1989). Sanford and colleagues (1985) found 0.23, 0.35, and 0.70 kg C/m^2 of charcoal in the profile of three soils in the Brazilian *tierra-firma* rainforest. Each value represents only a small fraction of the typical pool of soil organic carbon in rainforests (10.4 kg C/m^2; Schlesinger, 1977).

Soils in arid and semiarid climates typically contain pedogenic carbonate, which contains 800–1,700 Pg C globally (Schlesinger, 1982; Eswaran et al., 1993). In many areas this carbonate appears to originate from the deposition of aeolian carbonate dusts. In these areas or where the underlying geology consists of carbonate rocks, pedogenic carbonate merely represents a transfer and reprecipitation of inorganic carbon ($CaCO_3$) from the parent materials of soil formation. However, in areas of noncarbonate terrain, the pool of carbon in pedogenic carbonate represents a net storage of carbon that might otherwise be found in the atmosphere/ocean system (Schlesinger, 1985).

Pedogenic carbonate typically accumulates near the depth of the mean maximum annual percolation of soil water (Arkley, 1963); in many areas, buried carbonate layers show a record of past periods of greater rainfall (e.g., Schlesinger, 1985). Although the isotopic record of pedogenic carbonates can be of unusual value in unraveling past shifts in the distribution of vegetation (e.g., Quade et al., 1989), this pool of carbon in soils is sufficiently refractory (i.e., mean residence time

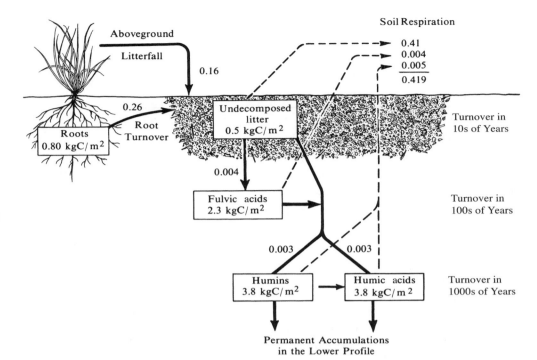

Figure 6.1: Dynamics of soil organic matter in the 0–20-cm layer of a hypothetical grassland soil, showing estimated carbon pools (kg C/m²) and annual transfers (kg C/m²/yr). (Data from Schlesinger, 1977.) Reproduced with permission from W. G. Schlesinger, *Annual Review of Ecology and Systematics* 8:51–81, © 1977, Annual Reviews Inc.

> 10,000 yr) to be ignored in all scenarios of the global carbon cycle during the next few centuries.

6.3 The Dynamics of the Soil Carbon System

Soil organic matter is ultimately derived from net primary production (NPP), which results in the deposition of plant debris in and on the upper layers of the soil profile. Globally, NPP appears to be approximately 50 Pg C/yr (Melillo et al., 1993; Potter et al., 1993). Since fractions of NPP are eaten and respired by animals (3 Pg C/yr; Whittaker and Likens, 1973) and consumed by fire (2–5 Pg C/yr; Crutzen and Andreae, 1990), we may assume that slightly less than 47 Pg C/yr enters the soil, assuming a steady state in terrestrial biomass.

Plant debris decomposes in the soil, releasing CO_2 to the soil atmosphere. Nearly all of this CO_2 diffuses to the surface of the soil and escapes to the atmosphere in a process called soil respiration. Only a tiny fraction (i.e., 0.3%; Edwards and Harris, 1977) appears to dissolve in the soil solution and escape from ecosystems in groundwater. In nearly all cases, the measured escape of CO_2 from the soil surface is larger than the input of organic carbon from NPP as a result of the respiration of live roots and mycorrhizae. Raich and Schlesinger (1992) estimated the global CO_2 in soil respiration at 68 Pg C/yr, suggesting that about one-third of the efflux is derived from root respiration. The mean residence time (mass/flux) for carbon in soil organic matter is approximately 32 years (Raich and Schlesinger, 1992).

We can assume that, prior to the widespread human disturbance of soils, the soil carbon pool was near steady-state. The rate of formation of refractory humic substances that might add to the long-term storage of carbon in soils is relatively slow (0.2–12 g C/m²/yr; Schlesinger, 1990). Some of the highest rates of accretion of soil organic matter are seen in areas covered by the last continental glaciation (Harden et al., 1992). A compilation of data from many ecosystems suggests that the formation of refractory soil organic matter results in a maximum long-term accretion of less than 0.4 Pg C/yr in soils globally (Schlesinger, 1990). This small net accumulation of organic matter in soils reflects the efficiency of terrestrial decomposition, which allows only approximately 0.7% of NPP to accumulate in the soil carbon pool each year. It is of interest that the annual transport of organic carbon in world rivers is also approximately 0.4 Pg C/yr (Schlesinger and Melack, 1981; Meybeck, 1982; Ludwig et al., 1996); the net formation of refractory humic substances in soils is roughly balanced by their erosion and transport to the sea, where ultimately they may be added to ocean sediments (Berner, 1982; Lugo and Brown, 1986; Sarmiento and Sundquist, 1992).

The advent of mechanized agriculture in the late 1800s led to dramatic losses of organic matter from cultivated soils. Typically, 20–40% of the native soil organic matter is lost when virgin lands are converted to agriculture (Schlesinger, 1986; Mann, 1986; Detwiler, 1986; Davidson and Ackerman, 1993). The losses are greatest during the first few years of land conversion, and slow after approximately 20 years of cultivation. These percentage losses are consistent with independent estimates of the relative portion of soil organic matter that exists in labile versus refractory pools (Spycher et al., 1983). Schlesinger (1984) calculated that 36 Pg C was lost from soils between 1860 and 1960, with a current rate of loss of approximately 0.8 Pg C/yr. Thus, the loss of carbon from soils is a sig-

Table 6.1. *Accumulation of soil organic matter in abandoned agricultural soils and in other disturbed sites, which are allowed return to native vegetation*

Ecosystem Type	Previous Land Use	Period of Abandonment (yr)	Rate of Accumulation (g C/m2/yr)	Reference
Subtropical forest	Cultivation	40	30–50	Lugo et al. (1986)
Temperate deciduous forest	Cultivation	100	45	Jenkinson (1990)
Temperate coniferous forest	Cultivation	50	21–26	Schiffman and Johnson (1989)
Temperate coniferous forest	Diked soils	100	26	Beke (1990)
Temperate deciduous forest	Mine spoils	50	55	Leisman (1957)
Temperate grassland	Mine spoils	28–40	28	Anderson (1977)

nificant component of the biotic flux of CO_2 to the atmosphere (Houghton et al., 1987).

Humans have also increased the extent of forest fires, especially in the tropics (Crutzen and Andreae, 1990). Seiler and Crutzen (1980) estimated that the annual production of charcoal from forest fires could be as large as 0.5–1.7 Pg C/yr, adding a significant amount of refractory material to the soil carbon pool. We have only limited knowledge about the long-term persistence of charcoal in soils. Shneour (1966) reported substantial microbial decomposition of graphite incubated in nonsterile soils. Estimates of the amount of charcoal in Amazon rainforest soils, divided by its measured radiocarbon age (250–6,260 yr), indicate a net accretion of only 0.4–2.8 g C/m2/yr of charcoal as a result of the historical occurrence of fires in that biome (Sanford et al., 1985). Based on the small fraction of biomass that is converted to charcoal during fire (ca. 5%), Andreae (1993) offered a revised estimate of 0.3–0.7 Pg C/yr for the long-term accumulation of charcoal in soils worldwide. Recent field data (Fearnside et al., 1993) suggest that the conversion efficiency may even be as low as 2.7%. In any case, current patterns of increasing biomass burning globally appear to generate a much larger flux of CO_2 to the atmosphere (2–5 Pg C/yr) than a net sink in soils.

Several recent reports by the U.S. Environmental Protection Agency assess the potential for improved soil management to maintain, restore, and enlarge the pool of organic carbon in agricultural soils (Kern and Johnson, 1993). When natural land is converted to agriculture using "no-tillage" techniques, losses of soil carbon are often smaller than when traditional cultivation is practiced (Blevins et al., 1977; Dick, 1983). Small increases in soil organic matter are sometimes seen when existing cultivated fields are converted to no-till agriculture. For example, Wood and colleagues (1991) found that intensive no-till management resulted in accumulations of 7–16 g C/m2/yr in the 0–10 cm depth of prairie soils that were previously cultivated. Nevertheless, the most optimistic scenario for the widespread adoption of no-till agriculture in the United States offsets only 0.7–1.1% of the projected fossil fuel use in the United States during the next 30 years (Kern and Johnson, 1993).

In the past few decades, agricultural practice has been abandoned on substantial areas of the eastern United States and Europe; presumably, reforestation has partially restored the pre-

historic pool of carbon in vegetation and soils in these regions. The rate of accumulation of soil organic matter is often much greater during early soil development than in mature soils (Table 6.1); nevertheless, it is much smaller than the rate of biomass accumulation during reforestation, which dominates any globally significant sink for carbon in these areas (e.g., Wofsy et al., 1993). Houghton and colleagues (1987) show a net sink of 0.08 Pg C/yr in Europe as a result of reforestation (cf. 0.1 Pg C/yr, in Kauppi et al., 1992). Delcourt and Harris (1980) report a similar value for the southeastern United States. Globally, although the rate of new agricultural clearing has slowed in recent years, the total amount of cultivated land is likely to continue to increase in the future as a function of the world's rising population.

The sum of these various human activities appears to yield a net release of CO_2 from soils to the atmosphere, and soils are likely to have contributed a significant proportion of the total release from the terrestrial biosphere as a result of human disturbance since 1850. A net release of carbon from land is also supported by measurements of the trend in $\delta^{13}C$ in tree rings and ice cores during the last century (Leavitt and Long, 1988; Siegenthaler and Oeschger, 1987). Globally, the biotic flux appears to have been larger than the fossil fuel flux until about 1960 (Houghton et al., 1983); the net release from land may be as large as 1.8 Pg C/yr today (Houghton et al., 1987).

6.4 Anticipated Changes in Soil Organic Matter

Numerous authors have postulated that the size of the terrestrial biosphere – vegetation and soils – is now increasing as a result of CO_2 fertilization, providing an explanation for the "missing sink" of CO_2 that ought otherwise to be found in the atmosphere. There is ample evidence that in atmospheres with high levels of CO_2 and in soils with abundant water and nutrients, the growth of crops is greater than in atmospheres with ambient CO_2 (Strain and Cure, 1985; Bazzaz, 1990; Idso and Kimball, 1993), and there is some evidence that enhanced growth of crops increases the amount of soil organic matter in agricultural soils (Leavitt et al., 1994; Wood et al., 1994).

Although there is much interest in the CO_2 fertilization effect, there is little or no empirical evidence that it results in increased storage of carbon in most natural ecosystems. Working in montane forests of the Sierra Nevada, Graumlich (1991) found no statistically significant trend in tree ring growth that could be linked to a CO_2-induced stimulation of plant growth since 1900. D'Arrigo and Jacoby (1993) found no trend in the growth rate of boreal forest species in Canada that could be linked to CO_2 after removing the correlation to recent climatic warming in that region. West and colleagues (1993) report recent increases in the tree ring width of longleaf pine in southern Georgia, which, they suggest, in the absence of a correlation to other factors, may indicate a response to higher CO_2. However, Van Deusen (1992) found no upward trend in the growth of loblolly pine in the southeastern United States during the last several decades.

Field experiments in the tundra of Alaska showed that rapid acclimation of native plants to high CO_2 reduced their potential for enhanced carbon acquisition (Tissue and Oechel, 1987), and Grulke and colleagues (1990) concluded that there was "little if any long-term stimulation of ecosystem carbon acquisition by increases in atmospheric CO_2," (p. 485) Billings and colleagues (1984) found that the storage of carbon in tundra soils increased at high CO_2 only if additional nitrogen was provided as a plant nutrient. Tissue and colleagues (1993) found that loblolly pine in native soils showed a strong nutrient limitation when grown at high CO_2. Similarly, Norby and colleagues (1992) found that seedlings of *Liriodendron tulipifera* showed no net increase in size when grown at high CO_2 with ambient levels of water and nutrients. In the latter experiment, the turnover of fine roots increased, potentially increasing the input of plant residues to the pool of soil organic matter. Therefore, we do not know whether a higher net primary production of plants in natural ecosystems subjected to elevated CO_2 translates into a greater long-term storage of carbon on land; it is possible that it could lead to greater rates of carbon turnover and little net sequestration in soils (e.g., Korner and Arnone, 1992).

Global warming is also expected to have a dramatic effect on the carbon storage of vegetation and soils. Ten thousand years ago, 29.5×10^6 km² of the Earth's land area was covered with ice and presumably contained little or no soil organic matter (Flint, 1971; Bell and Laine, 1985). Much of this area now supports tundra and boreal forest ecosystems, with substantial pools of soil carbon. Adams and colleagues (1990) calculated that the global pool of soil organic matter increased 490 Pg C during the warming that occurred from the last glacial maximum (18,000 years ago) to the present. Similarly, Prentice and Fung (1990) estimate that the soil carbon pool increased by 13–148 Pg C during this interval.

Smith and colleagues (1992) derive a prediction of the future distribution of vegetation from each of four recent climate models, finding that associated changes in the soil carbon pool range from a loss of 19.5 Pg C to a gain of 57.3 Pg C under climatic shifts that include the response of both temperature and precipitation to future levels of atmospheric CO_2. Relative to soils, in each case, the vegetation offers a substantial sink for atmospheric CO_2 when it has fully equilibrated with the new anticipated climate (Kirschbaum, 1993).

These models provide a picture of the carbon pool in the biosphere at two points in time; they do not anticipate potential transient states that may yield dramatic fluxes of CO_2 from the

biosphere to the atmosphere (Smith and Shugart, 1993). For example, while global climate may change rapidly, changes in the distribution of vegetation are likely to take hundreds of years (Overpeck et al., 1991; MacDonald et al., 1993). Most soils require more than 3,000 years to achieve a steady state in soil organic matter with respect to new conditions of vegetation or climate (Birkeland, 1984; Almendinger, 1990). During this time, soil microbial communities are likely to show an immediate response to higher soil temperatures, increasing the rate of soil respiration (Trumbore et al., 1996).

Rates of CO_2 efflux increase as a function of soil temperature (Schleser, 1982; Raich and Schlesinger, 1992), and there is good reason to believe that rates of soil respiration will increase with global warming (Peterjohn et al., 1993; Kirschbaum, 1995; Figure 6.2). Losses of soil organic matter during the transient state, as soils adjust to new, warmer climatic conditions, may cause substantial releases of CO_2 to the atmosphere. In an experiment with soils collected from a pine forest in central North Carolina, Winkler and colleagues (1996) found that soil respiration increased when A-horizon soils were incubated at temperatures above their present-day mean annual temperature of 15.5°C. However, the effect was short-lived, and all soils showed the same respiration rate after 60 days (Figure 6.3). Global warming may affect the carbon in deeper soils in a similar manner (Figure 6.2).

Even a 1% increase in the rate of CO_2 evolution from soils globally (increasing to 68.7 Pg C/yr) would be equivalent to a 14% increase in the annual flux of CO_2 to the atmosphere from fossil fuels. Schleser (1982) suggests that at least some of the increase in atmospheric CO_2 during the past century may have been the result of globally increasing temperatures over that period. Using a simulation model of soil processes, Jenkinson and colleagues (1991) predicted that during the next 60 years, as much as 61 Pg C may be lost from the global pool of carbon in soils and released to the atmosphere as CO_2. Recent data

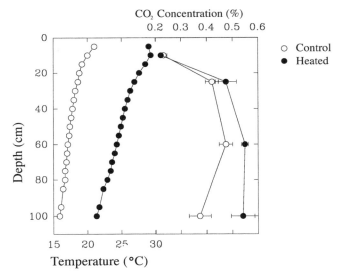

Figure 6.2: Soil temperature and pore-space CO_2 concentration (mean ± standard error of the mean) as a function of depth in control and experimental heated (+ 5°C) plots in a hardwood forest in Massachusetts (Data from J. P. Megonigal, unpublished data).

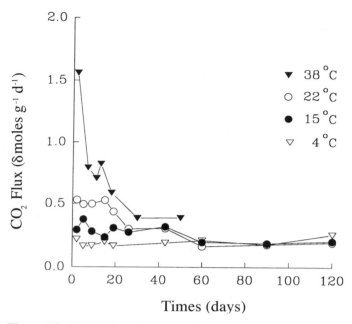

Figure 6.3: Respiration of A-horizon (0–28 cm) soils collected from a pine forest in central North Carolina and incubated at a range of temperatures for 120 days. The initial response of soil respiration showed a Q_{10} of 1.7–1.9, but after 60 days the respiration rates showed no significant differences. (Data from Winkler et al., 1996.)

from the arctic tundra of Alaska suggest that a net release of soil carbon may already be in progress (Oechel et al., 1993).

Ultimately, higher rates of decomposition may yield greater rates of nutrient turnover in the soil, allowing enhanced plant growth and carbon storage in terrestrial vegetation. In a soil-warming experiment in a black spruce forest of Alaska, Van Cleve and colleagues (1990) noted higher nutrient concentrations in the soil solution and in the foliage of spruce growing on heated plots (Table 6.2). Presumably, these nutrients were made available as a result of higher rates of decomposition when the soil was maintained at an elevated temperature. A greater mineralization of nutrients in globally warmer soils might help alleviate widespread nutrient deficiencies and allow vegetation to respond to higher levels of CO_2 (Oechel et al., 1994). Because the C/N ratio of the soil organic matter (ca. 12) is much lower than the C/N ratio of wood (ca. 160), a small increase in the rate of nitrogen mineralization in soils globally could potentially support a large increase in vegetation pro-

Table 6.2. *Effects of soil heating on tissue nutrient concentrations (%) in* Picea mariana

Plot	N	P	Ca	Mg	K
Control	0.76	0.099	0.28	0.08	0.56
Heated	1.05	0.136	0.28	0.09	0.80

Source: From Van Cleve et al. (1990).
Reprinted with permission K. Van Cleve, W. C. Oechel, and J. L. Hom (1990). *Canadian Journal of Forest Research 20,* 1530–1535.

duction and net carbon sequestration by the terrestrial biosphere (Rastetter et al., 1991; McGuire et al., 1992). Thus the response of vegetation and soils will be strongly linked during global climate change.

6.5 Conclusions

The soil carbon pool, per se, does not appear to be a likely candidate for housing the missing sink of carbon that plagues the current budgets for atmospheric CO_2. If less than 1% of global terrestrial NPP escapes decomposition and accumulates in soils (Schlesinger, 1990), enormous increases in NPP would be required to account for a significant proportion of the missing sink (2 Pg C/yr) in soils. At present, we see no strong indications of any CO_2-induced increase in NPP globally, presumably because of nutrient and moisture limitations over much of the Earth's land surface. Most models of future global climate, vegetation, and soils do not project a large sink of carbon in soils (e.g., Smith et al., 1992; Kirschbaum, 1993). In fact, during a greenhouse-induced climatic transient, substantial amounts of CO_2 are likely to be lost from soils, exacerbating the rate of greenhouse warming of the atmosphere. We know relatively little about whether increased rates of decomposition on a warmer Earth could alleviate the nutrient deficiencies that plague the growth of vegetation in many terrestrial ecosystems – allowing plants to respond to CO_2 enrichment in future global environments.

References

Adams, J. M., H. Faure, L. Faure-Denard, J. M. McGlade, and F. I. Woodward. 1990. Increases in terrestrial carbon storage from the last glacial maximum to the present. *Nature 348,* 711–714.

Almendinger, J. C. 1990. The decline of soil organic matter, total N, and available water capacity following the late-Holocene establishment of jack pine on sandy Mollisols, north-central Minnesota. *Soil Science 150,* 680–694.

Anderson, D. W. 1977. Early stages of soil formation on glacial till mine spoils in a semi-arid climate. *Geoderma 19,* 11–19.

Andreae, M. O. 1993. The influence of tropical biomass burning on climate and the atmospheric environment. In *Biogeochemistry of Global Change,* R. S. Oremland, ed., Chapman and Hall, New York, pp. 113–150.

Arkley, R. J. 1963. Calculation of carbonate and water movement in soil from climatic data. *Soil Science 96,* 239–248.

Bazzaz, F. A. 1990. The response of natural ecosystems to the rising global CO_2 levels. *Annual Review of Ecology and Systematics 21,* 167–196.

Beke, G. J. 1990. Soil development in a 100-year-old dike near Grand Pre, Nova Scotia. *Canadian Journal of Soil Science 70,* 683–692.

Bell, M., and E. P. Laine. 1985. Erosion of the Laurentide region of North America by glacial and glaciofluvial processes. *Quaternary Research 23,* 154–174.

Berner, R. A. 1982. Burial of organic carbon and pyrite sulfur in the modern ocean: Its geochemical and environmental significance. *American Journal of Science 282,* 451–473.

Billings, W. D., K. M. Peterson, J. O. Luken, and D. A. Mortensen. 1984. Interaction of increasing atmospheric carbon dioxide and soil nitrogen on the carbon balance of tundra ecosystems. *Oecologia 65*, 26–29.

Birkeland, P. W. 1984. *Soils and Geomorphology.* Oxford University Press, Oxford, U. K.

Blevins, R. L., G. W. Thomas, and P. L. Cornelius. 1977. Influence of no-tillage and nitrogen fertilization on certain soil properties after 5 years of continuous corn. *Agronomy Journal 69*, 383–386.

Bowden, R. D., K. J. Nadelhoffer, R. D. Boone, J. M. Melillo, and J. B. Garrison. 1993. Contributions of aboveground litter, belowground litter, and root respiration to total soil respiration in a temperate mixed hardwood forest. *Canadian Journal of Forest Research 23*, 1402–1407.

Campbell, C. A., E. A. Paul, D. A. Rennie, and K. J. McCallum. 1967. Factors affecting the accuracy of the carbon-dating method in soil humus studies. *Soil Science 104*, 81–85.

Crutzen, P. J., and M. O. Andreae. 1990. Biomass burning in the tropics: Impact on atmospheric chemistry and biogeochemical cycles. *Science 250*, 1669–1678.

D'Arrigo, R. D., and G. C. Jacoby. 1993. Tree growth-climate relationships at the northern boreal forest tree line of North America: Evaluation of potential response to increasing carbon dioxide. *Global Biogeochemical Cycles 7*, 525–535.

Davidson, E. A., and I. L. Ackerman. 1993. Changes in soil carbon inventories following cultivation of previously untilled soils. *Biogeochemistry 20*, 161–193.

Delcourt, H. R., and W. F. Harris. 1980. Carbon budget of the southeastern U. S. biota: Analysis of historical change in trend from source to sink. *Science 210*, 321–323.

Detwiler, R. P. 1986. Land use change and the global carbon cycle: The role of tropical soils. *Biogeochemistry 2*, 67–93.

Dick, W. A. 1983. Organic carbon, nitrogen, and phosphorus concentrations and pH in soil profiles as affected by tillage intensity. *Soil Science Society of America Journal 47*, 102–107.

Edwards, N. T., and W. F. Harris. 1977. Carbon cycling in a mixed deciduous forest floor. *Ecology 58*, 431–437.

Edwards, N. T., and P. Sollins. 1973. Continuous measurement of carbon dioxide evolution from partitioned forest floor components. *Ecology 54*, 406–412.

Eswaran, H., E. Van den Berg, and P. Reich. 1993. Organic carbon in soils of the world. *Soil Science Society of America Journal 57*, 192–194.

Fearnside, L. M., N. Leal, and F. M. Fernandes. 1993. Rainforest burning and the global carbon budget: Biomass, combustion efficiency and charcoal formation in the Brazilian Amazon. *Journal of Geophysical Research 98*, 16733–16743.

Flint, R. F. 1971. *Glacial and Quaternary Geology.* John Wiley and Sons, New York.

Graumlich, L. J. 1991. Subalpine tree growth, climate, and increasing CO_2: An assessment of recent growth trends. *Ecology 72*, 1–11.

Grulke, N. E., G. H. Riechers, W. C. Oechel, U. Hjelm, and C. Jaeger. 1990. Carbon balance in tussock tundra under ambient and elevated atmospheric CO_2. *Oecologia 83*, 485–494.

Harden, J. W., E. T. Sundquist, R. F. Stallard, and R. K. Mark. 1992. Dynamics of soil carbon during deglaciation of the Laurentide ice sheet. *Science 258*, 1921–1924.

Houghton, R. A., R. D. Boone, J. R. Fruci, J. E. Hobbie, J. M. Melillo, C. A. Palm, B. J. Peterson, G. R. Shaver, G. M. Woodwell, B. Moore, D. L. Skole, and N. Myers. 1987. The flux of carbon from terrestrial ecosystems to the atmosphere in 1980 due to changes in land use: Geographic distribution of the global flux. *Tellus 39B*, 122–139.

Houghton, R. A., J. E. Hobbie, J. M. Melillo, B. Moore, B. J. Peterson, G. R. Shaver, and G. M. Woodwell. 1983. Changes in the carbon content of terrestrial biota and soils between 1860 and 1980: A net release of CO_2 to the atmosphere. *Ecological Monographs 53*, 235–262.

Idso, S. B., and B. A. Kimball. 1993. Tree growth in carbon dioxide enriched air and its implications for global carbon cycling and maximum levels of atmospheric CO_2. *Global Biogeochemical Cycles 7*, 537–555.

Jenkinson, D. S. 1990. The turnover of organic carbon and nitrogen in soil. *Philosophical Transactions of the Royal Society of London, Biological Sciences 329*, 361–368.

Jenkinson, D. S., D. E. Adams, and A. Wild. 1991. Model estimates of CO_2 emissions from soil in response to global warming. *Nature 351*, 304–306.

Kauppi, P. E., K. Mielikainen, and K. Kuusela. 1992. Biomass and carbon budget of European forests, 1971 to 1990. *Science 256*, 70–74.

Kern, J. S., and M. G. Johnson. 1993. Conservation tillage impacts on national soil and atmospheric carbon levels. *Soil Science Society of America Journal 57*, 200–210.

Kirschbaum, M. U. F. 1993. A modelling study of the effects of changes in atmospheric CO_2 concentration, temperature and atmospheric nitrogen input on soil organic carbon storage. *Tellus 45B*, 321–334.

Kirschbaum, M. U. F. 1995. The temperature dependence of soil organic matter decomposition, and the effect of global warming on soil organic C storage. *Soil Biology and Biochemistry 27*, 753–760.

Korner, C., and J. A. Arnone. 1992. Responses to elevated carbon dioxide in artificial tropical ecosystems. *Science 257*, 1672–1675.

Leavitt, S. W., and A. Long. 1988. Stable carbon isotopic chronologies from trees in the southwestern United States. *Global Biogeochemical Cycles 2*, 189–198.

Leavitt, S. W., E. A. Paul, B. A. Kimball, G. R. Hendrey, J. R. Mauney, R. Rauschkolb, H. Rogers, K. Lewin, J. Nagy, P. J. Pinter, and H. B. Johnson. 1994. Carbon isotope dynamics of free-air CO_2-enriched cotton and soils. *Agricultural and Forest Meteorology 70*, 87–101.

Leisman, G. A. 1957. A vegetation and soil chronosequence on the Mesabi iron range spoil banks. *Ecological Monographs 27*, 221–245.

Ludwig, W., J.-L. Probst, and S. Kempe. 1996. Predicting the oceanic input of organic carbon by continental erosion. *Global Biogeochemical Cycles 10*, 23–41.

Lugo, A. E., and S. Brown. 1986. Steady state terrestrial ecosystems and the global carbon cycle. *Vegetatio 68*, 83–90.

Lugo, A. E., M. J. Sanchez, and S. Brown. 1986. Land use and organic carbon content of some subtropical soils. *Plant and Soil 96*, 185–196.

MacDonald, G. M., T. W. D. Edwards, K. A. Moser, R. Pienitz, and J. P. Smol. 1993. Rapid response of treeline vegetation and lakes to past climatic warming. *Nature 361*, 243–246.

Mann, L. K. 1986. Changes in soil carbon storage after cultivation. *Soil Science 142*, 279–288.

McGuire, A. D., J. M. Melillo, L. A. Joyce, D. W. Kicklighter, A. L. Grace, B. Moore, and C. J. Vorosmarty. 1992. Interactions between carbon and nitrogen dynamics in estimating net primary productivity for potential vegetation in North America. *Global Biogeochemical Cycles 6*, 101–124.

Melillo, J. M., A. D. McGuire, D. W. Kicklighter, B. Moore, C. J. Vorosmarty, and A. L. Schloss. 1993. Global climate change and terrestrial net primary production. *Nature 363*, 234–240.

Meybeck, M. 1982. Carbon, nitrogen, and phosphorus transport by world rivers. *American Journal of Science 282*, 401–450.

Nakane, K. 1976. An empirical formulation of the vertical distribution of carbon concentration in forest soils. *Japanese Journal of Ecology 26*, 171–174.

Norby, R. J., C. A. Gunderson, S. D. Wullschleger, E. G. O'Neill, and M. K. McCracken. 1992. Productivity and compensatory responses of yellow poplar trees in elevated CO_2. *Nature 357*, 322–324.

Oechel, W. C., S. Cowles, N. Grulke, S. J. Hastings, B. Lawrence, T. Prudhomme, G. Riechers, B. Strain, D. Tissue, and G. Vourlitis. 1994. Transient nature of CO_2 fertilization in arctic tundra. *Nature 371*, 500–503.

Oechel, W. C., S. J. Hastings, G. Vourlitis, M. Jenkins, G. Riechers, and N. Grulke. 1993. Recent change of Arctic tundra ecosystems from a net carbon dioxide sink to a source. *Nature 361*, 520–523.

Overpeck, J. T., P. J. Bartlein, and T. Webb. 1991. Potential magnitude of future vegetation change in eastern North America: Comparisons with the past. *Science 254*, 692–695.

Peterjohn, W. T., J. M. Melillo, F. P. Bowles, and P. A. Steudler. 1993. Soil warming and trace gas fluxes: Experimental design and preliminary flux results. *Oecologia 93*, 18–24.

Post, W. M., W. R. Emanuel, P. J. Zinke, and A. G. Stangenberger. 1982. Soil carbon pools and world life zones. *Nature 298*, 156–159.

Potter, C. S., J. T. Randerson, C. B. Field, P. A. Matson, P. M. Vitousek, H. A. Mooney, and S. A. Klooster. 1993. Terrestrial ecosystem production: A process model based on global satellite and surface data. *Global Biogeochemical Cycles 7*, 811–841.

Prentice, K. C., and I. Y. Fung. 1990. The sensitivity of terrestrial carbon storage to climate change. *Nature 346*, 48–51.

Quade, J., T. E. Cerling, and J. R. Bowman. 1989. Development of Asian monsoon revealed by marked ecological shift during the latest Miocene in northern Pakistan. *Nature 342*, 163–166.

Raich, J. W., and W. H. Schlesinger. 1992. The global carbon dioxide flux in soil respiration and its relationship to vegetation and climate. *Tellus 44B*, 81–99.

Rastetter, E. B., M. G. Ryan, G. R. Shaver, J. M. Melillo, K. J. Nadelhoffer, J. E. Hobbie, and J. D. Aber. 1991. A general biogeochemical model describing the response of the C and N cycles in terrestrial ecosystems to changes in CO_2, climate and N deposition. *Tree Physiology 9*, 101–126.

Sanford, R. L., J. Saldarriaga, K. E. Clark, C. Uhl, and R. Herrera. 1985. Amazon rain-forest fires. *Science 227*, 53–55.

Sarmiento, J. L., and E. T. Sundquist. 1992. Revised budget for the oceanic uptake of anthropogenic carbon dioxide. *Nature 356*, 589–593.

Schiffman, P. M., and W. C. Johnson. 1989. Phytomass and detrital carbon storage during forest regrowth in the southeastern United States piedmont. *Canadian Journal of Forest Research 19*, 69–78.

Schleser, G. H. 1982. The response of CO_2 evolution from soils to global temperature changes. *Zeitschrift für Naturforschung 37a*, 287–291.

Schlesinger, W. H. 1977. Carbon balance in terrestrial detritus. *Annual Review of Ecology and Systematics 8*, 51–81.

Schlesinger, W. H. 1982. Carbon storage in the caliche of arid soils: A case study from Arizona. *Soil Science 133*, 247–255.

Schlesinger, W. H. 1984. Soil organic matter: A source of atmospheric CO_2. In *The Role of Terrestrial Vegetation in the Global Carbon Cycle: Measurement by Remote Sensing*, G. M. Woodwell, ed. John Wiley and Sons, New York, pp. 111–117.

Schlesinger, W. H. 1985. The formation of caliche in soils of the Mojave desert, California. *Geochimica et Cosmochimica Acta 49*, 57–66.

Schlesinger, W. H. 1986. Changes in soil carbon storage and associated properties with disturbance and recovery. In *The Changing Carbon Cycle: A Global Analysis*, J. R. Trabalka and D. E. Reichle, eds. Springer-Verlag, New York, pp. 194–220.

Schlesinger, W. H. 1990. Evidence from chronosequence studies for a low carbon-storage potential of soils. *Nature 348*, 232–234.

Schlesinger, W. H., and J. M. Melack. 1981. Transport of organic carbon in the world's rivers. *Tellus 33*, 172–187.

Schulten, H.-R., and M. Schnitzer. 1993. A state of the art structural concept for humic substances. *Naturwissenschaften 80*, 29–30.

Seiler, W., and P. J. Crutzen. 1980. Estimates of gross and net fluxes of carbon between the biosphere and the atmosphere from biomass burning. *Climatic Change 2*, 207–247.

Shneour, E. A. 1966. Oxidation of graphitic carbon in certain soils. *Science 151*, 991–992.

Siegenthaler, U., and H. Oeschger. 1987. Biospheric CO_2 emissions during the past 200 years reconstructed by deconvolution of ice core data. *Tellus 39B*, 140–154.

Smith, T. M., R. Leemans, and H. H. Shugart. 1992. Sensitivity of terrestrial carbon storage to CO_2-induced climate change: Comparison of four scenarios based on general circulation models. *Climatic Change 21*, 367–384.

Smith, T. M., and H. Shugart. 1993. The transient response of terrestrial carbon storage to a perturbed climate. *Nature 361*, 523–526.

Spycher, G., P. Sollins, and S. Rose. 1983. Carbon and nitrogen in the light fraction of a forest soil: Vertical distribution and seasonal patterns. *Soil Science 135*, 79–87.

Strain, B. R., and J. D. Cure (eds.). 1985. *Direct Effects of Increasing Carbon Dioxide on Vegetation*. DOE/ER-0238, National Technical Information Service, Washington, D. C.

Tissue, D. T., and W. C. Oechel. 1987. Response of *Eriophorum vaginatum* to elevated CO_2 and temperature in Alaskan tussock tundra. *Ecology 68*, 401–410.

Tissue, D. T., R. B. Thomas, and B. R. Strain. 1993. Long-term effects of elevated CO_2 and nutrients on photosynthesis and rubisco in loblolly pine seedlings. *Plant, Cell and Environment 16*, 859–865.

Trumbore, S. E., O. A. Chadwick, and R. Amundson. 1996. Rapid exchange between soil carbon and atmospheric carbon dioxide driven by temperature change. *Science 272*, 393–396.

Van Cleve, K., W. C. Oechel, and J. L. Hom. 1990. Response of black spruce (*Picea mariana*) ecosystems to soil temperature modification in interior Alaska. *Canadian Journal of Forest Research 20*, 1530–1535.

Van Deusen, P. C. 1992. Growth trends and stand dynamics in natural loblolly pine in the southeastern United States. *Canadian Journal of Forest Research 22*, 660–666.

Wardle, D. A. 1992. A comparative assessment of factors which influence microbial biomass carbon and nitrogen levels in soil. *Biological Reviews 67*, 321–358.

West, D. C., T. W. Boyle, M. L. Tharp, J. J. Beauchamp, W. J. Platt, and D. J. Downing. 1993. Recent growth increases in old-growth longleaf pine. *Canadian Journal of Forest Research 23*, 846–853.

Whittaker, R. H., and G. E. Likens. 1973. Carbon in biota. In *Carbon and the Biosphere*, G. M. Woodwell and E. V. Pecan, eds. CONF-720510, National Technical Information Service, Washington, D. C., pp. 281–302.

Winkler, J. P., R. S. Cherry, and W. H. Schlesinger. 1996. The Q_{10} relationship of microbial respiration in a temperate forest soil. *Soil Biology and Biochemistry 28*, 1067–1072.

Wofsy, S. C., M. L. Goulden, J. W. Munger, S.-M. Fan, P. S. Bakwin, B. C. Daube, S. L. Bassow, and F. A. Bazzaz. 1993. Net exchange of CO_2 in a mid-latitude forest. *Science 260*, 1314–1317.

Wood, C. W., H. A. Torbert, H. H. Rogers, G. B. Runion, and S. A. Prior. 1994. Free-air CO_2 enrichment effects on soil carbon and nitrogen. *Agricultural and Forest Meteorology 70*, 103–116.

Wood, C. W., D. G. Westfall, and G. A. Peterson. 1991. Soil carbon and nitrogen changes on initiation of no-till cropping systems. *Soil Science Society of America Journal 55*, 470–476.

7

Grasslands and the Global Carbon Cycle: Modeling the Effects of Climate Change

D. O. HALL, J. M. O. SCURLOCK, D. S. OJIMA, AND W. J. PARTON

Abstract

The potential effects of climate change on grassland carbon budgets have received much less attention than those on forests. Changes in productivity and soil carbon need to be addressed, as well as the role of grassland burning in the global carbon cycle. Under a recent Scientific Committee on Problems of the Environment (SCOPE) collaborative project, we analyzed the sensitivity of global grassland ecosystems to modified climate and atmospheric CO_2 levels, concentrating on ecosystem dynamics rather than redistribution of grasslands. Worldwide, 31 grassland sites, temperate and tropical, were modeled under doubled-CO_2 climates projected by two different general circulation models. Results for climate change alone (without CO_2 effect) indicate that simulated soil C losses occur in most grassland regions, ranging from 0 to 14% of current soil C levels for the surface 20 cm. Direct CO_2 enhancement effects on decomposition and plant production tended to reduce the net impact of climate alterations alone. Detecting all these impacts will require a minimum of 16% change in net primary production and a 1% change in soil carbon. It is unclear whether grassland carbon stocks will increase overall under the new regimes, since changes in land use and biome area probably will result in a net source of carbon.

7.1 Introduction

Grassland ecosystems are far from uniform. They support myriad different species, both above and below ground, and are crucial in maintaining the ecological balance of a large part of the world, supporting diverse human and animal populations. The importance of the grassland environment, once recognized only by ecologists and conservationists, is now accepted by the growing numbers of scientists and decision makers concerned about human impact on the ability of grasslands to sustain many communities and their livestock. Ranging from the savannas of Africa to the prairies and steppes of North America and Russia, and to the converted grasslands of Latin America and Southeast Asia, they are an important habitat for much wildlife, many people, and many domestic animals. Approximately 20% of the world's population depends upon grasslands for their livelihood; they are subject to both commercial and noncommercial grazing as well as conversion to both dryland and irrigated agriculture.

Natural grasslands, both in the tropics and in temperate regions, play a major but poorly defined role in the global carbon cycle. Grasslands are one of the most widespread vegetation types worldwide, covering nearly one-fifth of the Earth's land surface, more than one-half of this area being within the tropics. At the regional level, for example, on the African continent, tropical grasslands may be more extensive than tropical forests. Worldwide, grasslands are likely to remain roughly constant in area for the near future; although some marginal parts are becoming desertified, conversion of forest to grazing land continues elsewhere. It is important to obtain more information about grassland ecosystems in their present form be-fore they are modified, either by global climate change or by increasingly intense management.

Change in management practices under current climate conditions has a greater influence on ecosystem carbon than does projected climate change (Burke et al., 1991). The changes in net primary production (NPP) are the result of increased decomposition or mineralization rates, but net ecosystem production (NEP) often decreases as a result of losses of carbon stocks in soils. Thus, a small increment in NPP of the plant system is counterbalanced by losses of soil carbon (Schimel et al., 1990).

It is possible that concurrent physiological changes resulting from increasing CO_2 levels (i.e., increased nitrogen use efficiency, NUE, and water use efficiency, WUE) will compensate for the negative NEP response. The observations of Owensby and colleagues (1993) suggest that both NUE and WUE are enhanced with increasing CO_2 levels, and that this enhanced plant production is manifest in greater root production where carbon can be more readily stabilized. In addition, the increased C:N ratio of the plant material will result in slower decomposition, stabilizing more soil carbon during the transient period.

7.2 Definition

Grasslands occur where seasonal drought prevents the development of extensive tree cover, as well as where our predecessors or contemporaries have cleared away forest to create grazing land. Natural or *climatically determined* grasslands include the savanna grasslands, which occur between the deserts of northern and southern Africa and the forests of the Congo Basin; also included are the North American prairie grasslands, and the dry steppes of Russia and Ukraine. The tropical savannas include a range of tree/grass systems whose dry season causes the grasses to die back and woody plants to shed their leaves. Grasslands may be defined as those with less than 10% tree cover; savannas have 10–50% woody plant cover and a well-developed grass layer in their unexploited state (Scholes and Hall, 1996).

Seminatural or *derived* grasslands, where forest has been cleared away and grass cover maintained by grazing, cutting, or burning, tend to be wetter and more productive. Many of the grasslands of South and Southeast Asia fall into this category. A third category of *agricultural* grasslands or pastures occurs where new grass species or varieties have been introduced and maintained by fertilization, irrigation, and so on. European lowland grazing lands are such an example.

7.3 Grassland Productivity, Burning, and Soil Carbon

The carbon in terrestrial vegetation and soils worldwide outweighs the amount found in the atmosphere and the ocean surface layers, all of which cycle carbon on annual to millennial time scales. Grassland ecosystems store most of their carbon in soils, where turnover times are relatively long (10–10,000

Table 7.1. *Net primary production, plant biomass, and area coverage for major world ecosystems (in Gt dry matter)*

	Total Land Only[a]	Tropical Forests	Temperate Forests	Boreal Forests	Tropical Woodlands	Temperate Woodlands	Tropical Grasslands	Temperate Grasslands	Agriculture	Marine
NPP										
Atjay (1979)	133 (9.0)[b]	30	8	7	13	5	26	4	15	—
Olson et al. (1983)	134 (8.9)	21	16	7	19	5	16	5	27	—
Whittaker & Likens	118 (7.8)	49	15	10	—[c]	6	14[d]	5	9	55
Plant biomass										
Atjay	1,244 (33)	542	174	205	88	54	58	20	10	—
Olson et al.	1,242 (34)	364	281	318	177	49	50	26	48	7
Whittaker & Likens	1,837 (40)	1,025	385	240	—[c]	50	60	14	14	4
Area (10^4 km^2)										
Atjay	149	15	6	9	10	5	12	13	16	—
Olson et al.	151	13	12	12	17	5	17	17	16	360
Whittaker & Likens	149	25	12	12	—[c]	9	15	9	14	361

Note: Many authorities still fail to agree upon the major categories of world ecosystems and their respective areas. For example, the area of "tropical forests" cited here is larger than recent estimates of approximately 8–10 × 10^6 km^2 for "tropical moist forests." Similarly, the distinction among "tropical woodlands," "tropical savannas," and "tropical grasslands" is not always clear.

[a]Also included, but not separately listed, are tundra, deserts, wetlands, polar, and extreme desert areas.

[b] = NPP in t/ha/yr or biomass in t/ha.

[c]Included in tropical forests.

[d]Categorized as savanna.

Source: Modified from Bolin (1986) and Woodwell (1986) and original references therein.

years); consequently, changes, though they may occur slowly, will be of considerable duration. Changes in grassland C storage can thus have a significant and long-lived effect on global C cycles, acting as a net source or sink for atmospheric CO_2.

Grassland carbon stocks, and the effects of projected climate changes on grasslands, have received relatively little attention compared with forests (Hall and Scurlock, 1991). For example, Melillo and colleagues (1993) suggest that tropical forests would account for about one-half of the global increase in NPP under elevated CO_2. However, various authors estimate that tropical grasslands alone occupy approximately 15 million km² and are nearly equal to tropical forests in terms of both land area and productivity, although the tropical forests contain 7–10 times as much standing biomass as tropical grasslands (Table 7.1). According to Scholes and Hall (1996), tropical grasslands, savannas, and woodlands, in the broad sense, occupy at least 16.1 million km², and the inclusion of 9 million km² of temperate grasslands totals a larger area than the 13–17 million km² of tropical forests alone.

Previous studies have suggested that the productivity of tropical savanna grasslands is 6.75 Pg C/yr (13.5 Pg dry matter) (Lieth, 1978), approximately 9% of total global terrestrial production. Scholes and Hall (1993) independently estimate NPP of between 3.2 and 10.8 Pg C/yr for tropical savannas and woodlands, or about one-half the net carbon fixation attributed to tropical forests. However, the proportion attributable to tropical grasslands could rise to more than 25% of total global terrestrial production if the minimum threefold underestimate of NPP found for three tropical grassland sites is generally applicable (Long et al., 1989). These findings have wide implications for the prediction of global carbon cycling and grassland ecosystem responses to climate changes. Taking into account that production in temperate grass ecosystems may also have been underestimated, owing to the difficulty of accounting for losses such as root exudation, the significance of natural grass ecosystems to total terrestrial production becomes apparent.

Fire plays an important role in the maintenance of many natural grasslands, both in temperate regions such as North America and in the tropics. Estimates of the widespread burning of tropical grasslands suggest that gross emissions of carbon from this source are substantial, on the order of 1.7 Pg/yr (Hao et al., 1990) or 2.4–4.2 Pg/yr (Hall and Scurlock, 1991). Scholes and Hall (1996) recently suggested that the fraction of aboveground biomass that is burned may have been overestimated in some cases, but agree that fire in tropical savannas and grasslands remains an important source of gross CO_2 emissions (0.9–1.7 Pg C) as well as trace gases. Frequent burning may result in a net loss of carbon from the soil, and volatilization of soil nitrogen may have feedback effects through primary productivity. The extent of carbonization of the standing biomass and annual NPP that is converted to long-lived charcoal depends on many factors and remains controversial. However, given the large areas of grasslands burned, especially in Africa (500 Mha or more annually), it is important to estimate the amount of carbon sequestered via this route (Scholes and Hall, 1996).

Estimates of the amount of carbon stored in grassland soils range from more than 10% of the world total (Eswaran et al., 1993) to 20% (Jenkinson et al., 1991) or even 30% (Anderson, 1991), that is, a range of 150–450 Pg out of a total of 1,500 Pg. Tropical savannas alone may account for 108 Pg of soil C, with a further 27 Pg C in aboveground woody plant biomass.

These figures may be compared with estimates in the range of 185–200 Pg for carbon in tropical forest soils (Jenkinson et al., 1991; Eswaran et al., 1993) and approximately 100 Pg for carbon in tropical forest biomass (Brown and Lugo, 1984). The effect of climate change and elevated CO_2 on grassland carbon cycling and soil C storage clearly merits the kind of effort that has been devoted to tropical forests.

It has been suggested that the majority of the "missing sink" for carbon in the global CO_2 balance may be accounted for by terrestrial vegetation rather than by oceanic uptake, and that such a sink may be located in both tropical and temperate regions (Tans et al., 1990; Enting and Mansbridge, 1991). It is therefore imperative to include grasslands with other major terrestrial ecosystems in attempts to model the global carbon cycle, in order to arrive at a well-founded policy toward climate change. It should be emphasized that the actual rate of climate change over the coming years will be crucial in determining whether ecosystems such as grasslands are able to adapt to climate changes.

7.4 Grassland Links with Climate and Climate Change

The seasonal distribution of rainfall is a major determinant of plant production in many semiarid and arid regions, where grasslands naturally occur. For example, in the North American Great Plains, most of the rainfall occurs during the growing season, from April to September. This permits greater biological productivity, more evapotranspiration (ET), and less runoff than a rainfall pattern more evenly distributed throughout the year, and accounts for a productive system despite the modest annual input of rain (250–800 mm). Many of the grasslands regions of the world share this characteristic. For example, it is widely assumed that potential changes in seasonal rainfall and temperature patterns in central North America and the African Sahel will have a greater impact on biological response and feedback to climate than changes in the overall amount of annual rainfall. Ecosystem model simulations of responses to climate change in the Great Plains demonstrated sensitivity of soil carbon storage and grassland biogeochemistry processes to seasonal distribution of precipitation changes (Schimel et al., 1990).

Grasslands may be among the earliest systems to exhibit the effects of climatic change (OIES, 1991). Sensitivity to climatic change may be a reflection of inadequate reserves of water or soil nutrients. Temperature increases will modify ecosystem process such as ET, decomposition, and photosynthesis. The combined effect of temperature and precipitation changes may alter the rates of ecosystem processes to offset one another or, alternatively, to act synergistically to amplify a positive or negative effect on overall C storage. Grassland ecosystem responses to increased atmospheric concentrations of CO_2 will be determined by the changes of increased plant production inputs, increased WUE, and modified nutrient availability.

Biospheric control over terrestrial ecosystem–atmosphere feedbacks is strong in grassland ecosystems because of the relationship between weather variations and biotic responses (Pielke and Avissar, 1990). Ecosystem changes in these regions resulting from changes in weather patterns or land use can alter processes controlling terrestrial C fluxes, such as plant production and decomposition (Schimel et al., 1990; Burke et al., 1991), as well as other key ecosystem processes, such as ET and trace gas production and consumption (Schimel et al., 1991; Ojima et al., 1993). The semiarid nature of the grassland climate regime and the extensive land-use changes that have taken place have accentuated the degree of susceptibility of the natural and managed grassland ecosystems.

The effect of changes in environmental variables such as temperature and water stress are relatively well understood; however, our understanding of nutrient constraints and the long-term effects of CO_2 fertilization are only now sufficiently well developed to allow evaluation of the changes in carbon fluxes in these grassland ecosystems (Schimel et al., 1991; Long, 1991; Owensby et al., 1993). The objective of the modeling study described in Section 7.5 is to characterize the factors controlling carbon dynamics in grassland ecosystems worldwide, and to evaluate the potential future response of

plant production and soil carbon to changes in climate and atmospheric CO_2.

7.5 Model Studies of Ecosystem Response

In order to study the impact of climate and atmospheric perturbations on soil organic matter and ecosystem dynamics, the Century model was employed. Century is a general model of plant–soil interactions that incorporates simplified representations of key processes relating to carbon assimilation and turnover (Figure 7.1). It has been described previously in detail (Parton et al., 1987, 1988, 1993).

Century simulates the dynamics of C and N for different plant–soil systems. Plant production in grasslands is a function of soil temperature and available water, limited by nutrient availability and a self-shading factor. The model includes the impact of fire and grazing on grassland ecosystems (Ojima et al., 1990; Holland et al., 1992). This model is linked to a soil organic matter submodel, which simulates the flow of C and nutrients through the different inorganic and organic pools in the soil, running on a monthly time step. Version 3.0 of Century was used for these model runs (Parton et al., 1993).

7.5.1 Modeling CO_2 Impact

The plant production parameters for both C_3 and C_4 grasslands under a doubled atmospheric CO_2 concentration were modified by changing production relative to potential evapotranspiration

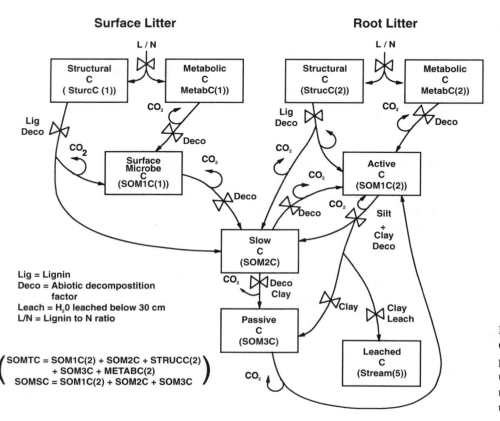

Figure 7.1: General structure of the Century model. Various management practices (e.g., fire and grazing) were used in the grassland simulations to represent more accurately actual land-use practices for the various sites.

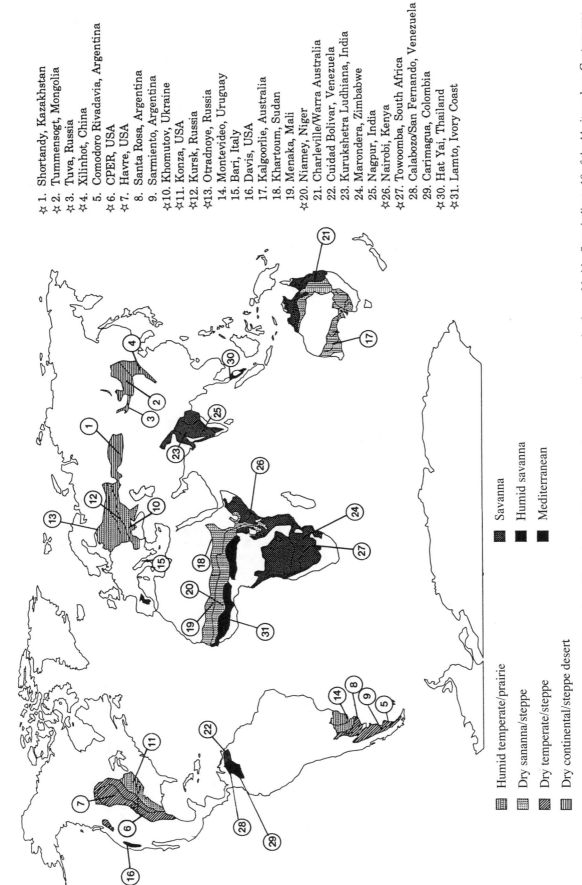

☆ 1. Shortandy, Kazakhstan
☆ 2. Tummensogt, Mongolia
☆ 3. Tuva, Russia
☆ 4. Xilinhot, China
☆ 5. Comodoro Rivadavia, Argentina
☆ 6. CPER, USA
☆ 7. Havre, USA
8. Santa Rosa, Argentina
9. Sarmiento, Argentina
☆10. Khomutov, Ukraine
☆11. Konza, USA
☆12. Kursk, Russia
☆13. Otradnoye, Russia
14. Montevideo, Uruguay
15. Bari, Italy
16. Davis, USA
17. Kalgoorlie, Australia
18. Khartoum, Sudan
19. Menaka, Mali
☆20. Niamey, Niger
21. Charleville/Warra Australia
22. Cuidad Bolivar, Venezuela
23. Kurukshetra Ludhiana, India
24. Marondera, Zimbabwe
25. Nagpur, India
☆26. Nairobi, Kenya
☆27. Towoomba, South Africa
28. Calabozo/San Fernando, Venezuela
29. Carimagua, Colombia
☆30. Hat Yai, Thailand
☆31. Lamto, Ivory Coast

▦ Humid temperate/prairie
▨ Dry sananna/steppe
▧ Dry temperate/steppe
▥ Dry continental/steppe desert

■ Savanna
■ Humid savanna
■ Mediterranean

Figure 7.2: Locations of the 31 grassland sites in this study and boundaries of Bailey ecoregions representing grasslands worldwide. Stars indicate 10 of the 11 sites where Century was validated.

Table 7.2. *Grassland sites modeled in this study*

Site, Country	Ecoregion (Bailey, 1989)	Lat./Long. (approx.)	Land Area Represented (10^6 km^2)	Annual Precip. (mm)	Mean Annual Temp. (°C)	Soil C (kg/m^2)	Above-ground C (g/m^2)	PET	DEC	N_{min}
Dry domain (300)										
1. Cold desert steppe division (331/333)										
Shortandy, Kazakhstan	331	52°N 71°E	0.804	351	1.4	7.37	73.1	96.3	0.003	3.34
Tummensogt, Mongolia	333	46°N 113°E	0.740	269	1.5	4.15	35.3	90.6	0.102	1.92
Tuva, Russia	333	52°N 94°E	0.109	214	−3.4	4.02	46.2	107.8	0.074	2.88
Xilinhot, China	333	44°N 117°E	0.441	360	−0.1	6.56	84.3	89.3	0.102	4.30
2. Dry steppe division (330) and temperate desert division (340)										
Comodoro Rivadavia, Argentina	342	49°S 68°W	0.454	222	12.7	1.13	19.1	90.5	0.172	1.74
CPER, USA	311/315	40°N 105°W	0.581	300	8.7	2.31	45.9	107.8	0.127	2.30
Havre, USA	331/332	49°N 110°W	1.450	312	5.5	3.23	33.9	107.0	0.087	1.74
Santa Rosa, Argentina	311/312	37°S 64°W	0.335	532	13.6	1.80	66.5	133.4	0.286	5.89
Sarmiento, Argentina	331/332	46°S 69°W	0.124	141	10.8	1.63	15.9	81.9	0.120	1.66
Humid temperate domain (200)										
3. Warm continental division (210) and prairie division (250)										
Khomutov, Ukraine	332	47°N 38°E	0.860	441	9.5	7.56	112.3	112.1	0.110	3.04
Konza, USA	251/255	39°N 97°W	0.781	818	12.9	5.78	184.8	137.8	0.271	5.77
Kursk, Russia	252	52°N 37°E	0.607	560	5.5	9.95	175.8	82.5	0.137	5.22
Otradnoye, Russia	212	61°N 30°E	1.181	543	3.8	6.09	106.3	72.4	0.104	3.27
Uruguay	254/255	35°S 56°W	0.529	936	15.1	6.48	89.5	93.5	0.346	6.76
4. Mediterranean division (260)										
Bari, Italy	262	41°N 17°E	0.084	574	15.8	3.17	107.6	118.0	0.233	4.53
Davis, USA	262	39°N 122°W	0.076	420	15.6	2.84	50.1	132.6	0.136	2.35

Dry domain (300) – dry savanna
5. Tropical/subtropical steppe division (310)

Kalgoorlie, Australia	312/315	31°S 122°E	1.813	255	17.2	3.70	40.1	141.8	0.143	2.12
Khartoum, Sudan	314	16°N 33°E	1.006	138	25.8	2.15	23.2	220.3	0.101	1.68
Menaka, Mali	314	16°N 2°E	0.846	290	29.7	1.23	35.5	225.3	0.156	1.93
Niamey, Niger	415	14°N 2°E	1.444	467	23.6	2.03	119.5	203.5	0.290	4.60

Humid tropical domain (400)
6. Savanna division (410)

Charleville/Warra, Australia	411/416	26°S 146°E	1.138	489	210.5	4.88	92.6	169.6	0.199	3.84
Ciudad Bolivar, Venezuela	416	8°N 64°W	0.129	981	27.5	4.51	330.6	143.7	0.506	8.86
Kurukshetra/Ludhiana, India	412	31°N 76°E	0.353	715	24.4	3.33	143.4	214.2	0.305	5.10
Marondera, Zimbabwe	411/415	18°S 31°E	2.995	819	18.3	4.60	221.8	106.7	0.294	6.54
Nagpur, India	413/415	21°N 79°E	1.184	1203	26.9	5.12	228.6	191.2	0.398	7.51
Nairobi, Kenya	413/416	1°S 36°E	1.565	680	19.0	6.24	209.5	87.1	0.322	4.14
Towoomba, South Africa	314	25°S 29°E	0.636	630	26.5	2.47	119.9	131.2	0.339	3.91

7. Humid savanna (414/415) and rainforest division (420)

Calabozo/San Fernando, Venezuela	414/415	9°N 67°W	0.044	1318	28.1	7.52	366.5	138.6	0.505	9.07
Carimagua, Colombia	414	4°N 72°W	0.185	2338	26.6	1.58	226.11	26.6	0.737	3.77
Hat Yai, Thailand	423	6°N 101°E	0.797	1540	27.6	2.22	461.6	144.7	0.703	7.76
Lamto, Ivory Coast	414	6°N 5°W	0.543	1170	27.9	1.78	305.0	151.8	0.628	5.69

Note: PET = potential evapotranspiration; DEC = modeled decomposition factor; N_{min} = nitrogen mineralization. Each site is assumed to be representative of its surrounding area on the Bailey (1989) ecoregion map. Refer to Ojima et al. (1996) for individual site references.

(PET) and to NUE. The C_3-specific modifications related to CO_2-induced changes of the assimilation efficiency of the rubisco C uptake pathway were not simulated. For the current analysis we uniformly implemented the impact of increased atmospheric CO_2 concentrations on all grasslands. The magnitude of the CO_2 effect causes a maximum of 20% increase in plant production with a doubling of atmospheric CO_2 concentration. The effect of modified NUE and PET on plant production is a simple linear effect of these processes on CO_2-stimulated production. In addition, changes in NUE affect litter quality, consequently, higher NUE results in a slower decomposition of this material under a given temperature and moisture regime.

Thus increasing CO_2 results in increased NUE and WUE in Century. Although Century shows increased NPP and carbon storage when CO_2 is increased, the N cycle interactions are again critical to understanding the model's response. In theory, NPP can increase by 40% at doubled CO_2; in reality, however, it increases by approximately 20%. As NPP increases, carbon storage rises, resulting in sequestration of N. Thus, increasing NUE results in wider tissue C:N ratios in detritus and a larger sink for N in soils. The steady-state increase in NPP reflects largely the change in NUE. However, this results in an additive effect when climate change and CO_2 are considered together, because the N released by warming can be used at the higher CO_2-fertilized efficiencies.

The synergism of these two effects suggests that CO_2 would reduce the losses of soil carbon caused by warming, and in some systems could even cause carbon to be sequestered (Parton et al., 1995). Responsiveness to CO_2 is also influenced by N inputs, as systems with high atmospheric N inputs can also show sustained responses to CO_2. Similar results from the G'-DAY model (Comins and McMurtrie, 1993) support this analysis of mechanisms, and highlight the importance of understanding the N cycle and ecosystem N budgets in modeling the response of terrestrial ecosystems to climate and CO_2.

7.5.2 Regional Ecosystem Modeling

We divided the grassland regions of the world into seven ecoregions (Figure 7.2) based on a scheme developed by Bailey (1989). Certain modifications have been made to Bailey's original ecoregion classification to accommodate differences in temperature (T) and rainfall (PPT) among the 31 grassland sites used in this analysis (Table 7.2). Site characteristics include mean annual T, mean annual PPT, aboveground net primary production (ANPP), and soil C levels to 20 cm depth.

A summary of current annual PPT and change in annual PPT as simulated by the general circulation model (GCM) is given in Table 7.2 and Figure 7.3. The global network of grassland sites had a range of mean annual PPT of less than 100 to greater than 2,000 mm. The mean annual T ranged from less than $-2°C$ to greater than $25°C$. The climates ranged from cold and dry to wet and warm (Table 7.3). For regional simulations, we selected representative sites and used site-specific climate and soil characteristics to simulate equilibrium current grassland conditions (Table 7.4). These initial conditions were then used to simulate ecosystem levels of soil C and plant production under perturbed CO_2 and climate patterns.

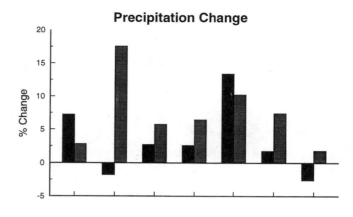

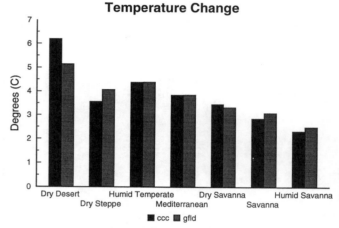

Figure 7.3: Changes in temperature and precipitation for each ecoregion under CCC and GFDL scenarios.

Site-specific soil texture information was used for the 11 sites where Century had been validated (Figure 7.2). For the other 20 sites, and for extrapolation of the regional estimates of plant and soil C, zonal estimates of soil texture were used. Land management practices, such as grazing intensity and fire frequency, were applied to a site according to historical land use. The management scenarios are described by Parton and colleagues (1993) for the 11 sites used in that study; the remaining sites used similar land management practices characteristic of the region. Century simulations for each site were run to equilibrium (approximately 5,000 model years) based on current climatologies.

7.5.3 Climatology

Ecosystem sensitivity to the temporal and regional resolution of climate change was evaluated by modifying the monthly weather record for the past 25 years using output from two high-resolution GCMs: the Canadian Climate Center model (CCC; Boer, 1988) and the Geophysical Fluid Dynamics Laboratory model (GFDL; as used by IPCC in 1990; Mitchell et al., 1990). We tested the sensitivity of grassland ecosystems to the following global change effects:

• Climate change effects (+CC, i.e., alterations to monthly mean temperature and precipitation)

Table 7.3. *Characteristics of the seven grassland regions*

Biome Type	Mean Annual Precipitation (mm)	Mean Annual Temperature (°C)	Mean Annual PET (mm)	PPT/PET[a]	Global Area (10^6 km²)	Vegetation Type
Cold desert steppe	299	−0.3	960	0.334	2.095	C_3
Temperate steppe	301	10.2	1,041	0.294	2.943	C_3/C_4
Humid temperate	700	9.3	997	0.680	3.958	C_4
Mediterranean	497	15.7	1,253	0.401	0.161	C_3/C_4
Dry savanna	387	24.1	1,977	0.165	5.109	C_4
Savanna	788	23.4	1,491	0.554	7.990	C_4
Humid savanna	1,555	27.4	1,404	1.195	1.708	C_-

[a]Ratio of annual precipitation to PET rate.

- Doubled-CO_2 response (+CO_2, i.e., increases in plant production resulting from changes in WUE, such as modified PET) and NUE resulting from elevated atmospheric CO_2 levels
- Combined effect of climate change and doubling of atmospheric CO_2 (CC + CO_2)

For each grassland site, a current weather file of monthly precipitation and monthly mean maximum and minimum temperatures was created using existing weather station data from the site itself or from a nearby meteorological station. When data were not available from a researcher at the site, we relied on the World Weather Disk (Weather Disk Associates, Inc., National Climate Data Center, USA). We used a 25-year weather record as the base climate to simulate the equilibrium grassland for each site. In order to generate the doubled-CO_2 climate, we spatially interpolated GCM grid values of projected doubled-CO_2 climate changes in monthly temperature and PPT for each site, based on the actual model output made for the 1990 IPCC report. The climate change inputs were based on the business-as-usual scenarios (see appendix of Houghton et al., 1990) for the CCC and the GFDL high-resolution runs. We applied these projected monthly values in a linear fashion

in a 50-year ramp (Figure 7.4). We used a 50-year time span, although the time to achieve an equivalent CO_2 doubling may be significantly greater than this (see, e.g., Kattenberg et al., 1996, fig. 6.20). The 50-year ramp was generated by taking the projected monthly climate changes and dividing these by 50. This incremental change then was added to each respective month during the 50-year ramp. We applied the projected monthly temperature changes equally to the minimum and the maximum mean monthly temperature values used by the Century model.

7.5.4 Regional Climate Changes and CO_2 Perturbations

The following section describes the ecosystem results for the grassland ecoregions (Parton et al., 1993). Overall, changes in ANPP in the climate change runs were correlated to changes in rainfall. Cases where rainfall amounts declined resulted in a lower simulated ANPP. Differences between the two GCM scenarios were observed primarily in the differences between projected rainfall amounts.

The impact of modifying both the climate and atmospheric CO_2 (CC + CO_2) resulted in enhancement of ANPP over simu-

Table 7.4. *Simulated annual plant production, soil C, abiotic decomposition, and N mineralization for all seven grassland regions prior to climate changes*

Biome Type	Aboveground Production (g/m2)	Belowground Production (g/m2)	Total Soil C (0–20 cm) (g/m2)	Net N Mineralization g N/m2/y	Annual Abiotic Decomposition Rate (0–1)
Cold desert steppe	59.7	92	5,530	3.11	0.093
Temperate steppe	36.4	59	2,020	2.67	0.158
Humid Temperate	133.8	161	7,171	4.81	0.194
Mediterranean	78.8	103	3,005	3.44	0.184
Dry savanna	54.6	85	2,275	2.58	0.173
Savanna	190.9	222	4,306	5.73	0.339
Humid savanna	339.6	334	3,273	6.61	0.644

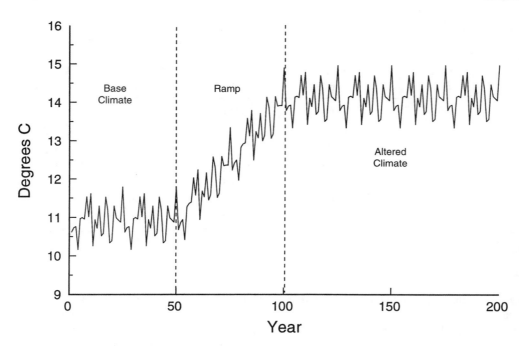

Figure 7.4: Example of climate change weather data used in the Century climate change model runs, for Sarmiento, Argentina. A repeating 25-year pattern is used throughout the Century simulations. Two 25-year cycles of weather data are followed by 50 years (two cycles) of climate change up to the doubled-CO_2 weather, which then continues in similar 25-year cycles.

lations of CC alone. The tropical simulations (dry savanna, savanna, and humid savanna) all resulted in increased ANPP with CC + CO_2 scenarios, regardless of GCM. Most of the temperate sites had increased ANPP, with the exception of nearly all the cold desert steppe sites and the Santa Rosa site in the temperate steppe region.

Plant production varied according to modifications in rainfall under the altered climate and according to altered N mineralization rates. The two GCMs differed in predictions of rainfall for a doubling of CO_2, and this difference was reflected in plant production. Soil decomposition rates responded most predictably to changes in temperature. Direct CO_2 enhancement effects on soil C loss and plant production tended to reduce the net impact of climate alterations alone.

In most cases, the simulations that combined CO_2 and CC modifications resulted in a net enhancement of soil C levels over those in the CC simulations alone. This trend did not occur in the extremely dry sites in both the temperate and the tropical regions.

Globally, the impact of climate changes alone on grassland soil C is estimated to result in a loss of 3–4 Pg C over a 50-year period. When a doubled atmospheric CO_2 concentration is modeled under current climatic conditions, the effect is to enhance plant production and to increase soil C storage by 2 Pg. When both climate change and direct CO_2 effects are considered, the resulting estimate is a soil C loss of 1–2 Pg C over the 50-year simulation. These estimates were based on current grassland areas and did not include any modifications related to potential land cover or land management changes.

Even though these losses are relatively small, the C pool in these ecosystems is relatively large. The potential for losses can be large if desertification or conversion to cropland occurs. Such losses can happen rapidly and take a long time to recover, especially if changes take place in plant community structure.

7.6 Conclusions

Two major controls related to climate factors are ET and decomposition, which affect not only rates of plant productivity but also soil C levels. The net result of the CO_2 and climate change impacts will depend on the differential changes in decomposition rates resulting from climate changes, any litter quality changes, and on the changes to decomposition owing to altered soil water dynamics resulting from plant-mediated changes in ET rates. Soil C levels will then be determined by the net changes in decomposition rates, plant production changes resulting from climate and CO_2-mediated effects, and rates of plant residue inputs into the soil system.

However, it remains unclear whether grassland carbon stocks may be increasing or decreasing, since opposing processes are taking place. Model results suggest that increasing atmospheric CO_2 will result in modest additional carbon storage, predominantly in certain tropical savannas where projected changes in climate will neither limit primary production nor excessively stimulate decomposition. However, it is precisely these areas that are most subject to management pressures of overgrazing and overfrequent burning, which can lead to degradation, reduced productivity, and loss of carbon through erosion and oxidation.

By the time atmospheric CO_2 has doubled, changes in biome distribution and changes in land use brought about by climate change will also be important, although grasslands are likely to remain constant or even to increase in area. However, the replacement of one biome type by another will probably result in a source rather than a sink of carbon. For example, the loss of carbon from forest biomass will take place much more rapidly than the accumulation of soil C in a grassland that replaces it. Future studies with Century need to address these issues and place them in the context of past, present, and future conversions among forests, grasslands, and agriculture. What is required is a series of links between process-based models such as Century and models of vegetation redistribution such as BIOME (Prentice et al., 1992).

Maintenance of grasslands worldwide through sustainable management is therefore imperative, both for carbon sequestration and in order to maintain biodiversity and human capacity. Many millions of agropastoralists already depend on grasslands for their livelihood, yet they are under threat from degradation and desertification resulting from overexploitation. These marginal environments, which also support a surprisingly high biological diversity, will be under increasing stress as climate changes in the future.

Acknowledgments

Data synthesis and model validation were made possible by the many collaborators involved with the Scientific Committee on Problems of the Environment (SCOPE) Project "Effects of Climate Change on Production and Decomposition in Coniferous Forests and Grasslands," the UN Environment Programme, the U.S. National Institute for Global Environmental Change, and the U.S. Department of Energy.

Current model development was funded by the NASA Earth Observing System project NACW-2662 and the U.S. Department of Energy. We extend our thanks to the many research staff at the Natural Resource Ecology Laboratory, Colorado State University, who have worked on the model, and to the National Center for Atmospheric Research for access to GCM output data.

References

Ajtay, G. L., P. Ketner, and P. Duvigneaud. 1979. Terrestrial primary production and phytomass. In *The Global Carbon Cycle*, B. Bolin, E. T. Degens, S. Kempe, and P. Ketner, eds. SCOPE 13, John Wiley and Sons, Chichester, U.K., 129–181.

Anderson, J. M. 1991. The effects of climate change on decomposition processes in grassland and coniferous forests. *Ecological Applications 1*, 326–347.

Bailey, R. G. 1989. Explanatory supplement to ecoregions map of the continents. *Environmental Conservation 16*, 307–309.

Boer, G. J. 1988. Some results concerning the effect of horizontal resolution and gravity-ware drag on simulated climate. *Journal of Climate 1*, 789–806.

Bolin, B. 1986. How much CO_2 will remain in the atmosphere? In *The Greenhouse Effect, Climate Change, and Ecosystems*, B. Bolin, E. T. Degens, S. Kempe, and P. Ketner, eds. SCOPE 29, John Wiley and Sons, Chichester, U.K., 93–155.

Brown, S., and A. E. Lugo. 1984. Biomass of tropical forests: A new estimate based on forest volumes. *Science 223*, 1290–1293.

Burke, I. C., T. G. F. Kittel, W. K. Lauenroth, P. Snook, C. M. Yonker, and W. J. Parton. 1991. Regional analysis of the Central Great Plains: Sensitivity to climate variability. *Bioscience 41*, 685–692.

Comins, H. N., and R. E. McMurtrie. 1993. Long-term response of nutrient-limited forests to CO_2 enrichment; equilibrium behavior of plant-soil models. *Ecological Applications 3*, 666–681.

Enting, I. G., and J. V. Mansbridge. 1991. Latitudinal distribution of sources and sinks of CO_2: Results of an inversion study. *Tellus 43B*, 156–170.

Eswaran, H., E. van den Berg, and P. Reich. 1993. Organic carbon in soils of the world. *Soil Science Society of America Journal 57*, 192–194.

Hall, D. O., and J. M. O. Scurlock. 1991. Climate change and productivity of natural grasslands. *Annals of Botany 67* (suppl.), 49–55.

Hao, W. M., M. H. Liu, and P. J. Crutzen. 1990. Estimates of annual and regional releases of CO_2 and other trace gases to the atmosphere from fires in the tropics. In *Fire in the Tropical Biota*, J. G. Goldammer, ed. Ecological Studies, Vol. 84., Springer-Verlag, Berlin, Germany, pp. 440–462.

Holland, E. A., W. J. Parton, J. K. Detling, and D. L. Coppock. 1992. Physiological responses of plant populations to herbivory and their consequences for ecosystem nutrient flows. *American Naturalist 140*, 685–706.

Houghton, J. T., G. J. Jenkins, and J. J. Ephraums (eds.). 1990. *Climate Change: The IPCC Scientific Assessment*. Cambridge University Press, Cambridge, U.K.

Jenkinson, D. S., D. E. Adams, and A. Wild. 1991. Model estimates of CO_2 emissions from soil in response to global warming. *Nature 351*, 304–306.

Kattenberg, A., F. Giorgi, H. Grassl, G. A. Meehl, J. F. B. Mitchell, R. J. Stouffer, T. Tokioka, A. J. Weaver, and T. M. L. Wigley. 1996. Climate models – Projections of future climate. In *Climate Change 1995: The Science of Climate Change, Contribution of Working Group I to the Second Assessment Report of the Intergovernmental Panel on Climate Change*, J. T. Houghton, L. G. Meira Filho, B. A. Callander, N. Harris, A. Kattenberg, and K. Maskell, eds. Cambridge University Press, New York, pp. 285–357.

Lieth, H. F. H. (ed.). 1978. *Patterns of Primary Productivity in the Biosphere*. Hutchinson Ross, Stroudsberg, Penn.

Long, S. P. 1991. Primary production in grasslands and coniferous forests with climate change: An overview. *Ecological Applications 1*, 139–156.

Long, S. P., E. Garcia Moya, S. K. Imbamba, A. Kamnalrut, M. T. F. Piedade, J. M. O. Scurlock, Y. K. Shen, and D. O. Hall. 1989. Primary productivity of natural grass ecosystems of the tropics: A reappraisal. *Plant and Soil 115*, 155–166.

Melillo, J. M., A. D. McGuire, D. W. Kicklighter, B. Moore, C. J. Vorosmarty, and A. L. Schloss. 1993. Global climate change and terrestrial net primary production. *Nature 363*, 234–240.

Mitchell, J. F. B., S. Manabe, T. Tokioka, and V. Meleshko. 1990. Equilibrium climate change. In *Climate Change: The IPCC Scientific Assessment*, J. T. Houghton, G. J. Jenkins, and J. J. Ephraums, eds. Cambridge University Press, Cambridge, U.K., pp. 131–172.

Office for Interdisciplinary Earth Studies (OIES). 1991. *Arid Ecosystems Interactions: Recommendations for Drylands Research in the Global Change Research Program*. OIES Report 6, University Corporation for Atmospheric Research, Boulder, Colorado.

Ojima, D. S., W. J. Parton, M. B. Coughenour, J. M. O. Scurlock, T. B. Kirchner, T. G. F. Kittel, D. O. Hall, D. S. Schimel, T. Seastedt, E. Garcia Moya, T. G. Gilmanov, T. R. Seastedt, A. Kamnalrut, J. I. Kinyamario, S. P. Long, J.-C. Menaut, O. E. Sala, R. J. Scholes, and J. A. van Veen. 1996. Impact of climate and atmospheric CO_2 changes on grasslands of the world. In *Effect of Climate Change on Production and Decomposition in Coniferous Forests and Grasslands*, A. Breymeyer, D. O. Hall, J. M. Melillo, and G. Agren, eds. Chap. 12, SCOPE Vol. 56, John Wiley and Sons, Chichester, U.K. (in press).

Ojima, D. S., W. J. Parton, D. S. Schimel, and C. E. Owensby. 1990. Simulated impacts of annual burning on prairie ecosystems. In *Fire in the North American Prairies*, S. L. Collins and L. Wallace, eds. University of Oklahoma Press, Norman, Okla.

Ojima, D. S., D. W. Valentine, A. R. Mosier, W. J. Parton, and D. S. Schimel. 1993. Effect of land use change on methane oxidation in temperate forest and grassland soils. *Chemosphere 26*, 675–685.

Olson, J., J. A. Watts, and L. J. Allison. 1983. *Carbon in Live Vegetation of Major World Ecosystems.* Report TR 004, DOE/NRB-0037, U. S. Department of Energy, Washington, D.C.

Owensby, C. E., P. I. Coyne, J. M. Ham, T. M. Aven, and A. K. Knapp. 1993. Biomass production in a tallgrass prairie ecosystem exposed to ambient and elevated levels of CO_2. *Ecological Applications 3*, 644–653.

Parton, W. J., D. S. Schimel, C. V. Cole, and D. S. Ojima. 1987. Analysis of factors controlling soil organic matter levels in great plains grasslands. *Soil Science Society of America Journal 51*, 1173–1179.

Parton, W. J., J. M. O. Scurlock, D. S. Ojima, T. G. Gilmanov, R. J. Scholes, D. S. Schimel, T. Kirchner, H.-C. Menaut, T. Seastedt, E. Garcia Moya, A. Kamnalrut, and J. L. Kinyamario. 1993. Observations and modeling of biomass and soil organic matter dynamics for the grassland biome worldwide. *Global Biogeochemical Cycles 7*, 785–809.

Parton, W. J., J. M. O. Scurlock, D. S. Ojima, D. S. Schimel, D. O. Hall, M. B. Coughenour, E. Garcia Moya, T. G. Gilmanov, A. Kamnalrut, J. I. Kinyamario, T. Kirchner, S. P. Long, J. C. Menaut, O. E. Sala, R. J. Scholes, and J. A. van Veen. 1995. Impact of climate change on grassland production and soil carbon worldwide. *Global Change Biology 1*, 13–22.

Parton, W. J., J. W. B. Stewart, and C. V. Cole. 1988. Dynamics of C, N, P and S in grassland soils: A model. *Biogeochemistry 5*, 109–131.

Pielke, R. A., and R. Avissar. 1990. Influence of landscape structure on local and regional climate. *Landscape Ecology 4*, 133–135.

Prentice, I. C., W. Cramer, S. P. Harrison, R. Leemans, R. A. Monserud, and A. M. Solomon. 1992. A global biome model based on plant physiology and dominance, soil proportion and climate. *Journal of Biogeography 19*, 117–134.

Schimel, D. S., W. J. Parton, T. G. F. Kittel, D. S. Ojima, and C. V. Cole. 1990. Grassland biogeochemistry: Links to atmospheric processes. *Climatic Change 17*, 13–25.

Schimel, D. S., T. G. F. Kittel, and W. J. Parton. 1991. Terrestrial biogeochemical cycles: Global interactions with the atmosphere and hydrology. *Tellus 43*, 188–203.

Scholes, R. J., and D. O. Hall. 1996. The carbon budget of tropical savannas, woodlands and grasslands. In *Effect of Climate Change on Production and Decomposition in Coniferous Forests and Grasslands*, A. Breymeyer et al., eds. Chap. 4, SCOPE Vol. 56, John Wiley and Sons, Chichester, U.K.

Tans, P. P., I. Y. Fung, and T. Takahashi. 1990. Observational constraints on the global atmospheric CO_2 budget. *Science 247*, 1431–1438.

Whittaker, R. H., and G. E. Likens. 1975. Primary production: The biosphere and man. In *Primary Productivity of the Biosphere*, H. Lieth and R. H. Whittaker, eds. Ecological Studies, Vol. 14, Springer-Verlag, New York.

Woodwell, G. M. 1986. On the limits of nature. In *The Global Possible*, R. Repetto, ed. Yale University Press, New Haven, Conn.

8

Constraints on the Atmospheric Carbon Budget from Spatial Distributions of CO_2

I. G. ENTING

Abstract

Extensive sampling networks have been established to determine the space–time distributions of greenhouse gases so that these data can be used to provide information about the sources and sinks. However, the problem of deducing sources and sinks from concentration data is an ill-conditioned (and often underdetermined) problem and as such is subject to large amplification of errors in observations or models. Various techniques that have been introduced to address this problem are reviewed. Particular attention is given to techniques of Bayesian synthesis inversion that can provide estimates of the uncertainty for the sources that are deduced.

8.1 The Context

The atmospheric budgets of radiatively active gases such as carbon dioxide (CO_2), methane (CH_4), and nitrous oxide (N_2O) remain subject to very considerable uncertainty. This uncertainty exists despite a wide range of observational and theoretical approaches that have been used in attempts to resolve the ambiguities. One particularly important approach has been to estimate the strengths of the various source and sink processes by using trace gas concentrations from global sampling networks. The principle is that the spatial distributions of concentrations constrain the spatial distributions of sources and sinks. These constraints in turn imply constraints on the possible source and sink processes.

In principle, determining sources and sinks from the spatial distributions of trace gas concentrations (technically called *inversion*) provides a "snapshot" of the distribution of sources, regardless of the processes that produce the sources. This should make it possible to detect short-term variability on time scales of approximately one year. Such variability is not captured by other approaches such as global carbon cycle models and ocean general circulation models (GCMs). This "snapshot approach" may make it possible to detect previously unexpected processes such as those arising from feedbacks on the carbon cycle. In practice, these desirable characteristics may apply on regional scales. On the global scale, however, the uncertainties are so large that the interpretation of spatial distributions of CO_2 plays only a minor role in reducing the uncertainties in the global carbon budget.

Inversion requires the use of an atmospheric transport model. The simplest example is the one-dimensional diffusion model used by Bolin and Keeling (1963). More complicated models are the two-dimensional advective-diffusive models used in studies such as those by Pearman and colleagues (1983), Pearman and Hyson (1986), Enting and Mansbridge (1989, 1991), and Tans and colleagues (1989). A more sophisticated approach is the use of three-dimensional models with transport fields either based on observed winds (Keeling et al., 1989) or derived from GCMs (Fung et al., 1983).

8.2 The Theory

8.2.1 Representation

A key issue in the discussion of inversion techniques is the way in which the distributions of sources and sinks are represented. One possibility is to consider spatial distributions, $S(x,y)$, as the basis representations. In a numerical calculation, the discretization of such distributions is in terms of grid points. An alternative is to consider a spectral representation with coefficients S_{mn} multiplying spherical harmonics $P_{nm}(x)$ $\exp(imy)$ where

$$S(x,y) = \Sigma_{mn} S_{mn} P_{nm}(x) \exp(imy) \qquad (8.1)$$

Numerical calculations will involve a truncation of the spectral expansion.

An alternative to the above representations, which are linked specifically to the geometry of the Earth, is a representation related to actual processes. The source distribution specified in terms of processes is linked to the spatial distribution through the spatial distributions of the various source components

$$S(x,y) = \Sigma_a S_a(x,y) \qquad (8.2)$$

where the sum is over distinct processes, indicated by a. When using this approach, it must be remembered that the results are conditional on the assumptions embodied in the choices of spatial distributions of source components.

8.2.2 Mass-Balance Inversions

There are two broad classes of techniques for actually calculating the source/sink distributions implied by concentration distributions. These can be denoted *deterministic* and *statistical*. The main deterministic technique is known as the *mass-balance* technique. It requires that the concentration distribution be specified as a function of time at precisely the set of locations at which the sources and sinks are to be deduced. The main application has been in deducing the surface distributions of sources from the surface distributions of CO_2 concentrations.

The mass-balance equation can be written as

$$\frac{d(C(x,y,z,t))}{dt} = T[C(x',y',z',t)] + S(x,y,z,t) \qquad (8.3)$$

where C is the tracer concentration and the operator T defines the effect of atmospheric transport. (Generally, we assume T is linear in the concentrations. This greatly simplifies many other inversion techniques but is largely irrelevant for the mass-balance technique.)

When considering the problem of deducing surface sources from surface concentrations, the mass-balance equation is applied in two different ways. It is used throughout the free atmosphere (usually with $S = 0$) to deduce dC/dt, which is integrated numerically to define the time evolution of concentrations in the free atmosphere. At the surface (i.e., $z = 0$), $C(x,y,0,t)$ is determined and therefore $d/dt\ C(x,y,0,t)$ is also known. Thus at any time, a knowledge of the full concentration distribution enables us to determine the transport contribution, $T[(x',y',z',t)]$, and to use the mass-balance equation to determine $S(x,y,0,t)$.

Since the mass-balance equation is applied on a point-by-point basis, the technique is invariably implemented by using a grid-point representation. The method is inherently time-dependent. Periodic solutions are obtained only for the special case of concentrations that have a periodic component and, optionally, a globally uniform trend.

Although the method has been applied primarily to deduce surface sources, it can also be applied in more general situations. Some possibilities are:

1. The method could be applied to deduce sources at all points in the atmosphere if the data were available.
2. For deducing sources at the surface, given surface concentration data, the method requires that the source in the free atmosphere be known. The method is not restricted to the case where this source is known to be zero. Thus, if the source of CO_2 from oxidation of carbon monoxide in the free atmosphere is calculated within the model, then the mass-balance approach can be used to deduce the surface sources.
3. In principle, the method could be applied to some subset of surface sites. Enting and Mansbridge (1991) excluded the Antarctic region from their surface inversion in order to ensure zero sources there; however, this result was achieved at the expense of an undesirable degree of instability in the estimated sources for nearby southern regions.

8.2.3 Inversion by Fitting Distributions

The *distribution fitting* methods work by expressing the sources as a sum of components:

$$S(x,y,z,t) = \Sigma_a S_a(x,y,z,t) = \Sigma_a c_a s_a(x,y,z,t) \quad (8.4)$$

Each normalized source component, $s_a(x,y,z,t)$, has a characteristic atmospheric response, $r_a(x,y,z,t)$. An atmospheric transport model is used to calculate these responses. Then the source strengths, c_a, are estimated by fitting a linear combination of responses to the observed data as

$$C(x_j,y_j,z_j,t_j) = \Sigma_a c_a r_a(x_j,y_j,z_j,t_j) \quad (8.5)$$

These methods rely on the linearity of the transport operator. The variations between methods of this type arise from differences in how the sources are discretized, how the fitting is performed, and how the time variation is handled.

One approach to the time variation is to consider cases that are periodic apart from a uniform global trend in the concentration and then decompose the problem as a Fourier series in terms of frequencies f_n, giving

$$C(x_j,y_j,z_j,f_n) = \Sigma_a c_a r_a (x_j,y_j,z_j,f_n) \quad (8.6)$$

This form was used by Enting and colleagues (1993, 1995). An alternative is to consider a fully time-dependent problem. We express the sources as

$$S(x,y,z,t) = \Sigma_{a,k} c_a(t_k) s_a(x,y,z,t_k) \quad (8.7)$$

and define the generalized responses of the form $r_a(x,y,z,t_1,t_2)$, which is the response at time t_1 to a type "a" source at time t_2.

These responses are used to express the concentrations as a sum over time t_k

$$C(x_j,y_j,z_j,t) = \Sigma_{a,k} c_a(t_k) r_a(x_j,y_j,z_j,t,t_k) \quad (8.8)$$

This type of approach was described by Prather and Bloomfield (1992) and Bloomfield and colleagues (1994), with time discretized at one-year intervals.

8.2.4 Statistical Approaches

The approaches described by equations (8.6) and (8.8) can be expressed in statistical terms as regression analyses. A weighted least-squares fit to the particular equation is required. This corresponds to the assumption that the observations are subject to independent errors. Equation (8.8) has the additional characteristic of being expressed in the time domain. In this case, the regression equations can be solved recursively. (As pointed out by Young, 1984, simple examples of the recursive formulation of regression were investigated by Gauss.) Such recursive regression formulations can be expressed in the more general terminology of Kalman filtering.

In a Kalman filtering formulation, the estimation problem becomes one of estimating the elements of a state vector that describes the system. In translating formulation (8.8) into Kalman filtering terms, the state vector at each time is represented by the strengths of each source component at *every* past time. An alternative form of state vector that would give a complete description would incorporate the current source strengths and the current concentrations throughout the whole of the atmosphere, not merely at those points where observations are made.

Some applications of Kalman filtering to atmospheric inversion problems have used reduced state-space representations. Hartley (1992) used only the current source strengths. Mulquiney (1992) used the combination of current sources plus current values of those concentrations that are observable.

These Kalman filtering inversions have several limitations:

1. There has been no evaluation of the approximations involved in reducing the state-space to a subset of what is required for a complete representation.
2. The statistical modeling of the sources has been extremely limited. The Kalman filtering formalism allows, indeed requires, a Bayesian formulation, since a stochastic expression for the source evolution must be specified. In the studies to date, this has not been related to actual source characteristics; indeed, Hartley used a source equation appropriate for constant sources when investigating a time-varying problem.
3. The Kalman filtering formalism is a one-sided process and so is suboptimal relative to more general procedures that use information from all times.

There would appear to be scope for making significant improvements in each of these three problem areas. In particular, the third problem can be addressed by using Kalman smoothers. It must be noted that in Bayesian inversion techniques (including the Kalman filtering approaches), a statistical characterization of the prior information about the sources is also required.

8.3 Inversion

8.3.1 The Source Deduction Problem

The problem of deducing sources and sinks of CO_2 and other trace gases from observations of surface concentrations is an ill-conditioned inverse problem. Figure 8.1 illustrates the origin of the difficulty. The causal chain moving down the left-hand side of the figure involves an attenuation of small-scale information because of the dissipative nature of atmospheric mixing. The "reconstruction" problem moving up the right-hand side of the figure involves amplifying the small-scale details; however, this implies that the errors introduced along the way are similarly amplified.

The difficulty was known to Bolin and Keeling (1963). Their assessment was that "no details of the sources and sinks can be considered reliable" (p. 3918). The inversion problem was analyzed using a simple diffusive representation of atmospheric transport (Newsam and Enting, 1988; Enting and Newsam, 1990). For the problem of deducing surface sources from surface concentration data, modes of variation of wavenumber n are subject to an amplification proportional to n. In other words, the ratio of implied source to measured concentration grows as n, the wavenumber, and this amplification factor also applies to "noise" in the measured concentrations. Newsam and Enting (1988) pointed out that the degree of error amplification will be determined by the most dissipative process acting and, therefore, that error amplification proportional to n is to be expected on both purely diffusion models and advective-diffusive models. This is confirmed numerically in the two-dimensional advective-diffusive model of Enting and Mansbridge (1989) (who misinterpreted the significance of the

result). The linear dependence on n is confirmed by the behavior of the Goddard Institute of Space Studies (GISS) three-dimensional tracer transport model, as illustrated in Figure 8.2.

Bolin and Keeling (1963) found an n^2 amplification because they were using a vertically averaged model. The errors involved in equating surface observations to the vertical averages needed by the model led to an overestimate of the degree of ill-conditioning.

8.3.2 General Inverse Problem Theory

Ill-conditioned inverse problems occur in most of the branches of science that involve indirect measurements; therefore, the problems have been widely studied. Many of the accounts are of quite general relevance. Many of the key issues are described by Jackson (1972). The book by Tarantola (1987) is particularly relevant to the work described here because it is closely based on Bayesian approaches. The Bayesian approach involves the use of additional prior information to constrain the instabilities that arise from unbounded error amplification. In the synthesis studies described below, this additional information is derived from studies of the processes.

Some of the key results applicable to a wide range of inverse problems are:

- Only a finite number of modes of variability (often quite a small number) can be determined from the data. A specific data set defines an observable subspace of the possible model parameters, that is, a subspace of the possible source/sink distributions in this case.

- In many geometrical problems, it is the largest spatial scales that can be resolved. The inversions typically are inadequate to resolve small-scale details.

- There is often a characteristic degree of error amplification as a function of spatial scale, such as the case described above with amplification proportional to wavenumber n. These rates of error amplification can be used to characterize the relative difficulty of inverse problems.

- Such behavior implies a tradeoff between the spatial resolution that is sought and the variance of the estimates that are obtained.

The discussion in the following sections builds on these general principles. Enting (1993) presented a series of approximate calculations that indicated how the tradeoff between resolution and uncertainty operated for the problems of deducing zonal-mean CO_2 sources. However, the analysis still required the use of additional a priori smoothness constraints in order to produce meaningful results.

8.4 Practical Aspects of Inversions

8.4.1 Smoothing

One approach that has been used to prevent the unbounded amplification of errors is to fit smoothed concentration distrib-

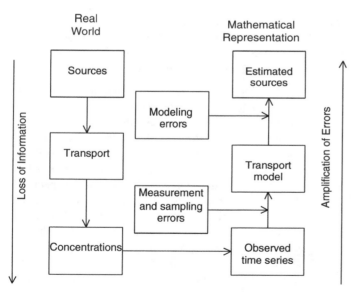

Figure 8.1: Origin of instability in inversions shown in the relation between the tracer transport in the atmosphere and the mathematical process involved in the inversion. The indirect nature of the inference means that there are a number of stages at which errors can enter. The atmospheric mixing processes act to lose small-scale information. The attempt to reconstruct this attenuated detail involves an amplification that will also amplify the errors.

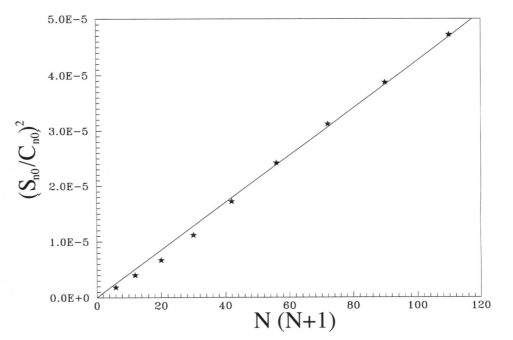

Figure 8.2: Amplification factors for GISS model, shown as *. For sources defined in terms of spherical harmonics (see Equation 8.1) the $m = 0$ components (S_{n0}) have concentration responses C_{n0}. The near-linear relation between wave number, n, and error amplification, S_{n0}/C_{n0}, confirms the point (Newsam and Enting, 1988) that the amplification factors for advective-diffusive transport will have the same behavior at larger wavenumbers as those for purely diffusive systems (solid line).

utions, thus suppressing the small-scale components whose errors are subject to unbounded amplification. This approach has been particularly important in the deterministic mass-balance inversions described above. Since this inversion process has no statistical component that can be used parametrically to control the effects of the error amplification, the stabilization of the source estimates must be achieved by either postprocessing the output or preprocessing the input. Most applications have used the latter choice.

Several forms of spatial smoothing have been considered. Enting and Mansbridge (1989) smoothed the latitudinal distributions of concentration by regression-fitting a truncated spectral expansion. Law and colleagues (1992) used a polynomial regression over latitude and assumed no longitudinal variation of concentration. However, such regression fits are defined globally so that, for example, the smoothed Southern Hemisphere concentrations (and source estimates) depend on both northern and southern concentration data. It is probably preferable to have fits that depend primarily on local concentration data. Such local fits were achieved by the spline-fitting procedures used by Tans and colleagues (1989) and Enting and Mansbridge (1991).

8.4.2 Low-Resolution Synthesis Inversion

Keeling and colleagues (1989) presented a synthesis analysis of the atmospheric carbon budget based on the use of a three-dimensional transport model using one year of analyzed winds. The global totals of the oceanic and terrestrial sinks were constrained by the results of a global carbon cycle model.

The oceanic exchanges were expressed in terms of three components with overlapping spatial distributions:

- A uniform ocean sink
- A North Atlantic sink (with balancing outgassing south of 40°S)

- An equatorial outgassing (with balancing sinks in the extratropical regions)

The terrestrial components were:

- The industrial component
- A biotic destruction component
- A biotic fertilization component
- A seasonal biotic component

As noted above, the representation of the source distribution in terms of only a small number of process components leads to stable estimates, but gives results that are conditional on the validity of the assumptions about the spatial distributions of sources and sinks within each of the components.

8.4.3 Higher-Resolution Synthesis Inversion

In order to avoid the potential "overspecification" from the use of small sets of components, researchers have performed several calculations using larger basis sets. The CH_4 study by Fung and colleagues (1991) used 12 components and presented a number of distinct combinations that fitted the observational data. For CO_2, Tans and colleagues (1990a) used 6 ocean source components, 3 terrestrial biotic components, plus a fossil source and a seasonal biotic flux.

For CO_2, Enting and colleagues (1993) used 14 distinct source components, $s_a(x,y,0,t)$, with additional components for isotopic disequilibria. The principle behind their approach was described by Wunsch and Minster (1982): Limited data can resolve only a small number of modes of variation in the system. However, the identification of these modes must come from the data characteristics and not be imposed a priori on the problem.

The initial study by Enting and colleagues used the following subdivision of CO_2 sources and sinks:

- Nine ocean regions
- A fossil carbon source
- A CO_2 source from the oxidation of CO
- A seasonal biotic source
- A source from land-use change
- A net biotic uptake

The inversion was stabilized by a Bayesian approach: In addition to the set of operations from Equation (8.6) as defined by the transport model, information in the form of independent prior estimates of the source strengths was included. These equations can be written as

$$c_a = c_a(\text{prior}) \tag{8.9}$$

and the coefficients were estimated on the basis of a combined weighted least-squares fit to the combined sets of Equations (8.6) and (8.9). The weights for Equation (8.6) are the observational uncertainties, and the weights for the set of Equation (8.9) are the standard deviations of the prior estimates. The simple form of Bayesian analysis used here requires that the prior estimates be independent of the CO_2 data that are fitted. Thus the estimates are based on combinations for direct measurements and process modeling. The specific values are taken from Enting and colleagues (1993, 1995), where further details are given. The ranges of uncertainty assigned to these prior estimates were deliberately chosen to be large in order to avoid biasing the results. Therefore there is scope for refining the estimated source strengths by combining the synthesis inversion with more comprehensive measurements or more detailed process modeling.

As described in Section 8.6, subsequent calculations have used a further disaggregation of the oceans and regional disaggregations of the biotic components. In all studies involving ^{13}C, additional components have been included to describe the isotopic disequilibrium effects associated with gross fluxes.

8.5 Error Estimation

8.5.1 Statistical Representation

In order to assess the value of the results of atmospheric transport modeling in reducing the uncertainties in atmospheric budgets of greenhouse gases, some quantification of the uncertainties in the estimated source/sink distribution is required. In synthesis studies, the uncertainty can be assessed in terms of statistical approaches to solving Equation (8.6) or (more generally) (8.8), supplemented by Equation (8.9) when Bayesian approaches are used.

At least three statistical approaches can be applied:

1.   Deduce the error characteristics from the statistical characteristics of the input data, which are in turn deduced from the observational time series, possibly supplemented by estimated error contributions due to measurement limitations.

2.   Deduce the error characteristics from the statistical characteristics of the input data, deducing these from the residuals to the fit that is used to estimate the sources.

3.   Deduce the error characteristics of the source estimates from the extent to which these estimates are sensitive to the choice of input data.

The synthesis results presented by Enting and colleagues (1993) and the extensions of these calculations described below used the first of these approaches so that the estimates are obtained from a weighted least-squares solution of Equations (8.6) and (8.9), or equivalently by using Equations (8.6) and (8.9) as the basis of a weighted linear regression. However, the analysis by Enting and colleagues (1993, 1995) also involved checking that the results were broadly consistent with the other two approaches.

The statistical modeling of CO_2 time series has been quite limited. Some initial studies are presented by Surendran and Mulholland (1987), but most other modeling has been in terms of the regression model of specified functions plus white noise. In the absence of a suitable error model to characterize the uncertainties in the observations, Enting and colleagues (1993) resorted to the results of the simulated sampling experiments described by Tans and colleagues (1990b).

As noted above, Bayesian inversion techniques also require a statistical characterization of the prior information concerning the sources.

8.5.2 Applications of Error Analyses

The ability to express the results of atmospheric transport modeling in a form with quantified uncertainties makes it possible to assess, quantitatively, the role of spatial information in constraining the atmospheric carbon budget. A number of other advantages arise from a quantification of the uncertainties in the inversion:

- The roles of alternative data sets in reducing the uncertainties can be compared.
- Similarly, it is possible to explore the amount of improvement that could be expected from an improvement in measurement precision.
- This intercomparison of the utility of alternative data sets can be extended to encompass the problem of designing sampling networks for specific objectives.

8.5.3 Model Error

The Bayesian synthesis inversion described above represents an important advance because it quantifies the effects of the measurement errors indicated in Figure 8.1. However, it does not address the issue of model error. Figure 8.1 indicates the potential importance of model error, since these errors are also subject to the potentially unbounded error amplification that occurs for data errors. The issue of model error has proved to be of considerable difficulty in many inverse problems. The two general issues (which also apply in the type of transport modeling inversions described here) are:

- The difficulty of formulating the problem of translating estimated model errors into uncertainties in the results
- The difficulty of obtaining numerical estimates for model error

Enting and Pearman (1993) presented a brief formal analysis of the role of model error in atmospheric constituent modeling problems. In particular, they pointed out that there were areas of overlap between "data error" and "model error," since many contributions to data error simply reflect the exclusion of small-scale processes from the models that are used to interpret the data. Tarantola (1987) addressed the general problem of formulating the analysis of model error in inverse problems by transforming model errors into "effective data errors." However, this approach still leaves the difficulty of obtaining realistic numerical estimates for the model errors and the additional difficulty of discerning the extent to which model errors translate into independent contributions to the "effective data error." These independence questions seem likely to be particularly important for the effects of large-scale errors in atmospheric transport models. Clearly, much more research is needed on this topic.

8.6 Results

In discussing the extent to which the use of atmospheric transport modeling helps to reduce the uncertainties in the atmospheric carbon budget, it is important to note that the most important constraints come from the global-scale rates of change of CO_2 and $^{13}CO_2$ and not from their spatial distributions. The global totals of sources (wavenumber $n = 0$) are determined primarily by the global mean rates of change (i.e., the $n = 0$ component) of CO_2 and ^{13}C.

To illustrate the issues, we present results using a disaggregation into 12 ocean regions, 4 land-use components, and 8 biotic regions, each with net uptake, seasonality, and isotopic disequilibrium. In addition there is a single component each from fossil carbon and CO oxidation. The main data used are the mean concentrations and seasonal cycles for 1986–87 for

20 sampling sites from the National Oceanic and Atmospheric Administration (NOAA) Geophysical Monitoring for Climatic Change (GMCC). These data items were assigned an uncertainty of ±0.3 ppmv. These data are supplemented by a global CO_2 trend of 1.5 ± 0.15 ppmv/yr.

In Table 8.1 we show the way in which additional data act to reduce the uncertainties. The most important comparison is between lines 3 and 4. In line 3, inclusion of the spatial distributions has reduced the regional uncertainties, relative to line 2, with little effect on the global total, whereas in line 4, inclusion of $\delta^{13}C$ data has reduced the uncertainties in the global total, relative to line 2, with little effect on the regional uncertainties.

In view of the very large uncertainties in these global estimates, Enting and colleagues (1995) also analyzed cases with additional constraints imposed on the total oceanic uptake, representing information from global carbon cycle modeling. (This study arose from a suggestion by Jorge Sarmiento when the present chapter was presented at the GCI.) Line 5 of Table 8.1 shows the results for the case when additional prior information is included, specifying the net global oceanic uptake as 1.4 ± 0.3 Gt C/yr. (This calculation uses the higher trend estimate of 1.8 ± 0.15 ppmv/yr, following Enting et al., 1993.)

All these estimates refer to CO_2 fluxes. Because of the fluxes of CO, these CO_2 fluxes differ from carbon fluxes, which in turn differ from carbon storage rates because of carbon transported from the terrestrial biota to the oceans via rivers.

The fact that (in the absence of additional constraints) the global net fluxes and the partition between oceanic and biotic uptake are determined primarily by the CO_2 and ^{13}C trends led Keeling and colleagues (1989) to determine the atmospheric budget by using a global carbon cycle model with a single atmospheric reservoir. The atmospheric transport model was used to determine the spatial distributions of sources within the global constraints on the net terrestrial and ocean fluxes.

The main results obtained from previous atmospheric transport modeling studies are:

- After subtracting the fossil source, there is a strong net sink in the midnorthern latitudes. Data on the partial

Table 8.1. *The effect of data on reducing uncertainties in global and regional estimates of net ocean sources*

C(t)	C(x,t)	δ(t)	δ(x)	O_{Total}	O_{Sth}	O_{Eq}	O_{Nth}
N	N	N	N	−1.50 ± 3.43	−2.40 ± 2.29	1.50 ± 2.00	−0.60 ± 1.58
Y	N	N	N	−2.76 ± 2.20	−2.96 ± 2.09	1.07 ± 1.87	−0.87 ± 1.52
Y	Y	N	N	−2.15 ± 2.13	−2.42 ± 0.83	1.87 ± 1.35	−1.60 ± 0.88
Y	N	Y	N	−0.06 ± 1.40	−1.76 ± 1.81	1.99 ± 1.69	−0.30 ± 1.43
Y	Y	Y	N	−0.25 ± 1.30	−1.97 ± 0.74	2.77 ± 1.08	−1.05 ± 0.72
Y	Y	Y	Y	−0.66 ± 1.16	−1.68 ± 0.70	2.99 ± 1.02	−1.97 ± 0.62
Constrained				−1.32 ± 0.29	−1.78 ± 0.63	2.58 ± 0.88	−2.11 ± 0.56

Note: Values shown are estimate sources in Gt C/yr with 1 standard deviation uncertainty. Y and N = presence and absence of groups of data in the fit; C(t) and δ(t) = trends in CO_2 and $\delta^{13}C$; C(x,t) = spatial and seasonal variations of CO_2; δ(x) denoting the spatial variation of $\delta^{13}C$. The first line corresponds to the prior estimates. The final line uses the same data set as the previous line (apart from using a higher CO_2 trend), plus an additional constraint of 1.4 ± 0.3 GtC/yr on the total oceanic uptake.

pressure of CO_2 (P_{CO_2}) suggest that the sink is terrestrial. The ^{13}C data are less definitive but have been interpreted as implying that the sink is oceanic, as implied by the last line of Table 8.1.

- The net equatorial source is smaller than expected from the combination of ocean outgassing and tropical deforestation.

- The southern sink (which is presumably oceanic, given the small proportion of land) is weaker than expected in comparison to northern ocean sinks.

- Most studies indicate a source at high southern latitudes.

As noted above, inversion results such as those of Table 8.1 (and the more detailed set of estimates in Table 8.2 in the appendix to this chapter) represent CO_2 fluxes and not carbon storage. The 2.28 Gt C/yr net biotic sink must be reduced by approximately 0.56 to account for the effect of carbon that enters terrestrial ecosystems as CO_2 and leaves as CO (or compounds that oxidize to CO). This gives a net carbon flux into the biota. To convert to the net rate of increase of biomass (i.e., storage rate), this influx must be reduced by approximately 0.6 Gt C/yr, representing carbon that enters the biota as CO_2 and is then transported directly to the oceans via rivers (Sarmiento and Sundquist, 1992).

8.7 Conclusions

The use of atmospheric transport modeling to determine sources and sinks of CO_2 can lead to useful information on regional scales. However, the uncertainty analysis presented here indicates that such an analysis cannot, on its own, achieve sufficient precision to give useful constraints on the global net fluxes, at a precision better than that implied by the global-scale trends. Thus the separation of oceanic from biotic fluxes must rely primarily on the interpretation of the trends of CO_2 and ^{13}C. At present, there are two conflicting approaches that give quite distinct results (Quay et al., 1992; Tans et al., 1993). The unconstrained fit used in these synthesis studies corresponds to the approach taken by Tans and colleagues, so the net ocean fluxes are lower than for other approaches.

The analysis above indicates several possibilities for refining the estimates:

- There is a need for improved ocean mixed layer ^{13}C data in order to resolve the discrepancy noted above.

- The analysis presented by Enting and colleagues (1993, 1995) and the example given above both use values of the uncertainty of the prior estimates that are generally conservative, in order to avoid biasing the results. In some cases, a more detailed study may indicate that particular processes can be determined, independent of atmospheric transport modeling, to a precision greater than has been assumed in the inversions thus far. Refinements in the precision of any one component will reduce the consequent uncertainties in all other components.

- A more comprehensive analysis of the statistics of the CO_2 time series may suggest ways in which regionally

representative signals can be extracted with a precision better than the ±0.3 ppmv, which is assumed here on the basis of the study by Tans and colleagues (1990b). However, repeating the calculations presented in Table 8.1 with the network precision reduced from ±0.3 to ±0.1 ppmv only reduces the uncertainty in the southern sink from 0.70 to 0.49 Gt C/yr.

- As noted above, the quantification of uncertainties makes it possible to design optimal sampling networks that minimize particular uncertainties.

- The time-dependent approaches need to be refined to address the questions of interannual variability.

Of course, major improvements over the uncertainties quoted above can be expected if any of the source strengths can be determined independent of any knowledge of atmospheric distributions. The most obvious example is the oceanic sink, where the net global uptake can be determined from ocean transport models. (Determining the spatial distribution of ocean fluxes using ocean models requires a description of the total ocean carbon cycle, including marine biology.) However, such modeling will generally determine long-term average fluxes and omit short-time variations, such as the effects of the El Niño–Southern Oscillation, and modeling of feedback processes cannot yet be regarded as sufficiently advanced to obviate the need for independent monitoring of ocean fluxes. Thus, despite the considerable uncertainties inherent in the inversion process, the use of atmospheric transport modeling to interpret spatial distributions of trace gases continues to play an important role in the study of radiatively active trace gases.

Appendix: Synthesis Inversion for 1986–87

The body of this chapter reviews a range of inversion techniques. The three-dimensional Bayesian synthesis inversion technique of Enting and colleagues (1993) has the particular characteristic that it produces estimates and associated uncertainties. Discussions at the Global Change Institute indicated the need for a more detailed presentation of the results of applying the technique.

The requirements for the synthesis analysis are:

- The space–time distributions $s_a(x,y,z,t)$. These are as defined in the initial study by Enting and colleagues (1993), with the exception that the equatorial ocean is subdivided into Atlantic, Indian, and East and West Pacific.

- The input data. These are the 20 1986–87 CO_2 concentrations and seasonal cycles from NOAA sites, the global CO_2 trend, the annual mean $\delta^{13}C$ from 5 CSIRO sites, and the $\delta^{13}C$ trend based on the Cape Grim record.

- Prior estimates of the source strengths (with standard deviations). Some of these are shown in Table 8.2.

Table 8.2 gives some of the source components used in the inversion, together with the estimates derived using the full set of CO_2 and ^{13}C data. The inversion included additional components for the seasonal fluxes and for the isotopic disequilib-

Table 8.2. *Details of Bayesian synthesis inversion, corresponding to the case shown in the final two lines of Table 8.1, using the trends and spatial distributions of both CO_2 and d13C, together with the seasonal cycles of CO_2 (units in GtC/yr)*

Source	Prior	Estimate	Constrained
Fossil	5.3 ± 0.3	5.24 ± 0.29	5.23 ± 0.29
CO oxidation	0.9 ± 0.2	0.89 ± 0.20	0.90 ± 0.20
Unbalanced uptake[a]			
Tropical America	0.0 ± 1.5	−2.24 ± 1.14	−1.98 ± 1.11
Tropical Africa	0.0 ± 1.5	−1.41 ± 1.25	−1.16 ± 1.23
Tropical Asia	0.0 ± 1.5	0.30 ± 1.25	0.49 ± 1.24
Evergreen	0.0 ± 1.5	0.70 ± 1.08	0.91 ± 1.07
Deciduous	0.0 ± 1.5	−1.95 ± 0.93	−1.98 ± 0.93
Boreal	0.0 ± 1.5	1.37 ± 0.69	1.33 ± 0.69
Grassland	0.0 ± 1.5	−0.68 ± 1.36	−0.51 ± 1.35
Other	0.0 ± 1.5	0.18 ± 0.98	0.27 ± 0.97
Total unbalanced uptake		−3.74 ± 1.39	−2.63 ± 1.01
Land-use change			
Tropical America	0.6 ± 0.5	0.34 ± 0.49	0.37 ± 0.49
Tropical Africa	0.4 ± 0.5	0.25 ± 0.49	0.28 ± 0.49
Tropical Asia	0.6 ± 0.5	0.60 ± 0.49	0.64 ± 0.49
Other	0.2 ± 0.3	0.25 ± 0.30	0.26 ± 0.30
Total from land-use		1.45 ± 0.88	1.55 ± 0.87
Net ocean fluxes[b]			
Far N. Atlantic	−0.2 ± 0.5	−1.23 ± 0.30	−1.22 ± 0.30
Far N. Pacific	0.2 ± 0.5	0.10 ± 0.21	0.09 ± 0.21
N. Atlantic	−0.3 ± 1.0	−0.08 ± 0.38	−0.15 ± 0.37
N. Pacific	−0.3 ± 1.0	−0.76 ± 0.46	−0.83 ± 0.44
Eq. W. Pacific	0.3 ± 1.0	1.25 ± 0.67	1.16 ± 0.66
Eq. E. Pacific	0.6 ± 1.0	0.65 ± 0.71	0.55 ± 0.69
Eq. Atlantic	0.3 ± 1.0	0.23 ± 0.64	0.08 ± 0.62
Eq. Indian	0.3 ± 1.0	0.87 ± 0.58	0.79 ± 0.57
S. Atlantic	−0.5 ± 1.0	−0.07 ± 0.87	−0.10 ± 0.87
S. Pacific	−1.0 ± 1.5	−1.59 ± 1.09	−1.63 ± 1.08
S. Indian	−0.7 ± 1.0	−0.63 ± 0.78	−0.68 ± 0.78
Southern	−0.2 ± 1.0	0.61 ± 0.46	0.64 ± 0.46
Total ocean		−0.66 ± 1.16	−1.32 ± 0.29

Notes: The rightmost column also uses a global constraint on the net oceanic uptake. The table lists the source components used (excluding those involved in seasonality and isotopic disequilibrium) together with the prior estimates and the estimates obtained from the inversions.
[a]Divided on the basis of region and ecosystem type.
[b]Divided by latitudes 16°N and °S and 48°N and °S.

rium effects of oceans and biota. The totals over processes are included in the table for convenience and were not part of the inversion, although (following a suggestion by Jorge Sarmiento) we have modified the procedure described by Enting and colleagues (1993) to allow for constraints on such totals to be included. The rightmost column in Table 8.2 illustrates the use of this option. Further details and additional calculations are given by Enting and colleagues (1995).

Acknowledgments

The Commonwealth Scientific and Industrial Research Organisation (CSIRO) atmospheric transport modeling program was funded by the State Electricity Commission of Victoria. The GISS tracer transport model was kindly supplied by Inez Fung. Calculations with the model were performed by Cathy Trudinger. Valuable comments from the referees are greatly appreciated.

References

Bloomfield, P., M. Heimann, M. Prather, and R. Prinn. 1994. Inferred lifetimes. In *Report on Concentrations, Lifetimes and Trends of CFCs, Halons and Related Species*, J. A. Kaye, S. A. Penkett, and F. M. Ormond, eds. NASA Reference Publication 1339, Washington, D.C.

Bolin, B., and C. D. Keeling. 1963. Large-scale atmospheric mixing as deduced from the seasonal and meridional variations in carbon dioxide. *Journal of Geophysical Research 68*, 3899–3920.

Enting, I. G. 1993. Inverse problems in atmospheric constituent studies: III. Estimating errors in surface sources. *Inverse Problems 9*, 649–665.

Enting, I. G., and J. V. Mansbridge. 1989. Seasonal sources and sinks of atmospheric CO_2: Direct inversion of filtered data. *Tellus 41B*, 111–126.

Enting, I. G., and J. V. Mansbridge. 1991. Latitudinal distribution of sources and sinks of CO_2: Results of an inversion study. *Tellus 43B*, 156–170.

Enting, I. G., and G. N. Newsam. 1990. Inverse problems in atmospheric constituent studies: II. Sources in the free atmosphere. *Inverse Problems 6*, 349–362.

Enting, I. G., and G. I. Pearman. 1993. Average global distribution of CO_2. In *The Global Carbon Cycle*, M. Heimann, ed. Proceedings of the NATO ASI at Il Ciocco, Italy, September 8–20, 1991. Springer-Verlag, Berlin, Germany, pp. 31–64.

Enting, I. G., C. M. Trudinger, and R. J. Francey. 1995. A synthesis inversion of the concentration and $\delta^{13}C$ of atmospheric CO_2. *Tellus 47B*, 35–52.

Enting, I. G., C. M. Trudinger, R. J. Francey, and H. Granek. 1993. *Synthesis Inversion of Atmospheric CO_2 Data Using the GISS Tracer Transport Model*. Technical Report No. 29, Division of Atmospheric Research, CSIRO, Australia.

Fung, I., J. John, J. Lerner, E. Matthews, M. Prather, L. P. Steele, and P. J. Fraser. 1991. Three dimensional model synthesis of the global methane cycle. *Journal of Geophysical Research 96D*, 13033–13065.

Fung, I., K. Prentice, E. Matthews, J. Lerner, and G. Russell. 1983. Three-dimensional tracer model study of atmospheric CO_2: Response to seasonal exchanges with the terrestrial biosphere. *Journal of Geophysical Research 88C*, 1281–1249.

Hartley, D. E. 1992. *Deducing Trace Gas Emissions Using an Inverse Method in Three-Dimensional Chemical Transport Models*. Ph.D. dissertation. Massachusetts Institute of Technology, Cambridge, Massachusetts.

Jackson, D. D. 1972. Interpretation of inaccurate, insufficient and inconsistent data. *Geophysical Journal of the Royal Astronomical Society 28*, 97–109.

Keeling, C. D., S. C. Piper, and M. Heimann. 1989. A three-dimensional model of atmospheric CO_2 transport based on observed winds: IV. Mean annual gradients and interannual variations. In *Aspects of Climate Variability in the Pacific and Western Americas*, D. H. Peterson, ed. Geophysical Monograph 55, American Geophysical Union, Washington, D.C., pp. 305–363.

Law, R., I. Simmonds, and W. F. Budd. 1992. Application of an atmospheric transport model to high southern latitudes. *Tellus 44B*, 358–370.

Mulquiney, J. E. 1992. Calculating the sources of CFCs using a Lagrangian transport model and Kalman filtering. *EOS 73 (Fall Meeting Supplement)*, 93.

Newsam, G. N., and I. G. Enting. 1988. Inverse problems in atmospheric constituent studies: I. Determination of surface sources under a diffusive transport approximation. *Inverse Problems 4*, 1037–1054.

Pearman, G. I., and P. Hyson. 1986. Global transport and inter-reservoir exchange of carbon dioxide with particular reference to stable isotope distribution. *Journal of Atmospheric Chemistry 4*, 81–124.

Pearman, G. I., P. Hyson, and P. J. Fraser. 1983. The global distribution of atmospheric carbon dioxide: I. Aspects of observations and modeling. *Journal of Geophysical Research 88C*, 3581–3590.

Prather, M., and P. Bloomfield. 1992. The "observed" lifetimes of $CFCl_3$ and CH_3CCl_3: Coupling statistical optimization with 3-D atmospheric models (abstract). *EOS 73 (Fall Meeting Supplement)*, 94.

Quay, P. D., B. Tilbrook, and C. S. Wong. 1992. Oceanic uptake of fossil fuel CO_2: Carbon-13 evidence. *Science 256*, 74–79.

Sarmiento, J. L., and E. T. Sundquist. 1992. Revised budget for the oceanic uptake of anthropogenic carbon dioxide. *Nature 356*, 589–593.

Surendran, S., and R. J. Mulholland. 1987. Modeling the variability in measured atmospheric CO_2 data. *Journal of Geophysical Research 92D*, 9733–9739.

Tans, P. P., J. A. Berry, and R. F. Keeling. 1993. Oceanic $^{13}C/^{12}C$ observations: A new window on ocean CO_2 uptake. *Global Biogeochemical Cycles 7*, 353–368.

Tans, P. P., T. J. Conway, and T. Nakazawa. 1989. Latitudinal distribution of the sources and sinks of atmospheric carbon dioxide derived from surface observations and an atmospheric transport model. *Journal of Geophysical Research 94D*, 5151–5172.

Tans, P. P., I. Y. Fung, and T. Takahashi. 1990a. Observational constraints on the global atmospheric CO_2 budget. *Science 247*, 1431–1438.

Tans, P. P., K. W. Thoning, W. P. Elliott, and T. J. Conway. 1990b. Error estimates of background atmospheric CO_2 patterns from weekly flask samples. *Journal of Geophysical Research 95D*, 14063–14070.

Tarantola, A. 1987. *Inverse Problem Theory: Methods for Data Fitting and Model Parameter Estimation*. Elsevier, Amsterdam, Netherlands.

Wunsch, C., and J.-F. Minster. 1982. Methods for box models and ocean circulation tracers: Mathematical programming and nonlinear inverse theory. *Journal of Geophysical Research 87*, 5647–5662.

Young, P. 1984. *Recursive Estimation and Time Series Analysis*. Springer-Verlag, Berlin, Germany.

9

Estimating Air–Sea Exchanges of CO_2 from pCO_2 Gradients: Assessment of Uncertainties

INEZ FUNG AND TARO TAKAHASHI

Abstract

The gradients in partial pressure of CO_2 across the air–sea interface provide a starting point for estimating regional and global CO_2 fluxes between the atmosphere and ocean. They also are critical constraints on global atmospheric and oceanic models used to infer the land–sea partitioning of CO_2 uptake. Here, we assess the factors that contribute to uncertainties in the estimated CO_2 fluxes.

We estimate measurement precision in pCO_2 to be ± 2 μatm, and extrapolation of the data to regions with no measurements yields uncertainties of ± 0.8 μatm. The short duration of spring blooms in the North Atlantic diminishes the uncertainties arising from sparse seasonal coverage in the measurements. We estimate an oceanic uptake of 0.3 Gt C/yr due to spring blooms in the North Atlantic. It is difficult to quantify the extent to which pCO_2 gradients may change by correcting the pCO_2 measurements to skin instead of bulk temperatures, as skin–bulk temperature differences may be positive (negative) with strong surface heating (cooling), or may vanish under high wind conditions. Uncertainties in fluxes associated with gas exchange rates cannot be separated from the yet unknown flux contributions from the covariance between high-frequency wind and pCO_2 fluctuations.

In addition to expanded spatial coverage, high-resolution and high-frequency sampling of the meteorology, hydrography, and carbon system in the atmospheric and oceanic boundary layers at a few locations is needed for improving estimates of air–sea CO_2 fluxes.

9.1 Introduction

A mass budget for atmospheric CO_2 over the past decade requires a total ocean and terrestrial sink for anthropogenic CO_2 (from fossil fuel combustion and land-use modification) that averages 3.0–5.0 Gt C/yr over that period, with the wide range stemming from uncertainties in the estimates of the CO_2 release associated with land-use modification. Although there is general agreement that the ocean should absorb a significant portion of anthropogenic CO_2 from the atmosphere (e.g., Broecker et al., 1979), there is considerable debate about the magnitude and location of the uptake.

Most of the estimates of ocean uptake have been derived from models that represent circulation and transport characteristics with varying degrees of sophistication. One- and two-dimensional models rely on the temporal and/or spatial distribution of tracers for determining flow rates. Three-dimensional ocean general circulation models (OGCMs) calculate time-dependent circulation responses to surface wind stress and buoyancy fluxes according to physical laws of fluid dynamics. OGCMs are validated by comparing modeled circulation statistics to those observed; tracers are used to evaluate, not to determine, the simulated flow characteristics. The OGCMs estimate, for fossil fuel releases alone, CO_2 sinks of 1.2–1.9 Gt C/yr (e.g., Maier-Reimer and Hasselmann, 1987; Bacastow and Maier-Reimer, 1990; Sarmiento et al., 1992; Siegenthaler and Sarmiento, 1993). These model results suggest that the ter-

restrial surface remains a net sink of fossil fuel CO_2, albeit with a magnitude smaller than the ocean sink. The amount of anthropogenic CO_2 (from fossil fuels and deforestation) absorbed by the land surface thus depends on the magnitude of the land-use source. Because of the variability of the ocean carbon system and the very nature of ocean measurements, the ocean signature of anthropogenic CO_2 has not been demonstrated unambiguously from direct measurements.

An indirect method of obtaining information about exchanges of CO_2 between the atmosphere and the underlying surface is via atmospheric models. In this approach, the inferred CO_2 exchanges with the land and sea are those required by the model to match the observed north–south gradient of atmospheric CO_2, after the atmospheric signatures of releases from fossil fuel combustion and land-use modification have been accounted for. This approach yields information about the location and magnitudes of surface exchanges, but it does not identify the mechanisms for the exchange. Keeling and colleagues (1989) assumed that 1.8 Gt C/yr was released from land-use modification, and estimated a net ocean CO_2 uptake of approximately 2.7 Gt C/yr in 1984, about one-half of industrial CO_2 emissions. On the other hand, Tans and colleagues (1990) assumed a land-use source of 0.3 Gt C/yr and concluded that the ocean uptake was approximately 1 Gt C/yr, so that the global ocean is a much smaller CO_2 sink than the terrestrial biosphere. The terrestrial sinks (after the land-use source has been accounted for) are of comparable magnitudes in the two studies. The different partitioning between the land sink and ocean sink as obtained by these studies resulted from the different ancillary information used to constrain the calculations, and it illustrates the uncertainties in this approach. Neither of the two studies succeeded in matching simultaneously both the observed atmospheric $\delta^{13}C$ gradient (used by Keeling and colleagues) and the air–sea fluxes of CO_2 derived from shipboard measurements (used by Tans and colleagues). Furthermore, these inferred fluxes represent the combined preindustrial (natural) and perturbed exchanges. Because the preindustrial exchanges with the ocean are geographically varying, and may be nonzero globally (Sarmiento and Sundquist, 1992), the inferred fluxes cannot be compared directly with those obtained from OGCM estimates.

The compilation of the CO_2 partial pressure differences between the surface water and overlying air (ΔpCO_2) by Tans and colleagues (1990) is the only direct global estimate of the air–sea exchange of CO_2 based on available pCO_2 measurements. The spatial and temporal coverage of the measurements is sparse; consequently, it has been necessary to fill the gaps in order to produce a global ΔpCO_2 distribution and an estimate of total carbon uptake by the world oceans. The resultant ΔpCO_2 and fluxes show weak fluxes of CO_2 into the northern oceans, compensated for by the outgassing from the equatorial oceans. These empirically derived fluxes in the oceans north of 15°S were used in the calculations of Tans and colleagues (1990).

Below, we reevaluate the ΔpCO_2 compilation in the Northern Hemisphere in light of both new data and issues raised since the publication of the 1990 study by Tans and colleagues (1990). The dearth of pCO_2 measurements in the Southern

Hemisphere reduces confidence in the direction of the estimated annual mean CO$_2$ flux and, therefore, any meaningful reevaluation. This data gap is the principal impediment to reliable estimates of the magnitude of the net global air–sea CO$_2$ flux. A concerted measurement program needs to be mounted. The estimates available remain useful as regional constraints for atmospheric models in the manner used by Tans and colleagues (1990) or as regional cross-checks for ocean carbon models such as those of Maier-Reimer and Hasselmann (1987) and Bacastow and Maier-Reimer (1990).

9.2 Measurement Precision

A principal source of data for the global compilation by Tans and colleagues (1990) is the seawater pCO$_2$ measurements made by the Lamont–Doherty group. The pCO$_2$ in the surface mixed layer of the oceans has been measured using both a continuous method and a discrete method.

In the continuous method, surface ocean water was continuously pumped through a ship's water sampling line into a gas–water equilibration chamber through a spray head (Broecker and Takahashi, 1966). Small droplets of water and air bubbles formed in the equilibrator caused a rapid exchange of CO$_2$ between two phases. The equilibrated air in the chamber is passed through a CO$_2$ analyzer (an infrared, IR, or gas chromatograph) for the determination of pCO_2. An IR CO$_2$ analyzer may be operated continuously, whereas a gas chromatograph allows measurements only at regular intervals (e.g., every 10 minutes). This method has been used for the surface water studies during the 1972–78 Geochemical Ocean Sections (GEOSECS) program by Takahashi and during the Transient Tracers in the Ocean/North Atlantic Study (TTO/NAS), South Atlantic Ventilation Experiment (SAVE), and other programs conducted by Weiss of the Scripps Institution of Oceanography and by Gammon and Feely of the National Oceanic and Atmospheric Administration/Pacific Marine Environmental Laboratory (NOAA/PMEL) – these programs supplied major sources of data used by Tans and colleagues (1990).

The discrete method was developed to determine pCO$_2$ when the volume of water is limited (e.g., deep water samples) or when portability and compactness of equipment are required. A fixed volume water sample is placed in a closed gas circulation system, in which a fixed volume of gas (normally one-tenth the volume of the water sample) is recirculated through the water sample by a pump for equilibration for approximately 15 minutes. When high precision (approximately ±0.1%) is required, the equilibrator system has been thermostated, and the equilibrated gas is analyzed using a gas chromatograph (Chipman et al., 1993). The pCO$_2$ values measured at a constant temperature are corrected to in situ temperatures using a temperature effect on pCO_2 (δ ln pCO$_2$/δT) of 0.0423°C^{-1} (Takahashi et al., 1993). This method has been used in TTO/NAS, SAVE, and other programs. During our 1984–89 North Pacific study of seasonal variability in surface waters (Takahashi et al., 1991), a modified discrete method was used extensively. A 20-l surface water sample was collected and equilibrated with 1.5 l of recirculating air for 15 minutes.

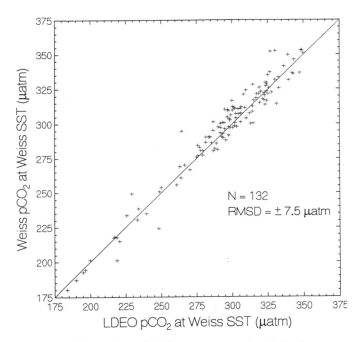

Figure 9.1: Comparison of pCO$_2$ measurements made by the Lamont–Doherty Earth Observatory (LDEO) and Weiss systems on the TTO/NAS cruises.

An aliquot (200 ml) of the equilibrated air was isolated in a gas sampling flask, which was sealed and shipped back to our land-based laboratories for pCO_2 determination. Because of the large amount of the seawater sample used, its temperature was maintained within a few tenths of a degree during the equilibration without a thermostated water bath. The precision of this method has been estimated to be ±2 μatm (or ±0.7%).

The compatibility of pCO_2 data obtained by different investigators using different methods has been tested using the data obtained by the continuous shower equilibrator system of Weiss and by the Lamont–Doherty discrete equilibrator (500 ml) system (Chipman et al., 1993) during the TTO/NAS expeditions in the North Atlantic Ocean. Gas chromatographs were used by both groups to obtain CO$_2$ measurements. Figure 9.1 shows that, over a pCO_2 range of 180–350 μatm, the Scripps data are consistent with the Lamont–Doherty data, with a root mean squared deviation (RMSD) of ±7.5 μatm ($N = 137$) or a standard deviation of the mean of ±0.7 μatm (7.5/$\sqrt{137}$). Systematic errors between two sets of measurements appear to be negligibly small.

9.3 An Estimate of Uncertainty in the Regional Distribution of Surface Water pCO_2

The largest source of uncertainty in the estimated air–sea flux of CO$_2$ comes from the inadequate sampling density of seawater pCO_2 in space and time. With a limited number of measurements made over a vast extent of ocean, estimates of air–sea CO$_2$ flux are affected by the means employed for interpolations and extrapolations. In the North Pacific, we have enough data to estimate errors associated with the interpolation scheme em-

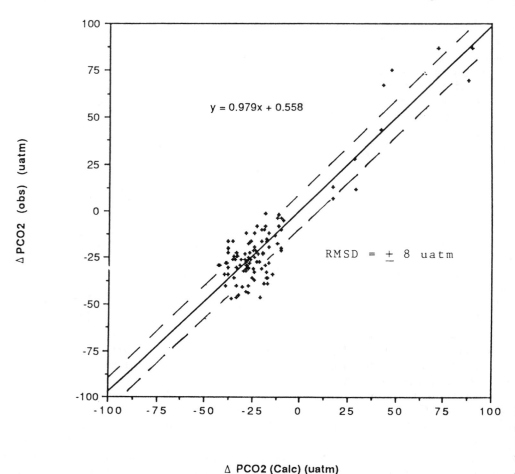

$$y = 0.979x + 0.558$$

RMSD = $\pm$ 8 uatm

△ PCO2 (obs) (uatm)

△ PCO2 (Calc) (uatm)

Figure 9.2: A comparison of the observed air–sea pCO₂ difference (ΔpCO_2) values with the computed using the interpolation method described in the text.

ployed for obtaining the regional pCO_2 distribution by Tans and colleagues (1990). In the North Pacific seasonal study program, the Pacific Ocean was sampled during 23 east–west crossings between Asian and U.S. ports. During each crossing, temperature, salinity, pCO_2, total carbon (TCO_2), phosphate, and mixed layer depth were measured (Takahashi et al., 1991) at approximately 40 stations located approximately 2° longitude apart. Regional distribution estimates are made by dividing the North Pacific into 2° × 2° pixel areas, and the data are sorted for two seasonal periods, January–April (winter) and July–October (summer). The mean value of ΔpCO_2 in each pixel has been computed by averaging all the values within 150 mi in a given seasonal period, with weighting factors inversely proportional to the square of distance from the center of the pixel. When there was no measured value in a pixel, a pixel value was obtained as the arithmetic mean of two interpolated values, one interpolated proportionally as a function of latitude and the other as a function of longitude. Errors for this interpolation scheme have been tested by comparing the measured ΔpCO_2 values with those computed for corresponding pixels. The data set (approximately 40 measurements) for one crossing was randomly selected and removed from the whole data set, and the pixel ΔpCO_2 values for the removed data set were computed using the scheme described above. The interpolated values computed when spring, summer, and winter transect data are removed one at a time, correlate linearly with the observed values with a slope of 0.98 and an RMSD of ±8 μatm, or a standard deviation of the mean of ±0.8 μatm (8/√110; Figure 9.2).

This gives an estimate for the precision of the ΔpCO_2 values interpolated over the North Pacific.

9.4 Spring Bloom

Another major source of uncertainty in the estimated air–sea flux of CO₂ is the shape of the pCO_2 variations over the course of a year. Ocean pCO_2 and other parameters of the ocean carbon system exhibit high-frequency fluctuations, the most dramatic documented example being the significant pCO_2 drawdown during spring plankton blooms in the North Atlantic (Watson et al., 1991). To obtain an annual mean value of the time series, it is necessary to have information about the high-frequency fluctuations themselves.

Near Iceland, a pCO_2 drawdown of approximately 160 μatm in a period of two weeks (Figure 9.3) was observed from measurements made in 1983–91 (Takahashi et al., 1993). This was accompanied by the complete removal of nutrient salts (i.e., nitrate, phosphate, and silicate) from the surface waters and a reduction of as much as 150 μmol/kg of TCO_2. The pCO_2 drawdown was consistent with that anticipated from the TCO_2 decrease. In the North Atlantic, observations made in the vicinity of 47°N 20°W during the Joint Global Ocean Flux Study (JGOFS) North Atlantic Bloom Experiment in 1989 also showed a drawdown on pCO_2 during the spring (Figure 9.4). Here, the spring bloom is less severe, with a pCO_2 and TCO_2 drawdown of 60 μatm and 60 μmol/kg, respectively, in a pe-

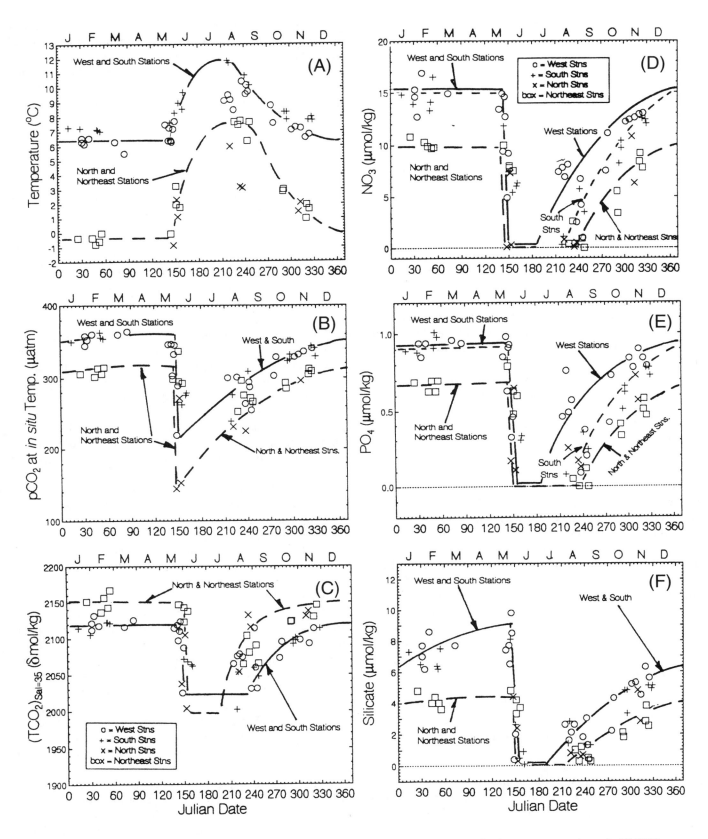

Figure 9.3: Seasonal variations of (A) surface water temperature; (B) pCO$_2$ in surface water at in situ temperature (μatm); (C) TCO$_2$ concentration (μml/kg); (D) nitrate concentration (μmol/kg); (E) phosphate concentration (μmol/kg); and (F) silicate concentration (μmol/kg) made in 1983–91 at four groups of stations located within 120 mi of Iceland. The northeast group is located at about 68°N 13°W, the North group at about 68°N 19°W, the West group at about 64.5°N 28°W, and the South group at about 63°N 22°W. (Data from Takahashi et al., 1993.) Reprinted with permission from T. Takahashi, J. Olafsson, J. Goddard, D. Chipman, and S. Sutherland, *Global Biogeochemical Cycles* 7:843–78, © 1993, American Geophysical Union.

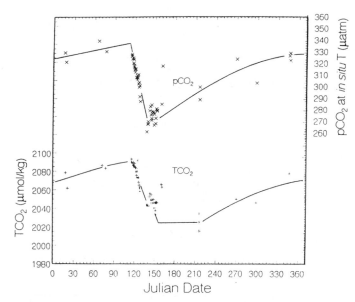

Figure 9.4: Seasonal changes in pCO₂ and TCO₂ concentration observed in the North Atlantic, 45°N–49°N and 15°W–25°W, during 1973–89. The cluster of data points between Julian days 115 and 152 represents data obtained during the JGOFS/NABE study at 47°N 20°W in April–June 1989 by Chipman and Goddard of LDEO. The solid curves represent the general seasonal trend observed. Reprinted with permission from T. Takahashi, J. Olafsson, J. Goddard, D. Chipman, and S. Sutherland, *Global Biogeochemical Cycles* 7:843–78, © 1993, American Geophysical Union.

riod of 45 days (Chipman et al., 1993; Takahashi et al., 1993). In the North Pacific, on the other hand, no dramatic spring blooms have been observed (Takahashi et al., 1993).

Tans and colleagues (1990) made a sinusoidal fit to the data available in the North Atlantic prior to 1989. Comparison of the newer observations and the sinusoidal fit shows that the sinusoidal fit gives a reasonable mean *pCO₂* for most of the year. Although the sinusoidal fit fails to capture the extreme rapid spring pCO₂ drawdown, the impact of the drawdown on the CO₂ flux is small because of the short duration of the spring bloom. The *pCO₂* averaged over May and June is approximately 300 μatm from observations and approximately 320 μatm from the sinusoidal fit, yielding an underestimate of approximately 20 μatm over the two months. A similar comparison of *pCO₂* curves near Iceland yields a comparable underestimate. Applying the underestimate of 20 μatm for two months to the North Atlantic gives a total annual underestimate of 0.03 Gt C/yr.

An independent estimate is provided by data of the satellite instrument Coastal Zone Color Scanner (CZCS), which measures pigment concentrations of photosynthetically active biomass in the ocean. Seasonal composites of CZCS data have been made from seven years of observations (cf. Esaias et al., 1986). The compositing, although necessary to maximize spatial coverage as individual scenes may be obscured by clouds, limits the usefulness of the data for evaluating bloom conditions in individual seasons. Nevertheless, the CZCS maps can be used as a general guide to regions of high surface produc-

tivity. The CZCS data are mapped to yield 1024 × 1024 arrays for the whole globe, and are given as nondimensional indexes with values typically between 0 and 5. We chose CZCS greater than 2 to indicate intense chlorophyll density. We multiplied the area where CZCS greater than 2 in the spring by the underestimate of 20 μatm for that season and obtained a global underestimate of 0.03 Gt C/yr.

9.5 Skin Temperature

The ocean gains heat from the atmosphere in the equatorial regions and returns the heat to the atmosphere at middle to high latitudes. The net heat flux, *Q*, is given by

$$Q = SW + LW\uparrow - LW\downarrow - SH - LH \tag{9.1}$$

where *SW* is the shortwave (solar) radiation absorbed; *LW*↑ and *LW*↓ are the downward and upward longwave radiation, respectively; and *SH* and *LH* are the sensible and latent heat fluxes, respectively.

Solar irradiance penetrates the ocean with an e-folding depth of approximately 10 m, while net longwave radiation and sensible and latent heat are lost from a layer less than 1 mm thick, referred to as the skin. Shipboard measurements of sea surface temperature ("bulk" temperatures) are done on water samples drawn from approximately 2–5 m. In the absence of wind stirring, regions of heat loss to the atmosphere would have a skin temperature that is cooler than the bulk temperature. Similarly, in equatorial regions, where the heat gain is much greater than the heat loss, skin temperature would be higher than bulk temperature. The bulk-minus-skin-temperature difference (ΔT) has been observed to be as large as −3 K in the "warm pool" in the western equatorial Pacific (Cechet, 1993).

Cechet (1993) also found that, in the western equatorial Pacific, ΔT decreases with increasing surface wind speeds, and disappears for wind speeds greater than 10 m/s. A similar decrease of ΔT at high wind speeds has been noted by Robertson and Watson (1992).

For shipboard measurements of *pCO₂*, water is brought from approximately 5 m depth to the surface. The *pCO₂* is measured at the temperature of the equilibrator and then corrected to in situ (bulk) temperature to yield the *pCO₂* of the surface waters. For estimation of CO₂ exchanges between the atmosphere and ocean, it has been argued that correction to skin rather than bulk temperatures would be appropriate (Robertson and Watson, 1992). Because skin temperatures are likely to be less than bulk temperatures, this correction would result in an additional invasion of 0.7 Gt C/yr to the ocean (Robertson and Watson, 1992).

In their calculation, Robertson and Watson used the Hasse's (1971) formula, which predicts ΔT as a function of incident solar radiation, net heat flux, and wind speed at the ocean surface. The Hasse equation was chosen because of the availability of global fields of the predictors. Without ΔT observations with large enough spatial and temporal coverage, it is impossible to evaluate the uncertainties or errors incurred by using a particular formula. Robertson and Watson surmise that ΔT is accurate to within 50%.

Schluessel and colleagues (1990) reported observations of ΔT during October–November 1984 at 20°N–50°N in the North Atlantic. In general, ΔT ranges between -1 and 1 K, with a mean of 0.1–0.2 K. The probability of negative ΔTs is higher when the heating processes (solar radiation and downward longwave radiation) exceed the cooling processes (upward longwave radiation, sensible and latent heat losses), such as under clear skies in the daytime when there is strong insolation, or under cloudy skies at night as downward longwave radiation increases with cloudbase temperature. Schluessel and colleagues (1990) found that predicting ΔT for the North Atlantic involved separate equations for day and night. The Hasse equation as used by Robertson and Watson (1992), which did not include the day–night discrimination, was not an accurate predictor for nighttime ΔTs or, by inference, winter ΔTs when solar irradiance is not dominating the surface heat budget.

Here we do not carry out a new estimate of the effect of skin temperature on pCO_2, because data on diurnal variations in the maritime boundary layer are limited; however, but we do wish to illustrate the magnitudes of the effects that wind speed and net heat flux can have on the pCO_2 flux correction. Esbensen and Kushnir's (1981) global distribution of monthly mean wind speed and net heat flux at the ocean surface is used. These data show that monthly mean wind speeds exceed 10 m/s throughout the year south of approximately 40°S and in the winter north of approximately 40°N, regions where Robertson and Watson obtained the largest skin temperature correction. Assuming $\Delta T = 0$ in these regions would decrease the Robertson and Watson (1992) flux correction by approximately 55%. The equatorial region from 5°S to 5°N, where there is strong net heating of the surface ocean, contributes approximately 5% to the flux correction. Inclusion of a warmer skin than bulk temperature under conditions of net surface heating (e.g., in the tropics and in midlatitude summers) would further reduce the magnitude of the correction. Other effects not considered in the estimate, such as the effect of temperature on carbonate chemistry in the skin layer, may act to increase the magnitude of the correction.

We note that the estimates of air–sea fluxes of heat, water, CO_2, and other trace substances are calculated as products of an exchange coefficient and the gradient of temperature, vapor pressure, or pCO_2 across the interface. The parameterization of an exchange coefficient is intimately tied to the thickness of the layer across which the gradients are estimated. Hence, although we recognize that a skin temperature correction on pCO_2 may be necessary in principle, it is equally important to establish whether the skin or bulk pCO_2 is employed in a particular parameterization of the gas exchange coefficient.

9.6 Gas Exchange Rate

At any moment in time t, the flux of CO_2 across the air–sea interface, $F(t)$, is expressed as the product of the instantaneous gas exchange coefficient $k(t)$ and the difference in pCO_2 across the interface, $\Delta pCO_2(t)$.

$$F(t) = k(t)\,\Delta pCO_2(t) \tag{9.2}$$

To obtain the averaged flux over a period, such as a month, we can write

$$k(t) = <k> + k'(t) \quad \text{and} \quad \Delta pCO_2(t) = <\Delta pCO_2> + \Delta p'(t) \tag{9.3}$$

where $\langle\ \rangle$ is the averaged value over the period and $(\ )'$ is the departure from the period mean. Hence:

$$<F> = <k(t)\,\Delta pCO_2(t)> = <k><\Delta pCO_2> + <k'(t)\,\Delta p'(t)> \tag{9.4}$$

It is clear that the mean flux does *not* equal the product of the mean gas exchange coefficient and the mean ΔpCO_2, but includes a contribution from the covariance of the fluctuations. It is difficult to estimate the sign of the covariations. Strong winds ($k'(t) > 0$) are likely to be accompanied by entrainment of colder and higher-TCO_2 water from below the mixed layer; however, whether $\Delta p'(t) > 0$ depends on the combination of biology, temperature, and buffer effects on pCO_2. Similarly, chemical enhancement of CO_2 exchange at low wind speeds may modify $\Delta p'(t)$ (Wanninkhof, 1992).

The dependence of the instantaneous $k(t)$ on the instantaneous wind $u(t)$ is established using data from wind tunnel experiments as well as measurements in lakes and open ocean (e.g., Liss and Merlivat, 1986; Smethie et al., 1985; Wanninkhof, 1992). The long history of surface marine observations and the regularity of satellite observations yield long time series of wind speeds, from which $<u>$ and higher-order moment statistics can be calculated (e.g., Etcheto and Merlivat, 1988; Boutin and Etcheto, 1991). The situation is deplorable for temporal statistics on ΔpCO_2. Typically, only a single measurement is available, and the measurement is used as the mean value over a period, such as a year. In other words, there are insufficient data to determine $<\Delta pCO_2>$; therefore, $<k'(t)\,\Delta p'(t)>$ certainly cannot be determined. The use of a gas exchange formula based on instantaneous winds together with a single distribution of ΔpCO_2 is therefore not appropriate.

In the ^{14}C method (Broecker et al., 1980, 1985, 1986), $<F14>$, the flux, and $<\Delta\Delta^{14}C>$, the time-averaged gradient in $\Delta^{14}C$ concentration across the air–sea interface, are determined independently, so that an equivalent exchange coefficient can be defined:

$$k_{eq} = <F14> / <\Delta\Delta^{14}C> \tag{9.5}$$

In this way, k_{eq} is the exchange coefficient required to yield the "correct" $<F14>$ for a given mean air–sea gradient. Note that $k_{eq} \neq <k>$.

For the calculation of $<F>$ where the ΔpCO_2 maps are treated as $<\Delta pCO_2>$, an equivalent gas exchange coefficient may be more compatible. This was done by Tans and colleagues (1990) and, as expected, it yielded a higher uptake than the use of the instantaneous exchange coefficients. We recognize that uncertainties in $\langle k_{eq}\rangle$ necessarily remain, because variability in $\Delta\Delta^{14}C$ is different from variability in pCO_2 (e.g., Thomas et al., 1988; Heimann and Monfray, 1989). This uncertainty may introduce a systematic error in the air–sea CO_2 flux of as much as 50%.

9.7 Concluding Remarks

Tans and colleagues (1990) obtained a flux of 0.59 Gt C/yr into the oceans north of 15°N. The above assessment suggests an addition of 0.03 Gt C/yr from spring blooms. Without additional information, it is difficult to quantify the adjustments of the flux resulting from skin temperature effects and gas exchange coefficients. Nevertheless, it seems unlikely that these adjustments would be large enough to yield an uptake large enough to eliminate a terrestrial sink in the northern midlatitudes.

Major uncertainties abound in the estimation of the global distribution of air–sea CO_2 fluxes. A first step toward removing the uncertainties would be an expanded program to measure pCO_2 variations in time and space, especially in the large expanse of the southern oceans. Other uncertainties may be addressed with in situ measurement programs and laboratory studies:

- The statistics of high-frequency fluctuations of pCO_2 in the surface waters
- The covariance between high-frequency fluctuations in wind and pCO_2
- The variations of temperature and pCO_2 within the upper 5 m of the ocean boundary layer
- The dependence of the gas exchange coefficient on the height and depth of pCO_2 measured, and on the stability of the atmospheric boundary layer

There are now several promising avenues for estimating the changes in carbon inventory in the ocean (e.g., Quay et al., 1993). These could provide cross-checks for the fluxes estimated from ΔpCO_2 distributions. We emphasize that there is no substitute for the continued monitoring of pCO_2 in the surface waters – at the very least for constraining atmospheric and ocean models, which are the ultimate tools for extrapolating the sparse measurements globally.

Acknowledgments

We thank P. Liss, M. Heimann, and A. Watson for their helpful comments in revising the manuscript. The support of the U.S. Department of Energy Carbon Cycle Program and National Aeronautics and Space Administration Mission to Planet Earth programs is acknowledged.

References

Bacastow, R., and E. Maier-Reimer. 1990. Ocean circulation model of the carbon cycle. *Climate Dynamics 4*, 95–125.

Boutin, J., and J. Etcheto. 1991. Intrinsic error in the air–sea CO_2 exchange coefficient resulting from the use of satellite wind speeds. *Tellus 43B*, 236–246.

Broecker, W. S., J. R. Ledwell, T. Takahashi, R. Weiss, L. Merlivat, T.-H. Peng, B. Jahne, and K. O. Munnich. 1986. Isotopic versus micrometeorologic ocean CO_2 fluxes: A serious conflict. *Journal of Geophysical Research 91*, 10517–10527.

Broecker, W. S., T.-H. Peng, and R. Engh. 1980. Modeling of the carbon system. *Radiocarbon 22*, 565–598.

Broecker, W. S., T.-H. Peng, G. Ostlund, and M. Stuiver. 1985. The distribution of bomb radiocarbon in the ocean. *Journal of Geophysical Research 90*, 6953–6970.

Broecker, W. S., and T. Takahashi. 1966. Calcium carbonate precipitation on the Bahama Banks. *Journal of Geophysical Research 71*, 1575–1602.

Broecker, W. S., T. Takahashi, H. J. Simpson, and T.-H. Peng. 1979. Fate of fossil fuel carbon dioxide and the global carbon budget. *Science 206*, 409–418.

Cechet, R. P. 1993. Diurnal SST cycle in the western equatorial Pacific: Measurements made aboard the R/V *Franklin* during TOGA-COARE and pre-COARE cruises. *AMOS Bulletin 6*, 26–32.

Chipman, D. W., J. Marra, and T. Takahashi. 1993. Primary production at 47°N and 20°W in the North Atlantic Ocean: A comparison between the ^{14}C incubation method and the mixed layer carbon budget. *Deep Sea Research II 40*, 151–169.

Esbensen, S. K., and Y. Kushnir. 1981. *The Heat Budget of the Global Ocean: An Atlas Based on Estimates from Surface Marine Observations.* Climatic Research Institute Report No. 29, Oregon State University, Corvallis, Oregon.

Esaias, W., G. Feldman, C. McClain, and J. Elrod. 1986. Monthly satellite-derived phytoplankton pigment distribution for the North Atlantic Ocean Basin. *EOS 67*, 835–837.

Etcheto, J., and L. Merlivat. 1988. Satellite determination of the carbon dioxide exchange coefficient at the ocean–atmosphere interface: A first step. *Journal of Geophysical Research 93*, 15669–15678.

Hasse, L. 1971. The sea surface temperature deviation and the heat flow at the sea–air interface. *Boundary Layer Meteorology 1*, 368–379.

Heimann, M., and P. Monfray. 1989. *Spatial and Temporal Variation of the Gas Exchange Coefficient for CO_2: I. Data Analysis and Global Validation.* Report No. 31, Max-Planck-Institut für Meteorologie, Hamburg, Germany.

Keeling, C. D., S. C. Piper, and M. Heimann. 1989. A three-dimensional model of atmospheric CO_2 transport based on observed winds: IV. Mean annual gradients and interannual variations. In *Aspects of Climate Variability in the Pacific and the Western Americas*, D. H. Peterson, ed. Geophysical Monograph 55, American Geophysical Union, Washington, D.C., pp. 305–363.

Liss, P., and L. Merlivat. 1986. Air–sea gas exchange rates: Introduction and synthesis. In *Role of Air–Sea Exchange in Geochemical Cycling*, P. Buat-Ménard, ed. D. Reidel, Hingham, Mass. pp. 113–129.

Maier-Reimer, E., and K. Hasselmann. 1987. Transport and storage of CO_2 in the ocean – An inorganic ocean circulation carbon cycle model. *Climate Dynamics 2*, 63–90.

Quay, P. D., B. Tilbrook, and C. S. Wong. 1993. Oceanic uptake of fossil fuel CO_2: Carbon-13 evidence. *Science 256*, 74–79.

Robertson, J. E., and A. J. Watson. 1992. Thermal skin effect of the surface ocean and its implications for CO_2 uptake. *Nature 358*, 738–740.

Sarmiento, J. L., J. C. Orr, and U. Siegenthaler. 1992. A perturbation simulation of CO_2 uptake in an ocean general circulation model. *Journal of Geophysical Research 97*, 3621–3645.

Sarmiento, J. L., and E. T. Sundquist. 1992. Revised budget for the oceanic uptake of anthropogenic carbon dioxide. *Nature 356*, 589–593.

Schluessel, P., W. J. Emery, H. Grassl, and T. Mammen. 1990. The bulk-skin temperature difference and its impact on satellite remote sensing of sea surface temperature. *Journal of Geophysical Research 95*, 13341–13356.

Siegenthaler, U., and J. L. Sarmiento. 1993. Atmospheric carbon dioxide and the ocean. *Nature 365*, 119–125.

Smethie, W. M., T. Takahashi, and D. W. Chipman. 1985. Gas exchange and CO$_2$ flux in the tropical Atlantic Ocean determined from ^{222}Rn and pCO$_2$ measurements. *Journal of Geophysical Research 90*, 7005–7022.

Takahashi, T., J. Goddard, D. Chipman, S. Sutherland, and G. Mathieu. 1991. *Assessment of Carbon Dioxide Sink/Source in the North Pacific Ocean: Seasonal and Geographic Variability, 1984–1989.* Final Technical Report for Contract 19X-SC428C to the U. S. Department of Energy, Lamont–Doherty Geological Observatory, Palisades, New York.

Takahashi, T., J. Olafsson, J. Goddard, D. Chipman, and S. Sutherland. 1993. Seasonal variation of CO$_2$ and nutrients in the high-latitude surface oceans: A comparative study. *Global Biogeochemical Cycles 7*, 843–878.

Tans, P. P., I. Y. Fung, and T. Takahashi. 1990. Observational constraints on the global atmospheric CO$_2$ budget. *Science 247*, 1431–1438.

Thomas, F., C. Perigaud, L. Merlivat, and J. F. Minster. 1988. World scale monthly mapping of the CO$_2$ ocean–atmosphere gas transfer coefficient. *Philosophical Transactions of the Royal Society of London, Series A 325*, 71–83.

Wanninkhof, R. 1992. Relationship between wind speed and gas exchange over the ocean. *Journal of Geophysical Research 97*, 7373–7382.

Watson, A. J., C. Robinson, J. E. Robertson, P. J. Williams, and M. H. R. Fasham. 1991. Spatial variability in the sink for atmospheric carbon dioxide in the North Atlantic. *Nature 335*, 50–53.

10

Atmospheric Oxygen Measurements and the Carbon Cycle

RALPH F. KEELING AND JEFF SEVERINGHAUS

Abstract

Precise measurements of atmospheric O_2 concentration can provide constraints on several aspects of the global carbon cycle. Seasonal variations in O_2 concentration, driven in part by biological and physical cycles in the ocean, can be used to constrain seasonal net photosynthesis rates of marine biota. Interannual variations in O_2 concentration, driven largely by O_2 uptake by fossil fuel burning and O_2 exchanges with land biota, can be used to partition the net global uptake of anthropogenic CO_2 into oceanic and land biotic components. The latter application is potentially complicated, however, by interannual sources and sinks of O_2 from the ocean. Model simulations are presented that suggest that interannually driven air–sea O_2 exchanges may be several times larger on a mole-for-mole basis than interannually driven air–sea CO_2 exchanges.

10.1 Introduction

It has recently become feasible to measure the atmospheric oxygen concentration to a degree of precision that allows for the detection of variations in the remote atmosphere (Keeling and Shertz, 1992; Bender et al., 1994; Bender et al., 1996; Keeling et al., 1996). These variations primarily reflect changes in O_2 because N_2 is constant to a very high level. Variations in O_2 are caused primarily by production and consumption of O_2 by photosynthesis, respiration, and combustion, and are thereby tied to the rate at which carbon is transformed between organic and inorganic forms. This link between O_2 and organic carbon means that atmospheric oxygen measurements can be used to constrain the rates and patterns of ocean biological productivity and the distribution of the sink for anthropogenic CO_2 between the land and the ocean.

Early results from samples collected at three sea-level sites – Alert, Ellesmere Island (82.5°N, 62.3°W); La Jolla, California (32.9°N, 117.3°W); and Cape Grim, Tasmania (40.7°S, 114.7°E) – were reported by Keeling and Shertz (1992) and are shown in Figure 10.1. Clearly detectable seasonal variations in oxygen concentration are evident at all three sites. An interannual decrease in O_2/N_2 is clearly evident in the La Jolla data. Concurrent CO_2 data are also given. Changes in O_2/N_2 are reported as deviations in the O_2/N_2 ratio relative to reference:

$$\delta\left(O_2/N_2\right) = \left(\frac{(O_2/N_2)\ sample}{(O_2/N_2)\ reference} - 1\right) \qquad (10.1)$$

The resulting deviations $\delta\left(O_2/N_2\right)$ are multiplied by 10^6, and the result is expressed in ("per meg") units. In these units, $1/0.2095 = 4.8$ per meg is equivalent to 1 ppmv because O_2 comprises 20.95% of air by volume (Machta and Hughes, 1970). Additional measurements of atmosphere oxygen concentration have been reported more recently by Keeling and colleagues (1996), Bender and colleagues (1996), and Keeling and colleagues (1998).

10.2 Seasonal Variations

The seasonal variations in oxygen concentration can be accounted for on the basis of three processes (Keeling and Shertz, 1992; Keeling et al., 1993). The first is the seasonal uptake and release of O_2 resulting from photosynthesis and respiration of terrestrial ecosystems. These exchanges of O_2 are closely tied to exchanges in CO_2 with an O_2/CO_2 exchange ratio of approximately $-1.05{:}1$ (Keeling, 1988). The seasonal variations in CO_2 in the Northern Hemisphere are almost entirely caused by these terrestrial exchanges, and they can be used to correct for the effects of terrestrial exchange on the O_2 concentration variations. The residual variations in O_2 concentration must be nonterrestrial, that is, oceanic, in origin. In the Southern Hemisphere, the lack of a large seasonal variation in CO_2 indicates that the O_2 concentration variations are almost entirely of oceanic origin.

The second process leading to seasonal variations is the warming and cooling of the upper oceans, which leads to seasonal ingassing and outgassing of O_2 and N_2 as determined by the solubility–temperature relations of these gases. Estimates of these exchanges based on seasonal changes in heat storage in the upper ocean suggest that this component of the seasonal variations is quite small, accounting for approximately 15% of the oceanic contribution to the seasonal variations in O_2 ratio (Keeling and Shertz, 1992).

The third process is the seasonal variations in photosynthesis, respiration, and vertical mixing in the upper ocean. Generally, there is net production of O_2 in the euphotic zone (approximately the upper 100 m of the ocean) where, on average, gross photosynthesis exceeds gross respiration, and there is a net consumption of O_2 below the euphotic zone by respiration. Both the net O_2 production rate in the euphotic zone and the rate of vertical mixing between the euphotic zone and deeper waters undergo large seasonal variations. At middle and high latitudes, the highest production rates tend to occur in the spring and summer when the waters above 100 m are well stratified (Peng et al., 1987; Asper et al., 1992). A considerable fraction of the O_2 produced at this time escapes into the atmosphere. Comparable amounts of O_2 are removed from the atmosphere in the fall and winter when stratification breaks down and oxygen-depleted waters from below 100 m depth are brought back into contact with the surface. The air–sea O_2 fluxes associated with this seasonal cycle are linked to the rate at which organic material is produced and exported from the euphotic zone (Jenkins and Goldman, 1985; Keeling et al., 1993), and they are linked to changes in dissolved inorganic carbon (DIC) in the water. The seasonal air–sea fluxes of CO_2 associated with these changes are much smaller than the O_2 fluxes because most of the dissolved inorganic carbon is in the form of carbonate and bicarbonate ions, which are not exchanged across the air–sea interface.

The marine biological component of the seasonal variations can be used to constrain estimates of the annual net photosynthetic production of organic carbon in the euphotic zone if O_2 transport within the atmosphere and between the euphotic zone of the oceans and the deeper waters is taken into account.

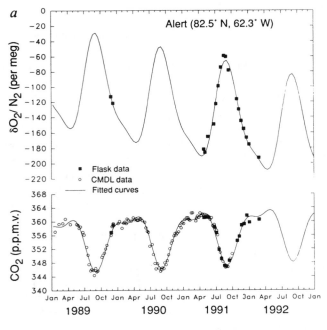

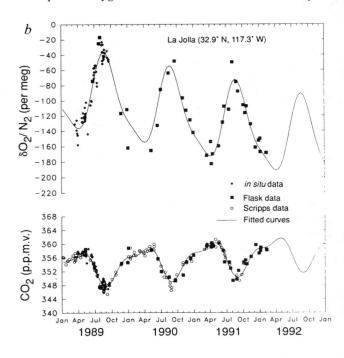

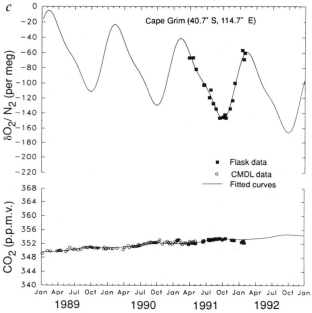

Figure 10.1: Measurements of $\delta(O_2/N_2)$ and CO_2 mole fraction at (a) Alert, (b) La Jolla, and (c) Cape Grim from the oxygen program directed by R. Keeling, now at Scripps Institution of Oceanography in La Jolla, California and formerly of the National Center for Atmospheric Research in Boulder, Colorado. The axes are scaled (5 per meg $\propto$ 1 ppm) so that changes in $\delta(O_2/N_2)$ and CO_2 are directly comparable on a mole O_2 to mole CO_2 basis. The data are shown against curve fits based on the sum of a four-harmonic seasonal function and a stiff-spline interannual trend, where the spline stiffness is adjusted to suppress frequencies higher than about 1 year. Reprinted with permission from *Nature*. R. F. Keeling and S. R. Shertz, *Nature* 358:723–27, © 1992, MacMillan Magazines Limited.

Using preliminary estimates of these quantities combined with the data shown in Figure 10.1, and assuming a Redfield ratio of 1.4:1 for O_2/C in marine organic matter, Keeling and Shertz (1992) estimated the global rate of net euphotic zone production to be 19 Gt C/yr.

Measurements of seasonal variations in O_2/N_2 spanning many years could provide an index of interannual variations in marine productivity on large spatial scales. Such data would be valuable for monitoring the sensitivity of marine productivity to climate change, decreasing stratospheric ozone, eutrophication, or other perturbations.

10.3 Long-Term Oxygen Depletion

It has long been noted that there is a mismatch between the combined sources of CO_2 from fossil fuel burning and land-use changes and the combined sinks in the ocean and the atmosphere. An additional missing sink for carbon is needed to balance things out (Keeling et al., 1989; Tans et al., 1990). This missing sink is of the same order of magnitude as the annual increase in atmospheric CO_2; consequently, identifying this sink is clearly important for estimating future CO_2 increases.

The most likely possibility for this sink is some sort of undocumented uptake of CO_2 by the land biosphere. At present, however, we cannot rule out the possibility that the oceans might be taking up more CO_2 than current models allow. This question of whether the missing sink is on land or in the ocean, or some combination of both, would be resolved if we could directly determine the net uptake of CO_2 by the oceans or the land biosphere. One method of obtaining this information involves using measurements of the long-term trend in atmospheric O_2.

Over the long term, we can represent the global budget for atmospheric CO_2 according to

$$\Delta CO_2 = F + C - O + B \qquad (10.2)$$

where ΔCO_2 is the annual averaged change in atmospheric CO_2, F is the source of CO_2 from burning fossil fuels, C is the CO_2 source from cement manufacturing, O is the oceanic CO_2 sink, and B is the net source of CO_2 from terrestrial ecosystems (B can be positive or negative), all in units of mol/yr. Likewise, we can represent the budget for atmospheric oxygen according to

$$\Delta O_2 = -F - H - \alpha_B B \qquad (10.3)$$

where ΔO_2 is the change in atmospheric oxygen, H is the O_2 sink owing to the oxidation of hydrogen and other elements besides carbon in fossil fuels, and α_B represents the O_2/C exchange ratio for terrestrial biomatter. There is no term in the long-term O_2 budget corresponding to the oceanic term in the CO_2 budget. This is because oceanic uptake of CO_2 essentially proceeds through reaction between dissolved CO_2 and carbonate ions, and therefore does not involve O_2. Also, the oceans do not significantly buffer the decrease in atmospheric O_2 because the amount of O_2 dissolved in the oceans is less than 1% of the amount in the atmosphere.

Adding Equations (10.2) and (10.3) and solving for O yields

$$O = -(\Delta O_2 + H) - (\Delta CO_2 - C) - (\alpha_B - 1)B \qquad (10.4)$$

The last term on the right-hand side of Equation (10.4) can be evaluated by solving Equation (10.3) for B, although this term is small, since $\alpha_B \approx 1$. Solving Equation (10.3) for B yields

$$B = -(1/\alpha_B)(\Delta O_2 + F + H) \qquad (10.5)$$

These equations show how observations of the change in atmospheric O_2 combined with estimates of fossil fuel CO_2 production and O_2 consumption can be used to directly calculate the net exchange of CO_2 with the oceans and with the land biosphere.

Using measurements of the O_2 and CO_2 trends from 1991 to 1994, Keeling and colleagues (1996) derive an estimate of global oceanic uptake of 1.7 ± 0.9 Gt C/yr and a net terrestrial carbon sink of 2.0 ± 0.9 Gt C/yr for 1991–94. The primary source of uncertainty comes from uncertainty in the O_2 trend, and this will decrease as longer records are obtained.

Assuming that the O_2 trend were measured sufficiently accurately, what would then limit our ability to resolve terrestrial and oceanic carbon fluxes using O_2 measurements? One limi-

tation arises because Equations (10.4) and (10.5) are both subject to uncertainty in the fossil fuel terms F and H. These quantities are estimated using statistics compiled by the United Nations for the production of different fuel categories (solids, liquids, gases), estimates of the fraction of each category oxidized, and estimates of the average elemental composition of the different fuel categories (Marland and Boden, 1991; Keeling, 1988). This approach leads to estimates of $F + H = 6.86 \times 10^{14}$ mol/yr and $H = 2.01 \times 10^{14}$ mol/yr for the year 1989. The errors in $F + H$ are estimated to lie between 6% and 10%, and the error in H to lie between 8% and 14%, depending on whether the errors for different fuel categories are assumed to be correlated (high estimate) or uncorrelated (low estimate) (Keeling, 1988). These error estimates are actually only rough guesses. Taking these figures, the uncertainty in $F + H$ alone contributes to an uncertainty of $\pm 4.0 \times 10^{13}$ to $\pm 7.0 \times 10^{13}$ mol/yr (0.5–0.8 Gt C/yr) in the net CO_2 source derived from Equation (10.5) and the uncertainty in H alone contributes to an uncertainty of $\pm 1.6 \times 10^{13}$ to $\pm 2.9 \times 10^{13}$ mol/yr (0.2–0.3 Gt C /yr) in the net CO_2 sink into the ocean derived from Equation (10.4).

It appears, therefore, that uncertainty in fossil fuel production ultimately limits the accuracy of determining the net terrestrial carbon source to approximately ± 0.5–± 0.8 Gt C/yr using the oxygen trend as displayed in Equation (10.5). Although this estimate is perhaps not as accurate as one would like, it is still much more accurate than current estimates of the net CO_2 source based on land-use changes (Houghton, 1991; Houghton et al., 1987); consequently, O_2 measurements indeed have the potential to contribute significantly to our understanding of terrestrial carbon exchanges. Even more valuable, perhaps, is the possibility of constraining the net oceanic uptake of CO_2 to the level of ± 0.3 Gt C /yr using Equation (10.4). If the current model estimates of the uptake of CO_2 by the oceans could be validated to this level, this would strengthen the case that the missing CO_2 sink is on land.

10.4 Interannual Variability

A limitation on the use of O_2 data to constrain terrestrial and oceanic carbon sources and sinks is uncertainty in the exchanges of O_2 between the oceans and the atmosphere on direct interannual time scales. Although uptake of anthropogenic CO_2 by the oceans has no effect on atmospheric O_2, it is still possible that variability in ocean dynamics or biological activity could lead to interannual air–sea exchange of O_2. Even a relatively small imbalance between the seasonal ingassing and outgassing fluxes of O_2 across the air–sea interface could lead to such variability. Equations (10.3) to (10.5) totally ignore any possible effects of the oceans on atmospheric O_2, so we must either find another way to estimate air–sea exchanges of O_2 or be able to argue that these exchanges are very small.

We have simulated the effects of interannual variability in biological productivity in the oceans using the high-latitude exchange/inferior diffusion-advection (HILDA) model (Joos et al., 1991; Shaffer and Sarmiento, 1995) coupled with a one-box model of the atmosphere. The purpose of this simulation is to explore the link between O_2 and CO_2 variations in the air arising from variations in oceanic productivity. We force the biological new production in the model to vary sinusoidally in time, and we calculate the variations in dissolved O_2 and dissolved inorganic carbon (DIC) in the ocean and variations in O_2 and CO_2 in the air arising from air–sea gas exchange. We express the results in terms of the amplitude ratio $\Delta O_2/\Delta CO_2$ (mol/mol) of the variations in the air and in terms of the phase differences between the atmospheric O_2 and CO_2 variations. These quantities are not sensitive to the magnitude of the imposed variations in new production; however, they are sensitive to periodicity of the forcing. Figure 10.2 shows the sensi-

tivity of these quantities to variations in new production with periods ranging from 1 year to 1,000 years.

These simulations use a fixed Redfield O_2/C ratio of 1.43:1 for photosynthesis and respiration in the oceans. If O_2 and CO_2 partitioned themselves identically between the atmosphere and oceans, then the changes in atmospheric O_2 and CO_2 would occur in this ratio. In fact, this ratio is not obtained in the atmosphere at any time scale. The reason involves differences in the kinetics of gas exchange for O_2 and CO_2 and differences in the equilibrium chemical capacity of seawater to absorb excess O_2 and CO_2.

When new production is forced with a period of one year, the atmospheric $\Delta O_2/\Delta CO_2$ ratio is 8 (see Figure 10.2). Such a high

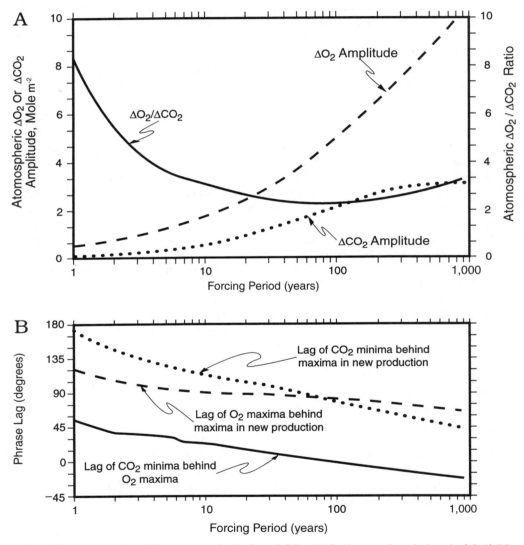

Figure 10.2: The effects of variable ocean productivity on atmospheric O_2 and CO_2. (A) Peak-to-peak variations in $\Delta O_2/\Delta CO_2$ and the $\Delta O_2/\Delta CO_2$ ratio; and (B) relative phasing of the atmospheric O_2 and CO_2 variations. The results were simulated using HILDA ocean model coupled to a well-mixed atmospheric reservoir. In this application of the HILDA model, the mixing in the low-latitude box was set using CFC data ($k = 465 + 7096e^{(z/220)}$ in m²/yr). The model was modified by changing the ratio of $CaCO_3$ to organic C production to 0.10 and adopting a remineralization scale of 800 m for organic particles and 500 m for $CaCO_3$. All other model parameters were used in Siegenthaler and Joos (1992). The new production in the model was varied sinusoidally between zero and twice the mean value. The mean new production was initially set at 1.3 mol C/m²/yr for forcing with a period of one year. The mean was decreased as the square root of the period as the period was increased. This was done to prevent nutrients from being depleted in the upper boxes and to prevent anoxia from developing in the deeper waters.

ratio can be expected for seasonal variations (Keeling et al., 1993) because exchange of CO_2 is heavily suppressed on this short time scale. Exchange of CO_2 is slow because most DIC is in the form of carbonate and bicarbonate ions, which are not exchanged across the air–sea interface. As the period of the forcing is increased from 1 year to 80 years, the $\Delta O_2/\Delta CO_2$ ratio decreases because CO_2 now has more time to equilibrate with the atmosphere. The $\Delta O_2/\Delta CO_2$ ratio achieves a minimum of approximately 2.2 at a period of 80 years and then increases for forcing with longer periods. This minimum ratio is still much larger than the Redfield ratio used. This minimum is probably quite sensitive to the remineralization depth range for organic matter and the mixing rates used in the model. A more realistic ocean model might produce lower $\Delta O_2/\Delta CO_2$ ratios in the air (Sarmiento and Orr, 1991). Beyond 80 years, the $\Delta O_2/\Delta CO_2$ ratio increases because perturbations in CO_2 in the upper ocean have time to communicate with the deep sea, which has a large capacity for buffering atmospheric CO_2 changes.

The possibility that there are net annual fluxes of O_2 between the ocean and atmosphere may be especially relevant on the two- to five-year time scale on which El Niños occur. There is evidence that atmospheric CO_2 is modulated at the ppm level on these time scales by the oceans (Whorf et al., 1993; Keeling et al., 1996). To the extent that these variations are related to variations in ocean biological productivity, we may expect, based on the above model simulations, that atmospheric O_2 may vary by an amount 3 to 4 times larger. If such large fluxes of O_2 do indeed occur, then we are likely to have serious difficulty in applying Equations (10.4) and (10.5) on this two- to five-year time scale. On the other hand, the atmospheric O_2 variations would still impose valuable constraints on any such changes.

Variability in biology or circulation may also produce variability in atmospheric O_2 on time scales longer than 5 years, although currently we have very little information that can be used to estimate the magnitude of such changes. In simulations that invoke relatively drastic alterations in oceanic productivity, Sarmiento and Orr (1991) found that although large air–sea fluxes of CO_2 and O_2 are possible on short time frames, the fluxes decrease significantly after approximately 10 years. High fluxes cannot be sustained over the long term because the O_2 deficit or C excess of the water transported to the surface eventually adjusts itself to cancel out the effect of changing productivity.

To close the gap in the atmospheric O_2 budget in decadal time scales, it may be necessary to develop independent constraints on decadal air–sea O_2 exchanges. One approach would be to conduct surveys of the concentrations of preformed nutrients – that is, those nutrients that originated with the water masses when they left the surface – since any change in atmospheric O_2 by marine biota must be accompanied by changes in the preformed nutrient distribution in the ocean (Keeling et al., 1993).

One application uses the seasonal cycles in O_2 concentration to impose constraints on estimates of marine biological productivities. Another application uses the long-term decrease in O_2 concentration to constrain terrestrial and oceanic sources and sinks of CO_2. The latter application may be complicated by interannual sources and sinks of O_2 from the oceans. To address this complication, it may be necessary to complement atmospheric O_2 measurements with independent constraints on long-term exchanges of O_2 between the atmosphere and the oceans.

Acknowledgments

We thank M. Bender, G. Marland, and J. Sarmiento for helpful reviews. This work was supported by the U.S. Environmental Protection Agency Global Change Research Program under IAG# DW44935603-01-2 and by the National Science Foundation under grants ATM-8720377 and ATM-9309765.

References

Asper, V. L., W. G. Deuser, G. A. Knauer, and S. E. Lohrenz. 1992. Rapid coupling of sinking particle fluxes between surface and deep ocean waters. *Nature 357*, 670–672.

Bender, M. L., J. T. Ellis, P. P. Tans, R. J. Francey, and D. Lowe. 1996. Variability of the O_2/N_2 ratio of southern hemisphere air, 1991–1994, implications for the carbon cycle. *Biogeochemical Cycles 10*, 9–21.

Bender, M. L., P. P. Tans, J. T. Ellis, J. Orchardo, K. Habfast. 1994. High precision isotope ratio mass spectrometry method for measuring the O_2/N_2 ratio of air. *Geochimica et cosmochimica acta 58*, 4751–4758.

Houghton, R. A. 1991. Tropical deforestation and atmospheric carbon dioxide. *Climate Change 19*, 99–118.

Houghton, R. A., R. D. Boone, J. R. Fruci, J. E. Hobbie, J. M. Melillo, C. A. Palm, B. J. Peterson, G. R. Shaver, and G. M. Woodwell. 1987. The flux of carbon from terrestrial ecosystems to the atmosphere in 1980 due to changes in land use: Geographic distribution of the global flux. *Tellus 39B*, 122–139.

Jenkins, W. J., and J. C. Goldman. 1985. Seasonal oxygen cycling and primary production in the Sargasso Sea. *Journal of Marine Research 43*, 465–491.

Joos, F., J. L. Sarmiento, and U. Siegenthaler. 1991. Possible effects of iron fertilization in the Southern Ocean on atmospheric CO_2 concentration. *Global Biogeochemical Cycles 5*, 135–150.

Keeling, C. D., R. B. Bacastow, A. F. Carter, S. C. Piper, T. P. Whorf, M. Heimann, W. G. Mook, and H. Roeloffzen. 1989. A three dimensional model of atmospheric CO_2 transport based on observed winds: I. Analysis of observational data. In *Aspects of Climate Variability in the Pacific and Western Americas*, D. H. Peterson, ed. American Geophysical Union, Washington, D.C., pp. 165–236.

Keeling, C. D., T. P. Whorf, M. Whalen, and J. van der Plicht. 1996. Interannual extremes in the rate of rise of atmospheric carbon dioxide since 1980. *Nature 375*, 666–670.

Keeling, R. F. 1988. *Development of an Interferometric Oxygen Analyzer for Precise Measurement of the Atmospheric O_2 Mole Fraction*. Doctoral dissertation, Harvard University, Cambridge, Massachusetts.

Keeling, R. F., R. G. Najjar, M. L. Bender, and P. P. Tans. 1993. What atmospheric oxygen measurements can tell us about the global carbon cycle. *Global Biogeochemical Cycles 7*, 37–67.

10.5 Summary

Measurements of the atmospheric O_2 concentration can be applied to improve our understanding of the global carbon cycle.

Keeling, R. F., S. C. Piper, and M. Heimann. 1996. Global and hemispheric CO_2 sinks deduced from changes in atmospheric O_2 concentration. *Nature 381*, 218–221.

Keeling, R. F., and S. R. Shertz. 1992. Seasonal and interannual variations in atmospheric oxygen and implications for the global carbon cycle, *Nature 358*, 723–727.

Keeling, R. F., B. B. Stephens, R. G. Najjaar, S. C. Doney, D. Archer, and M. Heimann. 1998. Seasonal variations in atmospheric O_2/N_2 ratio to the kinetics of air-sea gas exchange. *Global Biogeochemical Cycle*, in press.

Machta, L., and E. Hughes. 1970. Atmospheric oxygen in 1976 and 1970. *Science 168*, 1582–1584.

Marland, G., and T. Boden. 1991. CO_2 emissions: Modern record. In *Trends 91: A Compendium of Data on Global Change*, T. A. Boden, R. J. Sepanski, T. A. Boden, R. J. Sepanski, and F. W. Stoss, eds. Carbon Dioxide Information Analysis Center, Oak Ridge National Laboratory, Oak Ridge, Tenn., pp. 386–389.

Peng, T.-H., T. Takahashi, W. S. Broecker, and J. Olafsson. 1987. Seasonal variability of carbon dioxide, nutrients, and oxygen in the northern North Atlantic surface water: Observations and a model. *Tellus 39B*, 439–458.

Sarmiento, J. L., and J. C. Orr. 1991. Three-dimensional simulations of the impact of Southern Ocean nutrient depletion on atmospheric CO_2 and ocean chemistry. *Limnology and Oceanography 36*, 1928–1950.

Shaffer, G., and J. L. Sarmiento. 1995. Biogeochemical cycling in the global ocean: I. A new, analytical model with continuous vertical resolution and high-latitude dynamics. *Journal of Geophysical Research 100*, 2659–2672.

Siegenthaler, U., and F. Joos. 1992. Use of a simple model for studying oceanic tracer distributions and the global carbon cycle. *Tellus 44B*, 186–207.

Tans, P. P., I. Y. Fung, and T. Takahashi. 1990. Observational constraints on the atmospheric CO_2 budget. *Science 247*, 1431–1438.

Whorf, T. P., C. D. Keeling, and M. Wahlen. 1993. Recent interannual variations in CO_2 in both hemispheres. In *Climate Monitoring and Diagnostics Laboratory, No. 21, Summary Report 1991*, E. E. Ferguson and R. M. Rosson, eds. U.S. Department of Commerce, Boulder, Colo.

11

A Strategy for Estimating the Potential Soil Carbon Storage Due to CO_2 Fertilization

KEVIN G. HARRISON AND GEORGES BONANI

Abstract

Soil radiocarbon measurements can be used to estimate soil carbon turnover rates and inventories. A labile component of soil carbon has the potential to respond to perturbations such as CO_2 fertilization, changing climate, and changing land use. Soil carbon has influenced past and present atmospheric CO_2 levels and will influence future levels. A model is used to calculate the amount of additional carbon stored in soil because of CO_2 fertilization.

11.1 Introduction

The Intergovernmental Panel on Climate Change estimates that doubling atmospheric CO_2 over preindustrial levels will lead to a global-mean temperature increase of 1.5–4.5°C (e.g., Mitchell et al., 1990). Predicting when or if this doubling will occur requires an improved understanding of the global carbon cycle. One key question is the role of soil humus, which contains approximately 3 times the amount of carbon present in the preindustrial atmosphere. Scientists need to know if the soil carbon is labile or inert. If labile, soil carbon can respond to perturbations, either adding or removing atmospheric CO_2. If inert, soil humus does not significantly influence atmospheric CO_2 levels.

The purpose of this chapter is to show how soil radiocarbon measurements can be used to estimate soil carbon turnover times. The results presented here can be thought of as providing an example of what could be done on a biome-to-biome basis were more soil radiocarbon data available. Background material discusses the greenhouse effect, the global carbon cycle, and CO_2 fertilization, which are linked with soil carbon turnover times.

11.2 Background

The important relation between soil humus and the greenhouse warming can be difficult to understand. The greenhouse effect is caused by gases that trap radiation. An improved understanding of the global carbon cycle will lead to more accurate predictions of atmospheric CO_2 (a greenhouse gas) levels, which in turn will improve estimates of global warming. Soil humus represents a large unknown in the global carbon cycle. Below, we discuss the greenhouse effect, the carbon cycle, and soil carbon in detail.

Greenhouse gases trap radiation that would otherwise be lost from the Earth. For example, if the Earth's atmosphere had no naturally occuring water vapor, carbon dioxide, nitrous oxide, and methane, the Earth's temperature would be 35°C cooler (Broecker, 1985). This study concentrates on carbon dioxide, whose atmospheric concentration is expected to double in the next 100 years (Schimel et al., 1996, fig. 2.3). Such a doubling will elevate global temperatures by 1.5–4.5°C (Mitchell et al., 1990). Dixon and colleagues (1994) suggest that the sources of C to the atmosphere are 1.1 Gt C greater than the sinks. Sources include fossil fuel combustion, cement manufacturing, and deforestation. Sinks include the buildup in

the atmosphere and absorption by the ocean. Some postulate that this missing carbon (sources minus sinks) may be sequestered in the terrestrial biosphere because of CO_2 fertilization, climate change (Dai and Fung, 1993), and anthropogenic nitrogen deposition (Schindler and Bayley, 1993).

Soil carbon, which contains approximately 1,500 Gt C, has many organic compounds with turnover times that range from years to millennia. Many researchers have estimated global carbon inventories of soil carbon by using a variety of techniques: Schlesinger (1977) used vegetation types to estimate an inventory of 1,456 Gt C; Post and colleagues (1982) estimated soil humus to hold 1,395 Gt C by using climatic life zones; and Eswaran and colleagues (1993) used soil orders to estimate an inventory of 1,576 Gt C. Although there are differences among the techniques, the results generally agree.

Estimating soil carbon turnover has proven more elusive. As presented in this chapter, the technique of using soil radiocarbon measurements to estimate soil carbon turnover builds on the approaches of others: the Century model (Parton et al., 1987, 1989, 1993) and Rothamsted model (Jenkinson, 1990), mass-balance approaches, and soil humus fractionation approaches. The Century and Rothamsted models use measurements of soil carbon decomposition as the foundation for sophisticated ecosystem models. The model structures are similar, dividing soil organic material into fast, active, and passive fractions. The Century model has an active carbon turnover time range of 20–50 years and passive carbon turnover range of 800–1,200 years. The Rothamsted model uses a 20-year turnover time for active carbon and a near-infinite turnover time for passive carbon.

O'Brien and Stout (1978) use a sophisticated model to interpret their New Zealand radiocarbon measurements. Their model includes carbon input, decomposition rates, and soil diffusivity, which are constrained by depth profiles of total carbon and radiocarbon. They assign a 50-year turnover time for active carbon and a near-infinite time for passive carbon.

Researchers have tried to separate active and passive components by using physical and chemical fractionation techniques (Paul et al., 1964; Cambell et al., 1967; Martel and Paul, 1974; Goh et al., 1976, 1977, 1984; Scharpenseel et al., 1968a,b, 1989; Trumbore et al., 1989, 1990). Trumbore (1993) summarizes the fractionation technique results. One way to test the effectiveness of these fractionation schemes is to see if they predict, from pre-bomb residence time measurements, the observed amount of bomb radiocarbon increase in the soil. To date, the available fractionation schemes cannot do this (Trumbore, 1993).

11.3 Estimating Soil Carbon Turnover Times by Using Bulk Radiocarbon Measurements

This research uses a time-step one-box model and bulk soil radiocarbon measurements to estimate turnover times and inventories of active and passive carbon. The model has atmospheric ^{14}C values and CO_2 concentrations for every year from 1800 until the present. The user selects the carbon inventory and the turnover time. The turnover time equals the carbon inventory divided by the exchange flux. The exchange flux equals the amount of carbon that is added to the box (from photosynthesis) or lost from the box (respiration). Losses through erosion and dissolution are thought to be small (Schlesinger, 1986) and are not considered. The model can be run in either a steady-state mode (where the flux in equals the flux out) or a non-steady-state mode (in which carbon is either accumulating or decreasing). This research uses this model and soil radiocarbon data to show that soil carbon has more than one component and to estimate the turnover time of the passive fraction, the proportions of active and passive carbon in surface soil, and the active soil carbon residence time.

Researchers have concluded that soil consists of a complex soup of organic molecules whose turnover times range from a few years to thousands of years and cannot be characterized by a single turnover time (O'Brien and Stout, 1978; Balesdent et al., 1988; Parton et al., 1987, 1989, 1993; Jenkinson and Raynor, 1977; and Van Breemen and Feijtel, 1990). Figure 11.1 shows that a single residence time fails to characterize soil humus. Five prebomb values of surface soil had an average radiocarbon content that was 90% modern, where modern is defined as the amount of radiocarbon relative to 1850 wood (Harrison et al., 1993a). This 90% modern value translates into an 850-year turnover time, which shows very little increase in bomb radiocarbon with time. These values may not represent a global average, and values will vary among tropical, temperate, and boreal ecosystems. Further, different types of vegetation and soil types within the same climate often will have different values. Nevertheless, the measurement does illustrate what could be done were these data available. The soil radiocarbon values increase in the 1960s and then level off (Figure 11.1). This increase suggests that soil organic material contains an active component with a turnover time of significantly less than 850 years. This active component must be diluted with a passive component having a turnover time of greater than 850 years.

The passive soil turnover time can be estimated from soil radiocarbon measurements made at depths where little or no active soil carbon is present. Both carbon content and radiocarbon soil values decrease with increasing depth (Harrison et al., 1993a). At some depth, the soil radiocarbon measurements approach a minimum, where values tend to decrease very slowly. These values and depths vary for different locations (see Table 11.1). For example, a site in China had a 49% modern value at a depth of 65–105 cm. In contrast, a site in Germany was 71% modern at 80–100 cm. The average value for the sites listed in Table 11.1 is 63% modern, which corresponds to a 3,700-year turnover time for passive soil carbon.

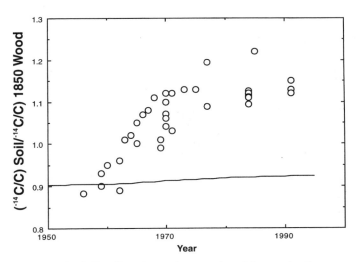

Figure 11.1: Soil radiocarbon values vs. time. Measured soil radiocarbon values for uncultivated soils are plotted against time from 1950 to 1991. The values tend to increase during the 1960s and then level off. The line indicates the model results for a theoretical carbon pool having an 850-year residence time. If soil carbon consisted of components having this single residence time, one would expect the observations to fall around this line. (Data from Table 11.2 and Harrison et al., 1993a.)

Using the limited available data, one can estimate the proportions of active and passive components in surface soil using the 63% modern passive soil radiocarbon value and the 90% modern prebomb surface soil value. Assume that the active component turns over quickly enough (<100 years) that its prebomb radiocarbon value is almost 100% modern (radiocarbon has a half-life of 5,700 years). A mixture of 25% passive and 75% active leads to the observed average radiocarbon value of 90% modern (Figure 11.2).

The postbomb increase in soil radiocarbon values can be used to estimate the active soil carbon turnover time. Figure 11.2 shows how a series of curves can be used to obtain the best fit. Upper and lower limits are formed by 10-year and 100-year turnover times, respectively. A 25-year turnover time produces the best fit to the available data. Most of the points are for temperate ecosystems, so warmer tropical ecosystems may have faster turnover times, and cooler boreal turnover times may be slower.

Table 11.1. *Deep soil radiocarbon values for cultivated and uncultivated soil*

% Modern	Depth (cm)	Reference	Soil Type	Location
49	65–105	Becker-Heidmann et al. (1988)	Mollisol	China
76	40–50	O'Brien (1986)	Mollisol	New Zealand
60	60–140	Scharpenseel & Becker-Heidmann (1989)	Vertisol	Israel
60	60–140	Becker-Heidmann (1989)	Udic	India
62	85–110	Tsutsuki et al. (1988)	Mollisol	Germany
71	80–100	Tsutsuki et al. (1988)	Mollisol	Germany

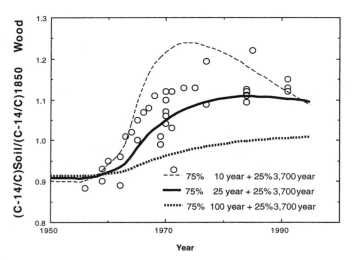

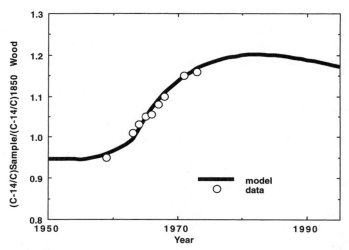

Figure 11.2: Soil radiocarbon model predictions and soil radiocarbon observations vs. time. We combine 75% of the fast-cycling components with 25% of the slow-cycling components to reproduce prebomb values. The prebomb values for the fast-cycling pools converge because their turnover time is much less than the mean life of radiocarbon. However, these curves diverge when bomb radiocarbon is released to the atmosphere. Soil radiocarbon for uncultivated soils is also plotted (open circles). We have superimposed our soil carbon modeling results for 10-, 25-, and 100-year residence times over these data. The 10- and 100-year cycling times almost bracket the experimental data, with the 25-year cycling time providing the best fit.

Figure 11.3: New Zealand test case. O'Brien and Stout (1978) published radiocarbon data for a New Zealand grassland site composed of a time series of surface soil values, including one prebomb and one deep soil radiocarbon value. We used this information to attempt to validate the model for a specific site. This site's surface soil had a slightly higher proportion of active to passive soil carbon and a slightly faster turnover time than the 25-year global average.

11.3.1 Determining the Global Inventory of Active Soil Organic Matter

Of the 1,500 Gt C in soil organic material, Schlesinger (1991) estimates that about one-third is present in wetland ecosystems. Although wetlands play an important role in regulating atmospheric CO_2 levels, they are beyond the scope of this study. The 3:1 active to passive proportions found in surface soil cannot be applied to the remaining 1,000 Gt C present in nonwetland soil because the proportion of active to passive carbon decreases with increasing depth. The integrated inventories are approximately 50% passive and 50% active (Harrison et al., 1993a). Extrapolating these distributions involves uncertainty (i.e., the proportions are likely to differ for other climates and types of vegetation); however, one can estimate the inventory of global soil carbon at 500 Gt C.

11.3.2 Turnover Time Model Validation

If these observations compared well with the modeled results, that finding would not rigorously prove that the modeled turnover times were correct; however, it would show that they were consistent with observations. Although these observations do not rigorously prove soil carbon turnover times, they do show whether the modeled turnover time is consistent with observations.

O'Brien and Stout (1978) reported measurements for a New Zealand grassland soil that included a deep soil value and a

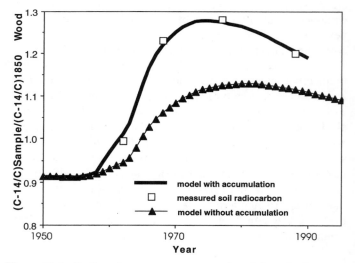

Figure 11.4: Radiocarbon measurements and model results for South Carolina. The model results that best fit the surface soil radiocarbon measurements were those from an accumulation model that took into account the increase in carbon inventory. The carbon increased from 0.52 g/cm² to 0.70 g/cm². The best model fit consisted of an initial mixture of 65% active carbon (having a turnover time of 12 years) and 35% passive carbon (having a turnover time of 2,300 years). The passive carbon turnover time was estimated from radiocarbon measurements of soil carbon in the upper 50 cm of the B horizon. The active reservoir turnover time is about twice as fast as the global average for native soils. (Data from Richter et al., 1995, and Table 11.4.)

Table 11.2. *Native soil radiocarbon values*

% Modern	Date	Depth (cm)	Reference	Vegetation/ Soil Type	Location
112	1991	0–10	Harrison (1994)	Hardwood	North Carolina
103	1991	10–20	Harrison (1994)	Hardwood	North Carolina
98	1991	20–30	Harrison (1994)	Hardwood	North Carolina
95	1991	30–40	Harrison (1994)	Hardwood	North Carolina
96	1991	40–50	Harrison (1994)	Hardwood	North Carolina
86	1991	70–80	Harrison (1994)	Hardwood	North Carolina
113	1985	0–5	This study	Boreal forest	Saskatchewan
123	1985	0–5	Harrison (1994)	Boreal forest	Saskatchewan
110	1985	0–7	Harrison (1994)	Boreal forest	Saskatchewan
109	1985	0–5	Harrison (1994)	Grass	Saskatchewan
115	1991	0–2	Harrison (1994)	Grass	Kansas
109	1984	0–23	Harrison (1994)	Grass	Lombardy
113	1982	0–2	Becker-Heidman (1989)	Hapludalf	Ohlendorf, Germany
107	1982	0–3	Becker-Heidman (1989)	Hapludalf	Ohlendorf, Germany
113	1983	3–5	Becker-Heidman (1989)	Hapludalf	Wohldorf, Germany
120	1983	0–2	Becker-Heidman (1989)	Hapludalf	Timmendor, Germany
96	1959	0–23	Trumbore (1993)	Temp. forest	California
111	1990	0–12	Trumbore (1993)	Temp. forest	California
90	1959	0–22	Trumbore (1993)	Trop. forest	Brazil
120	1986	4.7–7.5	Trumbore (1993)	Trop. forest	Brazil

time series of surface soil values that extended from prebomb times into the mid-1970s. The model that best fit the data consisted of a 23% passive, 77% active portion that turned over every 22 years. These values are very similar to those derived from the available global radiocarbon set. Further, Figure 11.3 shows the excellent agreement between the model and the data.

Another way to test the model is to see whether it can explain radiocarbon data from accumulating and cultivated ecosystems. An accumulating ecosystem increases its carbon stores. One example includes a recovering temperate forest located in the Calhoun National Forest (South Carolina) that is described by Richter and colleagues (1995) and Harrison (1994). This site contains a loblolly pine forest that was planted in 1959 on land that had been cultivated for 150 years. The soil has been accumulating carbon ever since the pines were planted. Figure 11.4 shows the agreement between the model and the data for accumulating ecosystems. Cultivated soil generally has lower radiocarbon values than native and recovering soil in the same time (Tables 11.2, 11.3, and 11.4). Harrison and colleages (1993b) showed that cultivated ecosystems can be modeled as well. They assumed that carbon lost from soil as a result of cultivation would be lost from the active carbon pool. When a correction was made for the mixing of deeper soil with the shallow surface soil because of cultivation (i.e., the plow mixes up the soil), the cultivation model produced good agreement with the available data.

Table 11.3. *Cultivated soil radiocarbon values*

% Modern	Date	Depth (cm)	Reference	Native Vegetation	Location
104	1985	0–7	Harrison (1994)	Boreal forest	Saskatchewan
104	1985	0–16	Harrison (1994)	Boreal forest	Saskatchewan
102	1985	0–17	Harrison (1994)	Boreal forest	Saskatchewan
99	1985	0–15	This study	Grassland	Saskatchewan
97	1985	0–15	This study	Grassland	Saskatchewan
104	1985	0–23	Harrison (1994)	Grassland	Lombardy
103	1991	0–2	Harrison (1994)	Grassland	Kansas
104	1985	0–15	Harrison (1994)	Grassland	Saskatchewan

Table 11.4. *Recovering soil radiocarbon values*

% Modern	Date	Depth (cm)	Reference	Vegetation	Location
115	1991	0–10	Harrison (1994)	Pine	North Carolina
98	1991	20–30	Harrison (1994)	Pine	North Carolina
113	1991	0–2	Harrison (1994)	Grassland	Kansas
99	1962	0–7.5	Harrison (1994)	Pine	South Carolina
123	1968	0–7.5	Harrison (1994)	Pine	South Carolina
128	1977	0–7.5	Harrison (1994)	Pine	South Carolina
121	1988	0–7.5	Harrison (1994)	Pine	South Carolina
77	1990	60–110	Harrison (1994)	Pine	South Carolina

Another way to validate the model is to see whether fluxes measured from the soil agree with values predicted by the model. A 500-Gt C pool turning over every 25 years emits 20 Gt C from the soil annually. This translates into a flux of 150 g C/m²/yr. The observed flux from a temperate forest soil ranges from 400 to 500 g C/m²/yr (Raich and Schlesinger, 1992), which is much higher than the predicted value. The measured values, however, include CO_2 sources besides microbial respiration of soil organic matter, such as root respiration and oxidation of litter and fine roots. It is impossible to separate these CO_2 sources from those being modeled. Also, land types having low organic carbon contents such as deserts make it difficult to compare global and regional values. These differences make impossible to achieve better agreement between the fluxes predicted by the model and measured fluxes for a temperate forest ecosystem.

11.3.3 Sensitivity Tests

The estimate that nonwetland soil contains a 500 Gt C pool that turns over every 25 years is a crude approximation that is based on many assumptions. Below, we test how sensitive these estimates are to the assumptions. The first generalization uses a 3,700-year turnover time (i.e., 63% modern) for passive soil carbon. The lowest radiocarbon value observed was 49% modern and translates into a mixture of 82% active and 18% passive carbon. If the highest value of 76% modern is used, the proportions become 65% fast and 35% slow. Measurements of deep soil radiocarbon average 63% modern, which corresponds to a 3,700 year turnover time. If typical values for deep soil contain less radiocarbon, they will only slightly increase the proportion of active to passive carbon in the surface soil. However, when the radiocarbon content rises above 63% modern, the proportions of active to passive begin to change more drastically, especially when they approach the mean prebomb value. Using a prebomb value of 90% modern as a global average can also lead to errors. This value and the passive carbon value determine the proportions of active and passive carbon. These values range from a low of 82% modern to a high of 96% modern, which translate into 51% active (49% passive) and 89% active (11% passive), respectively.

Another possible source of error comes from using a 25-year turnover time for active carbon. This estimate is based on

data that are from primarily temperate climes. Warmer tropical regions may have a faster turnover time. For example, Trumbore (1993) estimates tropical soil carbon to have about a 10-year residence time. Cooler climes may have slower turnover times.

11.4 Greening

Once the turnover time and inventory of fast cycling carbon has been estimated, it is possible to estimate the amount of carbon potentially stored in soil because of CO_2 fertilization. CO_2 fertilization occurs when plants increase their growth when exposed to elevated CO_2 levels (Strain and Cure, 1985; Bazzaz and Fajer, 1992). A convenient way of expressing CO_2 fertilization is with a CO_2 fertilization factor, that is, the percentage increase in growth for a doubling of CO_2 concentration.

For this study, a greening model has been developed to estimate the additional amount of carbon stored in soil because of CO_2 fertilization. The greening model builds on the carbon model described earlier. The flux of carbon into the box can be increased by adding the term $\beta (\Delta pCO_2)EF$, where β is the CO_2 fertilization factor, ΔpCO_2 is the fractional change in CO_2, and EF is the exchange flux. As the flux of carbon into the box increases, the decay flux (i.e., the decay constant times the amount of carbon in the box) also increases. If the level of atmospheric CO_2 stops increasing, the soil will attain a higher steady-state carbon content with an e-folding time of 25 years.

Table 11.5 lists the greening model results. About half the missing sink can be accounted for by using a CO_2 fertilization factor of 0.35 (i.e., plant growth increases 35% for a doubling of CO_2) and a carbon pool with a turnover time of 25 years and a size of 500 Gt C. By performing similar greening calculations for litter and short- and long-lived vegetation, the entire missing sink can be accounted for (Table 11.5).

11.4.1 Greening Model Validation

Although many indoor CO_2 fertilization experiments have shown increased growth at elevated CO_2 levels (Strain and Cure, 1985), extrapolating these results to natural vegetation is highly controversial (Bazzaz and Fajer, 1992). Further, applying these CO_2 fertilization factors to the carbon flux going into

Table 11.5. *Estimates of CO_2 fertilization for various carbon pools for an average year in the 1980s*

	τ	β	Inventory (Gt C)		
			350	500	650
			Annual Carbon Storage (Gt C/yr)		
Active soil carbon	15	0.60	0.7	1.0	1.4
		0.35	0.4	0.6	0.8
		0.15	0.2	0.3	0.3
	25	0.60	0.7	0.9	1.2
		0.35	0.4	0.5	0.7
		0.15	0.2	0.2	0.3
	40	0.60	0.5	0.7	0.9
		0.35	0.3	0.4	0.5
		0.15	0.2	0.2	0.3

	τ	β	Inventory (Gt C)
			500
			Annual Carbon Storage (Gt C/yr)
Wood	25	0.60	0.9
		0.35	0.5
		0.15	0.2

	τ	β	Inventory (Gt C)
			30
			Annual Carbon Storage (Gt C/yr)
Nonwoody tree parts	1.75	0.60	0.1
		0.35	0.1
		0.15	<0.05

	τ	β	Inventory (Gt C)
			70
			Annual Carbon Storage (Gt C/yr)
Litter	2.0	0.60	0.2
		0.35	0.1
		0.15	0.1

	τ	β	Inventory (Gt C)
			60
			Annual Carbon Storage (Gt C/yr)
Plants	4	0.60	0.2
		0.35	0.1
		0.15	0.05

	ASOC	W	NWTP	L	P		SUM
High guess (Gt C/yr)	1.4	0.9	0.1	0.2	0.2	=	2.8
Best guess (Gt C/yr)	0.5	0.5	0.1	0.1	01.	=	1.3
Low guess (Gt C/yr)	0.2	0.2	0.05	0.1	0.1	=	0.6–0.7

Notes: Five components used to simulate carbon uptake on land owing to CO_2 fertilization used in this study include active soil organic carbon (ASOC), wood (W), nonwoody tree parts (NWTP), and litter (L). Each of these pools has uncertainties associated with its turnover time, CO_2 fertilization factor, and inventory. For active soil carbon, the turnover times (τ) ranged from 15 to 40 years, with 25 years being the most likely value (Harrison et al., 1993a). The CO_2 fertilization factor (β) ranged from 60% to 15%, with 35% being the most likely value for all pools. The soil carbon inventories ranged from 350 to 650 Gt C, with 500 Gt C being the most likely value. Emanuel et al. (1984) and Post et al. (1990) estimated the inventories and turnover times for the remaining four pools. CO_2 fertilization has the potential to account for the entire missing sink.

Table 11.6. *Ecosystem division sensitivity test*

	Climate	Turnover time	Inventory (Gt C)	Annual carbon Storage (Gt C/yr)
Case 1: less tropical active carbon	Tropical	12	120	0.16
	Temperate	25	170	0.17
	Boreal	50	<u>210</u>	<u>0.16</u>
	Total		**500**	**0.49**
Case 2: equal amounts of active carbon	Tropical	12	170	0.23
	Temperate	25	170	0.17
	Boreal	50	<u>170</u>	<u>0.12</u>
	Total		**510**	**0.52**
Case 3: more tropical active carbon	Tropical	12	210	0.28
	Temperate	25	170	0.17
	Boreal	50	<u>120</u>	<u>0.09</u>
	Total		**500**	**0.54**

soil is speculation. Still, Zak and colleagues (1993) have shown that plants grown in doubled CO_2 in open-top chambers had more soil carbon than their nonelevated paired counterparts. Also, Norby and colleagues (1992) found evidence of increased fine root turnover for trees grown in elevated CO_2 concentrations, lending credibility to the belief that if plant growth is stimulated, soil carbon storage will be stimulated in turn.

11.4.2 Sensitivity Tests

Table 11.5 shows the sensitivity of estimates of soil carbon storage resulting from CO_2 fertilization. Specific tests are for the CO_2 fertilization factor, turnover time, and inventory size. In general, more carbon is sequestered for higher CO_2 fertilization factors, faster turnover times, and larger inventories. A further test is done to see the potential effect that splitting soil carbon spatially by climate would have on carbon sequestration (Table 11.6). If soil carbon turnover times for temperate forests fall between turnover times for boreal and tropical soil, then using a 25-year turnover time as an average for global ecosystems does not lead to huge errors, provided that other key assumptions also prove valid. (These assumptions, e.g., that the CO_2 fertilization factor of 35% can be applied to different ecosystems and that the inventories of active carbon are about equal in the different ecosystems, need validation with experimental work.) However, if the inventory and turnover of active carbon differ more dramatically than suggested by Table 11.6, the results will also differ. More soil radiocarbon measurements are needed to answer this question.

available radiocarbon data suggest that active soil carbon has a 25-year turnover time and a 500–Gt C inventory. Dividing the inventory by the turnover time leads to a 20–Gt C/yr exchange flux. Therefore, active soil carbon can respond significantly to perturbations such as CO_2 fertilization, changing climate, and anthropogenic nitrogen deposition. Even a small CO_2 fertilization effect will remove significant amounts of atmospheric CO_2 into the soil. For example, a greening model estimates that during the 1980s, 0.5 Gt C/yr may have been sequestered in soil humus, thus potentially explaining part of the missing sink. More soil radiocarbon data are needed to estimate the turnover time and inventories of active soil organic carbon on an ecosystem-by-ecosystem basis. Many of the soil radiocarbon data collected have remained unpublished, in part because they have been so difficult to interpret. It is hoped that these data can be coaxed out of musty file cabinets. Because radiocarbon values are time-sensitive, archived soil samples should be analyzed before they are cleaned out or their history is lost forever. Because radiocarbon measurements can cost as much as $500, publishing available data represents the most cost-effective approach toward expanding the available soil radiocarbon data base.

Researchers are working to reduce the uncertainties in CO_2 fertilization factors. By measuring inventories of active and passive carbon at the start of CO_2 fertilization experiments, it will be possible to observe directly soil carbon increases in response to CO_2 fertilization. Laboratory experiments are currently examining how plants respond to gradually increasing CO_2 concentrations (rather than simply to doubled CO_2 concentrations).

11.5 Conclusion and Future Research

This chapter presents a strategy for estimating the global carbon inventory and turnover time for nonwetland soil. The

Acknowledgments

I thank Wally Broecker, Inez Fung, and Jim Simpson for their careful guidance and helpful conversations, and I am grateful for the assistance given by Rick Fairbanks, Tyler Volk, Flip Froelich, Mac Post, Tony King, Charles Garten, and Tsung-Hung Peng. Jennifer Harden and Sue Trumbore carefully reviewed this manuscript and provided many carefully thought out suggestions. Mike Keene and Kim Schultz provided considerable help in the writing process. BethAnn Zambella offered support and encouragement. U.S. Department of Energy research grants, a National Aeronautics and Space Administration global change fellowship, and the U.S. Department of Energy distinguished global change postdoctoral fellowship funded this research. It was sponsored by the Carbon Dioxide Research Program, Environmental Sciences Division, Office of Health and Environmental Research, U.S. Department of Energy, under contract DE-AC05–84OR21400 with Martin Marietta Energy Systems, Inc.

References

Balesdent, J., G. H. Wagner, and A. Mariotti. 1988. Soil organic matter turnover in long-term field experiments as revealed by C-13 natural abundance. *Soil Science Society of America Journal 52,* 118–124.

Bazzaz, F. A., and E. D. Fajer. 1992. Plant life in a CO_2-rich world. *Scientific American 266,* 68–74.

Becker-Heidmann, P. 1989. *Die Teifenfunktionen der naturlichen Kohlenstoff-Isotopengehalte von vollstandig dunnschichtweise beprobten Parabraunerde und ihre Relation zur Dynamic der organischen Substanz in diesen Boden.* Ph.D. dissertation. Hamburg University, Hamburg, Germany.

Becker-Heidmann, P., L.-W. Liu, and H. W. Scharpenseel. 1988. Radiocarbon dating of organic matter fractions of a Chinese mollisol. *Zeitschrift für Pflanzenernahr. Bodenkunde 151,* 37–39.

Broecker, W. S. 1985. *How to Build a Habitable Planet.* Eldigio Press, New York.

Campbell, C. A., E. A. Paul, D. A. Rennie, and K. J. McCallum. 1967. Applicability of the carbon-dating method of analysis to solid humus studies. *Soil Science 104,* 217–223.

Dai, A., and I. Y. Fung. 1993. Can climate variability contribute to the "missing" CO_2 sink? *Global Biogeochemical Cycles 7,* 599–609.

Dixon, R. K., S. Brown, R. A. Houghton, A. M. Solomon, M. C. Trexler, and J. Wisniewski. 1994. Carbon pools and flux of global forest ecosystems. *Science 263,* 185–190.

Emanuel, W. R., G. G. Killough, W. M. Post, and H. H. Shugart. 1984. Modeling terrestrial ecosystems in the global carbon cycle with shifts in carbon storage capacity by land use change. *Ecology 65,* 970–983.

Eswaran, H., E. V. Den Berg, and P. Reich. 1993. Organic carbon in soils of the world. *Soil Science Society of America Journal 57,* 192–194.

Goh, K. M, T. A. Rafter, J. D. Stout, and T. W. Walker. 1976. Accumulation of soil organic matter and its carbon isotope content in a chronosequence of soils developed on aeolian sand in New Zealand. *Journal of Soil Science 27,* 89–100.

Goh, K. M., J. D. Stout, and J. O'Brien. 1984. The significance of fractionation dating the age and turnover of soil organic matter. *New Zealand Journal of Soil Science 35,* 69–72.

Goh, K. M., J. D. Stout, and T. A. Rafter. 1977. Radiocarbon enrichment of soil organic fractions in New Zealand soils. *Soil Science 123,* 385–390.

Harrison, K. G. 1994. *The Impact of CO_2 Fertilization, Changing Land Use, and N-deposition on Soil Carbon Storage.* Ph.D. dissertation. Columbia University, Palisades, New York.

Harrison, K. G., W. S. Broecker, and G. Bonani. 1993a. A strategy for estimating the impact of CO_2 fertilization on soil carbon storage. *Global Biogeochemical Cycles 7,* 69–80.

Harrison, K. G., W. S. Broecker, and G. Bonani. 1993b. The effect of changing land use on soil radiocarbon. *Science 262,* 725–726.

Harrison, K. G., W. M. Post, and D. D. Richter. 1995. Soil carbon turnover in a recovering temperate forest. *Global Biogeochemical Cycles 9,* 449–454.

Jenkinson, D. S. 1990. The turnover of organic carbon and nitrogen in soil. *Philosophical Transactions of the Royal Society of London B 329,* 361–368.

Jenkinson, D. S., and J. H. Raynor. 1977. The turnover of organic matter in some of the Rothamsted classical experiments. *Soil Science 123,* 298–305.

Martel, Y. A., and E. A. Paul. 1974. Use of radiocarbon dating of organic matter in the study of soil genesis. *Soil Science Society of American Proceedings 38,* 501–506.

Mitchell, J. F. B., S. Manabe, T. Tokioka, and V. Meleshko. 1990. Equilibrium climate change. In *Climate Change: The IPCC Scientific Assessment,* J. T. Houghton, G. J. Jenkins, and J. J. Ephraums eds. Cambridge University Press, Cambridge, U.K., pp. 131–172.

Norby, R. J., C. A. Gunderson, S. D. Wullschleger, E. G. O'Neill, and M. K. McCracken. 1992. Productivity and compensatory responses of yellow poplar trees in elevated CO_2. *Nature 357,* 322–324.

O'Brien, B. J. 1986. The use of natural and anthropogenic C-14 to investigate the dynamics of soil organic carbon. *Radiocarbon 28,* 358–362.

O'Brien, B. J., and J. D. Stout. 1978. Movement and turnover of soil organic matter as indicated by carbon isotope measurements. *Soil Biology and Biochemistry 10,* 309–317.

Parton, W. J., C. V. Cole, J. W. B. Stewart, D. S. Ojima, and D. S. Schimel. 1989. Stimulating regional patterns of soil, C, N and P dynamics in the U. S. central grasslands region. In *Ecology of Arable Land,* M. Clarholm and L. Bergström, eds. Kluwer Academic Publishers, Norwell, Mass., pp. 99–108.

Parton, W. J., D. S. Schimel, C. V. Cole, and D. S. Ojima. 1987. Analysis of factors controlling soil organic matter levels in Great Plains grasslands. *Soil Science America Journal 51,* 1173–1179.

Parton, W. J., M. O. Scurlock, D. S. Ojima, T. G. Gilmanov, R. J. Scholes, D. S. Schimel, T. Kirchner, J-C. Menaut, T. Seastedt, E. Garcia Moya, A. Kamnalrut, and J. I. Kinyamario. 1993. Observations and modeling of biomass and soil organic matter dynamics for the grassland biome worldwide. *Global Biogeochemical Cycles 7,* 785–809.

Paul, E. A., C. A. Campbell, D. A. Rennie, and K. J. McCallum. 1964. Investigations of the dynamics of soil humus utilizing carbon dating techniques. In *Transactions of the 8th International Soil Science Society,* Bucharest, Romania, pp. 201–208.

Post, W. M., W. R. Emanuel, P. J. Zinke, and A. G. Stangenverger. 1982. Soil carbon pools and world life zones. *Nature 298,* 156–159.

Post, W. M., T.-H. Peng, W. R. Emanuel, A. W. King, V. H. Dale, and D. L. DeAngelis. 1990. The global carbon cycle. *American Scientist 78,* 310–326.

Raich, J. W., and W. H. Schlesinger. 1992. The global carbon dioxide flux in soil respiration and its relationship to vegetation and climate. *Tellus 44B,* 81–89.

Richter, D. D., D. Markewitz, C. G. Wells, H. L. Allen, J. Dunscombe, K. G. Harrison, P. R. Heine, A. Stuanes, B. Urrego, and G. Bonani. 1995. Carbon cycling in an old-field loblolly pine forest: Implications for the "missing" carbon sinks and for the fundamental concept of soil. In *Proceedings of the Eighth North American Forest Soils Conference,* W. MacFee, ed. University of Florida, Gainesville, pp. 233–251.

Scharpenseel, H. W., and P. Becker-Heidmann. 1989. Shifts in C-14 patterns of soil profiles due to bomb carbon. *Radiocarbon 31*, 627–636.

Scharpenseel, H. W., P. Becker-Heidmann, H. U. Neue, and K. Tsutsuke. 1989. Bomb-carbon, C-14 dating and C-13 measurements as tracers of organic matter dynamics as well as of morphogenic and turbation processes. *The Science of the Total Environment 81/82*, 99–110.

Scharpenseel, H. W., C. Ronzani, and F. Pietig. 1968a. Comparative age determinations on different humic-matter fractions. In *Proceedings of the Symposium on the Use of Isotopes and Radiation in Soil Organic Matter Studies*, July 1968, International Atomic Energy Commission, Vienna, Austria, pp. 67–74.

Scharpenseel, H. W., M. A. Tamers, and F. Pietig. 1968b. Altersbestimmun von Boden durch die Radiokohlenstoffdatierungsmethode. *Zeitschr. Pflanzenemernahr Bodenkunde 119*, 34–52.

Schimel, D. S., D. Alves, I. G. Enting, M. Heimann, F. Joos, D. Raynaud, and T. M. L. Wigley. 1996. CO_2 and the carbon cycle. In *Climate Change 1995: The Science of Climate Change, Contribution of Working Group I to the Second Assessment Report of the Intergovernmental Panel on Climate Change*, J. T. Houghton, L. G. Meira Filho, B. A. Callander, N. Harris, A. Kattenberg, and K. Maskell, eds. Cambridge University Press, New York, pp. 65–86.

Schindler, D. W., and S. E. Bayley. 1993. The biosphere as an increasing sink for atmospheric carbon: Estimates from increased nitrogen deposition. *Global Biogeochemical Cycles 7*, 717–733.

Schlesinger, W. H. 1977. Carbon balance in terrestrial detritus. *Annual Review of Ecology and Systematics 8*, 51–81.

Schlesinger, W. H. 1986. Changes in soil carbon storage and associated properties with disturbance and recovery. In *The Changing Carbon Cycle: A Global Analysis*, J. R. Trabalka and D. E. Reichle, eds. Springer-Verlag, New York, pp. 194–220.

Schlesinger, W. H. 1991. *Biogeochemistry: An Analysis of Global Change*. Academic Press, New York.

Strain, B. R., and J. D. Cure. 1985. *Direct Effects of Increasing Carbon Dioxide on Vegetation.* DOE/ER-0238, U. S. Department of Energy, Washington, D. C.

Trumbore, S. E. 1993. Comparison of carbon dynamics in tropical and temperate soils using radiocarbon measurement. *Global Biogeochemical Cycles 7*, 275–290.

Trumbore, S. E., G. Bonani, and W. Wolfi. 1990. The rates of carbon cycling in several soils from AMS C-14 measurements of fractionated soil organic matter. In *Soils and the Greenhouse Effect*, A. F. Bouwman, ed. John Wiley and Sons, New York, pp. 407–414.

Trumbore, S. E., J. S. Vogel, and J. R. Southon. 1989. AMS C-14 measurements of fractionated soil organic mater. *Radiocarbon 31*, 644–654.

Tsutsuki, K., C. Suzuki, S. Kuwatsuka, P. Becker-Heidmann, and H. W. Scharpenseel. 1988. Investigation of the stabilization of the humus in mollisols. *Zeitschrift für Pflanzenernahr. Bodenkunde 151*, 87–90.

Van Breeman, N., and T. C. J. Feijtel. 1990. Soil processes and properties involved in the production of greenhouse gases, with special relevance to soil taxonomic systems. In *Soils and the Greenhouse Effect*, A. F. Bouwman and A. G. Bouwmann, eds. John Wiley and Sons, New York, pp. 195–223.

Zak, D. R., K. S. Pregitzer, P. S. Curtis, J. A. Teeri, R. Fogel, and D. L. Randlett. 1993. Elevated atmospheric CO_2 feedback between carbon and nitrogen cycles. *Plant and Soil 151*, 105–117.

Part III

Paleo-CO$_2$ Variations

12

Isotope and Carbon Cycle Inferences

MINZE STUIVER, PAUL D. QUAY, AND THOMAS F. BRAZIUNAS

Abstract

Various applications of carbon isotope (^{13}C and ^{14}C) records are described. The main data sources are dendrochronologically dated tree rings, ice cores, and ocean sediments (including corals). The representativeness and characteristics of these records are discussed. The history of atmospheric ^{14}C changes is determined by changes in oceanic upwelling rate and by solar and geomagnetic influences on upper atmosphere production rate. Separating these causal factors from the record is difficult, but analyses suggest interesting cyclic changes in North Atlantic deep water formation rates (periodicity around 500 years) and solar output (periodicity around 200 years). Isotopic data have provided valuable oceanic information regarding the current atmosphere-to-ocean flux of CO_2, deep water residence times, current upwelling rates, and glacial/interglacial changes in upwelling rate. This work is discussed and evaluated. Finally, the problems involved in interpreting radiocarbon dates in terms of calibrated (i.e., estimated calendar) dates are illustrated using the dating of the Mazama (U.S. Pacific Northwest) eruption as an example.

12.1 Introduction

Natural carbon contains the three carbon isotopes ^{12}C, ^{13}C, and ^{14}C. Of these isotopes, ^{12}C is by far the most abundant at 98.9% of total carbon. Thus, the carbon cycle in nature is essentially a ^{12}C cycle, with ^{13}C (1.1%) and ^{14}C (10^{-10}%) contributing only minor amounts. Nevertheless, ^{13}C and ^{14}C play a major role as tracers through which information on the physical and chemical properties of the carbon cycle can be obtained. The use of ^{13}C and ^{14}C in elucidating details of the carbon cycle is often tied to changes in the isotope abundance over time, as, for instance, the bicarbonate $^{13}C/^{12}C$ change of the past decades in the oceans or the $^{14}C/^{12}C$ change during the past 20,000 years in atmospheric CO_2. The ^{14}C isotope is used not only as a tracer but also routinely for age determinations. ^{14}C age determinations may have substantial uncertainties, and when ^{14}C dates are used, the precise translation of chronologies into calendar years can be quite cumbersome.

Isotope records can be representative of global, regional, or local carbon cycle change. Records of isotope change in the oceans (e.g., $^{13}C/^{12}C$ in marine sediments) and the biosphere (e.g., $^{14}C/^{12}C$ in soils) provide information on regional aspects. Many such regional records are needed for a global composite. However, owing to the atmosphere's fast mixing, true global records can be derived when the investigator accesses a library of fossil materials such as air or wood. The CO_2 concentration of trapped air in ice cores or the ^{14}C activities of dendrochronologically dated wood provide such global information. In the following sections we discuss a few specific carbon cycle details that are revealed by isotope records.

12.2 Atmospheric $\Delta^{14}C$ Change

A detailed atmospheric $^{14}C/^{12}C$ record can be derived from dendrochronologically dated wood. The procedure is to mea-

sure the remaining ^{14}C radioactivity of the wood and to correct for ^{14}C decay since its formation by using the ^{14}C half-life of 5,730 years. The decay-corrected measured ^{14}C content, after normalization on a fixed $^{13}C/^{12}C$ ratio to avoid biological fractionation effects, is expressed (Stuiver and Polach, 1977) as a relative deviation, $\Delta^{14}C$ (in percent or per mill), from the ^{14}C activity of the oxalic acid standard of the National Institute of Standards and Technology (NIST). Thus, atmospheric $^{14}C/^{12}C$ ratios are represented indirectly by the $\Delta^{14}C$ term.

In addition to dendrochronology, which yields the most precise time scale, other techniques can be used for the determination of ages of ^{14}C samples that were initially in equilibrium either with atmospheric CO_2 or with bicarbonate in portions of the major oceans. The resulting proxy records, however, are not as precise in age as the tree ring record, nor does the ^{14}C activity always represent global conditions. The counting of annual ice layers (cryo-counts) in selected (high accumulation rate) polar ice cores can yield ages with a surprisingly small error of 1–2% (Alley et al., 1993; Johnson et al., 1993; Taylor et al., 1993); however, the age of the gas bubbles differs from that of the ice because the gases diffuse through the snow and firn until polar ice forms, at which time the air-filled voids are isolated. Depending on the rate of accumulation, the fossil air represents a CO_2 concentration averaged over several decades, centuries, or even millennia. The average age of the trapped air also differs from that of the ice, and the estimation of the age difference adds to the age uncertainty.

Another proxy record involves the ^{14}C content of corals, for which uranium/thorium dates of high precision are available (Edwards et al., 1993; Bard et al., 1993). Here the precision of the time scale may approach the 1% range, but the resulting $\Delta^{14}C$ record is representative only of the regional ocean. The oceanic ^{14}C variations also will be smoothed relative to those in the atmosphere when production rate changes in the atmosphere cause the actual ^{14}C change. Furthermore, to estimate the atmospheric $\Delta^{14}C$ change from coral data, a correction must be made for the regional ^{14}C reservoir deficiency of the surface ocean. The time dependency of the reservoir deficiency is usually unknown, and only approximate atmospheric data can be obtained either by assuming a constant reservoir deficiency or by estimating its time dependency.

Ocean circulation and gas exchange-rate variations, tied to climatic variability, also result in a reapportioning of ^{14}C between the various earth carbon reservoirs. For instance, in the global ocean, a reduction in the rate of upwelling of older water (whose specific ^{14}C content is lower than that at the surface) will ultimately raise atmospheric $\Delta^{14}C$ levels. Unfortunately for carbon cycle adherents, the atmospheric $\Delta^{14}C$ record also reflects changes in the ^{14}C production rate that are tied to solar (Stuiver and Quay, 1980) and geomagnetic modulation of the cosmic ray flux. The geomagnetic field and the magnetic properties of the solar wind both limit the influx of galactic cosmic rays such that increases in geomagnetic field intensity and sunspot numbers cause lower ^{14}C production rates. For a proper assessment of carbon cycle–related $\Delta^{14}C$ change, the atmospheric $\Delta^{14}C$ perturbations induced by such change must be separated from those caused by geomagnetic and solar modulation of the cosmic ray flux.

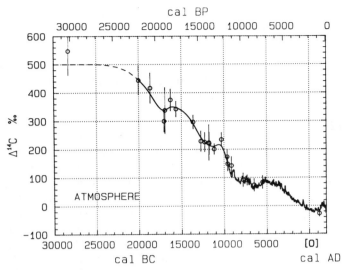

Figure 12.1: The atmospheric Δ14C record obtained from (1) dendrochronologically calibrated bidecadal tree ring 14C measurements (solid line from 9440 BC to AD 1950), (2) a cubic spline through 234U–230Th-calibrated coral 14C averages corrected by 400 14C years (solid line from 20,000 BC to 9440 BC), and (3) a smooth transition from an assigned pre-25,000 BC steady-state value of 500‰ (dashed line). The coral measurements are shown as circles with 2σ error bars. (References are given in the text.) Reprinted with permission from M. Stuiver and T. F. Braziunas (1993), *Radiocarbon* 35: 137–89.

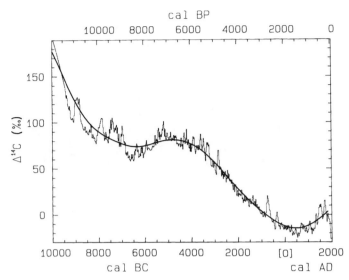

Figure 12.2: Atmospheric Δ14C (coral-derived for the earliest 500 years, tree ring–derived for the remaining part) for the past 12,000 years. The long-term Δ14C spline (smooth solid line), calculated from a 14C production spline (Stuiver and Braziunas, 1993b), resembles a 2,000-year moving Δ14C average. Most of the 20th-century Δ14C decline (up to AD 1952) is caused by fossil fuel CO_2 dilution. (Data from Stuiver and Quay, 1981.) Reprinted with permission from M. Stuiver and T. F. Braziunas (1993), *Radiocarbon* 35: 137–89.

Figure 12.1 depicts the long-term features of the atmospheric Δ14C record of the past 20,000 calibrated years (cal yr). Back to 11,390 cal yr prior to the present (BP), the data set (summarized in Stuiver and Reimer, 1993) is based on the 14C measurements (Kromer and Becker, 1993; Linick et al., 1986; Pearson and Stuiver, 1993; Stuiver and Pearson, 1993) of dendrochronologically dated wood samples (e.g., Becker, 1993), each spanning a bidecade. The 11,390 cal yr chronology has one remaining uncertainty near 9,900 cal yr BP, where the number of calibrated years covering the gap between the oldest part of the absolute German oak chronology and the youngest portion of the pine chronology (see Kromer and Becker, 1993, for details) must be estimated. More detail of the tree ring–derived bidecadal Δ14C profile is given in Figure 12.2.

The earlier part of the atmospheric Δ14C curve is based on the 14C measurements of corals, for which the calibrated ages are known from uranium/thorium dating (Bard et al., 1993). To generate atmospheric Δ14C information, the radiocarbon ages of the corals were reduced by 400 years (the assumed constant 14C deficiency of the local surface ocean reservoir). Of necessity, this part of the atmospheric curve has less detail than the part derived from the wood measurements because 14C variations generated by production-rate variations will be smoothed out in the large marine reservoir.

Heliomagnetic (solar), geomagnetic, and oceanic forcing all play a role in atmospheric Δ14C change. Major postulated changes in ocean circulation, such as a full turnoff of deep water formation, generate very large atmospheric Δ14C changes in global carbon reservoir models. (Examples of some calculations

are a 130‰ transient increase for a 1,300-year-long cessation of deep water formation, or a 280‰ increase for an infinitely long steady-state switch; Stuiver and Braziunas, 1993a.) Although deep water circulation changes must play a role in the long-term trend shown in Figure 12.1, most of the trend must be attributed to other causes. Consider, for instance, the 14C age differences between the deep and surface ocean (as derived from benthic-planktonic, or B P, age differences), which are very sensitive to the rate of deep water formation. The B P record (e.g., Andrée et al., 1986; Shackleton et al., 1988), although variable, does not encompass the large B P changes required for a full turnoff of deep water formation. (Calculated B P age differences range from approximately 4,000 years for the purely diffusive ocean to a more acceptable 1,000 years for the same ocean with the addition of current rates of deep water formation.)

Most of the Δ14C trend of the past 20,000 years could relate to production rate variation tied to geomagnetic dipole intensity change (Bard et al., 1990; Stuiver et al., 1991; Mazaud et al., 1991). As a first approximation, it is possible to remove the putative geomagnetic component of the Δ14C record by subtracting the long-term spline, as depicted in Figure 12.1. The residual Δ14C component of the record is depicted in Figure 12.3. When postulating the early part of this residual Δ14C record to be caused by ocean circulation change, Stuiver and Braziunas (1993a) calculate a reduction of approximately 20% in deep water formation during the 1,300-year-long Younger Dryas. Edwards and colleagues (1993), however, attribute the entire 150‰ Younger Dryas change to ocean circulation change, which, according to the model calculations, is compat-

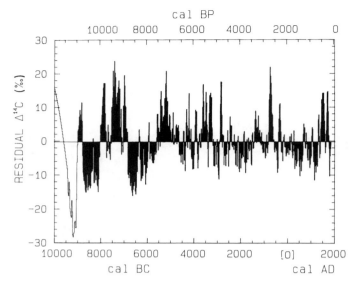

Figure 12.3: Residual $\Delta^{14}C$ obtained by deducting the Figure 12.2 long-term trend. Positive and negative values are shaded black, except for the oldest oscillation, which is (first 500 cal yr) derived in part, from coral determinations. The remaining $\Delta^{14}C$ values were derived from tree ring measurements and do not include the recent anthropogenic (post–AD 1850) portion of the atmospheric record. The lowest $\Delta^{14}C$ value for the post-YD (Younger Dryas) oscillation is at 9190 BC (11,140 cal yr BP); the subsequent rapid $\Delta^{14}C$ increase (duration ca. 150 yr) starts at 9060 BC (11,010 cal yr BP). The 18th-century $\Delta^{14}C$ maximum is associated with the Maunder sunspot minimum. Reprinted with permission from M. Stuiver and T. F. Braziunas (1993), *Radiocarbon* 35: 137–89.

ible with a full cessation of deep water formation over the entire interval. The latter scenario does not fully agree with the limitations imposed by the observed B P ^{14}C age difference record, whereas the former scenario may have overestimated the geomagnetic influence.

Stuiver and Braziunas (1993a,b) relate a 512-year $\Delta^{14}C$ periodicity found in the early Holocene bidecadal ^{14}C record to instabilities in North Atlantic thermohaline circulation. The $\Delta^{14}C$ record also suggests an increase in deep water formation near the start, instead of the termination, of the Younger Dryas interval (Stuiver and Braziunas, 1993a).

The 206-year $\Delta^{14}C$ cycle, persistent throughout the Holocene record, can be tied fairly convincingly to solar modulation. There is a possibility that the solar response is modified by climatic interaction (Stuiver and Braziunas, 1993a). There have been many recent attempts to link $\Delta^{14}C$ and climate variability, and a postulated climate–^{14}C relationship suggests a likelihood of solar-induced climate changes that augment the $\Delta^{14}C$ response as well, specifically of the Maunder minimum type (low sunspot numbers; Eddy, 1976). Single-year $\Delta^{14}C$ time series have periodicity in the 2–6-year range where the $\Delta^{14}C$ variance (Stuiver and Braziunas, 1993a) may be attributable to El Niño–Southern Oscillation (ENSO) perturbations. A 10–11-year component is partially tied to solar modulation of the cosmic ray flux, and multidecadal variability may relate to

either solar modulation or instability of North Atlantic thermohaline circulation.

12.3 The Role of the Oceans in the Carbon Cycle: Importance of Carbon Isotopes

12.3.1 Oceanic Uptake of Anthropogenic CO_2

Radiocarbon plays a critical role in the development of ocean models that are used to predict oceanic uptake rates of anthropogenically produced CO_2. The oceanic distribution of ^{14}C measured on the dissolved inorganic carbon during the Geochemical Ocean Sections Study (GEOSECS) expeditions (1972–78) provided the first look at the global distributions of the naturally occurring ^{14}C in the deep sea and the ^{14}C in the upper ocean derived from nuclear weapons testing. These ^{14}C measurements have been used to derive the rates of air–sea CO_2 gas transfer and upper-ocean mixing. These two rates are the foundation upon which ocean models have been built. Because CO_2 exchange between the atmosphere and oceans is at the crux of the Earth's CO_2 cycle, ^{14}C is the touchstone between models of CO_2 cycling and reality.

The first ocean model using ^{14}C in a comprehensive way was the box-diffusion model developed by Oeschger and colleagues (1975) at about the time when GEOSECS was beginning to map the oceanic ^{14}C distribution. Several modifications have been made to this one-dimensional box-diffusion model, such as incorporating the GEOSECS data and changing the geometry of ocean circulation to include mixing along isopycnal surfaces and deep sea exchange with the polar ocean (e.g., Siegenthaler and Joos, 1992). These models, generally, have air–sea CO_2 exchange rates of approximately 15–20 mol/m^2/yr and vertical diffusion rates between 1 and 1.5 cm^2/s in the upper 1,000 m of the ocean. When applied to the question of oceanic uptake of anthropogenic CO_2, these box-diffusion models yield estimates of 1.6–2.1 Gt C/yr (Siegenthaler and Joos, 1992).

$^{13}C/^{12}C$ is a tracer of anthropogenically produced CO_2 because the CO_2 added by fossil fuel combustion and biomass burning is approximately 20‰ depleted compared to atmospheric CO_2. The potential tracer capability of the $^{13}C/^{12}C$ of dissolved inorganic carbon in the ocean, measured primarily during GEOSECS, has been used less than ^{14}C by carbon modelers. Box-diffusion models that were used to predict atmospheric CO_2 levels but that also incorporated $^{13}C/^{12}C$ (e.g., Siegenthaler and Joos, 1992) predicted a surface ocean $^{13}C/^{12}C$ decrease of approximately 60% of the atmospheric $^{13}C/^{12}C$ decrease. These model predictions were similar to $^{13}C/^{12}C$ changes measured in corals.

Recently, Quay and colleagues (1992) measured a 0.4‰ decrease in the $\delta^{13}C$ of dissolved inorganic carbon (DIC) in the Pacific Ocean during the last 20 years. This ocean $^{13}C/^{12}C$ change, when coupled with the measured atmospheric CO_2 increase and $^{13}C/^{12}C$ decrease, was used to determine a net rate of oceanic CO_2 uptake of 2.1 ± 0.8 Gt C/yr. The accuracy of the mass-balance procedure used by Quay and colleagues is limited by our lack of oceanic $^{13}C/^{12}C$ measurements (in the near term) and by our incomplete knowledge of the biospheric carbon cycling time (in the long term). Tans and colleagues (1993) pointed out an apparent significant discrepancy between the air–sea $^{13}C/^{12}C$ disequilibrium (approximately 0.8‰) needed to account for the change in the oceanic $^{13}C/^{12}C$ inventory measured by Quay and colleagues, and the air–sea $^{13}C/^{12}C$ disequilibrium estimated from isotopic fractionation effects and surface ocean $\delta^{13}C$ values (0.4‰). The disequilibrium estimated by Tans and colleagues yields a net oceanic uptake of only 0.2 Gt C/yr; however, uncertainty in the Tans and colleagues calculation is substantial (1–2 Gt C/yr) and underscores the importance of precise measurements of the ^{13}C fractionation effects during CO_2 exchange. These fractionation effects have been measured, calculated, and reported by Mook and colleagues (1974) and Lesniak and Sakai (1989), for example. Both approaches used by Quay and colleagues and Tans and colleagues are hampered by the lack of high-quality oceanwide $^{13}C/^{12}C$ data.

12.3.2 Deep Sea Circulation Rates

One of the first oceanographic applications of radiocarbon was to determine the age of the water in the deep sea. Simple box models, using pre-GEOSECS ^{14}C data, yielded deep water residence times of approximately 1,000 years (e.g., Broecker, 1963). The ^{14}C depth trends in the deep sea, when described by one-dimensional vertical advection-diffusion models, yielded upwelling rates of approximately 2–4 m/yr. With GEOSECS data in hand, Stuiver and colleagues (1983) determined a deep water residence time of approximately 250 years for the Atlantic and Indian oceans and 500 years for the Pacific Ocean. Fiadeiro (1982) used a three-dimensional circulation model to simulate the ^{14}C distribution in the deep Pacific and found that an upwelling rate increasing with depth (from approximately 1–3 m/yr) best explained the mid–deep water ^{14}C minimum.

An extensive study of deep water circulation using ^{14}C was performed by Toggweiler and colleagues (1989a,b), using a primitive equation three-dimensional world ocean general circulation model (OGCM) from the Geophysical Fluid Dynamics Laboratory (GFDL). They found that the model's ventilation rates for the deep Atlantic, Pacific, and Indian oceans depend primarily on the rate of exchange across the Circumpolar Front (40–60°S). They found that the deep Pacific ^{14}C mid-depth minimum was adequately predicted by the model, as was the relatively young and well-mixed water of the Circumpolar Current. The model, however, underestimated the ^{14}C levels for the deep Atlantic because the Antarctic bottom water, with ^{14}C levels lower than the North Atlantic deep water (NADW), penetrated too far north. The GFDL model also predicted the deep ^{14}C levels more accurately for the prognostic version (without restoring temperature and salinity fields) than for the diagnostic model (with temperature and salinity restoring terms); this model response results from enhanced convection in the prognostic version. Toggweiler and colleagues compared their results to those from the three-dimensional Hamburg Large Scale Geostrophic OGCM and found that the Hamburg model underestimated the penetration depth of the overturning cell in the Circumpolar Current and

thus overestimated the ^{14}C levels in the polar Southern Ocean. They concluded that a more detailed evaluation of the three-dimensional OGCM was not possible with the coarse spatial resolution of the GEOSECS ^{14}C data set.

Paleoceanography has been a major beneficiary of the tracer capabilities of oceanic $^{13}C/^{12}C$ measurements. One application that stands out is the use of $^{13}C/^{12}C$ measured on benthic forams as a tracer of the strength of NADW formation during glacial–interglacial transitions (e.g., Boyle and Keigwin, 1987). This variability in NADW formation rates is a crucial component of Broecker's hypothesis that major ocean–atmosphere reorganizations occur during glacial–interglacial transitions (e.g., Broecker and Denton, 1989).

12.3.3 Upper-Ocean Mixing Rates

Bomb ^{14}C measured during GEOSECS and in corals has been used as a tracer of upper-ocean circulation rates. A simple meridional circulation scheme for the upper waters of the Atlantic, Pacific, and Indian oceans was determined from a comparison of the measured bomb ^{14}C inventories and the inventories expected from air–sea CO_2 exchange (Broecker et al., 1985). The underlying observation was that bomb ^{14}C was accumulating in the subtropical gyres faster than could be explained by CO_2 gas exchange. In the equatorial ocean ^{14}C was used to estimate rates of upwelling in the Pacific and Atlantic (Broecker et al., 1978; Wunsch, 1984; Quay et al., 1983). An oceanwide study of bomb ^{14}C as a tracer of upper-ocean circulation was done using the GFDL OGCM by Toggweiler and colleagues (1989b), who included some data from the Transient Tracers in the Ocean (TTO) experiment (1980–81) and from a French cruise in the Indian Ocean (INDIGO) (1985–87) in addition to GEOSECS. The GFDL model predicts successfully the latitudinal distribution of the bomb ^{14}C inventory. Significantly, it predicts that the post-GEOSECS buildup of ^{14}C will be primarily in the deeper thermocline layers of the Southern Ocean. The sub-Antarctic mode water forming at the northern edge of the Circumpolar Current in the GFDL model is the reason for the deep penetration of ^{14}C in the post-GEOSECS era.

12.3.4 Future Oceanic ^{14}C and $^{13}C/^{12}C$ Measurements

Radiocarbon is being measured during the World Ocean Circulation Experiment (WOCE), which began in 1990 in the north Pacific and will continue into the Indian Ocean and then the Atlantic Ocean. During the 10 or so years of the WOCE, $\Delta^{14}C$ will be measured by accelerator mass spectrometry (AMS) on approximately 2.5° latitude spacings and on approximately 30° longitude spacings with a precision of approximately ±7‰. The small sea water volume (0.5 l) sampling intensity for AMS initially applied to depths of less than 1,400 m where bomb ^{14}C resides. For deeper waters, which require higher $\Delta^{14}C$ precision (±3‰) to determine spatial gradients, some large-volume (250 l) ocean water samples have been collected in the Pacific for beta-counting measurement. As of 1994, the WOCE planned to collect only small-volume ^{14}C samples at all depths. The CO_2 extracted from these samples will be archived until AMS precision improves over the next few years.

A high-precision $^{13}C/^{12}C$ measurement program is currently not part of the WOCE program. Although the opportunity exists to obtain an oceanwide high-quality $^{13}C/^{12}C$ WOCE data set, via the AMS ^{14}C samples, the AMS samples are not being measured at this time with high precision ($\pm0.02‰$). Measuring the oceanic change in $^{13}C/^{12}C$ picture is complicated, unfortunately, by the analytical problems associated with the GEOSECS $^{13}C/^{12}C$ data (Kroopnick, 1985). The quality of the GEOSECS $^{13}C/^{12}C$ data must be evaluated on a station-by-station basis.

Currently, oceanic $^{13}C/^{12}C$ measurements are part of NOAA's Global Change Program, and six long-line transects have been measured during the past five years (four in the Pacific Ocean and two in the Atlantic).

12.4 ^{14}C Time Scale Calibration

^{14}C years are "model" years based on, among other things, the assumption of a constant atmospheric ^{14}C level during the past. ^{14}C years differ from "solar" years; therefore, to improve chronological detail of carbon cycle change, ^{14}C ages should be converted to calibrated years. "Calibration 1993," a recent issue of the journal *Radiocarbon* (Stuiver et al., 1993), gives the latest information on ^{14}C age–calibrated age conversion. Detailed ^{14}C age–calibrated age curves are given for Holocene atmospheric samples (derived from ^{14}C measurements of bidecadal wood samples) as well as decadal and single-year information for shorter time intervals.

Although the concept of ^{14}C age calibration is simple, complications arise because the events of interest are usually not represented by perfectly created ^{14}C samples. An example of this problem is a calibrated age determination of the Mazama eruption in the Pacific Northwest (Figure 12.4). Wood and Lake Gyttja samples, in contact with the ash layer boundaries, were ^{14}C dated. Samples QL-4659*, QL-4638*, and QL-4639 are tree branches from Heal Lake, Vancouver Island, whereas QL-1907* and QL-1908 are lake sediment samples from Carp Lake, Washington. The starred samples were collected above the ash layer, the others below. Age calibration of the ^{14}C dates yields the probability curves shown in Figure 12.4. The use of the atmospheric calibration curves is justified for the wood samples because these are fully equilibrated with atmospheric $^{14}CO_2$. The Gyttja ^{14}C ages, however, may be too old because the lake water ^{14}C level may have been lower than that of the atmosphere (^{14}C reservoir deficiency). The reservoir deficiency is unknown for this lake. Therefore, the Figure 12.4 probability plots for the lake sediment samples were derived by assuming zero age ^{14}C reservoir deficiency.

Another question concerns the geological setting of the samples. Are the materials indeed chronologically closely associated with the event of interest? The time it takes for branches to be transferred from the ecosystem surrounding the lake to the sediment bracketing the ash layer is variable, thus potentially increasing the "age" (and the age variability) of the event.

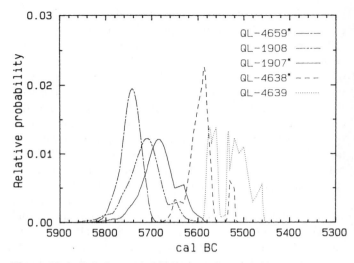

Figure 12.4: Relative probabilities per cal yr (total integrated probability over surface area of each curve is 1.0) obtained for five high-precision dates using Stuiver and Reimer's (1993) calibration program 3.0.3.

The biological ages (growth duration) of the branches and the interval it takes to form the sampled lake sediment also play a role, as the calibration curve must be adjusted for such intervals (e.g., by using the bidecadal data set for a branch containing 20 tree rings, and a five-point moving average for a branch with 100 tree rings). A smoothed calibration curve will give a better-defined midpoint probability profile for regions with major uncertainties about the ^{14}C calibrated age. (Averaging tends to reduce the "wiggliness" of the calibration curve.) For the lake sediment samples similar considerations apply, necessitating knowledge of an average sedimentation rate. Furthermore, because the probability curves derived for these samples are those of their midpoints and not their boundaries (e.g., the youngest growth ring of the branches), the ages had to be shifted by one-half the biological (or "geological") age for a better representation of the contact layer calibrated age.

For a reconciliation of the Figure 12.4 age differences, one can postulate (1) a ^{14}C reservoir deficiency for the sediment samples that resulted in a 150–cal yr shift, and (2) a 150–200–cal yr delay in the incorporation of the QL-4659 wood sample in the sediment. These considerations amount to rejecting the results for three out of five samples and yield a calibrated age in the 5500–5600 B.C. range (circa 6,650 ^{14}C yr BP) for the Mazama eruption.

Horizontal portions (plateaus) of the atmospheric calibration curve add to the calibrated age uncertainty, because a single ^{14}C age is compatible with a range of calibrated ages. These plateaus occur when atmospheric ^{14}C levels decline at the same rate as ^{14}C would decline through radioactive decay, that is, at 1% per 83 cal yr. Such atmospheric ^{14}C declines may relate to sunspot number increases or increased upwelling of older (^{14}C-deficient) waters.

Detailed measured oceanic calibration curves are, as yet, not available. An oceanic calibration curve has been calculated from the atmospheric $\Delta^{14}C$ profile (Stuiver and Braziunas, 1993b) by assuming that the $\Delta^{14}C$ change is solely related to

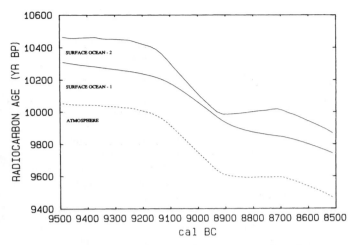

Figure 12.5: Smoothed (200-yr moving average) radiocarbon age profiles for the atmosphere, compared to calculated surface ocean curves 1 and 2. Curve 1 was calculated from a carbon reservoir model assuming atmospheric ^{14}C production rate change to be responsible for the observed atmospheric ^{14}C change, whereas curve 2 was calculated with oceanic circulation change as the causal agent.

production rate change in the atmosphere. In Figure 12.5 we compare a smoothed portion (200-year moving average) of the atmospheric calibration curve to the surface ocean curve (surface ocean – 1) calculated through such global carbon reservoir modeling. Here the $\Delta^{14}C$ decline of 1% per 83 cal yr in the atmosphere for the 8700–8900 B.C. interval is substantially modified in the surface ocean, and the 8700–8900 B.C. plateau disappears. However, when ocean circulation change is assumed to have been the causative factor, the surface ocean ^{14}C calculated from the carbon reservoir (surface ocean – 2) parallels that of the atmosphere because the ocean now forces the atmosphere. Thus, the presence or absence of ^{14}C age plateaus in detailed marine sediment profiles of ^{14}C age versus depth (high sedimentation rate areas only) is of crucial importance in separating forcing factors.

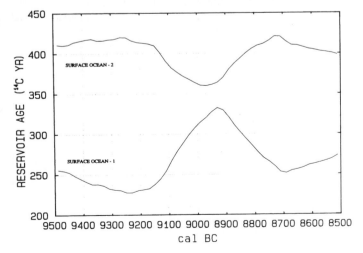

Figure 12.6: Reservoir deficiencies of the model ocean surface for radiocarbon age plateaus generated by production rate change (curve 1) or oceanic circulation change (curve 2).

Surface ocean reservoir ages (Figure 12.6) also differ substantially between the production rate and ocean circulation scenarios. Production-related atmospheric ^{14}C supply to the surface ocean results in concurrent fluxes to the deep ocean, whereas the atmosphere, when forced by the ocean, does not have such major losses to other reservoirs. As a result, the change in reservoir deficiency is larger for the production rate scenario. Reservoir deficiency perturbations also are opposite in sign because the ocean lags the atmosphere for the production rate scenario, whereas the atmosphere lags the ocean for a postulated oceanic increase.

Acknowledgments

The ^{14}C research of the Quaternary Isotope Laboratory was supported through the National Science Foundation and NOAA contract NA16RC0081. P. J. Reimer provided crucial technical support. For the Joint Institute for the Study of the Atmosphere and Ocean (JISAO), this is contribution number 236.

References

Alley, R. B., D. A. Meese, C. A. Shuman, A. J. Gow, K. C. Taylor, P. M. Grootes, J. W. C. White, M. Ram, E. D. Waddington, P. A. Mayewski, and G. A. Zielinski. 1993. Abrupt increase in Greenland snow accumulation at the end of the Younger Dryas event. *Nature 362*, 527–529.

Andrée, M., J. Beer, H. P. Loetscher, E. Moor, H. Oeschger, G. Bonani, H. J. Hofmann, M. Suter, W. Woelfli, and T.-H. Peng. 1986. Limits on the ventilation rate for the deep ocean over the last 12000 years. *Climate Dynamics 1*, 53–62.

Bard, E., M. Arnold, R. G. Fairbanks, and B. Hamelin. 1993. ^{230}Th-^{234}U and ^{14}C ages obtained by mass spectrometry on corals. *Radiocarbon 35*, 191–199.

Bard, E., B. Hamelin, R. G. Fairbanks, and A. Zindler. 1990. Calibration of the ^{14}C timescale over the past 30,000 years using mass spectrometric U-Th ages from Barbados corals. *Nature 345*, 405–410.

Becker, B. 1993. An 11,000-year German oak and pine dendrochronology for radiocarbon calibration. *Radiocarbon 35*, 201–213.

Boyle, E. A., and L. Keigwin. 1987. North Atlantic thermohaline circulation during the past 20,000 years linked to high-latitude surface temperature. *Nature 330*, 35–40.

Broecker, W. S. 1963. Radioisotopes and large scale oceanic mixing. In *The Sea*, M. N. Hill, ed. Vol. 2, 88–108. Interscience Publishers, New York, N.Y.

Broecker, W. S., and G. H. Denton. 1989. The role of ocean–atmosphere reorganizations in glacial cycles. *Geochimica et Cosmochimica Acta 53*, 2465–2501.

Broecker, W. S., T.-H. Peng, and M. Stuiver. 1978. An estimate of the upwelling rate in the equatorial Atlantic based on the distribution of radiocarbon. *Journal of Geophysical Research 83*, 6179–6186.

Broecker, W. S, T.-H. Peng, H. G. Ostlund, and M. Stuiver. 1985. The distribution of bomb radiocarbon in the ocean. *Journal of Geophysical Research 90*, 6953–6970.

Eddy, J. A. 1976. The Maunder Minimum. *Science 192*, 1189–1202.

Edwards, R. L., J. W. Beck, G. S. Burr, D. J. Donahue, J. M. A. Chappell, A. L. Bloom, E. R. M. Druffel, and F. W. Taylor. 1993.

A large drop in atmospheric $^{14}C/^{12}C$ and reduced melting in the Younger Dryas, documented with ^{230}Th ages of corals. *Science 260*, 962–968.

Fiadeiro, M. E. 1982. Three-dimensional modeling of tracers in the deep Pacific Ocean: II. Radiocarbon and the circulation. *Journal of Marine Research 40*, 537–550.

Johnson, S. J., B. Clausen, W. Dansgaard, K. Fuhrer, N. Gundestrup, C. U. Hammer, P. Iversen, J. Jouzel, B. Stauffer, and J. P. Steffenson. 1993. Irregular glacial interstadials recorded in a new Greenland ice core. *Nature 359*, 311–313.

Kromer, B., and B. Becker. 1993. German oak and pine ^{14}C calibration, 7200–9400 BC. *Radiocarbon 35*, 125–135.

Kroopnick, P. 1985. The distribution of ^{13}C of ΣCO_2 in the world oceans. *Deep Sea Research 32*, 57–84.

Lesniak, P. M., and H. Sakai. 1989. Carbon isotope fractionation between dissolved carbonate (CO_3^{2-}) and $CO_2(g)$ at 25°C and 40°C. *Earth and Planetary Science Letters 95*, 297–301.

Linick, W. L., A. Long, P. E. Damon, and C. W. Ferguson. 1986. High-precision radiocarbon dating of bristlecone pine from 6554 to 5350 BC. *Radiocarbon 28*, 943–953.

Mazaud, A., C. Laj, E. Bard, M. Arnold, and E. Tric. 1991. Geomagnetic field control of ^{14}C production over the last 80 ky: Implications for the radiocarbon time-scale. *Geophysical Research Letters 18*, 1885–1888.

Mook, W. G., J. C. Bommerson, and W. H. Staverman. 1974. Carbon isotope fractionation between dissolved bicarbonate and gaseous carbon dioxide. *Earth and Planetary Science Letters 22*, 169–176.

Oeschger, H., U. Siegenthaler, U. Schotterer, and A. Gugelmann. 1975. A box-diffusion model to study the carbon dioxide exchange in nature. *Tellus 27*, 168–192.

Pearson, G. W., and M. Stuiver. 1993. High-precision bidecadal calibration of the radiocarbon timescale, 500–2500 BC. *Radiocarbon 35*, 25–33.

Quay, P. D., M. Stuiver, and W. S. Broecker. 1983. Upwelling rates for the equatorial Pacific Ocean derived from the bomb ^{14}C distribution. *Journal of Marine Research 41*, 769–792.

Quay, P. D., B. Tilbrooke, and C. S. Wong. 1992. Oceanic uptake of fossil fuel CO_2: Carbon-13 evidence. *Science 256*, 74–79.

Shackleton, N. J., J.-C. Duplessy, M. Arnold, P. Maurice, M. A. Hall, and J. Cartlidge. 1988. Radiocarbon age of last glacial Pacific deep water. *Nature 335*, 708–711.

Siegenthaler, U., and F. Joos. 1992. Use of a simple model for studying oceanic tracer distributions and the global carbon cycle. *Tellus 44B*, 186–207.

Stuiver, M., and T. F. Braziunas. 1993a. Sun, ocean, climate and atmospheric $^{14}CO_2$: An evaluation of causal and spectral relationships. *The Holocene 3*, 289–305.

Stuiver, M., and T. F. Braziunas. 1993b. Modeling atmospheric ^{14}C influences and radiocarbon ages of marine samples to 10,000 BC. *Radiocarbon 35*, 137–189.

Stuiver, M., T. F. Braziunas, B. Becker, and B. Kromer. 1991. Climatic, solar, oceanic, and geomagnetic influences on late-glacial and Holocene atmospheric $^{14}C/^{12}C$ change. *Quaternary Research 35*, 1–24.

Stuiver, M., A. Long, and R. S. Kra (eds.). 1993. Calibration 1993. *Radiocarbon 35(1)*, special issue.

Stuiver, M., and G. W. Pearson. 1993. High-precision bidecadal calibration of the radiocarbon time scale, AD 1950–500 BC and 2500–6000 BC. *Radiocarbon 35*, 1–23.

Stuiver, M., and H. A. Polach. 1977. Discussion: Reporting of ^{14}C data. *Radiocarbon 19*, 355–363.

Stuiver, M., and P. D. Quay. 1980. Changes in atmospheric carbon-14 attributed to a variable sun. *Science 207*, 11–19.

Stuiver, M., and P. D. Quay. 1981. Atmospheric ^{14}C changes resulting from fossil fuel CO_2 release and cosmic ray flux variability. *Earth and Planetary Science Letters 53*, 349–362.

Stuiver, M., P. D. Quay, and H. G. Ostlund. 1983. Abyssal water carbon-14 distribution and the age of the world oceans. *Science 219*, 849–851.

Stuiver, M., and P. J. Reimer. 1993. Extended ^{14}C data base and revised CALIB 3. 0 ^{14}C age calibration program. *Radiocarbon 35*, 215–230.

Tans, P. P., J. A. Berry, and R. F. Keeling. 1993. Oceanic $^{13}C/^{12}C$ observations: A new window on ocean CO_2 uptake. *Global Biogeochemical Cycles 7*, 353–368.

Taylor, K. C., G. W. Lamorey, G. A. Doyle, R. B. Alley, P. M. Grootes, P. A. Mayewski, J. W. C. White, and L. K. Barlow. 1993. The "flickering switch" of late Pleistocene climate change. *Nature 361*, 432–436.

Toggweiler, J. R., K. Dixon, and K. Bryan. 1989a. Simulations of radiocarbon in a coarse resolution world ocean model: I. Steady state, pre-bomb distributions. *Journal of Geophysical Research 94*, 8218–8242.

Toggweiler, J. R., K. Dixon, and K. Bryan. 1989b. Simulations of radiocarbon in a coarse resolution world ocean model: II. Distributions of bomb-produced ^{14}C. *Journal of Geophysical Research 94*, 8243–8264.

Wunsch, C. 1984. An estimate of the upwelling rate in the equatorial Atlantic Ocean based on the distribution of bomb radiocarbon and quasi-geostrophic dynamics. *Journal of Geophysical Research 89*, 7971–7978.

13

Shallow Water Carbonate Deposition and Its Effect on the Carbon Cycle

BRADLEY N. OPDYKE

Abstract

Carbonate reefs and platforms have accumulated $CaCO_3$ at a rate of 8–9 $\times$ 10^{12} mol/yr over the last few million years. The Holocene rate of shallow water $CaCO_3$ deposition is approximately 17 $\times$ 10^{12} mol/yr. In order for the shallow water $CaCO_3$ flux to maintain its long-term average deposition rate, it must decline to below 8 $\times$ 10^{12} mol/yr during glacial intervals. Shallow water $CaCO_3$ sediments represent a large, dynamic carbon reservoir that rapidly affects the alkalinity of the surface ocean and hence the CO_2 content of the atmosphere. Shallow water carbonate deposition, while probably an important constraint on paleoatmospheric CO_2 concentrations, can only slightly influence the anthropogenically driven buildup of atmospheric CO_2.

13.1 Introduction

In order for the record of atmospheric CO_2 changes contained in glacial ice (Neftel et al., 1982; Barnola et al., 1987) to reflect changes in the deposition of marine carbonate, shallow water deposition during interglacial intervals must be significantly higher than the long-term average, and the global weathering of near–sea level carbonates must contribute an increased flux of dissolved calcium carbonate ($CaCO_3$) to the oceans during glacial low stands (Milliman, 1974). Review of research on the global Holocene shallow water carbonate flux reveals that the size of the shallow water carbonate reservoir is not well constrained. The recent improvement in numerical simulations of the carbon cycle using shallow water (neritic) carbonate deposition as a forcing parameter would allow a more precise estimate of the importance of the CO_2 input via this mechanism if the neritic carbonate flux were more accurately known.

13.2 Historical Background

For at least four decades, earth scientists have been interested in determining the mass of $CaCO_3$ deposition in the shallow and deep ocean. Traditionally, carbonate sedimentation has been divided into either shallow water (neritic) or deep water (pelagic) sedimentation. Although there have been many attempts to quantify the $CaCO_3$ depositional flux, a precise quantification has yet to be achieved.

Ironically, the mass of pelagic carbonate preserved in the deep ocean is better known than the neritic carbonate mass. In 1963, Turekian estimated that fluxes to neritic and pelagic environments were about equal. Concurrent with the publication of Turekian's paper, Livingstone (1963) estimated the riverine calcium flux to the global ocean to be approximately 1.1 $\times$ 10^{12} mol/yr. An early compilation of Atlantic and Pacific piston core data supplied the means to estimate the pelagic carbonate flux to the ocean, and Turekian (1965) calculated that it should account for the vast majority of the riverine calcium flux. Using the assumption that the calcium input must match

the rate of $CaCO_3$ deposition, he then concluded that neritic deposition must be negligible. This became the prevailing view for at least a decade, echoed by Garrels and Mackenzie (1971, pp. 216–218), who reported that "in shallow water areas the accumulation of carbonate is quantitatively unknown and of minor importance."

Although the geographical extent, though not the surface areas, of coral reef–related carbonate deposition had been known for some time (Wells, 1957; Figure 13.1), progress continued in the understanding of rates of deposition in neritic environments. Chave and colleagues (1972) compiled coral reef production rates published during the previous 100 years (a total of 19 articles) and made an estimate of present-day "reef system" net accumulation ($CaCO_3$ actually retained by the reef) of about 10^3 g/m²/yr. They allowed, however, that because the sea level is actively rising, net production may reach "up to more than 2 $\times$ 10^4 g $CaCO_3$/m²/yr" (p. 123).

Milliman (1974) estimated the global "reefal" area at double that of the Great Barrier Reef, the Bahamas, and all recorded atolls. His estimate for the area of neritic carbonate deposition was approximately 1.4 $\times$ 10^6 km². Milliman used an "average" deposition rate for these environments of 350 g/m²/yr, which corresponds to a global deposition rate of 0.5 $\times$ 10^{15} g/y, or 5 $\times$ 10^{12} mol/yr. Milliman noted that this is more than an order of magnitude higher than rates from ancient rock and proposed that the carbonate system is out of balance on a 10-kyr time scale, but is in balance over hundreds of thousands of years (Milliman, 1974, p. 248). He concluded that long-term equilibrium can be achieved by deposition of calcium carbonate on shallow shelves during interglacials and erosion of carbonate from those same shelves during glacial intervals (Figure 13.2).

The first semiquantitative assessment of the areas of "coral reef" complexes was completed by Smith (1978). He accomplished this by combining the data known about the distribution of coral reef complexes (Wells, 1957) with the average hypsometry of the continental shelves (Menard and Smith, 1966). He concluded that the total area of coral reef–related $CaCO_3$ accumulation was probably 6.2 $\times$ 10^5 km² ± 25%. After estimating that "average" net $CaCO_3$ production from the reef front to the lagoon was 10 mol $CaCO_3$ m²/yr (1,000 g/m²/yr), Smith calculated a global reef-related carbonate production rate of 6 $\times$ 10^{12} mol/yr.

Schlager (1981) compiled carbonate accumulation rate data for Holocene sediments using a slightly different technique, viewing sedimentation in terms of vertical growth. He states, "Average growth potential is in the 1 mm/yr range" (p. 201), which translates into slightly higher neritic $CaCO_3$ fluxes. Assuming an average porosity of 50% and density of 2.9 g/cm³ (aragonite), this yields a mass accumulation rate of 1,450 g/m²/yr and, using Smith's (1978) estimate of reefal area, gives an average global neritic accumulation of 8.1 $\times$ 10^{14} g/yr (8.1 $\times$ 10^{12} mol/yr).

The flux of Ca to the global ocean was also reassessed (e.g., Edmond et al., 1979, 1982). Owing to largely high-temperature reactions between seawater and basalt, magne-

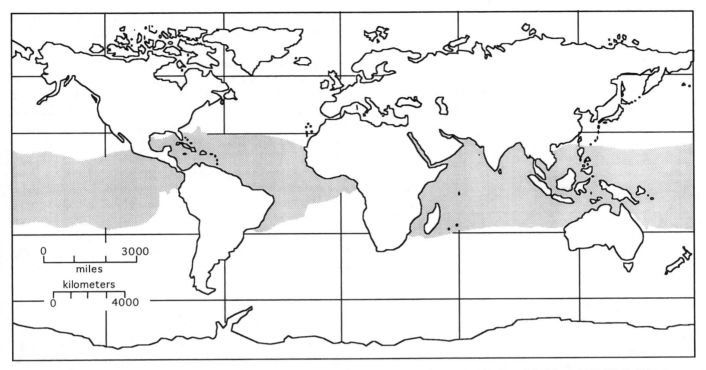

Figure 13.1: The stippled area encloses the shallow water carbonate reefs and banks of the world. Simplified from J. W. Wells (1957), *Coral Reefs*. In *Treatise on Marine Ecology,* ed. J. Hedgereth, Geological Society of America Memoir 67: 609–31.

sium (Mg) is removed from sea water to form a smectite or chlorite, while calcium (Ca) is released into seawater during the breakdown of Ca-plagioclase (Drever, 1982). The flux of Ca to the ocean as a result of this mechanism has been estimated to be between $2–3 \times 10^{12}$ mol/yr (Edmond et al., 1979) and 6×10^{12} mol/yr (Wilkinson and Algeo, 1989). These data, along with a new calculation of the riverine flux of Ca to the ocean of 13.7×10^{12} mol/yr (Meybeck, 1979), pushed the accepted total flux of Ca to the oceans to near 20×10^{12} mol/yr. The pelagic flux of $CaCO_3$ was taken to be approximately 13×10^{12} mol/yr (Davies and Worsley, 1981).

Some of the best data collected and published over the last decade have been related to work done on the Great Barrier Reef and in Papua New Guinea (e.g., Chappell and Polach, 1976; Davies and Hopley, 1983; Kinsey and Hopley, 1991). These include many radiometric dates, so that the depositional histories of entire reef systems can be documented (Davies and Hopley, 1983).

Kinsey and Hopley (1991) also assumed that Smith's (1978) global reef estimate of 6.2×10^5 km² is correct. Since the Great Barrier Reef totals 2.0×10^4 km², this estimate, combined with 11.5×10^4 km² for the atolls of the world, leaves 4.8×10^5 km² of reef area undocumented. The remaining total was estimated to be 1.6×10^5 km² of shelf or oceanic island barrier complexes (akin to the Great Barrier Reef) and 3.2×10^5 km² of fringing reefs. They calculated Holocene reef-related $CaCO_3$ accumulation to be approximately 11×10^{12} mol/yr (Table 13.1).

None of the above estimates of carbonate production, however, provided a systematic method with which to integrate $CaCO_3$ deposition from the fore reef across to the lagoon. What was needed was a more inclusive way to quantify the amount of $CaCO_3$ deposited in areas of neritic carbonate accumulation worldwide.

13.3 An Integrated Approach

Determination of the neritic carbonate accumulation rate, similar to that of other sediment, is strongly influenced by the period over which sedimentation is measured (Sadler, 1981; Anders et al., 1987). When available data on neritic carbonate accumulation rates from different neritic sedimentary environments are plotted against duration of sedimentation, the data follow a logarithmic distribution (Wilkinson et al., 1991; Opdyke and Wilkinson, 1993; Figure 13.3A). In general, accumulation rates measured at a 1,000-year time scale will be higher than when sedimentary sections that represent longer time intervals are measured. The longer the duration of sedimentation, the more hiatuses and erosional events are likely to be included in the record, and the uncertainty of the ^{14}C measurements (and U-Th measurements) increases in measurements made over shorter time intervals.

Describing all the various neritic sedimentary environments, such as reef-related deposits, mudbanks (e.g., Ginsburg

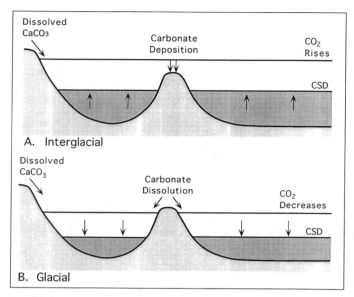

Figure 13.2: A qualitative (A) interglacial to (B) glacial comparison of CaCO₃ deposition and erosion to and from carbonate reefs and platforms. CSD is carbonate saturation depth. Reprinted with permission from J. D. Milliman (1974), Marine carbonates. In *Recent Sedimentary Carbonates*, vol. 1, eds. J. D. Milliman, G. Mueller, and U. Foerstner, Berlin: Springer-Verlag. © 1974, Springer-Verlag.

and James, 1974), and ooid shoals (e.g., Martin and Ginsburg, 1965), with a single equation allows a more quantitative calculation of the global neritic flux of $CaCO_3$. To this end, 5 kyr was chosen as the time interval to which each datum would be normalized (Opdyke and Wilkinson, 1993; Figure 13.3B). This calculation results in a global neritic carbonate accumulation rate of approximately 1.1 m/kyr (±0.5 m/kyr) (Figure 13.3B), which translates to 1.62×10^3 g/m²/yr (± 0.7 × 10³ g/m²/yr) as a global average deposition rate wherever shallow

water carbonate is accumulating. Normalizing to 5 kyr is appropriate because it is the time interval over which typical global carbon cycle models (at Holocene time scales) return to steady state after a perturbation is introduced.

This estimate of accumulation rate applies not only to the coral reef complexes, but also to mudbanks such as Florida Bay, ooid shoals such as those rimming parts of the Bahama Banks, and carbonate depositional areas such as those found in the Persian Gulf. Conservatively estimating these areas as an additional 2.0×10^5 km² brings the total area of neritic carbonate accumulation to approximately 8.0×10^5 km². If we accept the probable errors of ±25% (Smith, 1978) for these area estimates and use 1.1 m/kyr as the global average vertical flux of $CaCO_3$ sedimentation, the total neritic flux is between 10 and 16×10^{12} mol/yr of $CaCO_3$. Bank margin sedimentation may account for another 4×10^{12} mol/yr (e.g., Wilber et al., 1990).

13.4 The Coral Reef Hypothesis

The coral reef hypothesis (Berger, 1982; Berger and Killingley, 1982; Berger and Keir, 1984; Keir and Berger, 1985) was originally proposed to help explain the observed increase in atmospheric CO_2 content recorded in glacial ice during the last deglacial interval (e.g., Neftel et al., 1982; Barnola et al., 1987). $CaCO_3$ deposition in the surface ocean supplies CO_2 to the ocean–atmosphere system through the following reaction: $Ca^{2+} + 2HCO_3^-$ yields $CaCO_3 + CO_2 + H_2O$. The coral reef hypothesis requires an acceleration of deposition of neritic carbonates as sea level rises, reducing the alkalinity of surface water as well as adding CO_2 into the ocean–atmosphere system. This results in a net transfer of $CaCO_3$ deposition from the deep sea to the shallow shelves during the course of deglaciation and throughout the following interglacial. Early "coral reef" models simulated relatively short durations of time. They typically represented model runs of

Table 13.1. *Estimates of Carbonate Flux by Various Authors*

Author	Estimated Pelagic Flux (10¹² mol/yr)	Average Neritic Accumulation Rate (g/m²/yr)	Area of Neritic CaCO₃ (10⁶ km²)	Estimated Neritic Flux (10¹² mol/yr)
Turekian (1965)	11	Negligible	NA	NA
Garrels and Mackenzie (1971)	11	Negligible	NA	NA
Chave (1972)	NA	1,000–20,000	NA	NA
Milliman (1974)	12	350	1.4	5
Smith (1978)	NA	1,000	0.6	6
Schlager (1981)	NA	1,450	0.6	8.1
Davies and Worsley (1981)	13	NA	NA	NA
Kinsey and Hopley (1991)	NA	1,812	0.6	11
This study	13	1,600	0.8	14–20

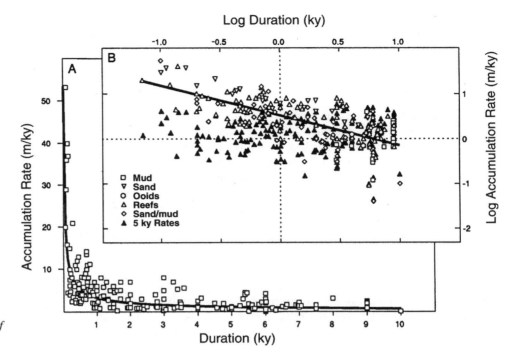

Figure 13.3: Holocene carbonate accumulation rates ($n = 220$) in platform settings. (A) Accumulation rate relative to time duration of accumulation. (B) Filled symbols, log accumulation rate vs. log duration for various carbonate facies; open symbols, 5 kyr duration values (A_5) calculated from duration (Dr) and rate (Ar) data as: $\log A_5 = (\log Ar) + (0.73 \log Dr) + 1.40$. (Adapted from Opdyke and Wilkinson, 1993.) Reprinted with permission of American Journal of Science, from B. N. Opdyke and B. H. Wilkinson (1993), *American Journal of Science* 293: 217–34.

10 kyr and envisioned neritic carbonate deposition in terms of depositional pulses or a single pulse on the order of 1,000–2,000 years in length. Oceanic alkalinity and weathering rates were held constant, and the maximum flux of neritic carbonate was typically assumed to be on the order of 10×10^{12} mol/yr.

Assuming that long-term shallow water carbonate accumulation rates are 8–9×10^{12} mol/yr and Holocene rates may be closer to 20×10^{12} mol/yr, Opdyke and Walker (1992) published a revision of the coral reef hypothesis. If we accept the long-term (million-year time scale) value of the shallow water carbonate flux and assume that Holocene deposition rates are approximately twice that rate, then lower neritic accumulation rates must accompany falling sea level and glacial periods. Weathering of newly exposed carbonate banks is assumed to increase as a falling sea level exposes larger areas of these deposits. By this mechanism, alkalinity in the deep ocean must increase as sea level falls. As this occurs, the carbonate saturation depth (CSD) must increase, allowing more pelagic $CaCO_3$ deposition in the deep sea, if noncarbonate productivity in the ocean remains the same.

There are testable limitations to how much the global CSD has varied from the Pleistocene into the Holocene. The distinct glacial–interglacial sedimentary cycles found in the central Pacific provide an ideal test site for measuring the extent to which deep sea alkalinity has changed (Arrhenius, 1952; Farrell and Prell, 1989, 1991). If one assumes that pelagic carbonate sediments perfectly record changes in the CSD, then the potential impact of changing alkalinity and hence changing CO_2 by this mechanism is probably limited to a maximum of 22 ppm (Broecker and Peng, 1993; Oxburgh and Broecker, 1993). However, if one accepts that the deep ocean saturation state is dynamic and rarely if ever reaches equilibrium with the

pelagic carbonate sediments, then atmospheric CO_2 change by the coral reef mechanism may account for more than 50 ppm of the observed glacial–interglacial change in CO_2 (Walker and Opdyke, 1995). This is a significant proportion of the 80–90 ppm change in atmospheric CO_2 observed in the Greenland and Antarctic ice cores during the last deglaciation (Neftel et al., 1982; Barnola et al., 1987).

13.5 "Coral Reef" Influence on the Future of Atmospheric CO_2 Contents

Neritic carbonate is likely to play a minor role in either mitigating or augmenting future atmospheric CO_2 contents while anthropogenic inputs continue. Two processes will influence how coral reef communities react to increased concentrations of atmospheric CO_2. First, if significant global warming does indeed occur and sea level does begin to rise, then accommodation space for the reef communities will increase, perhaps doubling their growth rate and subsequent output of CO_2 (Kinsey and Hopley, 1991) and therefore acting as a positive feedback for global warming. Additionally, the possibility exists that global warming will expand the habitats suitable for coral reef deposition to higher latitudes, possibly increasing the magnitude of this positive feedback. Second, it is possible that tropical coral reefs are at the upper limits of temperature tolerance and that any increase in surface water temperature will lead to bleaching of the coral heads and death of the coral reef community. This, in addition to direct human destruction of coral reefs through mining or pollution, would reduce the natural levels of CO_2 output from this system. The change in CO_2 by either one of these mechanisms is probably no more than $\pm 10 \times 10^{12}$ mol/yr, or approximately 2% of the present anthropogenic source of CO_2 to the atmosphere (5.4 Gt C/yr; Sundquist, 1993).

13.6 Summary

Although atmospheric CO_2 currently derived from neritic carbonate deposition represents only 2% of the present flux to the atmosphere, it may have been a very significant source of CO_2 in the past. Integrated global neritic carbonate sedimentation rates are approximately 1.1 m/kyr. If the approximate area for coral reefs, mudbanks, and ooid shoals is 800,000 km², then an intermediate estimate of the total Holocene flux of $CaCO_3$ to the shallow water zone (including offbank deposits) is 17×10^{12} mol/yr. Any estimate of the current neritic mass flux contains large uncertainties. Further refinement of the size of the source of atmospheric CO_2 from Holocene neritic carbonate requires a larger data base with respect to its areas, ages, and thicknesses of deposition. Given that the long-term (million-year) average neritic carbonate deposition is approximately 9×10^{12} mol/yr, then neritic carbonate deposition must have a substantial impact – perhaps as much as 50 ppm of the observed 80–90 ppm CO_2 rise as recorded in glacial ice during the last deglaciation. It is exciting to speculate that the "coral reef" response to glacially induced sea level rise may be an important positive feedback within Quaternary climate cycles.

Acknowledgments

Early drafts of this chapter were edited by Nancy Opdyke and Bruce Wilkinson. Later versions benefited from the observations of anonymous reviewers. Dale Austin and Dick Barwick aided with the drafting. Their help was greatly appreciated.

References

Anders, M. H., S. W. Krueger, and P. M. Sadler. 1987. A new look at sedimentation rates and the completeness of the stratigraphic record. *Journal of Geology 95,* 1–14.

Arrhenius, G. O. S. 1952. Sediment cores from the east Pacific. In *Reports of the Swedish Deep Sea Expedition, 1947–1948,* H. Pettersson, ed. Vol. 5, Elanders Boktr., Göteborg, Sweden, pp. 1–227.

Barnola, J. M., D. Raynaud, Y. S. Korotkevich, and C. Lorius. 1987. Vostok ice core provides 160,000 year record of atmospheric CO_2. *Nature 329,* 408–414.

Berger, W. H. 1982. Increase of carbon dioxide in the atmosphere during deglaciation: The coral reef hypothesis. *Naturwissenschaften 69,* 87–88.

Berger, W. H., and R. S. Kier. 1984. Glacial-Holocene changes in atmospheric CO_2 and the deep-sea record. In *Climate Processes and Climate Sensitivity,* J. E. Hansen and T. S. Takahashi, eds. AGU Monograph No. 29, American Geophysical Union, Washington, D.C., pp. 337–351.

Berger, W. H., and J. S. Killingley. 1982. The Worthington Effect and the origin of the Younger Dryas. *Journal of Marine Research 40,* supplement, 27–38.

Broecker, W. S., and T.-H. Peng. 1993. What caused glacial to interglacial CO_2 change? In *The Global Carbon Cycle,* M. Heimann,

ed. Proceedings of the NATO ASI at Il Ciocco, Italy, September 8–20, 1991. Springer-Verlag, Berlin, Germany, pp. 95–115.

Chappell, J., and H. A. Polach. 1976. Holocene sea-level change and coral-reef growth at Huon Peninsula, Papua New Guinea. *Geological Society of America Bulletin 87,* 235–240.

Chave, K. E., S. V. Smith, and K. J. Roy. 1972. Carbonate production by coral reefs. *Marine Geology 12,* 123–140.

Davies, P. J., and D. Hopley. 1983. Growth fabrics and growth rates of Holocene reefs in the Great Barrier Reef. *Journal of Australian Geology and Geophysics208,* 237–251.

Davies, T. A., and T. R. Worsley. 1981. Paleoenvironmental implications of oceanic carbonate sedimentation rates. *Society of Economic Paleontologists and Mineralogists Special Publication 32,* 169–179.

Drever, J. I. 1982. *The Geochemistry of Natural Waters.* Prentice-Hall, Englewood Cliffs, New Jersey.

Edmond, J. M., C. Measures, R. E. McDuff, L. H. Chan, R. Collier, B. Grant, L. I. Gordon, and J. B. Corliss. 1979. Ridge crest hydrothermal activity and the balances of the major and minor elements in the ocean: The Galapagos data. *Earth and Planetary Science Letters 46,* 1–18.

Edmond, J. M., K. L. Von Damm, R. E. McDuff, and C. I. Measures. 1982. Chemistry of hot springs on the East Pacific Rise and effluent dispersal. *Nature 297,* 187–191.

Farrell, J. W., and W. L. Prell. 1989. Climatic change and $CaCO_3$ preservation: An 800,000 year bathymetric reconstruction from the Central Equatorial Pacific Ocean. *Paleoceanography 4,* 447–466.

Farrell, J. W., and W. L. Prell. 1991. Pacific $CaCO_3$ preservation and $\Delta^{18}O$ since 4 Ma: Paleoceanic and paleoclimatic implications. *Paleoceanography 6,* 485–498.

Garrels, R. M., and F. T. Mackenzie. 1971. *Evolution of Sedimentary Rocks.* W. W. Norton, New York, pp. 216–218.

Ginsburg, R. N., and N. P. James. 1974. Holocene carbonate sediments of continental shelves. In *The Geology of Continental Margins,* C. A. Burke and C. C. Drake, eds. Springer-Verlag, Berlin, Germany, pp. 137–155.

Keir, R. S., and W. H. Berger. 1985. Late Holocene carbonate dissolution in the equatorial Pacific: Reef growth or neoglaciation? In *The Carbon Cycle and Atmospheric CO_2: Natural Variations Archean to Present,* E. T. Sundquist and W. S. Broecker, eds. American Geophysical Union Geophysical Monograph 32, American Geophysical Union, Washington, D.C., pp. 208–220.

Kinsey, D. W., and D. Hopley. 1991. The significance of coral reefs as global carbon sinks – Response to greenhouse. *Palaeogeography, Palaeoclimatology, Palaeoecology 89,* 1–15.

Livingstone, D. A. 1963. Chemical composition of rivers and lakes. In *Data of Geochemistry,* 6th edition, M. Fleischer, ed. USGS Professional Paper 440-G, U.S. Geological Survey, Washington, D.C.

Martin, E. L., and R. N. Ginsburg. 1965. Radiocarbon ages of oolitic sands on Great Bahama Bank. In *Proceedings, 6th International Conference of Radiocarbon and Tritium Dating,* U.S. Atomic Energy Commission Conference Report 650652, U.S. Atomic Energy Commission, Washington, D.C., pp. 705–719.

Menard, H. W., and S. M. Smith. 1966. Hypsommetry of ocean basin provinces. *Journal of Geophysical Research 71,* 4305–4325.

Meybeck, M. 1979. Concentrations des eaux fluvials en éléments majeurs et apports en solution aux oceans. *Revue de Geologie Dynamique et de Geographie Physique 21,* 215–246.

Milliman, J. D. 1974. Marine carbonates. In *Recent Sedimentary Carbonates,* Vol. 1, J. D. Milliman, G. Mueller, and U. Foerstner, eds. Springer-Verlag, Berlin, Germany.

Neftel, A., H. Oeschger, J. Schwander, B. Stauffer, and R. Zumbrunn. 1982. Ice core sample measurements give atmospheric CO_2 content during the past 40,000 years. *Nature 295,* 220–233.

Opdyke, B. N., and J. C. G. Walker. 1992. The return of the coral reef hypothesis: Glacial to interglacial partitioning of basin to shelf carbonate and its effect on Holocene atmospheric pCO_2. *Geology 20,* 733–736.

Opdyke, B. N., and B. H. Wilkinson. 1993. Carbonate mineral saturation state and cratonic limestone accumulation. *American Journal of Science 293,* 217–234.

Oxburgh, R., and W. S. Broecker. 1993. Pacific carbonate dissolution revisited. *Palaeogeography, Palaeoclimatology, Palaeoecology 103,* 31–39.

Sadler, P. M. 1981. Sediment accumulation rates and the completeness of stratigraphic sections. *Journal of Geology 89,* 569–584.

Schlager, W. 1981. The paradox of drowned reefs and carbonate platforms. *Geological Society of America Bulletin 92,* 197–211.

Smith, S. V. 1978. Coral-reef area and the contributions of reefs to processes and resources of the world's oceans. *Nature 273,* 225.

Sundquist, E. T. 1993. The global carbon dioxide budget. *Science 259,* 934–941.

Turekian, K. K. 1963. Rates of calcium carbonate deposition by deep-sea organisms, molluscs and coral-algae association. *Nature 197,* 277.

Turekian, K. K. 1965. Some aspects of the geochemistry of marine sediments. In *Chemical Oceanography,* Vol. 2, J. P. Riley and G. Skirrow, eds. Academic Press, New York, pp. 81–126.

Walker, J. C. G., and B. N. Opdyke. 1995. Influence of variable rates of neritic carbonate deposition on atmospheric carbon dioxide and pelagic sediments. *Paleoceanography 10–3,* 415–427.

Wells, J. W. 1957. Coral reefs. In *Treatise on Marine Ecology,* J. Hedgpeth, ed. Geological Society of America Memoir 67, Geological Society of America, Boulder, Colo., pp. 609–631.

Wilber, R. J., J. D. Milliman, and R. B. Halley. 1990. Accumulation of bank-top sediment on the western slope of the Great Bahama Bank: Rapid progradation of a carbonate megabank. *Geology 18,* 970–974.

Wilkinson, B. H., and T. Algeo. 1989. Sedimentary carbonate record of calcium-magnesium cycling. *American Journal of Science 289,* 1158–1194.

Wilkinson, B. H., B. N. Opdyke, and T. Algeo. 1991. Time partitioning in cratonic carbonate rocks. *Geology 19,* 1093–1096.

Part IV

Modeling CO$_2$ Changes

14

Future Fossil Fuel Carbon Emissions without Policy Intervention: A Review

JAE EDMONDS, RICHARD RICHELS, AND MARSHALL WISE

Abstract

This chapter surveys the literature regarding potential future fossil fuel carbon emissions in the absence of explicit control policies. We have assembled 30 base cases and uncertainty analysis trajectories from 18 separate analyses of fossil fuel carbon emissions for comparison to the Intergovernmental Panel on Climate Change (IPCC) 1991 *Integrated Analysis of Country Case Studies*. We discuss global forecasts of fossil fuel carbon emissions and associated energy consumption, regional forecasts of fossil fuel carbon emissions and associated energy production and consumption, analyses that have explicitly explored the uncertainty associated with global energy and fossil fuel carbon emissions, and differences in key assumptions among various base cases.

14.1 Introduction

In our survey of the literature on potential future fossil fuel carbon emissions in the absence of explicit control policies, we have assembled 30 base cases and uncertainty analysis trajectories from 18 separate analyses (Table 14.1, column 2) of fossil fuel carbon emissions for comparison to the Intergovernmental Panel on Climate Change (IPCC, 1991). A list of the studies, dates of publication, and models used is given in Table 14.1. Six of these trajectories have been drawn from the results of the 12th Energy Modeling Forum, "Global Climate Change: Energy Sector Impacts of Greenhouse Gas Emission Control Strategies" (EMF-12), and reflect a comparison of base cases with some standardization of assumptions. We have made no attempt to create an assessment of models. Several thorough literature reviews already perform that function. These include those by Nordhaus (1989), Hoeller and Wallin (1991), Rothman and Chapman (1991), Darmstadter (1991), Edmonds and Barns (1991), and Bradley and colleagues (1991).

14.2 Long-Term Projections of Global Fossil Fuel Carbon Emissions

A comparison of global fossil fuel carbon emissions baselines or base cases is given in Figures 14.1A and 14.1B, and a comparison of uncertainty cases is provided in Figure 14.1C. Figure 14.1A shows base cases for a variety of independently conducted studies; Figure 14.1B shows base cases for the seven studies included in EMF-12. Because the EMF studies shared a great number of key assumptions by virtue of the EMF-12 format, the variation in results, particularly in the post-2025 period, is not as great as that for the base cases shown in Figure 14.1A, for which no control over underlying assumptions was imposed, and which include cases developed over almost a decade of research.

Global fossil fuel carbon emissions in 1990 have been estimated to be 6.120Pg20C/yr as compared with 5.7 Pg C/yr in 1987 (Marland et al., 1994). Several observations regarding forecast emissions are worth making. Emissions increase over time in all cases except the IPCC IS92c case (see Table 14.1) and the extremes in uncertainty analysis results for Nordhaus and Yohe, Edmonds and colleagues, and Manne and Richels. For the year 2025, emissions ranged from 7.6 Pg C/yr (EPA SCW case) to 14 Pg C/yr (IIASA high case). The emissions in the EMF-12 cases span a far narrower range, from 9.4 Pg C/yr (Barns et al., 1991) to 11.6 Pg C/yr (Manne and Richels EMF). For the year 2025, the 25th and 75th percentiles span the range 5–13 Pg C/yr in Edmonds and colleagues UA. In comparison, the Nordhaus and Yohe UA estimated the same range to be 7–13 Pg C/yr.

By the year 2100, the range will have expanded substantially. All emissions in the base cases' are greater than the estimated emissions from the year 1990. The highest individual base case emission is 33 Pg C/yr (IPCC IS92e case); the lowest is 10.4 Pg C/yr (EPA SCW case in Table 14.1). This range does not encompass the EMF-12 base cases. These range between 20 Pg C/yr (Global-Macro) and 43 Pg C/yr (CETA). This range may seem large; however, Edmonds and colleagues' uncertainty analysis shows a range of 2–87 Pg C/yr for the 5th and 95th percentile cases, respectively, in 2075, and a range of 4–27 Pg C/yr between the 25th and 75th percentiles, a range wider than that spanned by the base cases in all other studies. The Nordhaus and Yohe uncertainty analysis spans the range 12–27 Pg C/yr for the year 2100 between their 25th and 75th percentiles, and 7–55 Pg C/yr between their 5th and 95th percentile cases.

14.3 Long-Term Projections of Global Energy Production and Use

Energy production and use trends that correspond to the carbon emissions described above have a similarly wide range, as shown in Figures 14.2A–C. "Energy" is defined here as "primary energy equivalent." As total global aggregate energy is considered, primary energy equivalent consumption equals primary energy production. (Changes in stocks are considered negligible.) This definition differs from total final energy consumption in that it includes all losses involved in energy transformation. Global energy production and consumption was approximately 344 EJ/yr in 1990.

Energy use rises with time in all baselines examined, with the exception of the extreme ranges of the uncertainty analyses conducted by Edmonds and colleagues (1986) and Nordhaus and Yohe (1983). In the year 2025 the range of base case energy consumption rates varies from 423 EJ/yr (IPCC, 1990 low) to 1,124/yr (Häfele, 1981; IIASA high case year 2030). This range easily encompasses the EMF-12 scenarios, which vary from 602 EJ/yr (CRTM-RD) to 706 EJ/yr (Global-Macro) in the year 2030. These ranges can be compared with the uncertainty analysis of Edmonds and colleagues, which gives a range of 690 EJ/yr to 849 EJ/yr between the 25th and 75th percentiles, and 251 EJ/yr to 1,742 EJ/yr between the 5th and 95th percentile cases.

Table 14.1. *Studies used in the comparison of baseline global fossil fuel carbon emissions*

Emissions Trajectory Identification	Source	Model Employed/Notes
Barns et al.	Barns et al. (1991)	Edmonds–Reilly model (ERM)
CETA	Weyant[a]	Carbon Emission Trajectory Assessment model, Peck and Teisberg (1991)
CRTM-RD	Weyant[a]	Carbon Rights Trade model, a static general equilibrium international trade model based on a data set from Global 2100, Perroni and Rutherford (1991)
Edmonds et al. UA[b]	Edmonds et al. (1986)	ERM, uncertainty analysis. 5th, 50th, and 95th percentile emissions trajectory results reported
EPA RCW EPA SCW	Lashof & Tirpak (1989)	Uses the Atmospheric Stabilization Framework modeling system. The energy module is a modified version of the ERM. RCW is rapidly changing world scenario; SCW is slowly changing world scenario
Global-Macro	Weyant[a]	Global energy market model with vintaged capital stocks for energy production and use, developed by ICF
IEA EMF	Weyant[a]	International Energy Agency model, IEA (1991)
IEA	IEA (1991)	
IIASA high IIASA low	Häfele (1981)	International Institute for Applied Systems Analysis (IIASA) model
IPCC 1990 high IPCC 1990 low	IPCC (1990)	Uses the Atmospheric Stabilization Framework modeling system; energy module is a modified version of ERM
IPCC IS92a IPCC IS92c IPCC IS92e	Leggett et al. (1992)	Uses the Atmospheric Stabilization Framework modeling system; energy module is a modified version of ERM
Manne & Richels	Manne & Richels (1991)	Global 2100
Manne & Richels EMF	Weyant[a]	Global 2100
Manne & Richels UA[b]	Manne & Richels (1993)	Global 2100
Mintzer	Mintzer (1987)	Modified ERM
Nordhaus & Yohe UA[b]	Nordhaus & Yohe (1983)	Nordhaus–Yohe model, uncertainty analysis. 5th, 50th, and 95th percentile emissions trajectory results reported
OECD Green	OECD (1991)	OECD Green model
OECD Green EMF	Weyant[a]	OECD Green model
WEC	WEC (1989)	Expert judgment, no formal computer model used

[a] J. P. Weyant, personal communication regarding EMF-12 models (1991).
[b] UA = uncertainty analysis.

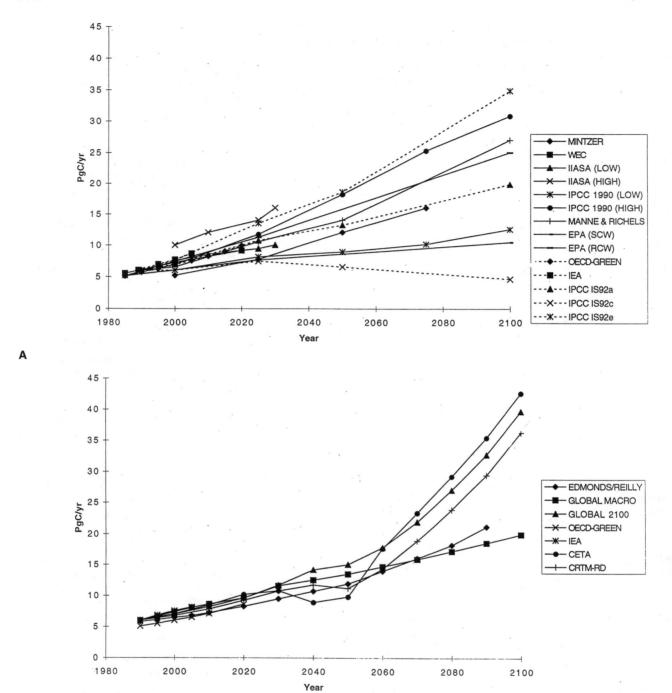

Figure 14.1: (A) Global fossil fuel carbon emissions baselines from (A) various studies and (B) the EMF-12 studies. (C) Global fossil fuel carbon emissions uncertainty analyses from three studies. All studies are identified in Table 14.1.

By the year 2100, the range of energy production and consumption will have expanded considerably relative to that found in the years 2025 and 2030. Primary equivalent energy use ranges from approximately 2,500 EJ/yr in the year 2100 (IPCC IS92e) to 650 EJ/yr (EPA SCW). Thus, even in the SCW case, energy use will have increased by almost a factor of 2 by the year 2100. The EMF-12 reference energy scenarios range from 1,282 EJ/yr (Global-Macro) to 2,813 EJ/yr (CETA). The uncertainty analysis of Edmonds and colleagues

gives a range of 454–1,933 EJ/yr between the 25th and 75th percentiles, and a range of 231–6,750 EJ/yr between the 5th and 95th percentile cases.

14.3.1 Oil

Global conventional oil consumption forecasts are shown in Figures 14.3A and 14.3B. These figures show that global oil consumption in most of the studies grows slightly for the first

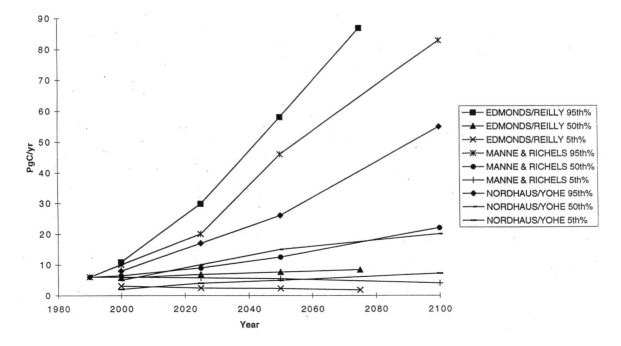

C

Figure 14.1: (*Continued*)

25–30 years, followed by a steady decline to the year 2100. Global oil consumption was approximately 130 EJ/yr in 1985 (IPCC, 1991).

Oil consumption in the year 2025 ranges from 68 EJ/yr (Mintzer, 1987), a decrease from the 1990 level, to 165 EJ/yr (EPA RCW case), an increase from 1990. The uncertainty analysis of Edmonds and colleagues (see Table 14.1) encompasses the extremes, with forecasts of 57 EJ/yr and 526 EJ/yr for the 5th and 95th percentiles, respectively. The EMF-12 scenarios all exhibit the pattern of an initial growth in global oil consumption that peaks around the year 2020 before declining. In the year 2030, the EMF-12 forecasts range from 127 EJ/yr (Manne and Richels EMF) to 143 EJ/yr (Barns et al., 1991).

Unlike the projections for total energy use and carbon emissions, the range of forecasts does not expand by the year 2100. In 2100, the EPA SCW and EPA RCW estimates of global oil consumption will range from 75 EJ/yr to 95 EJ/yr. Mintzer, however, shows consumption in 2075 at 30 EJ/yr. The EMF-12 forecasts are also much lower than those of the EPA, ranging from 26 EJ/yr (CRTM-RD) to 38 EJ/yr (Global-Macro). The uncertainty analysis of Edmonds and colleagues again gives the extreme forecasts for the year 2075 at 1 EJ/yr and 255 EJ/yr in their 5th and 95th percentile cases, although their 50th percentile or median case is within the range of the two EPA cases. The IPCC IS92e scenario is the sole case in which oil consumption continues to rise through the year 2100.

14.3.2 Gas

Figures 14.4A and 14.4B compare forecasts of global natural gas consumption for the various base cases. Global gas con-

sumption was approximately 60 EJ/yr in 1985 (IPCC, 1991). As depicted in the figures, most studies show global gas consumption increasing until the middle of the next century, reaching a peak level approximately 50–100% higher than the 1990 level. Gas consumption in the various cases then declines for the remainder of the century, nearing the 1990 level by the year 2100.

Gas consumption projections for the year 2025 range from 59 EJ/yr (Mintzer) to 156 EJ/yr (IPCC IS92e) for the non-EMF studies. In the EMF-12 studies, the projections for 2030 range from 106 EJ/yr (Manne and Richels EMF) to a high of 147 EJ/yr (ERM). Although the EMF-12 gas consumption values for 2025 are generally higher than those of the other studies, they are within the 5th and 95th percentile scenarios of the uncertainty analysis of Edmonds and colleagues.

For 2100, the range of gas consumption forecasts is narrowed considerably, and the projections from the EMF-12 studies are closer to the results of the other studies. The EMF-12 studies that extend to the year 2100 project gas consumption to be approximately 50–55 EJ/yr (J. P. Weyant, personal communication). Among the other studies, the IPCC 1990 high case provides the lowest forecast, at 28 EJ/yr; the EPA SCW scenario results in the highest forecast, at 50 EJ/yr.

14.3.3 Coal

Base case projections of global coal consumption from the various studies are provided in Figures 14.5A and 14.5B. Most of the cases considered here show a steady increase in coal consumption over the next century. Global coal consumption in 1985 was approximately 90 EJ/yr (IPCC, 1991).

Coal consumption projections for the year 2025 range from 160 EJ/yr (EPA SCW) to 277 EJ/yr (IPCC 1990 high) for the

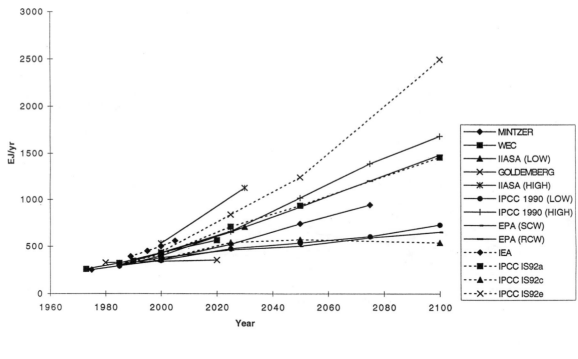

A

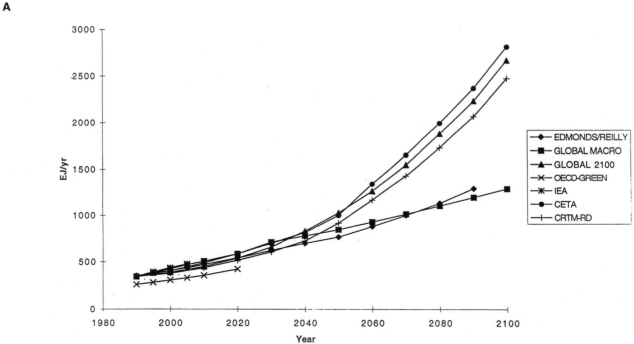

B

Figure 14.2: Global primary energy consumption baselines from (A) various studies and (B) the EMF-12 studies. (C) Global primary energy consumption uncertainty analyses from three studies.

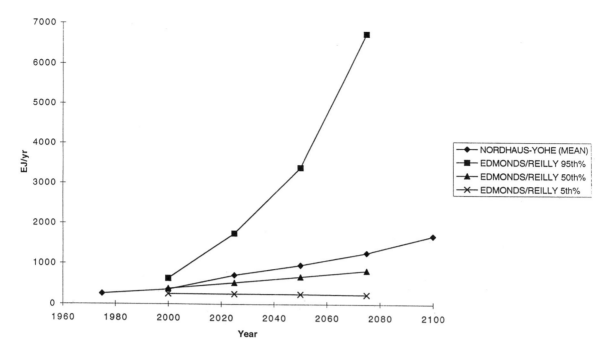

C

Figure 14.2: (*Continued*)

non-EMF base cases. In contrast to the oil and gas projections, the range of coal consumption values from the EMF-12 scenarios is relatively larger, though not by a great amount. For 2030, the CETA study shows global coal consumption at 170 EJ/yr, and the Manne and Richels EMF study gives the highest estimate at 311 EJ/yr.

As is the case for global carbon emissions projections, the range of coal consumption values is substantially larger for the year 2100. Again, the difference between the highest and lowest forecasts is greater in the EMF-12 cases, which range from 49 EJ/yr (CRTM-RD) to 1,755 EJ/yr (CETA). In the other studies, the range is from 335 EJ/yr (EPA SCW) to 1,127 EJ/yr (IPCC 1990 high). These projections are, however, well within the upper limit of the 95th percentile in the uncertainty analysis of Edmonds and colleagues, which shows global coal consumption in the year 2075 at 3,629 EJ/yr.

14.3.4 Nuclear Energy

Figure 14.6 shows projected global nuclear energy consumption for the non-EMF base cases. With the exception of one uncertainty analysis case showing a relatively large increase in nuclear consumption over the next 30 years, the trajectories depicted follow a slight growth for the first half of the next century, followed by a more rapid growth in some of the scenarios. Global nuclear energy consumption was 15 EJ/yr in 1985 (IPCC, 1991).

From Figure 14.6, it is apparent that the range of forecasts for nuclear consumption in the year 2025 is not as broad as the ranges for the fossil fuels. In the EPA SCW scenario, nuclear consumption is projected at 15 EJ/yr, compared to 40 EJ/yr in the EPA RCW scenario. The range between the two IPCC cases is lower, with 29.4 EJ/yr and 50.3 EJ/yr in IPCC 1990 low and high, respectively. Except for the projection of 160

EJ/yr in the 95th percentile case of Edmonds and colleagues' uncertainty analysis, forecasts of nuclear energy consumption from the other studies fall within the range set by the EPA SCW and IPCC 1990 high cases.

By the year 2100, the range of forecasts has become slightly larger. Again, the EPA SCW and IPCC 1990 high cases set the limits of the range. In the EPA SCW case, global nuclear energy consumption in 2100 is 40 EJ/yr, and the IPCC 1990 high case projects consumption in 2100 to be 188.3 EJ/yr.

14.3.5 Renewable Energy Sources

Figure 14.7 shows base case trajectories of global consumption of renewable energy sources – specifically, hydroelectric power, solar power, and biomass. In most of the studies, consumption of renewable energy is expected to grow steadily over the next century. Global renewable energy consumption in 1985 was approximately 21 EJ/yr (IPCC, 1991).

In 2025, projections of renewables consumption range from 50.7 EJ/yr (IPCC 1990 low) to 161 EJ/yr (Mintzer). The uncertainty analysis of Edmonds and colleagues gives projections ranging from 1.9 EJ/yr and 312 EJ/yr for the 5th and 95th percentile cases, respectively. Their falls below the range presented in the other studies, at 24.3 EJ/yr in 2025.

With the exception of the IPCC IS92a and IS92e cases, as well as those presented by Edmonds and colleagues, the range of projections grows little between the years 2025 and 2100. Figure 14.7 shows that the EPA SCW and RCW projections coincide fairly closely with the IPCC 1990 low and high cases, respectively, with the bottom of the range at 147 EJ/yr (IPCC 1990 low) and the top of the range at 250 EJ/yr (EPA SCW). In contrast, the IPCC IS92a and IS92e cases show significantly higher consumption, at 454 EJ/yr and 1,005 EJ/yr, respectively.

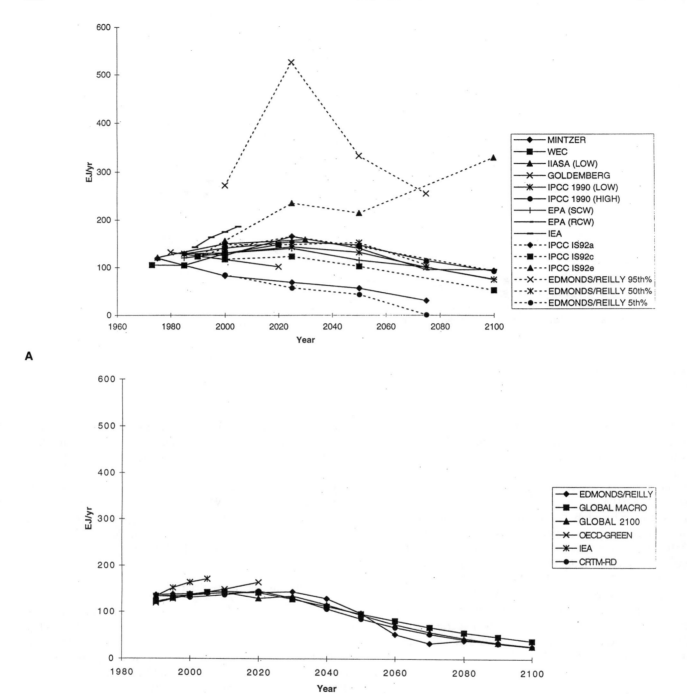

A

B

Figure 14.3: (A) Global oil consumption baselines and uncertainty analyses from various studies. (B) Global oil consumption baselines from the EMF-12 studies.

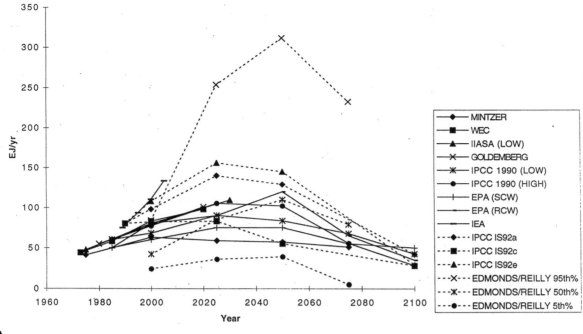

A

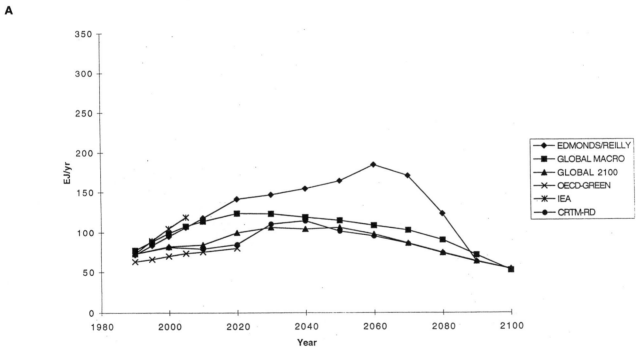

B

Figure 14.4: (A) Global gas consumption baselines and uncertainty analyses from various studies. (B) Global gas consumption baselines from the EMF-12 studies.

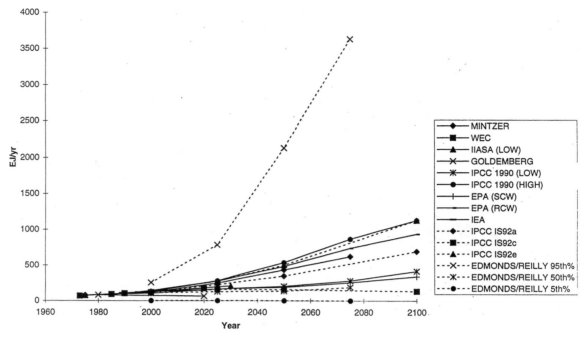

A

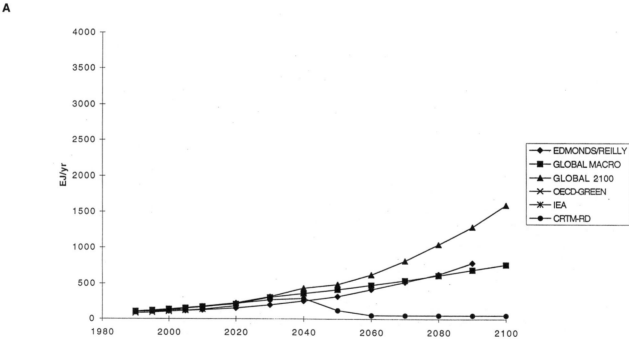

B

Figure 14.5: (A) Global coal consumption baselines and uncertainty analyses from various studies. (B) Global coal consumption baselines from the EMF-12 studies.

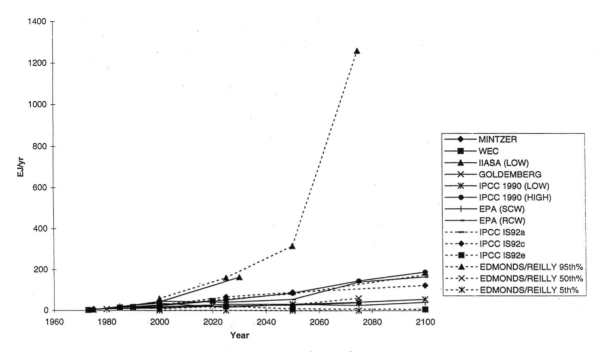

Figure 14.6: Global nuclear energy consumption baselines and uncertainty analyses.

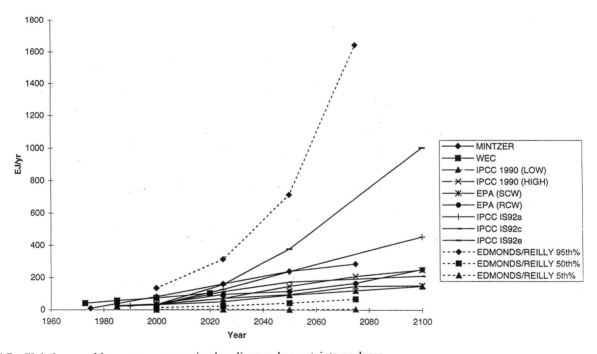

Figure 14.7: Global renewable energy consumption baselines and uncertainty analyses.

Table 14.2. *Definition of regions*

OECD Countries	Centrally Planned Countries	Rest of the World
United States	Soviet Union[b]	Africa
Austria	China	Afghanistan
Belgium	Kampuchea	Bangladesh
Canada	Laos	Brunei
Denmark	Mongolia	Burma
Finland	North Korean	India
France	Vietnam	Indonesia
West Germany[a]	Albania	Malaysia
Greece	Bulgaria	Nepal
Iceland	Cuba	Pakistan
Ireland	Czechoslovakia	Papua New Guinea
Italy	East Germany[a]	Philippines
Luxembourg	Hungary	South Korea
Netherlands	Poland	Sri Lanka
Norway	Rumania	Taiwan
Portugal	Yugoslavia	Thailand
Spain		Argentina
Sweden		Bolivia
Switzerland		Brazil
Turkey		Chili
United Kingdom		Colombia
Japan		Costa Rica
Australia		Dominican Republic
New Zealand		Ecuador
		Guatemala
		Guyana
		Haiti
		Honduras
		Mexico
		Nicaragua
		Panama
		Paraguay
		Peru
		Puerto Rico
		Surinam
		Trinidad & Tobago
		U.S. Virgin Islands
		Uruguay
		Venezuela
		Iran
		Iraq
		Syria
		Israel

[a]Although East and West Germany were unified in 1989, most studies treat them as separate.
[b]The Soviet Union is treated as a single nation in forecasts, as it was prior to 1991.

The base case from the Mintzer study, however, ends at the year 2075, with global renewable consumption of 283 EJ/yr, while the 95th percentile case in the uncertainty analysis of Edmonds and colleagues, gives the extreme value, at 1,646 EJ/yr for 2075.

14.4 Regional Long-Term Projections of Carbon Dioxide Emissions

To compare the base case or baseline projections at a more detailed level, this section looks at fossil fuel carbon dioxide emissions for different regions of the world. Not all the studies contain regional detail, so a smaller sample of cases is considered here. For purposes of comparison, we have constructed three regions: the United States and other countries of the Organization for Economic Cooperation and Development (OECD), the former Soviet Union and other currently and formerly centrally planned economies, and the rest of the world. The major countries included in these regions are listed in Table 14.2. This was the most common grouping of countries among all the studies. The relative ranking of emissions between these three groups changes completely. In 1990, the OECD has the greatest emissions of any of the three, and the rest of the world has the least emissions. By the year 2100, forecasts generally find that order to be reversed. That is, the OECD has the lowest emissions of the three regions, and the rest of the world has the highest emissions.

14.4.1 United States and Other OECD Countries

Fossil fuel carbon emissions for this region from various base cases are shown in Figure 14.8. For the year 2025, all the base cases show a growth in carbon emissions from the estimated 1990 level of 2.5 Pg C/yr (IPCC, 1991). The IPCC IS92c case

(Leggett et al., 1992) gives the lower bound of these cases, with carbon emissions in 2025 of 2.5 Pg C/yr. The highest baseline projection is from the IPCC IS92e case, with carbon emissions nearly 4.2 Pg C/yr for the year 2030. For the year 2100, the range of base case values has grown considerably. As Figure 14.8 shows, the IPCC IS92c and IPCC 1990 high cases provide the bottom and top of the range of projections, respectively, at 1.2 Pg C/yr and 7 Pg C/yr.

14.4.2 Centrally Planned Countries

Figure 14.9 shows base case fossil fuel carbon emissions for this region. The IPCC IS92c case offers the lowest projection for the period around 2025, with carbon emissions of 2.3 Pg C/yr for the year 2030; the highest projection is from the IPCC 1990 high case, at 4.7 Pg C/yr. By 2100, carbon emissions in the IPCC 1990 high remain at the top of the range, with emissions of 13 Pg C/yr. A slower growth in emissions is seen in the EPA SCW case, which projects emissions of 4.1 Pg C/yr in the year 2100. The IPCC IS92c case shows emissions decreasing to only 1.4 Pg C/yr by the year 2100.

14.4.3 Rest of the World

Base case fossil fuel carbon emissions for countries not included in the first two regions are provided in Figure 14.10. With the exception of the EPA SCW case and the IPCC IS92c case, which both show carbon emissions peaking in the middle of the next century and declining slightly afterward, all the cases here show growth in carbon emissions for this region through the year 2100. For the year 2025, carbon emissions range from 1.6 Pg C/yr (IPCC 1990 low) to 4.5 Pg C/yr (EPA RCW). By 2100, the range of emissions projections is from 1.9 Pg C/yr (IPCC IS92c) to 16.5 Pg C/yr (IPCC IS92e).

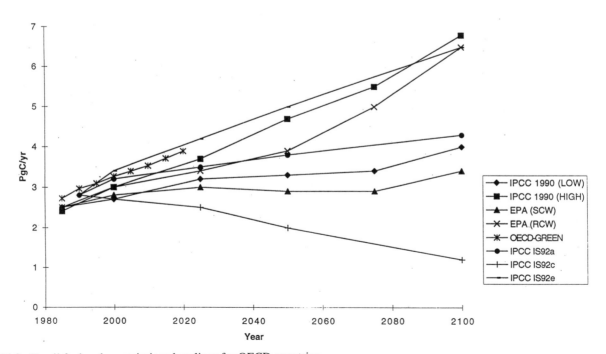

Figure 14.8: Fossil fuel carbon emissions baselines for OECD countries.

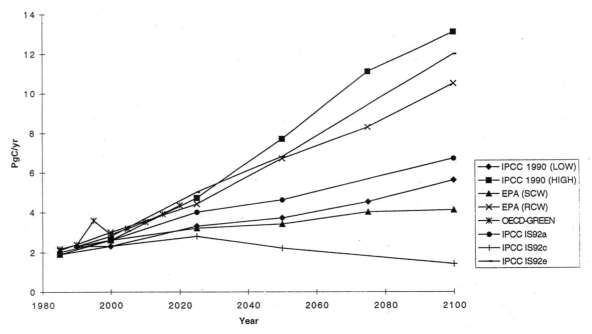

Figure 14.9: Fossil fuel carbon emissions baselines for currently and formerly centrally planned countries.

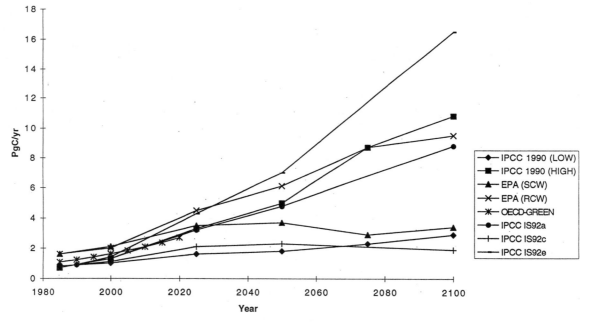

Figure 14.10: Fossil fuel carbon emissions baselines for the rest of the world.

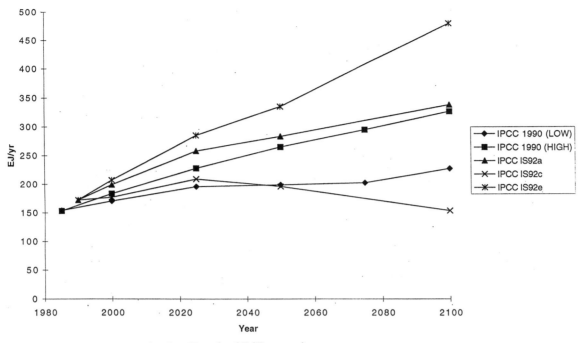

Figure 14.11: Primary energy consumption baselines for OECD countries.

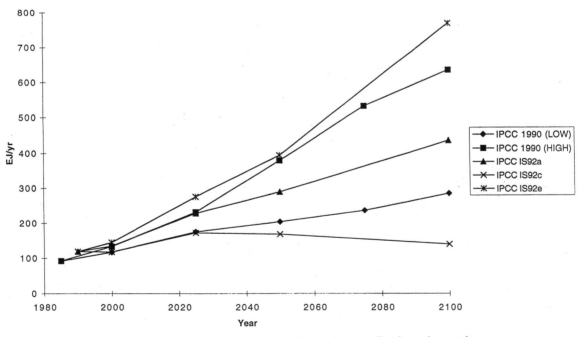

Figure 14.12: Primary energy consumption baselines for currently and formerly centrally planned countries.

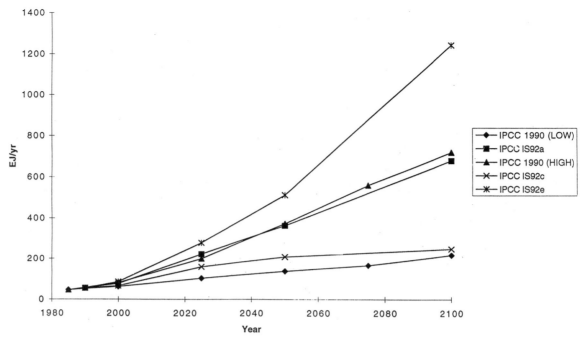

Figure 14.13: Primary energy consumption baselines for the rest of the world.

14.5 Regional Long-Term Projections of Energy Consumption and Production

In this section, we provide comparisons of energy consumption and production for the three regions discussed in Section 14.4.

14.5.1 United States and Other OECD

Base case primary energy production projections from various studies for the region comprised of the United States and other OECD countries are shown in Figure 14.11. The range of projections in 2025 is small, relative to the range in later years. For 2025, the range of regional primary energy consumption is from 196 EJ/yr (IPCC 1990 low) to 285 EJ/yr (IPCC IS92e). By 2100, the range has expanded to 153 EJ/yr (IPCC IS92c) to 480 EJ/yr (IPCC IS92e).

14.5.2 Centrally Planned Countries

Figure 14.12 shows base case primary energy consumption for this region from various studies. The trajectories here follow the carbon emissions trajectories for this region (shown in Figure 14.9) fairly closely. For the year 2025, the IPCC IS92c case gives the lowest projection at 172 EJ/yr, compared to the highest projection of 275 EJ/yr seen in the IPCC IS92e case. For 2100, the range of projections is from 139 EJ/yr (IPCC IS92c) to 770 EJ/yr (IPCC IS92e).

14.5.3 Rest of the World

Base case primary energy consumption projections for the remaining countries are depicted in Figure 14.13. All the studies considered here project a steady growth in primary energy consumption for this region over the next century. The range of projections for the year 2025 is bounded by the IPCC 1990 low case and the IPCC IS92e case at 102 EJ/yr and 277 EJ/yr, respectively. The IPCC 1990 low case also gives the lowest level of energy consumption for the year 2100, with a value of 216 EJ/yr. The top of the range is again the IPCC IS92e case, at 1,244 EJ/yr.

14.6 Why Forecasts Differ

In this section, we explain some of the causes for differences in energy and emissions results from the various studies. We begin by discussing the results of three uncertainty analyses that formally examined the relationship between uncertain parameters and corresponding carbon emissions trajectories. We then discuss some of the critical factors that account for much of the difference in carbon emission results from different studies. Finally, we provide a comparison of population and economic growth assumptions in the studies discussed here.

14.6.1 Uncertainty Analysis Results

Fossil fuel carbon emissions and related energy consumption results from three major uncertainty analysis research efforts were presented earlier in our discussion. The first such study was conducted by Nordhaus and Yohe (1983), followed by a similar effort by Edmonds and colleagues (1986) and a more recent study by Manne and Richels (1993). All these studies employed Monte Carlo simulation techniques to specify parameter sets for their models. In a Monte Carlo analysis, assumptions are made *ex ante* about the probability distributions of key model input parameters as well as any interdependencies between them. A systematic experimental design procedure is then used to exercise the model with randomly selected samples of parameter values. This procedure provides probability distributions of carbon emissions (and other results), which offer more insight into the uncertainty and precision of the results than is possible with a single "best guess" forecast of results.

It should be emphasized that this method of uncertainty analysis does not solve the problem of forecasting the future. It is only an exercise of a model and is limited by the quality and validity of its assumptions and model structure. However, a model is a reduced-form representation of our understanding of an actual system. Performing an uncertainty analysis on a model not only provides the probability description of alternative paths of emissions, but also – and perhaps more importantly – indicates which of the assumptions and parameters are most important to our forecasts of carbon emissions. The analysis does not resolve the uncertainties but instead serves to represent our understanding of the uncertainties in a logical and consistent manner.

Nordhaus and Yohe (1983) performed the first formal uncertainty analysis of future carbon emissions. They identified 10 key input parameters, to which they assigned probability distributions. Compared to later models, the model used by Nordhaus and Yohe is relatively simple in that it considered only two types of energy: fossil (carbon-emitting) energy, and nonfossil energy. Their analysis singled out the ease of substitution between fossil and nonfossil energy (which is also implicitly a function of the availability of nonfossil sources) as the most important parameter in explaining the uncertainty in carbon emissions. In addition, they deemed the rate of growth of total factor productivity, a measure that affects both labor and energy productivity, to be next most important parameter. Interestingly low on the list of factors affecting the baseline emissions was population growth. Edmonds and colleagues (1986) obtained a similar result, although the range of population growth assumptions used in both studies' uncertainty analyses is insufficient to include present population forecasts.

Edmonds and colleagues (1986) performed a similar uncertainty analysis; however, the model they employed contained much more detail in its description of the energy-producing and -consuming sectors. The carbon emissions forecasts from this study formed a nonnormal distribution with median values significantly below the mean values. As shown in Figure 14.1A, the range of trajectories required to bracket a 90% confidence interval for carbon emissions is from a 1.4% annual decline in emissions to a 3% annual growth. From their study, Edmonds and colleagues identified the four most important factors in explaining carbon emissions uncertainty: labor productivity in developing countries, labor productivity in developed countries, end-use energy efficiency improvement, and world income elasticity of energy demand. They did not confirm the significance of the interfuel substitution parameter, which was most important in the Nordhaus and Yohe (1983) study. This different result can be attributed to increased energy modeling detail in the Edmonds–Reilly model, which incorporated multiple sources of energy supply in addition to interfuel substitution options.

Manne and Richels (1993) performed an uncertainty analysis in which a poll of expert opinion was used to create the prior distributions of key model parameters. Specifically, they polled experts on driving parameters such as gross domestic product (GDP) growth rates, elasticity of substitution between labor and capital, rate of autonomous energy efficiency improvements, commercial year for economically competitive carbon-free electricity, and the cost of the nonelectric backstop. As Figure 14.1A shows, the range of carbon emissions projections obtained bounds all but the uncertainty cases of Edmonds and colleagues (1986).

14.6.2 Factor Analysis of Base Case Differences

An adequate understanding of the differences in base case fossil fuel carbon emissions is not possible without an examination of the differences in the component factors that drive the results. We mentioned some of the important factors in our discussion of the uncertainty analysis studies. In general, the list of major factors in determining fossil fuel carbon emissions includes population growth, economic growth, labor productivity, and various technological considerations such as energy efficiency improvement and the feasibility and cost of switching to nonfossil fuels. This list is not intended to be exhaustive, and these factors would usually not be considered independently.

Owing to differences in the structures of models, a thorough and consistent factor analysis across various studies is not always possible. Also, many of these factors (e.g., rate of technological improvement) may lack standard definitions. The EMF-12 suite of cases has been constructed with a consistent set of key factor input assumptions. As we have shown in the figures depicting EMF-12 case results, significant differences in results will still occur under consistent factor assumptions because of structural and conceptual differences in the models employed.

14.6.3 Comparison of Population and Economic Growth Assumptions

Population and economic growth are two important factors of carbon emissions that do have standard definitions and units of measure. Population is simply the number of people, and GDP is a standard measure of the size of an economy. When other things are equal, fossil fuel carbon emissions will tend to increase directly with population and with the size of the economy. Current energy models are sufficiently sophisticated to recognize that other things are not usually equal and that technological improvement and other factors are critical determinants of future carbon emissions. However, population and economic growth remain two critical factor assumptions that should be compared across the various base cases.

Figure 14.14 shows world population assumptions for various studies to the year 2100. Most of these studies forecast that population growth will slow after 2050 and reach approximately 10.5 billion in 2100. The tendency for population growth forecasts to approach zero in long-period analysis is a by-product of the modeling approach used. The present approach is to identify a time at which fertility rates decline to replacement rate levels. As a consequence of this formulation of the demographic problem, all forecasts eventually attain steady-state populations. Whether the world's major regions will have completed their demographic transitions over the course of this period in time is uncertain.

The EPA SCW case gives the highest projection for 2100 at 13.2 billion. Interestingly, the 95th percentile case of Edmonds and colleagues' uncertainty analysis lies below that of other

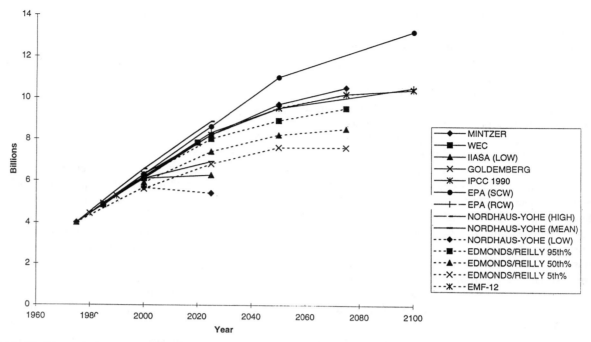

Figure 14.14: World population assumptions for various studies and uncertainty analyses.

studies. This difference not only reflects that the consensus about population growth may have changed in the last five years, but also reemphasizes the point that the results of an uncertainty analysis are functions of our model structures and our assumptions about parameter distributions.

A comparison of worldwide economic growth rate assumptions is provided in Figure 14.15. Most of the studies assume annual GDP growth rates between 2% and 3% over the next century. For the year 2100, the lowest projected growth rate is in the EPA SCW case, with a value near 1%. The median value of Ed-

monds and colleagues' uncertainty analysis is near the other studies' values, and the 95th percentile case bounds the range of projections with a growth rate of approximately 5% per year.

14.7 Conclusions

The present literature on potential future global fossil fuel use and associated carbon emissions explores a relatively wide range of potential emissions trajectories. Differing combina-

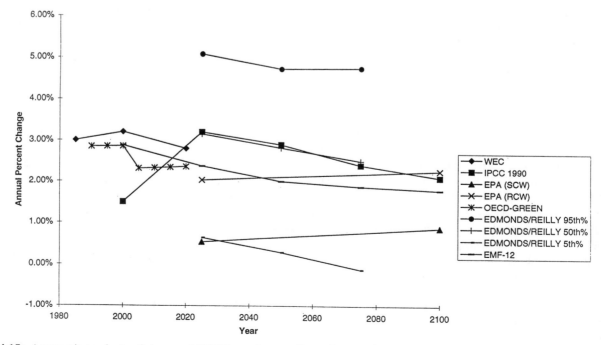

Figure 14.15: Assumptions of rate of change of GDP for various studies and uncertainty analyses.

tions of key assumptions lead to differing emissions trajectories. The range of plausible base case assumptions is sufficient to produce a very wide range of emissions by the year 2100. This result is consistent with intuition. The evolution of energy systems over the future century may evolve in a wide variety of ways, depending upon such assumptions as population growth, economic growth, and the evolution of technologies for producing and consuming energy. It is noteworthy that declining base case emissions to the year 2100 occur only in the tails of the ranges of uncertainty analyses. This does not mean that it is impossible to construct plausible scenarios in which long-term base case emissions decline even without any explicit actions in response to concerns regarding global climate change. Rather, it means that at present, such scenarios appear relatively unlikely in comparison to cases in which emissions are constant or rising with time. The range of base case future global fossil fuel carbon emissions tends to widen as the forecast horizon extends into the future. This general tendency seems also to be consistent with intuition. Not all factors contributing to models of future fossil fuel carbon emissions appear to be equally important. Assumptions about economic growth, population, technological change, and nongreenhouse environmental costs appear to have greater influence than, for example, assumptions regarding the fossil fuel resource base. Systematic exploration of the range of coupled social and natural science systems is a potentially fruitful avenue for future research efforts.

Acknowledgments

This work was prepared under contract DE-AC06–76RLO1830 for the U.S. Department of Energy and under contract DE-AC06–76RLO1831 for the Electric Power Research Institute.

References

Barns, D. W., J. Edmonds, and J. Reilly. 1991. *Use of the Edmonds/Reilly Model to Model Energy Sector Impacts of Greenhouse Gas Emissions Controls.* Draft paper prepared for the Energy Modeling Forum, study, Global Climate Change: Energy Sector Impacts of Greenhouse Gas Emissions Control Strategies (EMF 12), Stanford University, Stanford, California.

Bradley, R. A., E. C. Watts, and E. R. Williams (eds). 1991. *Limiting Net Greenhouse Gas Emissions in the United States.* U.S. Department of Energy, Washington, D.C.

Darmstadter, J. 1991. *The Economic Cost of CO_2 Mitigation: A Review of Estimates for Selected World Regions.* Discussion Paper ENR91–06, Resources for the Future, Washington, D.C.

Edmonds, J. A., and D. W. Barns. 1991. *Factors Affecting the Long-Term Cost of Global Fossil Fuel CO_2 Emissions Reductions.* PNL Global Studies Program Draft Working Paper, Pacific Northwest Laboratories, Richmond, Washington.

Edmonds, J. A., J. M. Reilly, R. H. Gardner, and A. Brenkert. 1986. *Uncertainty in Future Global Energy Use and Fossil Fuel CO_2 Emissions 1975 to 2075.* TR036, DO3/NBB-0081 Dist. Category UC-11, National Technical Information Service, U.S. Department of Commerce, Springfield, Virginia.

Goldemberg, J., T. B. Johansson, A. K. N. Reddy, R. H. Williams. 1988. *Energy for a Sustainable World.* John Wiley and Sons, New York.

Häfele, W. 1981. *Energy in a Finite World.* Ballinger Publishing Company, Cambridge, Mass.

Hoeller, P., and M. Wallin. 1991. *Energy Prices, Taxes and Carbon Dioxide Emissions.* Working Papers No. 106, Economics and Statistics Department, Organization for Economic Co-operation and Development, Paris, France.

IEA (International Energy Agency). 1991. *Carbon Taxes and CO_2 Emissions Targets: Results from the IEA Model.* Research report to the Policy Studies Branch, Organization for Economic Cooperation and Development, Paris, France.

IPCC (Intergovernmental Panel on Climate Change). 1990. *Climate Change: The IPCC Scientific Assessment,* J. T. Houghton, G. J. Jenkins, and J. J. Ephraums, eds. Cambridge University Press, Cambridge, U. K.

IPCC. 1991. *Integrated Analysis of Country Case Studies.* Report of the U.S./Japan Expert Group to the Energy and Industry Subgroup of the Response Strategies Working Group of the IPCC.

Lashof, D. A., and D. A. Tirpak. 1989. *Policy Options for Stabilizing Global Climate, Draft Report to Congress.* Office of Policy, Planning and Evaluation, U.S. Environmental Protection Agency, Washington, D.C.

Leggett, J. A., W. J. Pepper, and R. J. Swart. 1992. Emissions scenarios for IPCC: An update. In *Climate Change 1992: The Supplementary Report to the IPCC Scientific Assessment,* J. T. Houghton, B. A. Callander, and S. K. Varney, eds. Cambridge University Press, Cambridge, U. K., pp. 69–95.

Manne, A. S., and R. G. Richels. 1991. Global CO_2 emissions reductions – The impacts of rising energy costs. *The Energy Journal 12,* 87–108.

Manne, A. S., and R. G. Richels. 1993. *The Costs of Stabilizing Global CO_2 Emissions – A Probabilistic Analysis Based on Expert Judgments.* Electric Power Research Institute, Palo Alto, California.

Marland, G., R. J. Andres, and T. A. Boden. 1994. Global, regional, and national CO_2 emissions. In *Trends '93: A Compendium of Data on Global Change,* T. A. Boden, D. P. Kaiser, R. J. Sepanski, and F. W. Stoss, eds., Carbon Dioxide Information Analysis Center, Oak Ridge National Laboratories, Oak Ridge, Tenn., pp. 505–584.

Mintzer, I. M. 1987. *A Matter of Degrees: The Potential for Controlling the Greenhouse Effect.* World Resources Institute, Washington, D.C.

Nordhaus, W. D. 1989. *A Survey of Estimates of the Cost of Reduction of Greenhouse Gas Emissions.* Department of Economics, Yale University, New Haven, Conn.

Nordhaus, W. D., and G. W. Yohe. 1983. Future carbon dioxide emissions from fossil fuels. In *Changing Climate,* National Academy Press, Washington, D.C., pp. 87–153.

OECD (Organization for Economic Cooperation and Development). 1991. Draft Paper on the OECD Comparative Modelling Exercise. OECD, Washington, D.C.

Peck, S. C., and T. J. Teisberg. 1991. *CETA: A Model for Carbon Emissions Trajectory Assessment.* Electric Power Research Institute, Palo Alto, Calif.

Perroni, C., and T. F. Rutherford. 1991. *International Trade in Carbon Emission Rights and Basic Materials: General Equilibrium Calculations for 2020.* Department of Economics, Wilfrid Laurier University, Waterloo, Ontario, Canada.

Rothman, D. S., and D. Chapman. 1991. *A Critical Analysis of Climate Change Policy Research.* Draft paper, Department of Agricultural Economics, Cornell University, Ithaca, New York.

WEC (World Energy Conference, 14th Congress). 1989. *Global Energy Perspectives 2000–2020.* Conservation and Studies Committee, WEC, Montreal, Canada.

15

The Future Role of Reforestation in Reducing Buildup of Atmospheric CO$_2$

GREGG MARLAND

Abstract

Among the options proposed for mitigating the buildup of atmospheric CO_2 is planting new forest areas to sequester carbon from the atmosphere. One of the questions of interest in modeling the global carbon cycle is the extent to which reforestation is likely to succeed in providing physical removal of CO_2 from the atmosphere. There are many strategies for using forest land to mitigate the atmospheric buildup of CO_2: decreasing the rate at which forests are cleared for other land uses, increasing the density of carbon storage in existing forests, improving the rate and efficiency at which forest products are used in the place of other energy-intensive products, substituting renewable wood fuels for fossil fuels, improving management of forests and agroforestry, and increasing the amount of land in standing forest. Because increasing the area of forests has social, political, and economic limitations, in addition to physical limitations, it is hard to envision a large increase in forest area except where there are associated economic benefits. Our speculation is that, over the next several decades, (1) the forest strategies most likely to be pursued for the express purpose of CO_2 mitigation are those that provide more, or more efficient, substitution of forest products for energy or energy-intensive resources and that (2) the physical accumulation of additional carbon in forests will be of lesser importance.

15.1 Introduction

The idea of planting trees specifically to offset emissions of CO_2 from fossil fuel burning was first suggested by Dyson (1977) and Dyson and Marland (1979). Dyson and Marland described tree planting as an emergency measure that could be used to halt or reverse the growth of atmospheric CO_2 while society developed an energy system not based on fossil fuels. They recognized that carbon uptake in trees could be effective only over the time it took the trees to approach maturity, and they described tree planting more as a short-term emergency measure than as a long-term strategy. Dyson and Marland concluded that it is physically possible to plant enough new trees (while arresting forest clearing) to offset fossil fuel emissions for a few decades if growth in atmospheric CO_2 is viewed as a serious short-term threat to the global climate. In a later paper, Marland (1988) reexamined the magnitude of the effort that would be required to balance fossil fuel carbon emissions. This work showed that a complete offset would require new tree plantations on a scale comparable to the land area of Australia and growing at the rate of fast-growing American sycamore trees on short-rotation plantations in Georgia (7.5 tons C/ha/yr). Marland concluded that although tree planting was unlikely to provide a full offset (except as an emergency measure), it could play some role in a CO_2 mitigation strategy.

15.2 Potential for Reforestation

The idea of planting trees as part of an ongoing strategy to reduce current net CO_2 emissions is now broadly accepted, and several such projects have actually been implemented (see, e.g., Trexler et al., 1989; Dixon et al., 1993). Tree planting is being viewed as a way to slow the rate of atmospheric CO_2 increase so that society can either gain better understanding of what climate change portends or have more time for systems to adapt to changes that do occur. A number of papers have discussed the potential for tree planting to offset a portion of current CO_2 emissions from fossil fuels and the cost per ton of carbon for doing so (see, e.g., Sampson and Hair, 1992; Henderson and Dixon, 1993). Tree planting has been included as a part of the U.S. Climate Change Action Plan (Clinton, 1993).

The purpose of this chapter is not so much to analyze what is physically possible as it is to speculate on what is likely to be accomplished over the next several decades. Our intent is not to provide a technical review of the burgeoning literature, but to convey a sense of the prospects offered. As we examine scenarios for future energy use and their implications for the global carbon cycle and for atmospheric concentrations of CO_2, we query the extent to which emissions from fossil fuels are likely to be countered by new tree plantings. There are, of course, a variety of strategies for using forestry projects to mitigate the buildup of CO_2 in the atmosphere (e.g., Dixon et al., 1994; Sampson et al., 1993); however, our focus here is on growing trees physically sequester carbon to. Should scenarios of future CO_2 emissions include planned CO_2 sequestering? We are, obviously, no more able to anticipate future tree planting than we are able to anticipate future patterns of energy use; however, there is useful information that colors our expectations. We have some knowledge of the current situation: that forest clearing continues in many places around the world (see Houghton, Chapter 4, this volume), and that the global population continues to increase.

Figure 15.1 shows data on the historical relationship between population growth and the increase in arable land and the decrease in forest land in Southeast Asia. In addition to illustrating the monotonic decrease in forest land with increasing population, this plot raises the question of what we mean by "reforestation." In places where increasing the actual area of forests is not achievable, it may still be possible to increase average carbon storage in forests through accelerated replanting of harvested areas and through changes in forest management. Among others, Iverson and colleagues (1993) have pointed out that most tropical forests are not at their maximum C density and that C storage could be increased without increasing the area of forest land. Also, in the United States, for example, the area in forest has increased over recent decades as increasing agricultural productivity has allowed some marginal cropland return to forest.

A relationship between population growth and forest clearing should not be taken too rigidly. Meyer and Turner (1992) discussed the relationship and concluded that although population growth is an important factor in forest clearing, it is significantly modified by the natural and institutional context. Waggoner (1994) noted the continuing improvement in agricultural productivity globally and held that there are plausible scenarios whereby the world could feed twice the current population on less agricultural land than is now in use. Although these scenarios depend on human values, diet, economics, and

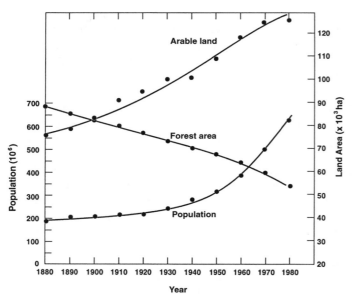

Figure 15.1: Changing land use and population since 1880 in Pakistan, Bangladesh, Burma, Malaysia, Brunei, and Northern India. (Data from Richards et al., 1987; see also Richards and Flint, 1994.)

technology, Waggoner believes that a world of 10 billion people can still spare land for nature.

As a matter of perspective, it is interesting to examine the prospects for reforestation in the United States and to start with the current situation. Turner and colleagues (1993) have published what they believe to be the carbon balance for forests in the contiguous United States. They have taken the net annual biologic C flux into forests, subtracted that which is harvested, estimated the portion of the harvest that ends up in either landfills or long-lived products, adjusted for wildfires, and estimated that 0.124 Pg C is taken up annually in U.S. forests and forest products (Table 15.1). This compares with

emissions of carbon from fossil fuels, which totaled 1.346 Pg C from the United States and 6.188 Pg C globally in 1991 (Andres et al., Chapter 3, this volume).

It has recently been observed that Northern Hemisphere temperate zone forests seem to be net sinks for carbon in many areas. Kauppi and colleagues (1992) described an increase in growing stocks in European forests during the 1970s and 1980s; they worried, however, that favorable development of forest resources is at risk in the near future. Plantinga and Birdsey (1993) found an increase in growing stock on private timberlands in the United States, but their base case projection showed the inventory peaking by 2010 and removals exceeding growth thereafter (including a decline in the area of timberlands). Kurz and Apps (1993) found that forest ecosystems in Canada were a sink for carbon in 1986, because of the increasing average age of forests – a process that cannot continue indefinitely.

In the first detailed analysis of land that might be reforested in the United States, Moulton and Richards (1990) estimated that up to 56% of U.S. CO$_2$ emissions might be offset with an aggressive reforestation strategy on 140 million ha of "economically marginal and environmentally sensitive" crop- and pasture land. Although the Moulton and Richards analysis is widely conceived to be very optimistic (e.g., NAS, 1992), several features of their analysis nonetheless provide very useful insight. Figure 15.2 illustrates that, for the maximum reforestation strategy, the bulk of the net carbon uptake would occur on croplands, with notably smaller contributions from pasture lands and improved planting and management of current forest lands. Again, emphasizing the shape of the curve rather than the numerical values on the ordinate, Figure 15.3 illustrates that the marginal cost of sequestering carbon can be expected to increase dramatically as the amount of carbon sequestering increases. This increase in cost can be attributed largely to increasing land costs. As the amount of carbon sequestering increases, reforestation begins to compete with

Table 15.1. *Carbon balance for forests in the contiguous 48 U.S. states in 1990 (in Pg C)*

Biologically driven C flux into forests			0.286
Harvest driven C flux (from forests)			−0.185
of which, carbon into forest products		0.139	
of which, C soon oxidized	0.103		
C in long-lived products	0.020		
C in landfills	0.016		
Net increase of C in standing forests			0.101
Sum of carbon stored in products			0.036
C removed by wildfires			−0.013
Credit for use as fuel			?
Total annual balance (forests and forest products)			+0.124

Reprinted with permission from D. P. Turner, J. F. Lee, G. J. Koerper, and J. R. Barker (1993), *The Forest Sector Carbon Budget of the United States: Carbon Pools and Flux Under Alternative Policy Options.* EPA/600/3-93/093, Environmental Protection Agency.

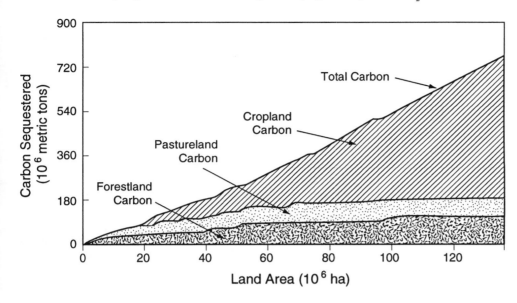

Figure 15.2: Carbon sequestration through tree planting and forest management in the United States by land type and the total land area involved for each level of sequestration. Sequestering is assumed to occur in order of increasing cost ($/ton C). (Data from Moulton and Richards, 1990.) Reprinted with permission from R. J. Moulton and K. R. Richards (1990), *Costs of Sequestering Carbon through Tree Planting and Forest Management in the United States.* GTR WO-8, U.S. Department of Agriculture Forest Service.

other land uses, and the cost per ton of carbon sequestering increases. Although some inexpensive offsets may be available, tree planting is likely to occur primarily when there is value in the trees or tree products.

Others have also speculated on the amount of land available for reforestation in the United States. Turner and colleagues presented some scenarios for carbon uptake in U.S. forests; their most aggressive scenario (5 million ha of new forestland) would result in 0.015 Pg/yr of net carbon uptake for 50 years (above the current-plans scenario). Parks and colleagues (1992) suggested that 47 million ha of marginal crop- and pastureland may be "physiologically suited" for conversion to forest in the United States, a prospect made possible by increasing agricultural productivity. Twenty percent of the conversion could involve net economic gain. Wright and colleagues (1992; see also Graham et al., 1992) offered a maximum tree-planting scenario involving 28 million ha in the United States.

In developing countries, particularly in the tropics, although deforestation continues in many areas (Dale et al.,

1991), it appears that there are significant areas of degraded lands that might be reforested without competition from higher land uses such as agriculture. Grainger (1988) defined degraded lands as those with "modification or substantial removal of the composition and/or structure of their vegetative cover and/or consequent depletion of soil fertility" (p. 32) and tabulated over 2 billion ha in the tropics (Table 15.2). The bulk of this degraded land, however, receives too little rainfall to sustain forests. Of the 758 million ha with theoretical potential, only part is suitable for reforestation in practice. According to Grainger, large-scale reforestation is justified for both economic and ecological reasons. Yet, looking at a specific part of the area, for example, Grainger found it "difficult to think of conserving or regenerating large areas of tropical moist deciduous forest. Population, commercial, and land tenure pressures would probably make expansion of forests in these areas difficult unless allied to major local needs for fuelwood or industrial wood" (p. 41). Regarding tropical rainforests, Grainger wrote that intensive tree planta-

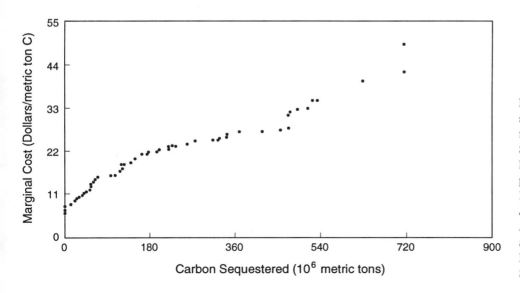

Figure 15.3: Marginal cost of carbon sequestering through tree planting and forest management in the United States. (Data from Moulton and Richards, 1990.) Reprinted with permission from R. J. Moulton and K. R. Richards (1990), *Costs of Sequestering Carbon through Tree Planting and Forest Management in the United States.* GTR WO-8, U.S. Department of Agriculture Forest Service.

Table 15.2. *Area of tropical degraded lands (millions of hectares)*

Region	Logged Forests	Forest Fallow	Deforested Watershed	Desertified Drylands	All Lands*
Africa	39	59	3	741	842
Asia	54	59	56	748	917
Latin America	44	85	27	162	318
Total	137	203	86	1,651	2,077

Source: Grainger (1988).
*Adapted from Grainger, 1988.

tions could be established to take the pressure off remaining natural forest areas. As in the United States, the suggestion is that there are real and opportunity costs (e.g., the opportunity to use the land in other ways) involved in tree planting, and that planting will be most successful where there is local economic benefit.

Houghton and colleagues (1993) identified 3,200 million ha in the tropics where woody biomass has been decreased and carbon might again be sequestered, with a potential accumulation of 160–170 Pg C. Houghton acknowledged, however, that this is biophysical potential and that some of the land is preempted by agriculture and infrastructure and unlikely to be reforested. The same is true everywhere; for social, political, and economic reasons the land available for reforestation is less than the land area with physical potential.

In an examination of the costs of sequestering carbon in forests, Swisher (1991) concluded that although the total potential for carbon storage is significant, it does not appear that forestry can offset more than a few percent of global CO$_2$ emissions from fossil fuel use. Swisher's examination of proposed projects in Central America found that they can offer socioeconomic benefits at the local level. Indeed, Swisher suggested, it is difficult to imagine any forest development achieving sufficient longevity for long-term carbon storage without providing such benefits. Winjum and Lewis (1993) argued that nonfinancial benefits and costs should be included in forest management considerations and that storing carbon has forest value. These benefits are often not apparent at the local level, although one can envision policy strategies that compensate local economies for providing regional or global-scale services.

Trexler was involved with Applied Energy Services in 1989, when that company committed to a forestry project in Guatemala to offset CO$_2$ emissions from a planned power plant in Connecticut. In 1992, Trexler and colleagues wrote, "There is also so much momentum built into the deforestation process in the form of population growth and rural settlement efforts that actually reversing current trends poses mammoth difficulties. . . . It would be a remarkable achievement to bring the terrestrial biota into a CO$_2$ balance. . . . It is even harder to conceive of an effort large enough to offset a significant portion of rapidly rising fossil fuel emissions of CO$_2$" (p. 92). Examination of the Guatemala project shows that 90% of stemwood to be grown will be harvested for various uses. The CO$_2$

benefit attributed to the project is largely through substitution of sustainably produced wood for unsustainable harvest of standing forest.

Sampson and colleagues (1993) concluded that the greatest potential for managing forests to affect the net sources and sinks of carbon lies in the opportunities to use biomass for energy production.

15.3 Uses of Forest Products

From this perspective, it appears that much of the CO$_2$ benefit of forestry projects undertaken for the express purpose of mitigating anthropogenic CO$_2$ emissions will come not from increasing the amount of land in mature forest or even from increasing the standing stock of forests, but from using renewable forest products in the place of other products. Thirty percent of the recent net carbon uptake of U.S. forests resides in landfills and long-lived products (Turner and colleagues 1993). Another use of tree products is as a substitute for fossil fuels, Turner and colleauges (1993) noted the possibility of taking a credit for tree products used as fuel; however, this did not fit into the structure of their analysis and was not actually included (see Table 15.1). To illustrate the importance of fuel substitution in the carbon balance, a credit for biomass used as fuel can be worked into the balance described by Harmon and colleagues (1990). In their analysis of the harvest of an old-growth Douglas fir forest (Table 15.3A), Harmon and colleagues showed that the bulk of carbon stored in the forest is released on harvest and that literally centuries are required before the total carbon sequestered in the forest and forest products returns to its initial state. We can, however, modify their carbon balance to acknowledge that some of the forest harvest was used as a fuel and that had it not been so used, some other, probably fossil, fuel would have been required. This modification recognizes that carbon is, in effect, sequestered both in long-lived wood products and in unburned fossil fuels (Table 15.3B) (Marland and Marland, 1992). For this forest management strategy, the net effect on CO$_2$ emissions is not simply the effect on carbon standing in the forest and the fate of long-lived wood products, but includes the displacement of other carbon-emitting processes. Additionally, the forest harvest has economic value as a source of energy services.

Table 15.3A. *Disposition of C on harvest of a 450-year-old Douglas fir forest*

Carbon in old-growth boles			325 Mg C/ha
Carbon in long-lived products		138	
Carbon in short-lived products, fuel, paper, and residue	187		
Carbon sequestered		138	
Carbon released to atmosphere	187		

Source: Harmon et al. (1990).

Cumulative net CO$_2$ emissions can be reduced by planting trees either to store carbon in standing forest or to displace emissions from fossil fuels; however, over the long term the latter strategy appears to have greater potential (Figure 15.4). Figure 15.4 distorts the details, however, because it fails to acknowledge specifics such as the inefficiencies of fuel substitution; for example, energy input is required to manage a tree plantation, and carbon standing in a forest cannot substitute one-for-one for carbon lying in a coal deposit. Figure 15.4 also fails to deal with the carbon stored in soils, a factor that can make a significant contribution to net carbon storage.

Marland and Marland (1992) have constructed a simple model of carbon flows to provide a more realistic appraisal of the relative merits of using trees for carbon storage as opposed to using them to recycle carbon through a woody fuel. In a series of illustrative plots (see, e.g., Figure 15.5), Marland and Marland suggested that (depending on time horizon, anticipated productivity, current standing stock, and substitution efficiency) at least three options should be considered. Under various combinations of these parameters, the optimal strategy for minimizing net emissions of CO$_2$ to the atmosphere may be (1) to manage existing forest for carbon storage in the forest (and forest soils), (2) to replace an existing land use with tree plantings intended to store carbon in the forest, or (3) to replace an existing land use with plantation biomass (trees or, perhaps, a herbaceous crop) intended as a substitute for fossil fuels. In general terms, Marland and Marland suggested that for areas with large standing biomass and low productivity, the most carbon-efficient strategy is to preserve the current forest

stand, whereas for areas with high productivity the most carbon-efficient strategy is to plant biomass for use as an energy crop (or to substitute for energy-intensive products otherwise). The boundary between what we characterize here as low productivity and high productivity is very site specific; however, it also depends on the time horizon over which planning occurs and on the efficiency with which the biomass is produced and used.

In the context of the present discussion, this means that the places where reforestation for carbon storage is the most carbon-efficient strategy are those areas where there is low standing biomass and low productivity. Although net carbon offset will be greater than that for other strategies that might be undertaken at these low-productivity sites, carbon uptake will be slow on a per hectare per year basis. On the other hand, there will also be some average accumulation of carbon in the standing biomass at areas chosen for energy plantations; however, it will be considerably less than in a mature forest.

Having raised the issue of site-specific characteristics, we note that Wright and colleagues (1992) identified 159 million ha in the United States capable of producing energy crops without irrigation and speculated that 90 million of these could achieve the estimated critical rate of 2.5 Mg C/ha/yr probably necessary to produce energy crops economically. After considering other factors, such as agricultural land needs, Wright and colleagues focused on a tree-planting scenario based on 28 million ha. Table 15.4 shows the geographic distribution of the 159 million ha of "capable" land in the United States, 78% of which is currently classified as cropland. As noted in the

Table 15.3B. *Net effect on CO$_2$ emissions to the atmosphere from harvest of the same 450-year-old Douglas fir forest when we acknowledge that part of the harvest was burned to displace fossil fuel combustion*

Carbon in old-growth boles			325 Mg C/ha
Carbon in long-lived products		138	
Carbon in wood-based fuels (total = 113)		68	
Fossil fuel carbon displaced	45		
Wood fuel C burned without fossil fuel displacement	74		
Total effective carbon sequestered		206	
Net carbon released to atmosphere	119		

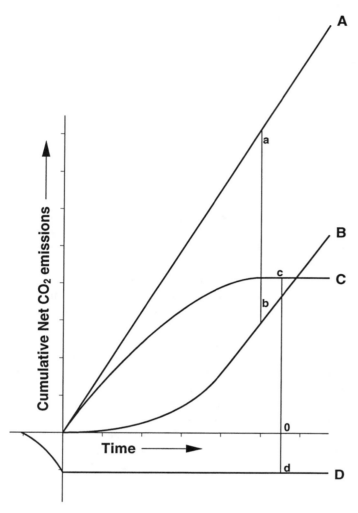

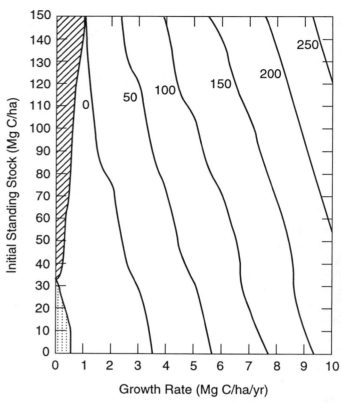

Figure 15.5: Net cumulative carbon sequestering as a result of tree growth plus fossil fuel displacement after 50 years is contoured as a function of site occupancy at time 0 (initial standing stock in Mg C/ha) and site productivity (initial growth rate in Mg C/ha/yr). Four regions are delineated by the contour lines. The contours are in Mg C/ha and show the excess of net sequestering for a system in which biomass is harvested and used efficiently to substitute for fossil fuels as compared to a system where the intent is long-term carbon storage in standing trees. For combinations of initial site occupancy and productivity to the right of the 0 contour, the biomass fuel scenario yields greater cumulative net C offset at the end of 50 years. For combinations to the left of the 0 contour but to the right of the shaded region, either fuel substitution or on-site sequestering will yield a net carbon offset, but on-site sequestering is more effective. For combinations in the upper left shaded region, i.e., with high initial site occupancy and low productivity, any forest harvest will result in net C emissions to the atmosphere because even after 50 years the forest has not recovered to its initial carbon storage, even when fossil fuel displacement is taken into account. In the stippled area on the lower left, it is assumed that the initial standing stock is so small that there can be no useful harvest (see Marland and Marland, 1992).

Figure 15.4: Schematic representation of cumulative net emissions of CO_2 as a function of time for various combinations of a coal-fired power plant and forest management strategy. Path A shows the uniform increase of cumulative net CO_2 emissions from the coal-fired power plant. Path B represents the cumulative net emissions if enough trees are planted so that CO_2 emissions from the power plant are exactly offset by the photosynthetic uptake of C in young, rapidly growing trees. The latter part of path B shows net emissions growing parallel to those of path A as the mature forest stand no longer has a net uptake of carbon. Path C represents net CO_2 emissions when a sustained-yield energy plantation is established after initial clearing and use of an existing forest stand. Path D represents the emissions from a power plant that burns wood from a sustained-yield energy plantation established on a site not previously occupied by forest. Path D envisions a tree plantation started some years prior to operation of the power plant so that the plantation can be partially harvested each year to fuel the power plant with no net CO_2 emissions. The distances ab and cd represent the amount of carbon held in mature forest; 0d is the amount held in plantation forest. Reprinted with permission from R. J. Moulton and K. R. Richards (1990). *Costs of Sequestering Carbon through Tree Planting and Forest Management in the United States.* GTR WO-58, U.S. Department of Agriculture Forest Service.

Moulton and Richards (1990) analysis discussed earlier, if large areas of fast-growing trees are to be planted in the United States as part of a strategy to offset a significant amount of fossil fuel–based CO_2 emissions, the trees will likely compete for land in the agricultural regions of the United States, including, initially, lands in those regions that are "economically marginal and environmentally sensitive" (Moulton and Richards, 1990, p. 1) or that currently represent excess agricultural ca-

Table 15.4. Quantity and location of land capable of producing wood energy crops in the U.S. without irrigation (millions of hectares)

Region	Land Area
Rocky Mountains	0
Pacific Coast	1
Southeast	14
Northeast	18
South Central	36
North Central	90
Total	159

Source: Wright et al. (1992).

pacity. These trees can provide an alternative to fossil fuel burning; however, average carbon storage will be modest.

15.4 Conclusions

In closing, we reiterate that there are multiple opportunities to reduce net CO$_2$ emissions with forestry projects and that these are complimentary in the sense that for any specific land area there are choices from which to seek the optimal strategy. Opportunities range from protecting existing forests, to improving forest management, to expanding the area in forest, to increasing the magnitude or efficiency of substitution of renewable forest products for more carbon-intensive products. For forest land with large standing biomass and low productivity, it may take a very long time to recapture the carbon once released, and the optimal strategy is likely to be protection of the standing forest.

Projects intended to mitigate the net emissions of carbon to the atmosphere take many forms. Winjum and colleagues (1993) summarized that improved forest management can enlarge carbon storage in existing forests, Swisher (1994) reminded us that low-intensity uses of land that cannot support production forestry can still provide local benefits and some long-term carbon storage, and Sampson and colleagues (1993) found considerable potential for using biomass fuels to avoid fossil fuel burning.

This discussion is intended to provide perspective on the extent to which reforestation projects undertaken to offset anthropogenic CO$_2$ emissions might increase the mass of carbon sequestered in the biosphere and, hence, need to be incorporated in scenarios for the future behavior of the global carbon cycle. Although the discussion is largely qualitative, it is clear that despite the existence of apparently large areas capable of supporting reforestation, the literature on reforestation recognizes social and economic limitations on what is physically possible. There are two fundamental strategies for reforestation projects to reduce net CO$_2$ emissions: growing trees for long-term carbon storage in the maturing forest, and sustainable management of forest land to produce energy and other

products to displace carbon-intensive activities such as fossil fuel burning. Land most likely to be replanted (or planted) to forest with the intent of long-term storage of carbon in the standing biomass is land with lower productivity, lower demand for other purposes, and lower rates of carbon accumulation per unit of area. Areas capable of higher productivity are more likely to be used to produce energy or other forest products and will be more important for their C emissions displacements than for carbon "sequestering." Although the opportunities for forestry to play a role in mitigation strategies seem to be significant, the primary focus of new planting is likely to be sustainable and efficient use of forest products rather than long-term accumulation of carbon.

References

Clinton, W. J. 1993. *The Climate Change Action Plan.* The White House, Washington, D.C.

Dale, V. H., R. A. Houghton, and C. A. S. Hall. 1991. Estimating the effects of land-use change on global atmospheric CO$_2$ concentrations. *Canadian Journal of Forest Research 21*, 87–89.

Dixon, R. K., K. J. Andrasko, F. G. Sussman, M. A. Lavinson, M. C. Trexler, and T. S. Vinson. 1993. Forest sector carbon offset projects: Near-term opportunities to mitigate greenhouse gas emissions. *Water, Air, and Soil Pollution 70*, 561–577.

Dixon, R. K., S. Brown, R. A. Houghton, A. M. Solomon, M. C. Trexler, and J. Wisniewski. 1994. Carbon pools and flux of global forest ecosystems. *Science 263*, 185–190.

Dyson, F. J. 1977. Can we control the carbon dioxide in the atmosphere? *Energy 2*, 287–291.

Dyson, F. J., and G. Marland. 1979. Technical fixes for the climatic effects of CO$_2$. In *Workshop on the Global Effects of Carbon Dioxide from Fossil Fuels*, W. P. Elliott and L. Machta, eds. Miami Beach, Florida, March 7–11, 1977. CONF-770385, U.S. Department of Energy, Washington, D.C., pp. 111–118.

Graham, R. L., L. L. Wright, and A. F. Turhollow. 1992. The potential for short-rotation woody crops to reduce U.S. CO$_2$ emissions. *Climatic Change 22*, 223–238.

Grainger, A. 1988. Estimating areas of degraded tropical lands requiring replenishment of forest cover. *International Tree Crops Journal 5*, 31–61.

Harmon, M. E., W. K. Ferrell, and J. F. Franklin. 1990. Effects on carbon storage of conversion of old-growth forests to young forests. *Science 247*, 699–701.

Henderson, S., and R. K. Dixon (eds.). 1993. Management of the terrestrial biosphere to sequester atmospheric CO$_2$. *Climate Research,* Special Issue 3 (1 and 2), 1–140.

Houghton, R. A., J. D. Unruh, and P. A. Lefebvre. 1993. Current land cover in the tropics and its potential for sequestering carbon. *Global Biogeochemical Cycles 7*, 305–320.

Iverson, L. R., S. Brown, A. Grainger, A. Prasad, and D. Liu. 1993. Carbon sequestration in tropical Asia: An assessment of technically suitable forest lands using geographic information systems analysis. *Climate Research 3*, 23–38.

Kauppi, P. E., K. Mielikainen, and K. Kuusela. 1992. Biomass and carbon budget of European forests, 1971–1990. *Science 256*, 70–74.

Kurz, W. A., and M. J. Apps. 1993. Contribution of northern forests to the global C cycle: Canada as a case study. *Water, Air, and Soil Pollution 70*, 163–176.

Marland, G. 1988. *The Prospect of Solving the CO$_2$ Problem Through Global Reforestation.* DOE/NBB-0082, U.S. Department of Energy, Washington, D.C.

Marland, G., and S. Marland. 1992. Should we store carbon in trees? *Water, Air, and Soil Pollution 64,* 181–195.

Meyer, W. B., and B. L. Turner II. 1992. Human population growth and global land-use/cover change. *Annual Review of Ecology and Systematics 23,* 39–61.

Moulton, R. J., and K. R. Richards. 1990. *Costs of Sequestering Carbon Through Tree Planting and Forest Management in the United States.* GTR WO-58, Forest Service, U.S. Department of Agriculture, Washington, D.C.

NAS (National Academy of Sciences, Panel on Policy Implications of Greenhouse Warming). 1992. *Policy Implications of Greenhouse Warming.* National Academy Press, Washington, D.C.

Parks, P. J., S. R. Brame, and J. E. Mitchell. 1992. Opportunities to increase forest area and timber growth on marginal crop and pasture land. In *Forests and Global Change. Vol. 1: Opportunities for Increasing Forest Cover,* R. N. Sampson and D. Hair, eds. American Forestry Association, Washington, D.C., pp. 97–121.

Plantinga, A. J., and R. A. Birdsey. 1993. Carbon fluxes resulting from U.S. private timberland management. *Climatic Change 23,* 37–53.

Richards, J. F., and E. P. Flint. 1994. A century of land-use change in south and southeast Asia. In *Effects of Land-Use Change on Atmospheric CO$_2$ Concentrations: South and Southeast Asia as a Case Study,* V. H. Dale, ed. Springer-Verlag, New York, pp. 15–66.

Richards, J. F., E. S. Haynes, J. R. Hagen, E. P. Flint, J. Arlinghaus, J. B. Dillon, and A. L. Reber. 1987. *Changing Land Use in Pakistan, Northern India, Bangladesh, Burma, Malaysia, and Brunei, 1880–1980.* U.S. Department of Energy, Washington, D.C.

Sampson, R. N., and D. Hair (eds.). 1992. *Forests and Global Change. Vol. 1: Opportunities for Increasing Forest Cover.* American Forestry Association, Washington, D.C.

Sampson, R. N., L. L. Wright, J. K. Winjum, J. D. Kinsman, J. Beneman, E. Kursten, and J. M. O. Scurlock. 1993. Biomass management and energy. *Water, Air, and Soil Pollution 70,* 139–159.

Swisher, J. N. 1991. Cost and performance of CO$_2$ storage in forestry projects. *Biomass and Bioenergy 1,* 317–328.

Swisher, J. N. 1994. Bottom-up comparisons of CO$_2$ storage and costs in forestry and biomass energy projects. In *Proceedings of the Biomass Conference of the Americas,* Burlington, Vermont.

Trexler, M. C., P. E. Faeth, and J. M. Kramer. 1989. *Forestry as a Response to Global Warming: An Analysis of the Guatemala Agroforestry and Carbon Sequestration Project.* World Resources Institute, Washington, D.C.

Trexler, M. C., C. A. Haugen, and L. A. Loewen. 1992. Global warming mitigation through forestry options in the tropics. In *Forests and Global Change. Vol. 1: Opportunities for Increasing Forest Cover,* R. N. Sampson and D. Hair, eds. American Forestry Association, Washington, D.C., pp. 73–96.

Turner, D. P., J. F. Lee, G. J. Koerper, and J. R. Barker. 1993. *The Forest Sector Carbon Budget of the United States: Carbon Pools and Flux Under Alternative Policy Options.* EPA/600/3–93/093, U.S. Environmental Protection Agency, Washington, D.C.

Waggoner, P. E. 1994. *How Much Land Can Ten Billion People Spare for Nature?* Task Force Report No. 121, Council for Agricultural Science and Technology, Ames, Iowa.

Winjum, J. K., and D. K. Lewis. 1993. Forest management and the economics of carbon storage: The nonfinancial component. *Climate Research 3,* 111–119.

Winjum, J. K., R. A. Meganck, and R. K. Dixon. 1993. Expanding global forest management: An "easy first" approach. *Journal of Forestry 91(4),* 38–42.

Wright, L. L., R. L. Graham, A. F. Turhollow, and B. C. English. 1992. The potential impacts of short-rotation woody crops on carbon conservation. In *Forests and Global Change. Vol. 1: Opportunities for Increasing Forest Cover,* R. N. Sampson and D. Hair, eds. American Forestry Association, Washington, D.C., pp. 123–156.

16

Simple Ocean Carbon Cycle Models

KEN CALDEIRA, MARTIN I. HOFFERT, AND ATUL JAIN

Abstract

Simple ocean carbon cycle models are constructed to reflect the interaction between the atmospheric and oceanic components of the global carbon cycle. In this chapter, (1) a two-box ocean model is used to demonstrate principles used in constructing simple ocean carbon cycle models, (2) a variety of simple ocean carbon cycle models are described, and (3) results of various models are shown and compared. Physical transport of carbon in simple ocean models is not based on first principles, but is accomplished using parameterizations calibrated with carbon isotopes and/or other tracer fields. Well-calibrated simple ocean carbon cycle models may yield CO_2 absorption predictions that are more accurate than predictions based on ocean general circulation models. Nevertheless, ocean general circulation models may be required to estimate the impact of climate and ocean circulation feedbacks on CO_2 fluxes between the ocean and atmosphere.

16.1 Introduction

The activities of *Homo sapiens* have been significantly perturbing the global carbon cycle for the past several hundred years. From 1750 to 1990, the burning of fossil fuels and the clearing of forests has released approximately 380 Gt C as CO_2 into the atmosphere, but only approximately 160 Gt C remains there (Sundquist, 1993); the rest has been absorbed by the oceans and terrestrial ecosystems. Ocean models indicate that the oceans have absorbed approximately 140 Gt C, leaving approximately 80 Gt C unaccounted for (perhaps indicating an additional sink in terrestrial ecosystems). A major focus of ocean carbon cycle modelers has been to try to develop models that will lead to more accurate predictions of the rate at which anthropogenic CO_2 has been and will be absorbed by the oceans. This chapter describes some simple ocean carbon cycle models that can be used for that purpose. (See Table 16.1 for a taxonomy of simple ocean carbon cycle models.)

The flux of CO_2 to the atmosphere from the burning of fossil fuels is about two orders of magnitude greater than that from natural geologic CO_2 sources. In the natural carbon cycle, unperturbed by human-induced CO_2 emissions, the primary sources of carbon to the oceans and atmosphere are volcanoes, midocean ridges, CO_2-rich hot springs, and the weathering of carbonate rocks and sedimentary organic carbon. Most of this carbon is subsequently sequestered in sediments as carbonate rock and organic carbon (Lasaga et al., 1985). Ocean chemistry is thought to adjust to two conditions: (1) The partial pressure of CO_2 (pCO_2) in the surface ocean approaches a value that is close to the atmospheric value (which is determined by considerations involving silicate-rock weathering and CO_2 degassing rates); and (2) the carbonate ion concentration in the deep ocean adjusts such that calcium carbonate ($CaCO_3$) sedimentation balances sources of calcium and carbon to the ocean. From these considerations, long-term ($> 10^6$ yr) values for ocean pH and for the oceanic concentrations of total dissolved inorganic carbon (ΣCO_2) and alkalinity can be calculated (Lasaga et al., 1985).

Table 16.1. *Taxonomy of ocean carbon cycle models with examples*

1. Classical multibox

Minimum number of boxes is 3: atmosphere, surface mixed layer and deep sea. Single-box ocean fails because downward mixing is limited by thermocline diffusion bottleneck, not gas exchange at the surface.

Revelle & Suess (1957); Bolin & Eriksson (1959); Broecker et al. (1971); Machta (1973); Hoffert (1974); Keeling (1973, 1977); Bacastow & Keeling (1973); Bjorkstrom (1979, 1986); Broecker & Peng (1982)

2. High-latitude surface box

Attempts to explain pCO_2 variations in ice cores though changes in circulation, solubility or biological carbon pumps leveraged by high-latitude gas exchange.

Sarmiento & Toggweiler (1984); Siegenthaler & Wenk (1984); Knox & McElroy (1984); Volk & Hoffert (1985); Toggweiler & Sarmiento (1985); Volk & Liu (1988)

3. Multibasin multibox

Pandora and related models calibrated on tracers to include many physically distinct water mass regimes and processes.

Broecker & Peng (1986, 1987); Bjorkstrom (1986); Keir (1988, 1989); Walker (1991); Walker & Kasting (1992)

4. Box-diffusion

Vertically stacked boxes equivalent to pure diffusion finite-difference formulation beneath a well-mixed layer; steady-state calibrated on prebomb radiocarbon.

Oeschger et al. (1975); Siegenthaler & Oeschger (1978); Hoffert et al. (1979); Enting & Pearman (1987); Keeling et al. (1989a,b); Caldeira & Kasting (1993)

5. Outcrop-diffusion

Combines diffusion with high-latitude connection to deep sea to simulate isopyncnal mixing outcrop.

Siegenthaler (1983)

6. Upwelling-diffusion and HILDA

Classical deep sea recipes for temperature and tracers updated by circulation and mixing connections to polar sea surface plus surface and high-latitude mixed layer.

Wyrtki (1962); Munk (1966); Hoffert et al. (1981); Volk (1984); Shaffer (1989); Khesghi et al. (1991); Siegenthaler & Joos (1992); Jain et al. (1993); Shaffer & Sarmiento (1995)

7. OGCM

Transport by three-dimensional internal current fields, but vertical mixing input either as subgrid diffusivity or numerical diffusion; solutions depend on mixing assumptions.

Sarmiento (1986); Maier-Reimer & Hasselmann (1987); Toggweiler et al. (1989a,b); Maier-Reimer & Bacastow (1990); Bacastow and Maier-Reimer (1990, 1992); Sarmiento et al. (1993)

8. Convolution integral

Calculational method to estimate ocean carbon cycle model results without performing full model calculations.

Maier-Reimer & Hasselmann (1987); Harvey (1989); Wigley (1991)

On the decade-to-century time scale, the removal of anthropogenic CO_2 from the atmosphere primarily reflects a redistribution of carbon among atmosphere, biosphere, and ocean reservoirs. On the time scale of several centuries or more, interactions between marine waters and sediments can be an important factor affecting atmospheric CO_2 content (Sundquist, 1990).

This chapter has three parts. The first part discusses processes and techniques common to many simple ocean carbon cycle models, using a simple two-box ocean model as an example. The second part presents a brief description of simple ocean carbon cycle models. The third part presents some significant results of these models.

16.2 A Taxonomy of Simple Ocean Models

16.2.1 A Two-Box Ocean Carbon Cycle Model

In this section, we use a two-box ocean carbon cycle model to demonstrate fundamental techniques and principles used in simple ocean carbon cycle models. We use a two-box model (1) to show the difference between models that use perturbation approaches and models that represent biological fluxes explicitly, (2) to demonstrate how Redfield ratios can be used to line the carbon cycle to ocean nutrient fields (such as phosphate), and (3) to estimate crudely the effect of changes in ocean mixing rates and biological productivity on CO_2 absorption by the oceans. (A one-box model cannot reproduce the time-dependent structure of the response of the carbon cycle to the anthropogenic CO_2 perturbation, because the time scale of the perturbation is short relative to the ocean mixing time scale.)

In a two-box model, the ocean is divided into two well-mixed compartments, a surface ocean and a deep ocean (Figure 16.1). The concentration of carbon is greater in the deep ocean than in the surface ocean, primarily owing to a *biological pump* and a *solubility pump* (Volk and Hoffert, 1985). The surface ocean provides the primary base of the marine food chain, because phytoplankton can grow only where there is sufficient light (Parsons et al., 1984). Some of the organic carbon generated in the surface ocean settles into the deep ocean where it may be oxidized into ΣCO_2. Thus, biological production at the ocean's surface transports ΣCO_2 to the deep ocean. This transport is known as the biological pump or the soft-tissue pump. CO_2 is more soluble in cold water than in warm water. Cold water is also more dense than warm water of the same salinity. Cold sinking water tends to have a higher ΣCO_2 concentration than warm, buoyant water; this tends to make the deep ocean ΣCO_2-rich relative to the surface ocean. This transport is known as the solubility pump. In steady state, the upward and downward fluxes of carbon must balance; the upward flux of carbon that balances the biological and solubility pumps is provided by fluid advection and ocean mixing processes, which tend to return carbon to the surface ocean where the ΣCO_2 concentration is lower.

16.2.1.1 Basic Two-Box Ocean Model Equations

In a two-box ocean model, the equation for the time rate of change of C_s, the ΣCO_2 concentration in the surface box, can be represented as

$$M_s \frac{dC_s}{dt} = F_{mix}(C_d - C_s) - F^C_{bio} - F^C_{sol} + F^C_{gasex} \quad (16.1)$$

(See Table 16.2 for a description of symbols used.)

A corresponding equation for the deep ocean can be written as

$$M_d \frac{dC_d}{dt} = F_{mix}(C_s - C_d) + F^C_{bio} + F^C_{sol} \quad (16.2)$$

CO_2 fluxes into the surface ocean (F^C_{gasex}) are typically modeled as the product of an air–sea gas exchange coefficient (k_{gasex}), the area of the ocean (A_{ocean}), and the difference between the partial pressure of CO_2 in the atmosphere ($pCO_{2,atm}$), and the partial pressure in the surface ocean ($pCO_{2,atm}$):

$$F^C_{gasex} = A_{ocean} k_{gasex}(pCO_{2,atm} - pCO_{2,s}) \quad (16.3)$$

In steady state, in this sort of two-box model, $pCO_{2,atm} = pCO_{2,s}$, and $F^C_{gasex} = 0$. Model results are relatively insensitive to variation in k_{gasex}, because the primary resistance to oceanic CO_2 absorption is transport within the ocean and not air–sea gas exchange.

16.2.1.2 Perturbation Model of CO_2 Absorption by the Ocean

If we make the assumptions that the preindustrial ocean was in steady state, and ocean mixing and the biological and solubility pumps are unaffected by the change in atmospheric CO_2 in the industrial age, we can develop a perturbation model corre-

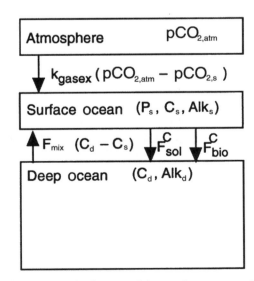

Figure 16.1: Schematic diagram of the two-box ocean carbon cycle model described in the text. Symbols are described in Table 16.2.

Table 16.2. *List of symbols used in the text*

Symbol	Description
A_{ocean}	Area of the oceans (= 3.5×10^{14} m²)[a]
Alk_d	Deep ocean alkalinity concentration (= 2,406 µeq kg-1; model computed value)
Alk_s	Surface ocean alkalinity concentration (= 2,344 µeq kg-1; model value calibrated to achieve preindustrial $pCO_{2,s}$ = 280 ppm)
C_d	The ΣCO_2 concentration in the deep ocean box (2,288 mol ΣCO_2/kg; Volk and Hoffert, 1985)
C_s	The ΣCO_2 concentration in the surface ocean box (2,016 mol ΣCO_2/kg; Volk and Hoffert, 1985)
E	The flux of anthropogenic CO_2 into the atmosphere (kg/yr)
F^C_{bio}	The flux of ΣCO_2 to the deep ocean produced by the biological pump (7.8×10^{14} mol ΣCO_2/yr; model computed value)
F^C_{gasex}	The net flux of CO_2 into the oceans due to ocean–atmosphere gas exchange (mol ΣCO_2/yr)
F^C_{human}	The flux of CO_2 into the atmosphere brought on by human activity (including net fluxes from the biosphere; mol CO_2/yr)
F^C_{sol}	The flux of ΣCO_2 to the deep ocean produced by the solubility pump (2.0×10^{14} mol ΣCO_2/yr; model computed value)
F_{mix}	The mass flux of water between the surface ocean and deep ocean boxes (3.6×10^{18} kg/yr)
F^P_{bio}	Phosphate flux to the deep ocean produced by the biological pump (5.9×10^{12} mol P/yr; model computed value)
G	The CO_2 signal as a function of time observed in the atmosphere for a δ-function atmospheric input of CO_2, or equivalently a unit step-function change in the initial atmospheric CO_2 concentration (dimensionless)
k	Eddy-diffusion coefficient for vertical transport in the ocean (m²/yr)
k_{bio}	A constant relating surface ocean phosphate concentration and the biological export of phosphate from the surface ocean (kg/yr)
k_{gasex}	Gas exchange coefficient for CO_2 (= 0.06 mol/m²/yr/ppm; Maier-Reimer and Hasselmann, 1987)
M_{atm}	The molar mass of the atmosphere (= 1.8×10^{20} mol)
M_d	The mass of the deep ocean (= 7E1.3 $\times 10^{21}$ kg)
M_s	Mass of the surface ocean (= ~3.6×10^{19} kg for a 100-m layer)
$pCO_{2,atm}$	The partial pressure of CO_2 in the atmosphere at the ocean surface (ppm)
$pCO_{2,s}$	The partial pressure in the surface ocean (ppm)
P_d	The phosphate concentration in the deep ocean (2.14 µmol P/kg)
P_s	The phosphate concentration in the surface ocean (0.5 µmol P/kg)
$R_{Alk:P}$	The ratio of alkalinity to phosphate ($R_{C:P}$) in the biogenic detrital "rain" out of the surface ocean and into the deep ocean (= 38 eq/(mol P))
$R_{C:P}$	The ratio of carbon to phosphate in the biogenic detrital "rain" out of the surface ocean and into the deep ocean (= 132.5 mol ΣCO_2/(mol P)), assuming that 80% of the detrital rain is composed of organic matter and that 20% is composed of $CaCO_3$
t	Time (yr)
w	Mean upwelling velocity for the ocean excluding the polar waters (m/yr)
z	Depth below the surface of the ocean (m)
0	Subscript 0 indicates a steady-state value
ΔC	Perturbation of the ΣCO_2 concentration from the steady-state value as a function of depth (mol C/m³)
ΔC_d	The perturbation to the steady-state deep ocean total dissolved carbon concentration (= $C_d - C_{d,0}$; mol ΣCO_2/kg)
ΔC_s	The perturbation to the steady-state surface ocean total dissolved carbon concentration (= $C_s - C_{s,0}$; mol ΣCO_2/kg)
ΔM	The CO_2 mass in the atmosphere above the preindustrial mass (kg)
$\Delta pCO_{2,s}$	The perturbation to the steady-state partial pressures of CO_2 in the surface ocean (= $pCO_{2,s} - pCO_{2,s,0}$; ppm)
ζ	A buffer factor expressing the fractional change in the partial pressure of CO_2 in the mixed layer ($\Delta P_{ocean}/P_{ocean,0}$) divided by the fractional change in mixed-layer total dissolved inorganic carbon ($\Delta C_s/C_{s,0}$)
ΣCO_2	Total dissolved carbon (= $CO_2 + H_2CO_3 + HCO_3^- + CO_3^=$)

[a]The model values for the two-box ocean developed here are presented for pedagogic purposes; a more sophisticated model should be used if accurate results are required.

sponding to this two-box ocean model. Two new variables, ΔC_d and ΔC_s, can be defined, representing the perturbation to the steady-state values of deep ocean $C_{d,0}$ and surface ocean $C_{s,0}$, that is,

$$C_d = \Delta C_d + C_{d,0} \tag{16.4}$$

and

$$C_s = \Delta C_s + C_{s,0} \tag{16.5}$$

where the subscript 0 indicates a steady-state value. Substituting Equations (16.4) and (16.5) into Equations (16.1) and (16.2) with the assumptions stated above, and subtracting the steady-state solution yields the equations

$$M_d \frac{d\Delta C_d}{dt} = F_{mix}(\Delta C_s - \Delta C_d) \tag{16.6}$$

and

$$M_s \frac{d\Delta C_s}{dt} = F_{mix}(\Delta C_d - \Delta C_s) + F_{gasex}^C \tag{16.7}$$

Note that, under these assumptions, the biological and solubility pumps "drop out" of the solution to the perturbation equations. This principle is used in many simple ocean models (e.g., Oeschger et al., 1975), as well as general circulation models (e.g., Sarmiento et al., 1992; Maier-Reimer and Hasselmann, 1987), to perform carbon cycle calculations without explicitly modeling ocean biology or the steady-state distribution of carbon in the oceans. (Nevertheless, it should be noted that the neglect of biological fluxes can introduce errors of up to 10% in the steady-state profile of ^{14}C ratios; Fiadeiro, 1982.)

Many simple models use a factor (ζ) known as the buffer factor or Revelle factor (Revelle and Suess, 1957) to express the fractional change in the partial pressure of CO_2 in the surface ocean ($\Delta pCO_{2,s}/pCO_{2,s,0}$) caused by a fractional change in mixed-layer total dissolved inorganic carbon ($\Delta C_s/C_{s,0}$):

$$\zeta = \frac{\Delta pCO_{2,s}/pCO_{2,s,0}}{\Delta C_s/C_{s,0}} \tag{16.8}$$

With the assumption that ζ is constant, Equation (16.8) can be rearranged to yield a linear equation relating $\Delta pCO_{2,s}$ to ΔC_s. (Because the partial pressure of mixed-layer CO_2 is a nonlinear function of total inorganic carbon concentration, this assumption is valid only for small changes in atmospheric CO_2 content; see Maier-Reimer and Hasselmann, 1987.)

If we add an equation for the change of mass of CO_2 in the atmosphere, such as

$$M_{atm} = \frac{dpCO_{2,atm}}{dt} = F_{human}^C - F_{gasex}^C \tag{16.9}$$

then this linear system can be solved analytically for the decay of a unit pulse of CO_2 into the atmosphere. The results of this computation for an instantaneous doubling of atmospheric CO_2 content, compared with those of Maier-Reimer and Hasselmann (1987), are shown in Figure 16.2A. This calculation demonstrates that even a model as simple as this two-box per-

turbation model can capture some of the basic features of CO_2 absorption by the oceans. The oceans initially absorb a pulse on a rapid time scale associated with the equilibration of the surface ocean waters with the atmosphere. Then the absorption process is dominated by the slower process of transporting carbon from the surface ocean to the deep ocean.

When there are significant changes in ΣCO_2 (or alkalinity) concentration, a constant buffer factor is no longer an acceptable approximation, and the results of carbonate chemistry calculations need to be considered. CO_2 dissolving into the ocean speciates into carbonic acid (H_2CO), bicarbonate ion (HCO_3^-),

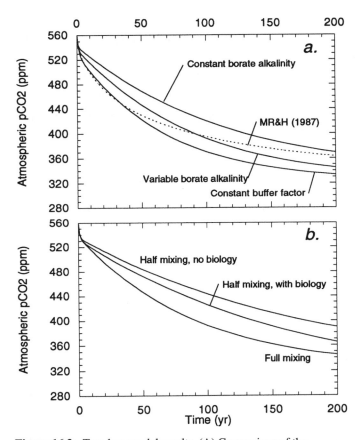

Figure 16.2: Two-box model results. (A) Comparison of the response of the two-box model described in the text with the results of Maier-Reimer and Hasselmann (1987) for a CO_2 pulse equal to 25% of the preindustrial atmospheric CO_2 content. The importance of using variable borate alkalinity to compute carbonate equilibria can be seen from the two-box model results that were computed using, alternatively, a constant surface ocean CO_2 buffer factor (bottom line), constant borate alkalinity (top line), and variable borate alkalinity (middle line). (B) Comparison of CO_2 absorption by the oceans computed using the two-box model for a CO_2 impulse. The "variable borate alkalinity" case from (A) is compared with two cases in which ocean circulation (F_{mix}) is instantaneously diminished by a factor of 2 at time $t = 0$. In the "no biology" case, CO_2 absorption rates are computed using the perturbation model described in the text. In the "with biology" case, a biological feedback is computed using the linear biology response model described in the text. Parameter values used are as specified in Table 16.2.

and carbonate ion ($CO_3^=$); most of the dissolved inorganic carbon in the ocean is in the form of HCO_3^-. (See Stumm and Morgan, 1981, for a detailed exposition of how to calculate the speciation of ΣCO_2 in seawater.) The roughly 25% increase in the pCO_2 of surface seawater that has occurred in the industrial era would be accomplished by a 2.5% increase in the ΣCO_2 concentration. The small percentage change in ocean total dissolved inorganic carbon concentration at any location is masked by natural variability. Hence, it is difficult to measure the direct uptake of carbon by the oceans. For this reason, ocean carbon cycle models calibrated on other tracers have been employed to estimate carbon uptake rates by the oceans.

Important qualitative behaviors of the carbonate chemistry are:

- More total dissolved inorganic carbon results in a higher CO_2 partial pressure.
- More alkalinity results in a lower CO_2 partial pressure.
- Warmer water results in a higher CO_2 partial pressure.

Marine biological productivity tends to diminish both the inorganic carbon and alkalinity concentrations of the surface ocean waters. However, the diminution of inorganic carbon concentration has the larger effect on the pCO_2 in the surface ocean; hence, enhanced biological productivity would tend to diminish the partial pressure of CO_2 in the surface ocean.

The dissolution of CO_2 in the oceans primarily occurs by two avenues. The first involves only carbonate chemistry and can be schematically represented as

$$CO_2 + CO_3^= + H_2O \rightarrow 2\,HCO_3^- \quad (16.10)$$

The second avenue for the CO_2 to dissolve in the oceans involves reaction with the borate ion, and can be represented as

$$CO_2 + B(OH)_4^- \rightarrow HCO_3^- + B(OH)_3 \quad (16.11)$$

Near present-day atmospheric CO_2 concentrations, approximately two-thirds of the CO_2 flux into the ocean dissolves by Reaction 10 and one-third by Reaction 11. The results of using a constant buffer factor (ζ), as opposed to using a value of ζ based on carbonate (or carbonate plus borate) equilibria calculations are also shown in Figure 16.2A. The consideration of borate results in approximately 30% more rapid oceanic absorption than the case in which borate is not considered; hence, borate must be considered in calculations of the speciation of total dissolved inorganic carbon.

16.2.1.3 Modeling the Effects of Marine Biology

The assumptions that ocean circulation, and the biological and solubility pumps, remain unchanged in a higher-CO_2 world allowed these pumps to be ignored in this perturbation calculation. However, if global warming were to alter ocean circulation significantly (Manabe and Stouffer, 1993), these assumptions might no longer be valid.

This two-box model can be used to demonstrate how the biological pump may be explicitly modeled. The flux of biogenic matter from the surface to deep ocean alters the oceanic carbon, nutrient, oxygen, and alkalinity fields. In such models,

phosphate is typically taken to be the nutrient limiting biological productivity (e.g., Broecker and Peng, 1982), although other macronutrients such as nitrate (Fasham et al., 1993), micronutrients such as iron (Martin, 1991), or zooplankton grazing (Fasham et al., 1993) may also limit productivity in the surface ocean.

In the simplest form of such models, surface phosphate concentrations are taken to be constant, and the F_{bio}^P is calculated to maintain a constant phosphate concentration in the surface ocean (e.g., Toggweiler and Sarmiento, 1985). At the next level of complexity, F_{bio}^P may be taken to be an increasing function of phosphate concentration. The simplest such statement would be

$$F_{bio}^P = k_{bio}P_s \quad (16.12)$$

where k_{bio} is a constant relating biological export from the mixed layer to mixed-layer phosphate concentration. More complicated formulation may involve Michaelis–Menton kinetics (e.g., Maier-Reimer, 1993) or represent ecosystem structure (Fasham et al., 1993). (See Totterdell, 1993, for a bibliography of marine biology models.)

The equations for the time rate of change of the surface phosphate concentration (P_s) may be written

$$M_s \frac{dP_s}{dt} = F_{mix}(P_d - P_s) - F_{bio}^P \quad (16.13)$$

and the deep ocean phosphate concentration (P_d) may be calculated by assuming conservation of oceanic phosphate mass.

On large spatial scales, the proportions of carbon, alkalinity, and phosphate in the organic matter transported out of the base of the mixed layer are thought to be relatively constant throughout the world ocean, although there may be considerable variability in these proportions on small spatial and temporal scales (Broecker and Peng, 1982). The phosphate and carbon cycles may be linked using these characteristic ratios, known as Redfield ratios (Redfield et al., 1963), of carbon to phosphate ($R_{C:P}$) and alkalinity to phosphate ($R_{Alk:P}$) in the detrital "rain" out of the mixed layer:

$$F_{bio}^C = R_{C/P}F_{bio}^P \quad (16.14)$$

(Equations for alkalinity parallel the equations for carbon, except that there is no exchange in the atmosphere, and alkalinity transport in the biogenic flux from the surface ocean to the deep ocean is the product of $R_{Alk/P}$ and F_{bio}^P.)

Changes in marine biological productivity affect surface ocean alkalinity concentrations and, hence, affect the partial pressure of CO_2 in the surface waters. Therefore, the alkalinity field must be calculated to estimate the distribution of carbon in the ocean and atmosphere for scenarios involving a change in ocean circulation or marine biological productivity.

With slower ocean mixing, the residence time for water in the mixed layer is greater; hence, nutrients might be expected to be stripped more efficiently from the mixed layer. Consequently, surface ocean nutrient utilization in the surface ocean may become more efficient at slower ocean mixing rates, even though overall productivity decreases. To demonstrate the effect this more efficient nutrient utilization can have on the absorption of CO_2 under conditions of altered ocean circulation, Figure 16.2B compares results from the perturbation version of the two-box model with a version incorporating Equations (16.12)–(16.14) for a scenario in which ocean mixing fluxes are instantaneously diminished by a factor of 2. In the model version incorporating biological effects, surface nutrient utilization is enhanced as ocean circulation slows; hence the drawdown of an atmospheric CO_2 perturbation is more rapid. This is true even though the biological transport of carbon from the surface to deep ocean is diminished. The biological feedback is opposite in sign but smaller in magnitude than the direct physical effects of diminished circulation on oceanic CO_2 absorption.

16.2.2 Multibox Ocean Carbon Cycle Models

Cold CO_2-rich water penetrates to the deep ocean, while warm CO_2-depleted water tends to stay near the surface; the temperature dependence of the solubility of CO_2 in seawater "pumps" CO_2 to the deep ocean. Consideration of this fact suggests a three-box model (e.g., Toggweiler and Sarmiento, 1985; Knox and McElroy, 1984; Siegenthaler and Wenk, 1984) in which there are warm and cold surface boxes and a deep water box. These models calculate the solubility of CO_2 in each of the surface boxes as a function of temperature. The solubility of CO_2 in the surface boxes affects the CO_2 fluxes between the atmosphere and surface ocean boxes, and hence the concentration of ΣCO_2 in the surface boxes. The cold surface box is specified as being the primary locus of deep water formation, and tends to have a greater ΣCO_2 concentration than the warm surface box. Hence, the formation of cold CO_2-rich deep water pumps ΣCO_2 to the deep ocean box.

Most organic carbon that falls out of the mixed layer is oxidized in the upper kilometers of the ocean, whereas most carbonate carbon dissolves in the bottom kilometer of the ocean. This consideration suggests the possible division of subsurface waters into an intermediate water box and a deep water box (Volk and Liu, 1988; Caldeira and Rampino, 1993). The inclusion of an intermediate water box permits four-box models to capture features such as the oxygen concentration minimum and a phosphate maximum.

Broecker and Peng (1987) and Keir (1988, 1989) have used box models to represent individual ocean basins and/or water masses. These models are able to represent interbasin differences in carbon-isotopic composition, and have been used generally to explore relations among changes in ocean circulation, tracer distributions, and atmospheric CO_2.

16.2.3 The Box-Diffusion Model

On the decade-to-century time scale, the infusion of anthropogenic CO_2 into the ocean may be primarily due to small-scale vertical transport in eddies, and not the large-scale ocean circulation. This suggests a pure diffusion approach to a perturbation model whereby the anthropogenic perturbation is modeled as penetrating into the ocean by purely diffusive mechanisms.

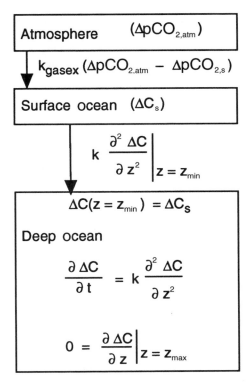

Figure 16.3: Schematic diagram of a box-diffusion model similar to that developed by Oeschger et al. (1975). Symbols are described in Table 16.2.

The box-diffusion model (Oeschger et al., 1975; Figure 16.3) represents the ocean as a vertical column of boxes, with neighboring boxes exchanging water parcels at a specified rate. The model is usually used to model perturbations to a steady-state carbon distribution; hence biological fluxes are not modeled. The governing equations for this model, with a finite number of vertical boxes, can be stated in terms of exchange of mass between neighboring boxes. However, in the limit of a large number of boxes, the governing equation for the box-diffusion model becomes

$$\frac{d\Delta C}{dt} = \frac{d}{dz}\left(k\frac{d\Delta C}{dz}\right) \tag{16.15}$$

where z is depth in meters. At the surface boundary, the flux of carbon into the ocean is specified as in the two-box model. At the bottom boundary, there is a zero-flux boundary condition.

Typically, box-diffusion models are calibrated using [14]C. There are two basic ways of doing this: One is to try to reproduce the steady-state distribution of ^{14}C produced by natural processes in the atmosphere; the other is to try to reproduce, as a function of time, the distribution of ^{14}C produced during atmospheric atomic bomb tests. Unfortunately, the diffusion coefficients are not the same for the two calibration methods. Box-diffusion models calibrated using natural ^{14}C absorb CO_2 more slowly than box-diffusion models calibrated using bomb [14]C. Natural ^{14}C has been penetrating into the oceans for as long as there have been oceans, whereas bomb ^{14}C has been penetrating into the oceans for only several decades. Hence, it is thought that natural ^{14}C calibrations may represent a mean ocean mixing rate on long time scales, whereas bomb ^{14}C calibrations may represent the mean ocean mixing rate in the upper several hundred meters of the ocean, where most of the bomb ^{14}C resides.

16.2.4 The Outcrop-Diffusion Model

There is considerable vertical density stratification in the ocean. In the ocean, warm water tends to sit above denser, cooler water. Similarly, less saline water tends to sit above denser, more saline water. However, largely because of the equator-to-pole surface temperature gradient, there is a surface ocean density gradient. Surface waters are less dense in equatorial regions and more dense near the poles. The ocean's dense deep and bottom waters tend to form in the densest polar regions.

Vertical eddies are damped by vertical density stratification, because buoyancy forces tend to counteract the movement of a parcel of water into waters of a different density. In contrast, these forces do little to damp eddies that evolve along surfaces of constant density. Because mixing across surfaces of constant density (diapycnal mixing) is damped, mixing tends to be much greater along surfaces of constant density (isopycnal mixing). It may be about as easy for a tracer to diffuse 1,000 km along a nearly horizontal isopycnal surface as it would for the tracer to diffuse 1 m in the nearly vertical diapycnal direction. Because the surface ocean is more dense at the poles than in tropical regions, isopycnal surfaces at depth in the tropics tend to outcrop at the poles. Hence, CO_2 absorbed by the surface ocean in polar regions may be efficiently transported to deeper, more equatorial waters, along surfaces of constant density.

Siegenthaler (1983) studied what would happen in the limiting case with infinitely rapid isopycnal mixing (and a finite diapycnal eddy-diffusion coefficient) by constructing the "outcrop-diffusion" model in which all deeper layers of the ocean were taken to be in direct contact with the atmosphere. When the vertical exchange coefficients in these box-diffusion and outcrop-diffusion models were tuned to produce the steady-state natural ^{14}C profile, it was found that the outcrop-diffusion model absorbed CO_2 faster than the box-diffusion model (and probably faster than the real ocean). This result demonstrates that although two models may be calibrated with the same tracer data, the difference in model architecture may lead to differing results.

16.2.5 The Upwelling-Diffusion Model

Vertical globally averaged concentration profiles for ocean tracers such as carbon, nitrate, phosphate, and oxygen show extrema in the first kilometer or so of the deep ocean. Such extrema at intermediate depths cannot be calculated in a one-dimensional pure diffusion model, as pure diffusion models tend toward constant or linear profiles. This is because not only does the real ocean diffuse, but also on the mean there is a large-scale overturning of the oceans.

Deep water forms in relatively limited areas in the polar regions, and, in the mean, water tends to upwell outside of the polar regions. (In fact, much of this upwelling is localized in certain coastal and equatorial regions; however, this geographic variability cannot be captured in a one-dimensional ocean carbon cycle model.) This fact suggests the construction of a model in which polar water is injected in the base of a one-dimensional column representing the nonpolar regions of the global ocean (Figure 16.4; Munk, 1966; Hoffert et al., 1981). This model is very similar to the box-diffusion model, except that, in addition to diffusing, tracers are advected toward the ocean surface. The governing equation for the perturbation to the ΣCO_2 concentration in this model can be represented as

$$\frac{d\Delta C}{dt} = \frac{d}{dz}\left(k\frac{d\Delta C}{dz}\right) + \omega\frac{d\Delta C}{dz} \qquad (16.16)$$

Calibration of this model using natural ^{14}C suggests an upwelling velocity (ω) of 3.5 m/yr and a diffusion coefficient (k) on the order of 4,700 m²/yr (Jain et al., 1995). Model results using these values are consistent with both natural and bomb ^{14}C data. The bottom boundary condition of this model reflects the composition of newly formed deep water. With the appropriate choice of the composition of newly formed deep water, this model reproduces observed tracer fields reasonably well

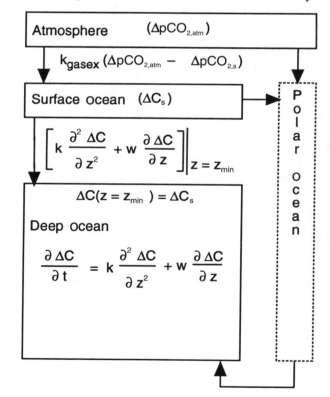

Figure 16.4: A schematic representation of an upwelling-diffusion model. The Hoffert et al. (1981) version considers a cold-water box, but does not explicitly resolve it as a separate reservoir. The HILDA model (Shaffer, 1989; Siegenthaler and Joos, 1992) divides the polar water box into two explicit reservoirs and considers mixing transport between the deep polar water box and all levels in the one-dimensional deep ocean compartment. Symbols are described in Table 16.2.

(e.g., Volk, 1984; Khesghi et al., 1991). Two objections that could be made to this model are: (1) there is no unique and obvious way to determine the composition of newly formed deep water, and (2) not all newly formed deep water is bottom water; some newly formed deep water leaves the deep water–forming regions at intermediate depth.

A form of upwelling-diffusion model (high latitude exchange/interior diffusion advection, or HILDA, model) that accounts for these possible objections to the classical upwelling-diffusion model has been developed by Shaffer (1989) and Siegenthaler and Joos (1992). This model appends two polar water boxes to the upwelling-diffusion column. In this model, the deep layers exchange water with the polar deep water box, so that polar water effectively enters the ocean at all depths. This feature serves to mimic both isopycnal mixing and deep convection in the polar seas. When calibrated using a diffusion coefficient that decreases with depth, this model is able to reproduce the vertical profiles of both bomb and natural ^{14}C (Siegenthaler and Joos, 1992).

16.2.6 Convolution Models

The ocean models discussed above all explicitly consider some kind of ocean buffer factor and explicitly calculate the change in concentrations of various chemical species. However, there is another type of model that is based on parameterized results of general circulation models (Maier-Reimer and Hasselmann, 1987; Harvey, 1989; Wigley, 1991). Maier-Reimer and Hasselmann (1987) produced exponential curve fits to the change in atmospheric CO_2 concentration as a function of time for a variety of CO_2 emission pulse sizes using a perturbation approach with an ocean general circulation model. In the appendix to their paper, Maier-Reimer and Hasselmann compared the full model results with results of a convolution integral calculation for a variety of emission scenarios.

In a convolution integral calculation of the ocean carbon cycle, the CO_2 mass in the atmosphere above the preindustrial mass (ΔM), given some CO_2 source flux [$E(t)$], is calculated by computing the integral:

$$\Delta M(t) = \int_0^t E(u)G(t-u)du = \int_0^t E(t-u)G(u)du \quad (16.17)$$

where $t = 0$ represents the time at the initial condition when $\Delta M = 0$, and $G(t)$ represents the response of the system to a unit impulse of CO_2. The impulse-response function, $G(t)$, is "defined as the CO_2 signal observed in the atmosphere for a δ-function atmospheric input at time $t = 0$ (or equivalently a unit step-function change in the initial atmospheric CO_2 concentration)" (Maier-Reimer and Hasselmann, 1987, p. 79).

In general, convolution integrals produce accurate results only for nearly linear systems. This modeling approach may be useful for small perturbations in which the linearizing assumption of a constant buffer factor and ocean circulation. All the impulse-response functions calculated by Maier-Reimer and Hasselmann represent injections of CO_2 into a 265-ppm preindustrial atmosphere. Because the response to a CO_2 impulse is

a function of the initial condition, no convolution of impulses into a 265-ppm preindustrial atmosphere will appropriately reflect the behavior of pulse injected into an atmospheric CO_2 content that differs significantly from that value. However, Harvey (1989) and Wigley (1991) have modified the pure impulse-response function calculation to mimic the effects of a changing buffer factor in a way that produces reasonable results for some common CO_2 scenarios.

16.3 Results of Simple Ocean Carbon Cycle Models

In Table 16.3, we compare the results of three simple ocean carbon cycle models (a pure diffusion model, Oeschger et al., 1975; an outcrop-diffusion model, Siegenthaler, 1983; an upwelling-diffusion model, Jain et al., 1995; and HILDA, Siegenthaler and Joos, 1992) and two ocean general circulation models (Toggweiler et al., 1989a,b; and Heimann, personal communication, 1993). The simple ocean carbon cycle models have been calibrated using the distribution of ^{14}C. When Siegenthaler (1983) calibrated the box-diffusion and outcrop-diffusion models using surface bomb ^{14}C concentration and bomb ^{14}C inventory, he found that the modeled increases in mean deep ocean ^{14}C concentration were approximately 30% greater than observed (Table 16.3). Siegenthaler and Joos (1992), using the HILDA model, found that in order to produce natural as well as bomb ^{14}C distribution in the ocean, the vertical diffusivity in their model must be greater in the upper ocean than in the deep ocean. In contrast, the Jain et al. (1995) model reproduces both the natural and bomb ^{14}C distribution in the ocean with a constant vertical eddy diffusivity. Nevertheless, Table 16.2 shows that the changes predicted by the models of both Siegenthaler and Joos (1992) and Jain and colleagues (1995) are quite close to the observed values.

Table 16.3 also shows that the results from these two ocean general circulation models (OGCMs) do not lie within the range of uncertainty of the observed data (Toggweiler et al., 1989a,b; Heimann, 1993), as did some of the simpler carbon cycle model results. The results from Toggweiler and colleagues (1989a,b) are determined from the Geophysical Fluid Dynamics Laboratory (GFDL) primitive equation OGCM. The Heimann (1993) results are based on the Hamburg Large Scale Geostrophic OGCM (Maier-Reimer et al., 1993). The GFDL model predicts a mean surface water bomb ^{14}C concentration of 160‰, in agreement with data from the Geochemical Sections Ocean Study (GEOSECS); however, the prediction of the bomb ^{14}C inventory as well as penetration depth are too low relative to the observations. The Hamburg model predicts very high values of bomb ^{14}C surface concentrations, approximately 23% higher than the observed value. The bomb ^{14}C inventory is higher by approximately 16% than the observed value, even though the model does reproduce the observed natural ^{14}C inventory. These results suggest that the current published OGCMs do not correctly simulate the atmosphere–ocean exchange and the vertical mixing in the upper ocean.

The models listed in Table 16.3 yield an ocean CO_2 uptake rate for the 1980s of between 1.45 Gt C/yr and 3.33 Gt C/yr.

Table 16.3. *Comparison of model results with observations*

| | Observation | Upwelling-Diffusion Model; Jain et al. (1995) | HILDA Model[a]; Siegenthaler & Joos (1993) | Box Diffusion Model; Siegenthaler (1983) | Outcrop-Diffusion Model; Siegenthaler (1983) | OGCM | |
						Toggweiler et al.[b,c]	Maier-Reimer et al.[c]
Preindustrial (1840) ocean surface $\Delta^{14}C$ (‰)	-49 ± 3[d]	-50	-55	-46	-46	—	—
Prebomb (1950) ocean surface $\Delta^{14}C$ (‰)	-58 ± 3[d]	-59	-62	-53	-53	-50	—
Bomb $\Delta^{14}C$ in ocean surface (‰)[e]	160 ± 15[f]	171	155	163	163	160	206
Mean ^{14}C for the deep sea $\Delta^{14}C$ (‰)	-160[g]	-163	—	-108	-110	—	—
Total CO_2 concentration, ΣC (mol/m³)[b]	2.15[i]	2.18	2.13	2.05	2.05	2.15	—
Bomb ^{14}C inventories (10^{13}atoms/m²)	8.00 ± 0.5[f]	8.30	7.95	8.42	8.42	6.90	9.27
Penetration depth (m)	328 ± 20[j]	308	338	355	355	283	320
Oceanic CO_2 uptake, 1980s (Gt C/yr)	2.0 ± 0.8[k]	2.10	2.15	2.45	3.33	1.67	1.45
Oceanic CO_2 uptake, 1980s, normalized by penetration depth (Gt C/yr)	2.0 ± 0.8[k]	2.24	2.09	2.33	3.08	1.94	1.49

[a]Values for HILDA model are area weighted mean from low- and high-latitude belts.
[b]Assumes prebomb atmosphere (1950) in steady state.
[c]Personal communication.
[d]Druffel and Suess (1983).
[e]Defined as the difference between the $\Delta^{14}C$ at the time of GEOSECS and the prebomb $\Delta^{14}C$.
[f]Broecker et al. (1985).
[g]Weighted mean of Atlantic and Pacific data (Oeschger et al., 1975).
[h]Average value for that part of the water column that is contaminated with bomb ^{14}C.
[i]Estimated by Toggweiler et al. (1989b) from the GEOSECS observations.
[j]Estimated from the formula of Broecker et al. (1985).
[k]Estimate (Sarmiento and Sundquist, 1992) of ocean carbon uptake based on measurements outlined by Tans et al. (1990).

Both GCMs (Toggweiler et al., 1989a; and Maier-Reimer et al., 1993) predict ocean CO_2 uptake rates at the low end of this range, whereas the diffusion model (Siegenthaler, 1983) may overpredict ocean CO_2 uptake. The models results of Siegenthaler and Joos (1992) and Jain and colleagues (1995) lie near the center of this range.

Schematic ocean models of the type discussed here predict similar present-day carbon uptake rates to models using the current distributions of three-dimensional ocean circulation models when calibrated to recover the global mean penetration depth of bomb radiocarbon measured by GEOSECS (Siegenthaler and Sarmiento, 1993). When so calibrated, both simple and more complicated ocean carbon cycle models predict recent (1980–89) oceanic carbon uptake rates in the range 2.0 ± 0.6 Gt C/yr.

In Figure 16.5 we present the results of a simple ocean carbon cycle model calculations (Jain et al., 1995) of the rates of CO_2 absorption for a number of CO_2 stabilization scenarios under consideration by the Intergovernmental Panel on Climate Change (IPCC). The results of other simple carbon cycle models are qualitatively similar. These calculations show that, for these scenarios, the rate of CO_2 absorption by the oceans will diminish in the long term, although the rate of ocean CO_2 absorption may increase for part of the next century for several of these scenarios. To stabilize atmospheric CO_2 content in the long term, for any of these scenarios, requires reducing diminishing total fossil fuel use plus net terrestrial biosphere CO_2 emissions by at least a factor of 3 over the present emission rate.

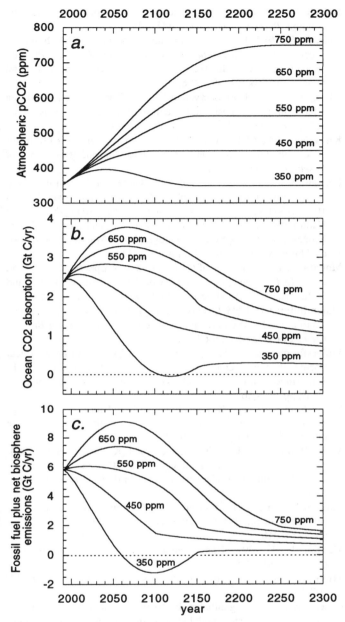

physics of fluid motion; however, OGCMs include highly parameterized representations of convective overturning, sub-grid-scale diffusion and, sometimes, biological carbon transport (e.g., Maier-Reimer, 1993, and Sarmiento et al., 1993). Of these carbon transport mechanisms, only fluid advection is based on first principles (Schlesinger and Jiang, 1990).

It would be more satisfying if the parameterization in simple ocean models were based on physical principles, instead of calibration to tracer fields. However, simple ocean carbon cycle models, relative to OGCMs, are easy to develop, constrain, and test; they execute rapidly and produce understandable output. Because they are calibrated using carbon isotopes, which are transported by the same mechanisms as the anthropogenic CO_2 perturbation, well-calibrated simple ocean carbon cycle models may produce CO_2 absorption predictions that are more reliable than more complex OGCMs. Nevertheless, complex scenarios involving the dynamic coupling of the atmosphere and ocean may need to be simulated using OGCMs.

Acknowledgments

We would like to thank Ian Enting, Danny Harvey, and Haroon Kheshgi for valuable reviews. This chapter is dedicated to Ulrich Siegenthaler. The work of Ken Caldeira and Atul Jain was performed under the auspices of the U.S. Department of Energy Environmental Sciences Division by the Lawrence Livermore National Laboratory under contract No. W-7405-ENG-48.

Figure 16.5: Results (A) for various IPCC CO_2 stabilization scenarios, from a simple ocean carbon cycle model (Jain et al., 1995), and (B) for ocean CO_2 absorption. Results are also shown (C) for the sum of oceanic absorption plus the change in CO_2 mass in the atmosphere (= fossil fuel plus net terrestrial biosphere emission). Reprinted with permission from A. K. Jain, H. S. Kheshgi, and D. J. Wuebbles, *Global Biogeochemical Cycles* 9: 153–66. © 1995, American Geophysical Union.

16.4 Conclusions and Discussion

Why would someone want to use a simple ocean carbon cycle model when complex three-dimensional ocean carbon cycle models such as those of Maier-Reimer (1993) and Sarmiento and colleagues (1992) are available? OGCMs have the satisfying property of being based, in part, on the fundamental

References

Bacastow, R. B., and C. D. Keeling. 1973. Atmospheric carbon dioxide and radiocarbon in the natural carbon cycle. In *Carbon and the Biosphere*, G. M. Woodwell and E. V. Pecan, eds. U.S. Atomic Energy Commission (available as CONF-720510 from National Technical Information Service, Springfield, Virginia), pp. 86–135.

Bacastow, R. B., and E. Maier-Reimer. 1991. Dissolved organic carbon in modeling oceanic new production. *Global Biogeochemical Cycles 5,* 71–85.

Bjorkstrom, A. 1979. A model of CO_2 interaction between atmosphere, oceans and land biota. In *The Global Carbon Cycle*, B. Bolin, E. T. Degens, S. Kempe, and P. Ketner, eds. SCOPE 13, John Wiley and Sons, New York, pp. 403–457.

Bjorkstrom, A. 1986. One-dimensional and two-dimensional ocean models for predicting the distribution of CO_2 between the ocean and the atmosphere. In *The Changing Carbon Cycle,* J. R. Trabalka and D. E. Reichle, eds. Springer-Verlag, New York, pp. 258–278.

Bolin, B., and E. Eriksson. 1959. Changes in the carbon content of the atmosphere and sea due to fossil fuel combustion. In *The Atmosphere and Sea in Motion,* B. Bolin, ed. Rockefeller Institute Press, New York.

Broecker, W. S., Y.-H. Li, and T.-H. Peng. 1971. Carbon dioxide – Man's unseen artifact. In *Impingement of Man on the Oceans,* D. W. Hood, ed. Wiley-Interscience, New York, pp. 287–324.

Broecker, W. S., and T.-H. Peng. 1982. *Tracers in the Sea.* Eldegio Press, Lamont-Doherty Geological Observatory, Palisades, New York.

Broecker, W. S., and T.-H. Peng. 1986. Carbon cycle 1985, glacial to interglacial changes in the operation of the global carbon cycle. *Radiocarbon 28,* 309–327.

Broecker, W. S., and T.-H. Peng. 1987. The role of $CaCO_3$ compensation in the glacial to interglacial atmospheric CO_2 change. *Global Biogeochemical Cycles 1,* 15–39.

Broecker, W. S., T.-H. Peng, G. Ostlund, and M. Stuiver. 1985. The distribution of bomb radiocarbon in the ocean. *Journal of Geophysical Research 90,* 6953–6970.

Caldeira, K., and J. F. Kasting. 1993. Insensitivity of global warming potentials to carbon dioxide emission scenarios. *Nature 366,* 251–253.

Caldeira, K., and M. R. Rampino. 1993. Biogeochemical stabilization of the carbon cycle and climate after the Cretaceous/Tertiary boundary mass extinction events. *Paleoceanography 8,* 515–525.

Druffel, E. M., and H. E. Suess. 1983. On the radiocarbon record in banded corals. *Journal of Geophysical Research 88,* 1271–1280.

Enting, I. G., and G. I. Pearman. 1987. Description of a one-dimensional carbon-cycle model calibrated by the techniques of constrained inversion. *Tellus 39B,* 459–476.

Fasham, M. J. R., J. L. Sarmiento, R. D. Slater, H. W. Ducklow, and R. Williams. 1993. Ecosystem behavior at Bermuda Station 'S' and ocean weather station 'India': A general circulation model and observational analysis. *Global Biogeochemical Cycles 7,* 379–415.

Fiadeiro, M. E. 1982. Three dimensional modeling of tracers in the deep Pacific Ocean. II: Radiocarbon and circulation. *Journal of Marine Research 40,* 537–550.

Harvey, L. D. D. 1989. Managing atmospheric CO_2. *Climate Change 15,* 343–381.

Hoffert, M. I. 1974. Global distributions of atmospheric carbon dioxide in the fossil fuel era: A projection. *Atmospheric Environment 8,* 1225–1249.

Hoffert, M. I., A. J. Callegari, and C.-T. Hseih. 1981. A box-diffusion carbon-cycle model with upwelling, polar bottom water formation and a marine biosphere. In *Carbon Cycle Modeling,* B. Bolin, ed. SCOPE 16, John Wiley and Sons, New York, pp. 287–305.

Hoffert, M. I., Y.-C. Wey, A. J. Callegari, and W. S. Broecker. 1979. Atmospheric response to deep-sea injections of fossil-fuel carbon dioxide. *Climate Change 2,* 53–68.

Jain, A. K., H. S. Kheshgi, and D. J. Wuebbles. 1995. Distribution of radiocarbon as a test of global carbon cycle models. *Global Biogeochemical Cycles 9,* 153–166.

Keeling, C. D. 1973. The carbon dioxide cycle: Reservoir models to depict the exchange of atmospheric carbon dioxide with the oceans and land plants. In *Chemistry of the Lower Atmosphere,* S. I. Rasool, ed. Plenum, New York, pp. 251–329.

Keeling, C. D. 1977. Impact of industrial gases on climate. In *Energy and Climate,* Geophysics Study Committee, National Academy of Sciences Press, Washington, D.C., pp. 72–91.

Keeling, C. D., R. B. Bacastow, A. F. Carter, S. C. Piper, T. P. Whorf, M. Heimann, W. G. Mook, and H. Roeloffzen. 1989a. A three-dimensional model of atmospheric CO_2 transport based on observed winds. I. Analysis of observational data. In *Aspects of Climate Variability in the Pacific and Western Americas,* D. H. Peterson, ed. Geophysical Monograph 55, American Geophysical Union, Washington, D.C., pp. 165–236.

Keeling, C. D., S. C. Piper, and M. Heimann. 1989b. A three-dimensional model of atmospheric CO_2 transport based on observed winds. IV: Mean annual gradients and interannual variations. In *Aspects of Climate Variability in the Pacific and Western Americas,* D. H. Peterson, ed. Geophysical Monograph 55, American Geophysical Union, Washington, D.C., pp. 305–363.

Keir, R. S. 1988. On the late Pleistocene ocean geochemistry and circulation. *Paleoceanography 3,* 443–445.

Keir, R. S. 1989. Paleoproduction and atmospheric CO_2 based on ocean modeling. In *Productivity of the Ocean: Past and Present,* W. H. Berger, V. S. Smetacek, and G. Wefer, eds. John Wiley and Sons, New York, pp. 395–406.

Khesghi, H. S., B. P. Flannery, and M. I. Hoffert. 1991. Marine biota effects on the compositional structure of the world oceans. *Journal of Geophysical Research 96,* 4957–4969.

Knox, F., and M. McElroy. 1984. Changes in atmospheric CO_2: Influence of the marine biota at high latitudes. *Journal of Geophysical Research 89,* 4629–4637.

Lasaga, A. C., R. A. Berner, and R. M. Garrels. 1985. An improved geochemical model of atmospheric CO_2 fluctuations over the past 100 million years. In *The Carbon Cycle and Atmospheric CO_2: Natural Variations Archean to Present,* E. Sundquist and W. S. Broecker, eds. Geophysical Monograph 32, American Geophysical Union, Washington, D.C., pp. 397–411.

Machta, L. 1973. The role of the oceans and biosphere in the carbon cycle. In *The Changing Chemistry of the Oceans,* D. Dryssen and D. Jagner, eds. Wiley-Interscience, New York, pp. 121–145.

Maier-Reimer, E. 1993. Geochemical cycles in an ocean general circulation model: Preindustrial tracer distributions. *Global Biogeochemical Cycles 7,* 645–677.

Maier-Reimer, E., and R. Bacastow. 1990. Modeling of geochemical tracers in the ocean. In *Climate-Ocean Interaction,* M. E. Schlesinger, ed. Kluwer Academic Publishers, Dordrecht, Netherlands, pp. 233–267.

Maier-Reimer, E., and K. Hasselmann. 1987. Transport and storage of CO_2 in the ocean – An inorganic ocean-circulation carbon-cycle model. *Climate Dynamics 2,* 63–90.

Maier-Reimer, E., U. Mikolajewicz, and K. Hasselmann. 1993. Mean circulation of the Hamburg LSG OGCM and its sensitivity to the thermohaline surface forcing. *Journal of Physical Oceanography 23,* 731–757.

Manabe, S., and R. J. Stouffer. 1993. Century-scale effects of increased atmospheric CO_2 on the ocean atmosphere system. *Nature 364,* 215–218.

Martin, J. H. 1991. Iron, Liebig's Law, and the greenhouse. *Oceanography 4,* 52–55.

Munk, W. H. 1966. Abyssal receipes. *Deep Sea Research 13,* 707–736.

Oeschger, H., U. Siegenthaler, U. Schotterer, and A. Guglemann. 1975. A box-diffusion model to study the carbon dioxide exchange in nature. *Tellus 27,* 168–192.

Parsons, T. R., M. Takahashi, and B. Hargrave. 1984. *Biological Oceanographic Processes.* Pergamon Press, Elmsford, N.Y.

Redfield, A. C., B. H. Ketchem, and F. A. Richards. 1963. The influence of organisms on the composition of sea-water. In *The Sea, Vol. 2,* M. N. Hill, ed. Interscience Publishers, New York, pp. 26–77.

Revelle, R., and H. E. Suess. 1957. Carbon dioxide exchange between the atmosphere and the ocean and the question of an increase of CO_2 during the past decades. *Tellus 9,* 18–27.

Sarmiento, J. L. 1986. Three-dimensional ocean models for predicting the distribution of CO_2 between the ocean and atmosphere. In *The Changing Carbon Cycle,* J. R. Trabalka and D. E. Reichle, eds. Springer-Verlag, New York, pp. 279–294.

Sarmiento, J. L., J. C. Orr, and U. Siegenthaler. 1992. A perturbation simulation of CO_2 uptake in an ocean general circulation model. *Journal of Geophysical Research 97,* 3621–3645.

Sarmiento, J. L., R. D. Slater, M. J. R. Fasham, H. W. Ducklow, J. R. Toggweiler, and G. T. Evans. 1993. A seasonal three-dimensional ecosystem model of nitrogen cycling in the North Atlantic euphotic zone. *Global Biogeochemical Cycles 7*, 417–450.

Sarmiento, J. L., and E. T. Sundquist. 1992. Revised budget for the oceanic uptake of anthropogenic carbon dioxide. *Nature 356*, 589–593.

Sarmiento, J. L., and J. R. Toggweiler. 1984. A new model for the role of the oceans in determining atmospheric $p\text{CO}_2$. *Nature 308*, 621–624.

Schlesinger, M. E., and X. Jiang. 1990. Simple model representation of atmosphere-ocean GCMs and estimation of the time scale of CO_2-induced climate change. *Journal of Climate 3*, 1297–1315.

Shaffer, G. 1989. A model of biogeochemical cycling of phosphorous, nitrogen, oxygen and sulfur in the ocean: One step toward a global climate model. *Journal of Geophysical Research 94*, 1979–2004.

Shaffer, G., and J. L. Sarmiento. 1995. Biogeochemical cycling in the global ocean 1. A new, analytical model with continuous vertical resolution and high latitude dynamics. *Journal Geophysical Research 100*, 2659–2672

Siegenthaler, U. 1983. Uptake of excess CO_2 by an outcrop-diffusion model of the ocean. *Journal of Geophysical Research 88*, 3599–3608.

Siegenthaler, U., and F. Joos. 1992. Use of a simple model for studying oceanic tracer distributions and the global carbon cycle. *Tellus 44B*, 186–207.

Siegenthaler, U., and H. Oeschger. 1978. Predicting future atmospheric carbon dioxide levels. *Science 199*, 388–395.

Siegenthaler, U., and J. L. Sarmiento. 1993. Atmospheric carbon dioxide and the ocean. *Nature 365*, 119–1125.

Siegenthaler, U., and T. Wenk. 1984. Rapid atmospheric CO_2 variations and ocean circulation. *Nature 308*, 624–626.

Stumm, W., and J. J. Morgan. 1981. *Aquatic Chemistry*, 2nd Edition. John Wiley and Sons, New York.

Sundquist, E. T. 1990. Steady- and non-steady-state carbonate-silicate controls on atmospheric CO_2. *Quaternary Science Review 10*, 283–296.

Sundquist, E. T. 1993. The global carbon dioxide budget. *Science 259*, 934–940.

Tans, P. P., I. Y. Fung, and T. Takahashi. 1990. Observational constraints on the global atmospheric CO_2 budget. *Science 247*, 1431–1438.

Toggweiler, J. R., K. Dixon, and K. Bryan. 1989a. Simulations of radiocarbon in a coarse resolution world ocean model: I. Steady state prebomb distributions. *Journal of Geophysical Research 94*, 8217–8242.

Toggweiler, J. R., K. Dixon, and K. Bryan. 1989b. Simulations of radiocarbon in a coarse resolution world ocean model: II. Distributions of bomb-produced carbon-14. *Journal of Geophysical Research 94*, 8243–8264.

Toggweiler, J. R., and J. Sarmiento. 1985. Glacial to interglacial changes in atmospheric carbon dioxide: The critical role of ocean surface water at high latitudes. In *The Carbon Cycle and Atmospheric CO_2*: Natural Variations Archean to Present, E. Sundquist and W. S. Broecker, eds. Geophysical Monograph 32, American Geophysical Union, Washington, D.C., pp. 163–184.

Totterdell, I. J. 1993. An annotated bibliography of marine biological models. In *Towards a Model of Ocean Biogeochemical Processes*, G. T. Evans and M. J. R. Fasham, eds. Springer-Verlag, New York, pp. 317–339.

Volk, T. 1984. *Multi-Property Modeling of the Marine Biosphere in Relation to Global Climate and Carbon Cycles*. Ph.D. dissertation. New York University, University Microfilms International, Ann Arbor, Michigan.

Volk, T., and M. I. Hoffert. 1985. Ocean carbon pumps: Analysis of relative strengths and efficiencies in ocean-driven atmospheric CO_2 changes. In *The Carbon Cycle and Atmospheric CO_2*: Natural Variations Archean to Present, E. Sundquist and W. S. Broecker, eds. Geophysical Monograph 32, American Geophysical Union, Washington, D.C., pp. 91–110.

Volk, T., and Z. Liu. 1988. Controls of CO_2 sources and sinks in the Earth scale surface ocean: Temperature and nutrients. *Global Biogeochemical Cycles 2*, 73–89.

Walker, J. C. G. 1991. *Numerical Adventures with Geochemical Cycles*. Oxford University Press, New York.

Walker, J. C. G., and J. F. Kasting. 1992. Effect of fuel and forest conservation on future levels of atmospheric carbon dioxide. *Palaeogeography, Palaeoclimatology, Palaeoecology 97*, 151–189.

Wigley, T. M. L. 1991. A simple inverse carbon-cycle model. *Global Biogeochemical Cycles 5*, 373–382.

Wyrtki, K. 1962. The oxygen minimum in relation to ocean circulation. *Deep Sea Research 9*, 11–23.

17

Very High Resolution Estimates of Global Ocean Circulation, Suitable for Carbon Cycle Modeling

ALBERT J. SEMTNER

Abstract

This chapter assesses the state of the art in modeling the physical conditions of the global ocean. Results of some new simulations are now becoming available for use by carbon cycle models for the case of modern climatic forcing. Future improvements in computer power should allow realistic physical and biogeochemical simulations for other climatic conditions, such as paleoclimates and anthropogenic climate change.

17.1 Introduction

About 10 years ago, Charles D. Keeling stopped by my office at the National Center for Atmospheric Research (NCAR) and inquired about the availability of ocean simulation fields for modeling the carbon cycle of the global ocean. I told him that, unfortunately, there was as yet no global model representation of ocean circulation that was physically realistic enough to be used for the purpose he envisioned. The global models of the time had been run for only short periods from simple initial conditions, and/or their grids were very coarse (approximately 5 degrees). I did recommend that he contact Jorge Sarmiento at the Geophysical Fluid Dynamics Laboratory (GFDL) in Princeton, New Jersey, to obtain their new modeling results for the North Atlantic, which were being produced on a relatively fine grid for that time (approximately 1 degree of latitude and longitude). Even those results had not achieved the same length scales as those for currents in the real ocean, which are typically small fractions of a degree in longitude.

Since the work of Sarmiento and Bryan (1982), a number of ocean studies have been conducted, both with the GFDL model (e.g., Toggweiler et al., 1989) and with other models (e.g., Maier-Reimer and Hasselmann, 1987). These models have accomplished long-term integrations with 3-degree grids and with some biogeochemical variables included as well. Until recently, however, there have been no truly high-resolution simulations that were also global. This chapter reports on the existence of some global model data sets that are now becoming available, and that can be useful for modeling the carbon cycle with the present-day ocean circulation. The new models portray the strong currents, which accomplish the property transports, and the current fluctuations, which carry out much of the horizontal and vertical mixing, at nearly their real scales and strengths. The new studies have the potential to be extended, via greatly increased computer power and added atmospheric models, to give climate predictions from first principles, such as for paleoclimates or climates produced by increasing greenhouse gases.

Even before such computational improvements, somewhat coarser resolution models can be used for long integrations, wherein the effects of the unresolved scales are "parameterized" using information from the higher-resolution simulations.

The premise of this chapter is that to maximize the likelihood that past and future climates can be represented successfully, including aspects of the carbon cycle, it is important to have adequate realizations of the global ocean circulation. Ideally, these should reproduce the time–mean currents at proper length scales and intensities, as well as the effects of eddies (either explicitly or by parameterization). Eddy parameterizations, when used, should not be ad hoc as in the past, but should be physically based (e.g., see Danabasoglu et al., 1994). Validation of simulated fields can be made by comparison with in situ hydrographic and current-meter data and with satellite estimates of eddy kinetic energy and the large-scale transports of heat and moisture.

This is a long-term effort. Currently, it is demanding to do eddy-resolving decadal simulations; however, millennial simulations can become routine early in the next century. The goal is to exploit "teraflop" computers with optimized ocean models. (A speed of a trillion operations per second is expected by the year 2005.) In the meantime, present-day ocean circulation can be computed on existing machines and validated using new in situ data from programs such as the World Ocean Circulation Experiment and satellite data from sources such as the ERS-1 and TOPEX/Poseidon missions. Furthermore, some near-term evaluation of model capabilities to reproduce the equilibrium thermohaline circulation from first principles may be realizable with coarser-grid models having parameterized eddies. Transient climate changes, such as those that might occur over multiple centuries in response to increasing greenhouse gases, could also be simulated in this way; the physical results would be of great utility for carbon cycle studies, particularly if grid sizes remained small enough to represent the true time–mean currents.

Once the ocean circulation has been simulated, both the time-step constraint to resolve internal gravity waves and the grid size criterion to resolve an inertial subrange below the mesoscale can be significantly relaxed in studies that involve tracers such as biogeochemical species. As a result, the prospects for improved understanding of the oceanic carbon cycle are not as computer-restricted as those for understanding the physical system. Fairly efficient advection-diffusion equations with sources/sinks are to be solved. In this regard, the community of carbon cycle modelers can also use time and/or space averaging, regional sampling, and other effective offline strategies. Such methods can cut their computer requirements drastically and facilitate further improvements in ocean carbon cycle models. To be consistent with models, one could compute Lagrangian diffusivity tensors from high-resolution studies and use these at coarser resolution. In one high-resolution simulation described in this chapter, complete three-day history tapes as well as derived statistical tapes are routinely generated and retained on the Mass Storage System at NCAR. Model output and extraction programs are made available upon request.

17.2 Model Formulations

The physical models that are discussed here are all descended from the Bryan–Cox multilevel primitive equation formulation. Other formulations, such as isopycnal ones, have promise; however, it has not yet been shown that they are superior in performance to the basic GFDL model of Bryan (1969). The GFDL model has been extensively used in ocean and climate modeling in various forms that were developed by Semtner (1974), Cox (1984), and Pacanowski and colleagues

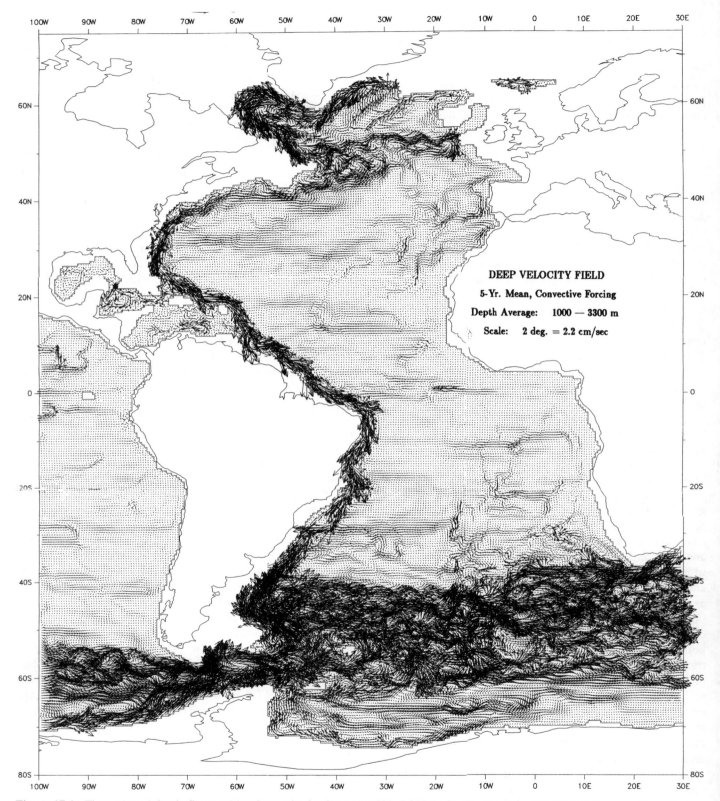

Figure 17.1: Time–mean Atlantic flow, averaged over the depth range 1,000–3,300 m, for the global model with 0.5-degree grid spacing. This improves on years 27.5–32.5 of the experiment by Semtner and Chervin (1992), now run with no deep restoring to observations except in polar latitudes and near Gibraltar. Southward transport of the deep western boundary current is 16 megatons/sec, very similar to that observed.

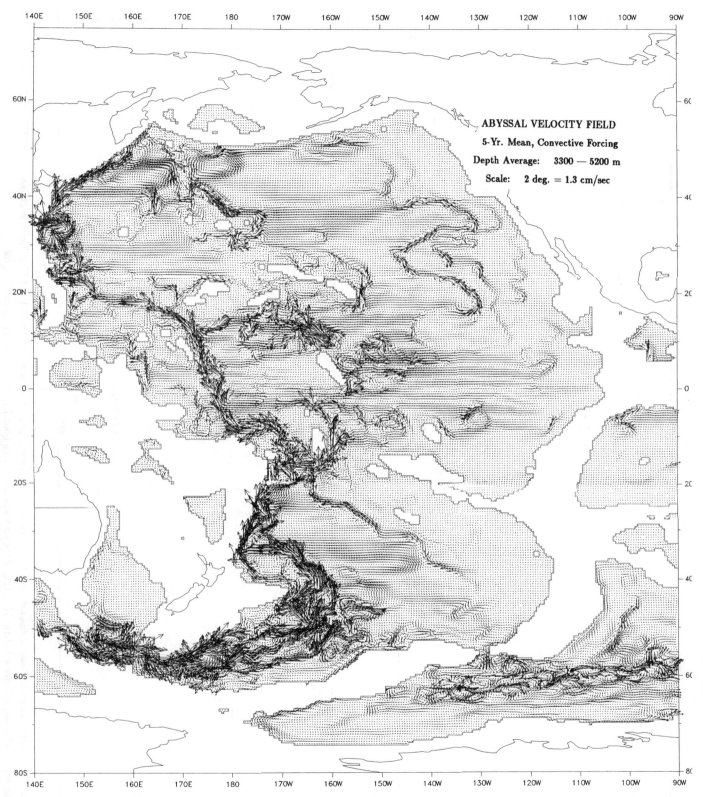

Figure 17.2: Time–mean Pacific flow, averaged over the depth range 3,300–5,200 m, for the global model with 0.5-degree grid spacing. This improves on years 27.5–32.5 of the experiment of Semtner and Chervin (1992), with essentially no Pacific restoring. Strong northward flow, as shown extending to at least 20°N, is known to exist on the basis of the observed distributions of properties in the deep Pacific.

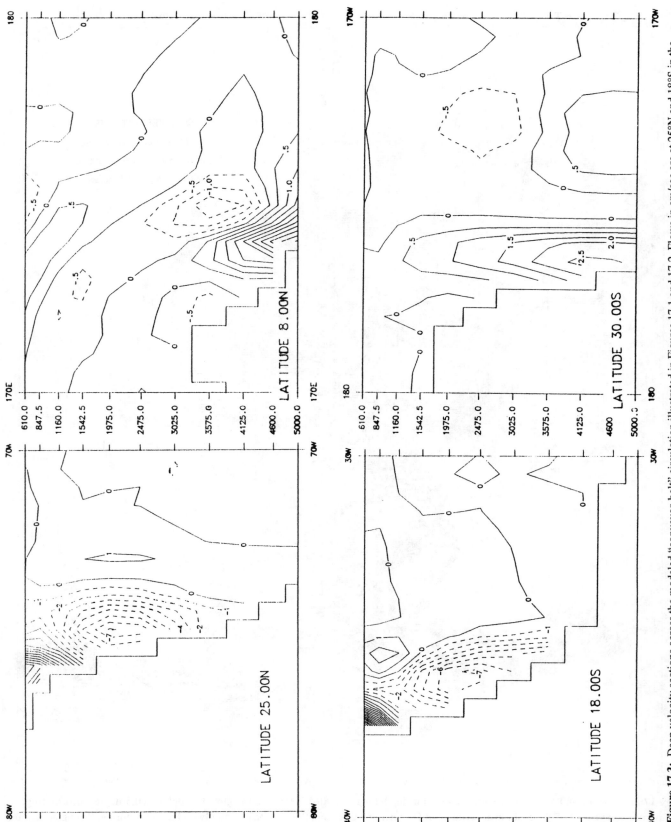

Figure 17.3: Deep velocity sections across the modeled "conveyor-belt" circulation illustrated in Figures 17.1 and 17.2. Flows in cm/sec are at 25°N and 18°S in the Atlantic and at 30°S and 8°N in the Pacific. Note that the scale width of the deep boundary currents is only about 1 degree.

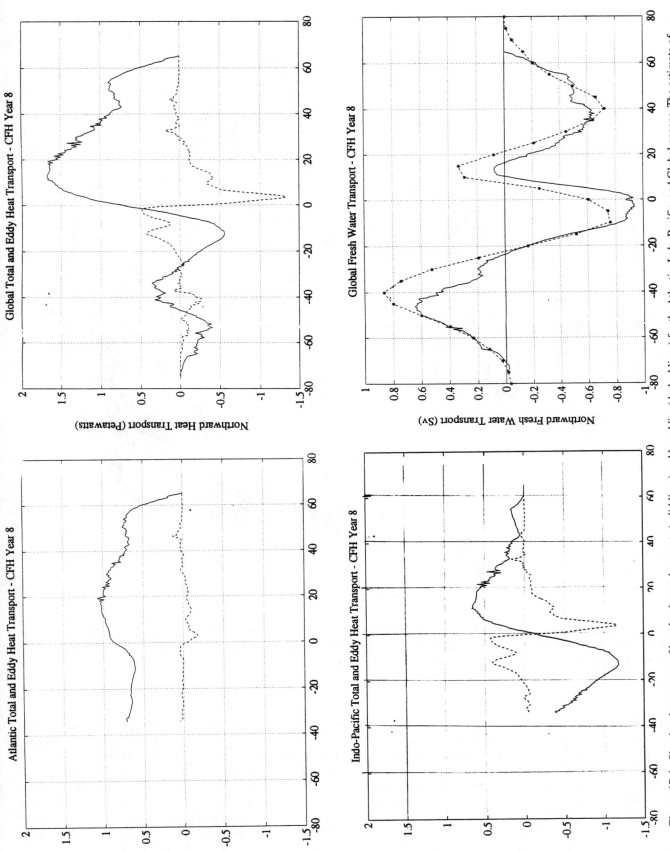

Figure 17.4: Simulated transports of heat by the total currents (solid lines) and by eddies (dashed lines) for the Atlantic, Indo-Pacific, and Global oceans. The estimate of global transport is in disagreement with observations of Hastenrath (1980) mainly in the band of midsouthern latitudes; this discrepancy is probably associated with eddies that are too weak. The fourth panel shows simulated fresh-water transport of the global ocean (solid), which agrees with an observed estimate (dashed) from Wijffels et al. (1992).

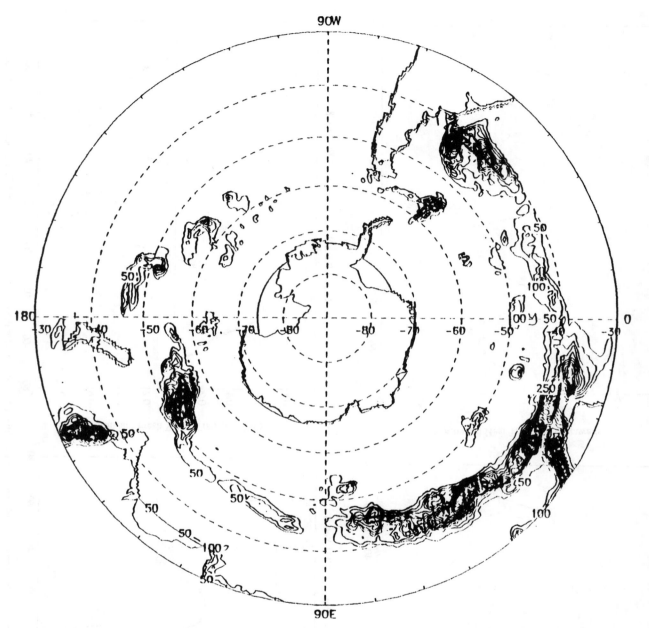

Figure 17.5: Eddy kinetic energy of simulated surface currents in the Southern Ocean for the case of 0.50-degree grid spacing. The contour interval is 50 cm²/sec². Instabilities of strong currents are occurring in the right regions, but with magnitudes that are too low relative to satellite estimates by a factor of 2–4. (Data from Wilkin and Morrow, 1994.)

(1990). Chervin and Semtner (1990) developed a multitasked version for parallel vector supercomputers, which includes bi-harmonic mixing parameterizations to allow better treatment of eddies with length scales of tens of kilometers.

A global configuration with 0.50-degree grid spacing was shown to be tractable by Semtner and Chervin (1988); the general circulation of a 32.5-year ocean simulation with marginally resolved eddies has already been reported in Semtner and Chervin (1992). That integration used continuous deep restoring to the observed temperature and salinity of Levitus (1982). Those constraints have subsequently been removed, except in

the polar latitudes and near Gibraltar, in order to allow a more prognostic integration. Results of a rerun of the final five years of the earlier integration will be reported in Section 17.3.

Two ongoing integrations will also be discussed in Section 17.3, for which resolution increases are the most prominent modification. Both of the new integrations use a Mercator-style grid, whereby a constant longitudinal spacing is chosen and then the latitudinal spacing is set equal to that constant times the cosine of latitude. This gives a square grid, with the desirable property that grid resolution improves with increasing latitude. Stammer and Boning have shown (1992) that increased resolution is needed at high latitudes in order to re-solve the eddy field observed by Geosat satellite altimetry.

The two grids with Mercator spacing use longitudinal spac-ing of 0.40 degree and 0.28 degree; their average latitudinal grid spacings over the latitude range 0–75 degrees are in fact 0.25 degree and 0.16 degree. Both the 0.25-degree and the 0.16-degree (average) grids are being evaluated to determine how adequately they can reproduce observed levels of eddy variability and eddy transports. A recent regional simulation of the Gulf Stream (Schmitz and Thompson, 1993) shows that a grid as small as 0.1 degree may be needed for full convergence; however, the properties of a 0.16-degree (average) model are similar to those of a 0.125-degree model poleward of 60 de-grees of latitude, and that may suffice. Also, it is possible to use satellite altimeter data to force the model to achieve realistic eddy variability through data assimilation techniques, and thus achieve adequate realizations of eddy mixing with somewhat lower resolution requirements (e.g., see Verron et al., 1992).

The 0.25-degree and 0.16-degree simulations differ in other respects from the 0.50-degree integration. They both use much more realistic geometry, which is made possible by free-surface height formulations. A time-splitting free-surface method of Killworth and colleagues (1991) is used with the 0.25-degree model, whereas an implicit method of Dukowicz

and Smith (1994) is used with the 0.16-degree model. The 0.25-degree model is being run on the CRAY Y-MP/8 at NCAR, and the 0.16-degree model is being run in collaboration with the Los Alamos National Laboratory's group of Malone, Smith, and Dukowicz on that laboratory's 1,024-node Connec-tion Machine 5. The new model developed by the Los Alamos group runs at about 10 gigaflops. It also extends to latitude 78°N, to permit more deep water formation. Both of the higher-resolution models use monthly wind data from the European Centre for Medium-Range Weather Forecasts (ECMWF) for the individual years from 1985 to 1994, as analyzed by a method of Trenberth and colleagues (1989). These winds re-place the climatological ship winds of Hellerman and Rosen-stein (1983), which were employed at 0.50 degree. Surface heat and moisture fluxes are computed using 30-day restoration to observed monthly Levitus (1982) ocean temperature and salin-ity; however, the fluxes will eventually include climatological terms from the observed surface energy balance.

17.3 Prognostic Results from the Various Models

Figures 17.1–17.5 show results of the 0.50-degree integration with prognostic deep circulation, with the exception of restora-

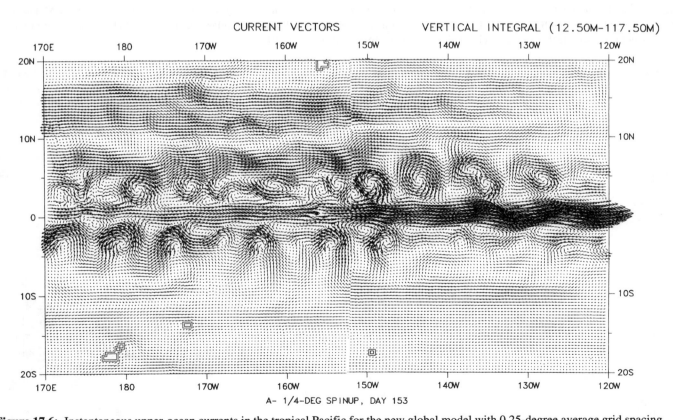

Figure 17.6: Instantaneous upper-ocean currents in the tropical Pacific for the new global model with 0.25-degree average grid spacing, fully realistic coastline and bathymetry, and ECMWF winds. The simulation was initialized from interpolated output of the 0.50-degree run. After 153 days with 1985 winds, well-developed 30-day waves have formed to mix cold upwelling water into the surroundings. (Eddy heat transports agree well with observations of Bryden and Brady, 1989.) The model's equatorial undercurrent rises into the depth range of the plot (0–135 m) on the eastern side of the figure.

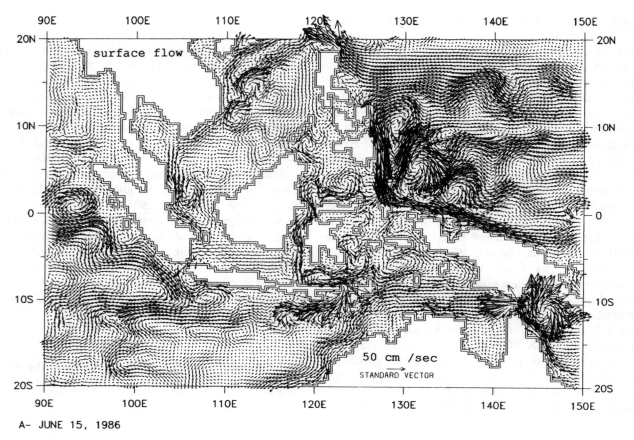

A– JUNE 15, 1986

Figure 17.7: Representative surface currents in the vicinity of Indonesia, for the improved global model. The flow through the Makassar Strait between Borneo and Sulawesi, and additional flows through the Molucca Sea east of Sulawesi, are known to be a major part of the warm-water route for the thermohaline circulation (Gordon, 1986; Godfrey et al., 1993).

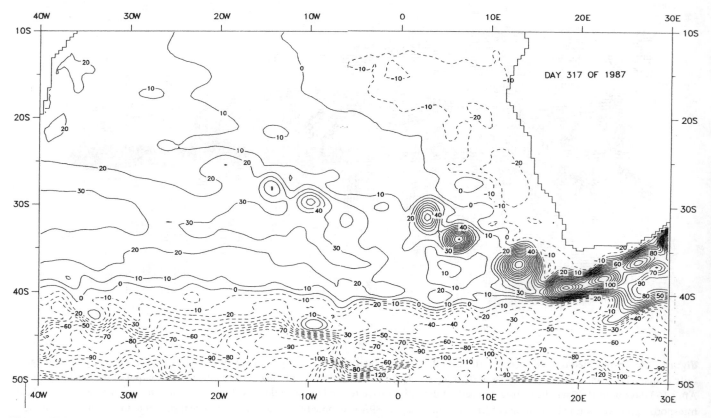

Figure 17.8: Surface height field in the southeast Atlantic, as simulated for late 1987. Large Agulhas eddies propagating northwest at speeds of approximately 8 km/day are similar to those observed during 1987 by satellite altimeter (Gordon and Haxby, 1990). This portion of the conveyor belt consists almost exclusively of warm-core eddies.

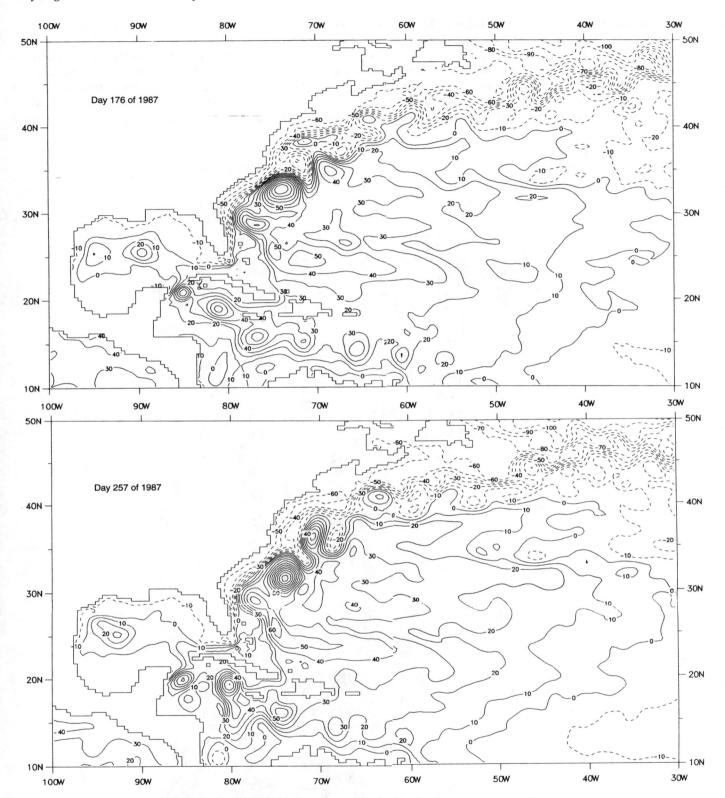

Figure 17.9: Simulated surface height in the northwest Atlantic during 1987. As in nature, the Gulf Stream in the new model separates near Cape Hatteras and develops large meanders and rings. Other transients, such as those in the Caribbean Sea and the Gulf of Mexico, are also well simulated.

tion in latitudes 55–65°N and 75–65°S, and near Gibraltar. The five-year mean currents of Figure 17.1 for the deep (1,000–3,300 m) flow in the Atlantic and Figure 17.2 for the abyssal (3,300–5,200) flow in the Pacific clearly show the deep "conveyor-belt" circulation that has been inferred on the basis of observations by Broecker (1991). As indicated in Figure 17.3, deep boundary currents are relatively strong and narrow;

a grid spacing of 0.50 degree is probably barely adequate to resolve them.

Heat transports in Figure 17.4 are within the errors bars for the Atlantic observations of Bryden and Hall (1980) and nearly within the error bars for the Pacific observations at 24°N of Bryden and colleagues (1991). Also, the large eddy heat transports near the equator in the Indo-Pacific are quite consistent

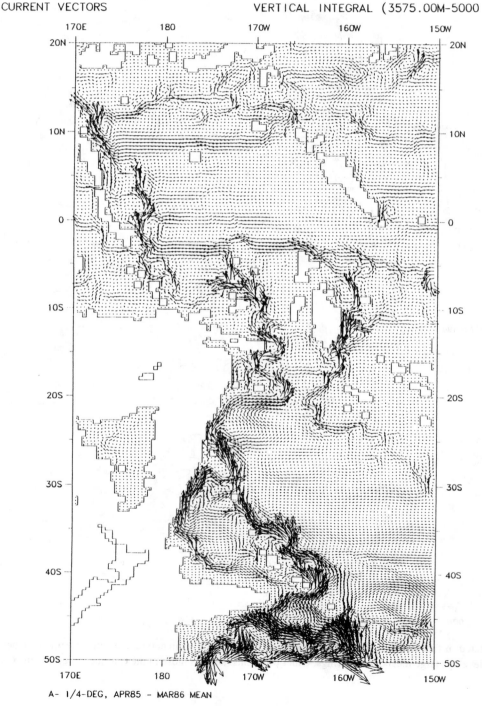

Figure 17.10: Annual averaged currents in the southwest Pacific for the depth range 3,300–5,200 m of the 0.25-degree (average) model. As a result of the improved representation of bathymetry, the flows along the Louisville Seamounts east of New Zealand and through the Samoan Passage near 13°S can be better compared with records of current meters that monitor the deep thermohaline flow.

with the observational estimates of Bryden and Brady (1989). However, global transports in the Southern Hemisphere are less realistic, that is, they are equatorward in midlatitudes. This discrepancy may be due to incorrect surface forcing in the Southern Hemisphere; however, it may also be due to the lack of adequate eddy resolution in middle and high latitudes, where direct observations of ocean heat transport by the time-mean currents indicate a near-zero contribution from that source (e.g., see Bennett, 1978). Indeed, Keffer and Holloway (1988) inferred from satellite estimates of eddy kinetic energy in the Southern Ocean that the poleward eddy heat transport could be as large as 1.3 petawatts. Consistent with that speculation, analyses of the 0.50-degree model output of Semtner and Chervin (1992) by Wilkin and Morrow (1994) have found that the simulated eddy kinetic energy in strong-current regions of the Southern Ocean is too low by a factor of 2–4, in comparison with the satellite observations. (The eddy kinetic energy of the present run is shown in Figure 17.5.)

It appears, therefore, that the 0.50-degree model is marginally adequate for portraying the circulations and transport of the Northern Hemisphere, but not those of the Southern Hemisphere, particularly at high latitudes. Therefore, the results of the 0.25-degree grid need to be examined. These are provided in Figures 17.6–17.13. Figures 17.6–17.9 show representative snapshots of flow for the upper-ocean part of the conveyor-belt circulation. With new geometry, resolution, and forcing, the "warm-water route" described by Gordon (1986) is depicted more accurately than in the 0.50-degree model. Much of the rising water in the Tropical Pacific flows through Indonesia, crosses the Indian Ocean, and passes into the southwest Atlantic in the form of intense Agulhas eddies (similar to those observed by Gordon and Haxby, 1990). Ultimately, the water joins the Gulf Stream, which is now better represented than previously in terms of its separation from the coast at Cape Hatteras and its eddying behavior after separation. Figures 17.10 and 17.11 indicate that the abyssal pathways of bottom water in the Southern Hemisphere are well resolved by the new model with its realistic bathymetry. A proper portrayal of interbasin connections is vital for overall realism. Figure 17.12 shows that global thermohaline circulation is more realistic than that described by Semtner and Chervin (1992). Figure 17.13 provides dramatic evidence of an increase in eddy kinetic energy over the Pacific and Southwest Indian oceans, as a result of improved grid resolution.

Although only a preliminary analysis of the 0.25-degree model has been conducted prior to completion of the 1985–94 simulation, it is clear that this new configuration performs much better than the 0.50-degree model. As mentioned above, good resolution of eddies will probably require at least the 0.16-degree version that is currently running on the Connection

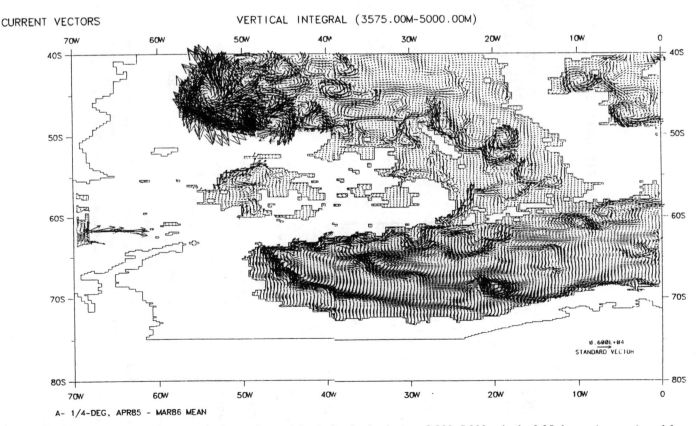

CURRENT VECTORS   VERTICAL INTEGRAL (3575.00M–5000.00M)

A– 1/4–DEG, APR85 – MAR86 MEAN

Figure 17.11: Annual averaged currents in the southwest Atlantic for the depth range 3,300–5,200 m in the 0.25-degree (average) model. Two well-defined circulations are shown to exit the Weddell Basin and to pass northward into the Argentine Basin or south of the mid-Atlantic Ridge. Valid representation of these currents is important, if correct transports are to be obtained for key properties that enter the ocean at the surface of the Weddell Sea.

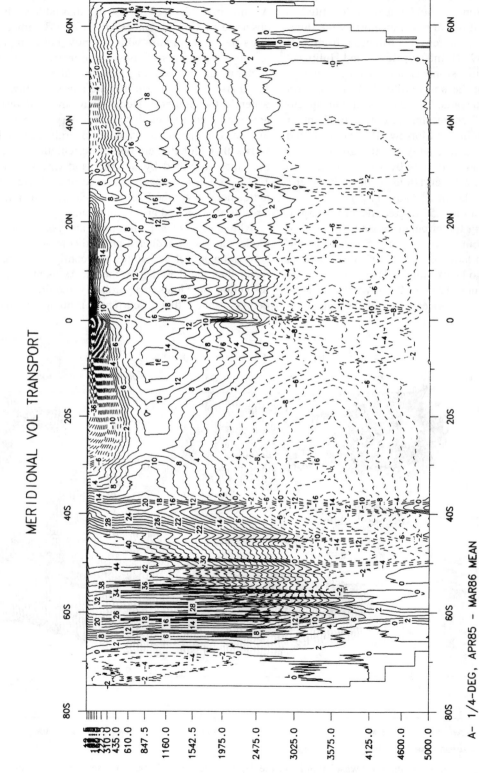

Figure 17.12: The global overturning stream function for a 1-year period. The production rates of North Atlantic deep water and of Antarctic bottom water in this 0.25-degree representation are now enhanced to approximately 18 megatons/sec and 5 megatons/sec, respectively. Relative to the 0.50-degree results of Semtner and Chervin (1992), there is also more penetration of abyssal Antarctic waters into the Northern Hemisphere and less concentrated upwelling at depth near the equator.

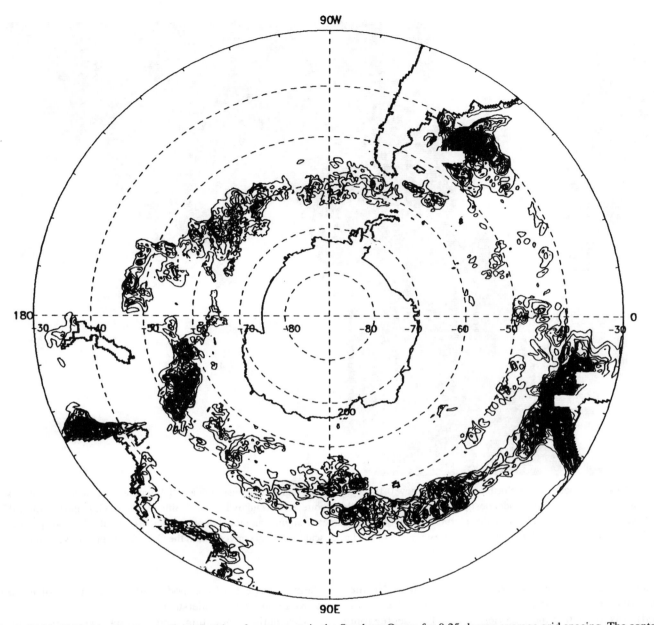

Figure 17.13: Eddy kinetic energy of simulated surface currents in the Southern Ocean for 0.25-degree average grid spacing. The contour interval is 50 cm²/sec². Instabilities of strong currents are much more pronounced throughout the Pacific and the southwest Indian oceans, relative to the 0.50-degree results of Figure 17.5. Associated eddy heat transports should be enhanced as a result, on the basis of the theoretical turbulence considerations of Keffer and Holloway (1988).

Machine 5 at Los Alamos. This will give resolution of the Antarctic circumpolar current at 65°S of 0.11 degree. Furthermore, the addition of the Norwegian and Greenland seas and more of the Weddell and Ross seas will greatly aid in portraying water-mass formation by deep convection and plume processes in those regions. (Figures 17.14 and 17.15 show the geometric representation of these domains.) It is significant that small grid sizes of approximately 7 km allow direct representation of some of the scales of convective features; this opens the door to parameterization of other subgrid-scale processes at even smaller scales. Once that is achieved, the 0.16-degree model should be capable of running to equilibrium, in order to produce thermohaline circulations from first principles. The primary advance required at that point will then be one of computing power.

17.4 Conclusions

It is clear from the above figures and discussion that high-resolution ocean models are becoming physically more realistic as resolution and physical formulations both improve. Because of the deficiencies of even a 0.50-degree model in

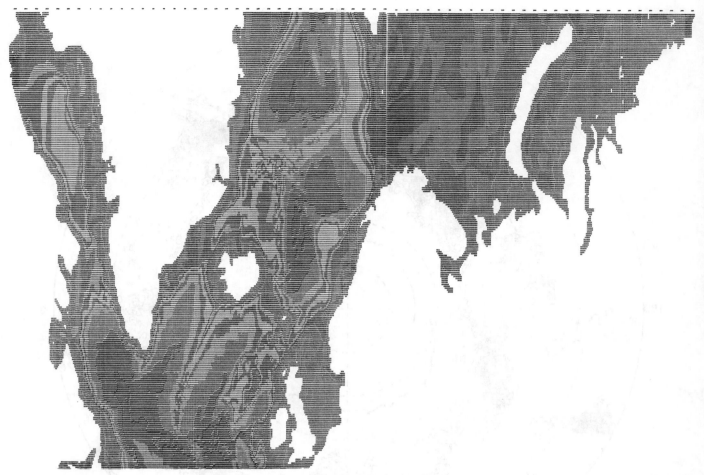

Figure 17.14: Geometry of the North Atlantic for a new 0.16-degree average grid being using in a collaborative effort with Los Alamos investigators. A version of the model designed for massively parallel architectures (Dukowicz and Smith, 1994) is being used on the 1,024-node Connection Machine 5. This shaded picture is composed of characters signifying model depth at each grid point. The grid spacing at 78°N is less than 7 km, which allows direct representation of the larger scales of deep convection in the Greenland and Norwegian seas.

representing eddy processes (even though the mean Northern Hemisphere conveyor-belt circulation is fairly well represented), it is likely that the 0.25-degree and 0.16-degree representations will be needed to give truly quantitative estimates of global ocean circulation. Fields from both of these models are being made available for use in carbon cycle studies as needed. Further progress in physical modeling can be expected as computer power increases to 1,000 times that available in 1990 by the early part of the next century. Meanwhile, confidence in simulation capability will grow as both fine-grid and eddy-parameterized coarse-grid models are evaluated and improved using the growing data base of oceanic observations.

Carbon cycle models that involve solving advection-diffusion equations for scalar variables are not as limited by computer power as are physical models. Consequently, the output from the 0.25-degree and 0.16-degree models can be used in an offline mode to help simulate the ocean carbon cycle, requiring much less computing power. It is hoped that this will allow progress in understanding the carbon cycle: To the extent that the physical fields are now more accurate, the biogeo-

chemical aspects of the model will be easier to isolate, to improve upon, and to understand.

Circulation modeling and carbon cycle modeling are clearly complementary to each other. Carbon cycle modeling helps validate the predictions of the physical models, and vice versa. Combined models will ultimately be necessary for reconstructing some climates of the past. The practical payoff from improving both types of models will be increased confidence in predictions of future climate regimes as well.

17.5 Acknowledgments

Support for this research was provided by the Physical Oceanography Program of the National Science Foundation and by the Computer Hardware, Advanced Mathematics, and Model Physics Program (CHAMMP) of the U.S. Department of Energy. Supercomputing resources were provided by the Scientific Computing Division of the National Center for Atmospheric Research. Parallel computing resources at the

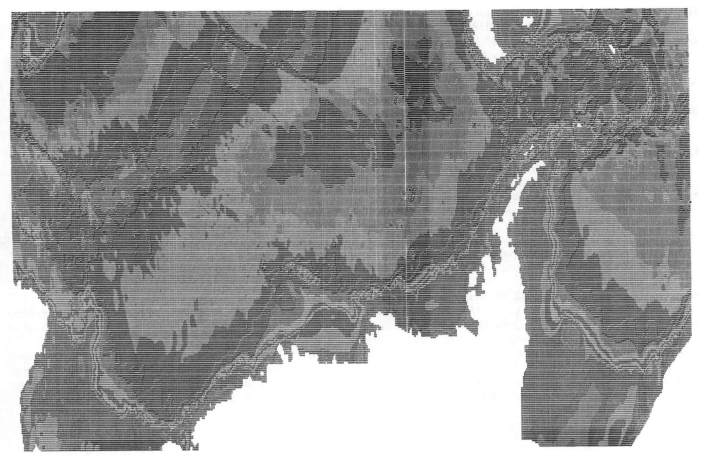

Figure 17.15: Geometry for the vicinity of the Weddell and Ross seas, constructed for use in the collaborative effort with Los Alamos investigators on the CM-5. The shaded picture is composed of 20 characters to show model depth at each grid point. Average grid spacing for the Antarctic circumpolar current region is approximately 0.11-degree. The grid spacing near 78°S is 7 km, which allows direct representation of the larger scales of deep convection in the Southern Hemisphere.

Los Alamos National Laboratory are being provided through CHAMMP.

References

Bennett, A. F. 1978. Poleward heat fluxes in Southern Hemisphere oceans. *Journal of Physical Oceanography 8*, 785–798.

Broecker, W. S. 1991. The great ocean conveyor. *Oceanography 4*, 79–89.

Bryan, K. 1969. A numerical model for the study of the world ocean. *Journal of Computational Physics 4*, 347–376.

Bryden, H. L., and E. C. Brady. 1989. Eddy momentum and heat fluxes and their effects on the circulation of the equatorial Pacific Ocean. *Journal of Marine Research 47*, 55–79.

Bryden, H. L., and M. M. Hall. 1980. Heat transports by currents across 25°N latitude in the Atlantic Ocean. *Science 207*, 884–886.

Bryden, H. L., D. H. Roemmich, and J. A. Church. 1991. Ocean heat transport across 24°N in the Pacific. *Deep-Sea Research 38*, 297–324.

Chervin, R. M., and A. J. Semtner. 1990. An ocean modelling system for supercomputer architectures of the 1990's. In *Climate-Ocean Interactions*, M. E. Schlesinger, ed. Kluwer Academic Publishers, Dordrecht, Netherlands, pp. 87–95.

Cox, M. D. 1984. *A Primitive Equation Three-Dimensional Model of the Ocean.* Geophysical Fluid Dynamics Laboratory Ocean Group Technical Report No. 1, Geophysical Fluid Dynamics Laboratory/National Oceanic and Atmospheric Administration, Princeton University, Princeton, New Jersey.

Danabasoglu, G., J. C. McWilliams, and P. R. Gent. 1994. The role of mesoscale tracer transports in the global ocean circulation. *Science 264*, 1123–1126.

Dukowicz, J. K., and R. D. Smith. 1994. Implicit free-surface method for the Bryan-Cox-Semtner ocean model. *Journal of Geophysical Research 99(C4)*, 7991–8014.

Godfrey, J. S., A. C. Hirst, and J. Wilkin. 1993. Why does the Indonesian Throughflow appear to originate from the North Pacific? *Journal of Physical Oceanography 23*, 1087–1098.

Gordon, A. 1986. Interocean exchange of thermocline water. *Journal of Geophysical Research 91*, 5037–5046.

Gordon, A., and W. Haxby. 1990. Agulhas eddies invade the South Atlantic: Evidence from Geosat altimeter and shipboard conductivity-temperature-depth survey. *Journal of Geophysical Research 95*, 3117–3125.

Hastenrath, S. 1980. Heat budget of tropical ocean and atmosphere. *Journal of Physical Oceanography 10*, 159–170.

Hellerman, S., and M. Rosenstein. 1983. Normal monthly wind stress over the world ocean with error estimates. *Journal of Physical Oceanography 13*, 1093–1104.

Keffer, T., and G. Holloway. 1988. Estimating Southern Ocean eddy flux of heat and salt from satellite altimetry. *Nature 332*, 624–626.

Killworth, P. D., D. J. Webb, D. Stainforth, and S. M. Paterson. 1991. The development of a free-surface Bryan-Cox-Semtner ocean model. *Journal of Physical Oceanography 21*, 1333–1348.

Levitus, S. 1982. *Climatological Atlas of the World Oceans.* National Oceanic and Atmospheric Administration Professional Paper No. 13. U.S. Government Printing Office, Washington, D.C.

Maier-Reimer, E., and K. Hasselmann. 1987. Transport and storage of CO_2 in the ocean – An inorganic ocean circulation carbon cycle model. *Climate Dynamics 2*, 63–90.

Pacanowski, R. C., K. W. Dixon, and A. Rosati. 1990. *GFDL Modular Ocean Model.* Unpublished report, National Oceanic and Atmospheric Administration/Geophysical Fluid Dynamics Laboratory, Princeton, New Jersey.

Sarmiento, J. L., and K. Bryan. 1982. An ocean transport model for the North Atlantic. *Journal of Geophysical Research 87*, 394–408.

Schmitz, W. J., and J. D. Thompson. 1993. On the effects of horizontal resolution in a limited-area model of the Gulf Stream system. *Journal of Physical Oceanography 23*, 1001–1007.

Semtner, A. J. 1974. *An Oceanic General Circulation Model with Bottom Topography.* Technical Report 9, Department of Meteorology, University of California, Los Angeles, California.

Semtner, A. J., and R. M. Chervin. 1988. A simulation of the global ocean circulation with resolved eddies. *Journal of Geophysical Research 93*, 15502–15522 and 15767–15775.

Semtner, A. J., and R. M. Chervin. 1992. Ocean general circulation from a global eddy-resolving model. *Journal of Geophysical Research 97*, 5493–5550.

Stammer, D., and C. W. Boning. 1992. Mesoscale variability in the Atlantic Ocean from Geosat altimetry and WOCE high resolution numerical modeling. *Journal of Physical Oceanography 22*, 732–752.

Toggweiler, J. R., K. Dixon, and K. Bryan. 1989. Simulations of radiocarbon in a coarse-resolution world ocean model: I. Steady state prebomb distributions. *Journal of Geophysical Research 94*, 8217–8242.

Trenberth, K. E., J. G. Olson, and W. G. Large. 1989. *A Global Ocean Wind Stress Climatology Based on ECMWF Analyses.* NCAR Technical Note NCAR/TN-338+STR, National Center for Atmospheric Research, Boulder, Colorado.

Verron, J., J.-M. Molines, and E. Blayo. 1992. Assimilation of Geosat data into a quasi-geostrophic model of the North Atlantic between 20 and 50°N: Preliminary results. *Oceanologica Acta 15*, 575–583.

Wijffels, S. E., R. W. Schmitt, H. L. Bryden, and A. Stigebrandt. 1992. On the transport of freshwater by the oceans. *Journal of Physical Oceanography 22*, 155–162.

Wilkin, J. L., and R. A. Morrow. 1994. Eddy kinetic energy and momentum flux in the Southern Ocean: Comparison of a global eddy-resolving model with altimeter, drifter, and current-meter data. *Journal of Geophysical Research 99(C4)*, 7903–7916.

18

Effects of Ocean Circulation Change on Atmospheric CO$_2$

ROBIN S. KEIR

Abstract

The effects of ocean circulation on steady-state atmospheric CO$_2$ concentration in ocean models pertaining to glacial climates are reviewed in this chapter. In this context, it appears that ocean circulation changes could provoke four basic effects: (1) Circulation-activated change in calcium carbonate (CaCO$_3$) production can change the deep ocean CO$_3$= concentration and (2) the rain ratio of organic C to CaCO$_3$ production; (3) change in thermohaline circulation or upper ocean mixing may alter the shape of the vertical gradient of dissolved CO$_3$=; and (4) changing thermohaline circulation may interact with both biological production and air–sea exchange in high-latitude deep water formation areas to effect change in atmospheric CO$_2$ through the solubility and biological pumps.

18.1 Introduction

The records of calcium carbonate (CaCO$_3$), carbon isotopes, Cd/Ca ratio, and benthic forminiferal speciation indicate that the ocean circulation varies with climate change during the Pleistocene (Crowley, 1985; Mix and Fairbanks, 1985; Boyle and Keigwin, 1985; Duplessy et al., 1988). In addition, atmo-

spheric CO$_2$ appears to decrease by approximately 90 ppm during ice ages as compared to relatively warm climates such as that of the Holocene (Barnola et al., 1987; Neftel et al., 1988). Various models of the ocean and atmosphere have been developed to simulate proposed mechanisms that might have produced the glacial–interglacial CO$_2$ change, and some of these models include changes in various circulation parameters (Knox and McElroy, 1984; Sarmiento and Toggweiler, 1984; Siegenthaler and Wenk, 1984; Broecker and Peng, 1987; Keir, 1988; Boyle, 1988; Heinze et al., 1991). These results are reviewed here with the purpose of clarifying mechanisms by which changes in circulation may alter the atmospheric concentration. Generally, the focus will be on the effect of a reduction in thermohaline circulation because the paleo-proxies indicate that the formation of North Atlantic deep water during ice ages decreased. However, some consideration will also be given to changes in vertical mixing structure, because greater trade winds during the glacial climates appear to have increased equatorial upwelling as well (Arrhenius, 1952).

Table 18.1 provides selection of previous model results in which one or more transport features are changed by a factor of 2. These features vary from model to model and, except for the simple two-box ocean model, generally include a merid-

Table 18.1. *Atmospheric CO$_2$ vs. circulation in various ocean models*

Model Type	Circulation Change	ΔpCO_2, ppm	Reference
2-box, closed	any	0	
2-box, CO$_3$= i/o	1/2 vertical exchange		
	(a) CaCO$_3$ and organic production proportional	−17	*a*
	(b) CaCO$_3$ production constant	+52	*a*
3-box, closed	1/2 conveyor belt	+25	*b*
constant polar	1/2 polar convection	−45	*b*
biological production	1/2 both	−30	*b*
5-box, fixed deep ocean CO$_3$=	2X upper-ocean exchange	−13 to −27	*c*
CaCO$_3$ α organic production	Switch from deep to intermediate water production	−8	*c*
10-box, closed	1/2 all transports	−36	*d*
13-box, CO$_3$= i/o	1/2 meridional loop		
	(a) CaCO$_3$ μ organic production	−26	*e*
	(b) CaCO$_3$ production constant	+4	*e*
	2X upper-ocean exchange		
	(a) CaCO$_3$ μ organic production	+31	*e*
	(b) CaCO$_3$ production constant	−13	*e*
Hamburg GCM	1/2 all transports	−26	*f*

*a*Calculated from model of Keir & Berger (1983).
*b*Sarmiento & Toggweiler (1984).
*c*Boyle (1988).
*d*Broecker & Peng (1987).
*e*Keir (1988).
*f*Heinze et al. (1991).

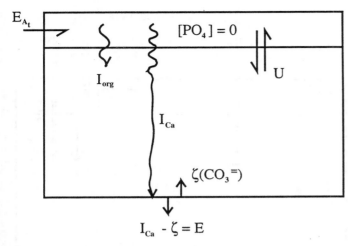

Figure 18.1: Two-box ocean model with vertical exchange transport, μ, and input of dissolved CO$_3^=$, E$_{At}$, to the surface ocean. Production of organic C, I$_{org}$, is limited by the upward flux of nutrients. The partial dissolution, ζ, of the particulate carbonate flux, I$_{Ca}$, in the deep ocean is dependent on its CO$_3^=$ concentration. (Data from Keir and Berger, 1983.) Reprinted with permission from R. S. Keir and W. H. Berger, *Journal of Geophysical Research* 88: 6027–28. ©1983, American Geophysical Union.

ional component and one or more vertical exchange features. The results from some of the simpler models are discussed in the following sections; however, it appears from Table 18.1 that a variety of atmospheric pCO$_2$ changes can result from the different types of circulation changes. Across models, some of the results appear fairly consistent; for example, three different ocean models give a decrease of approximately 30 ppm when all transport is reduced by one-half. However, it is also observed that the same circulation change in different models can result in differing atmospheric CO$_2$ changes, depending on the nature of the change in biological production of CaCO$_3$ that is assumed to occur. For example, doubling the vertical exchange of the upper-ocean boxes in a five-box ocean (Boyle, 1988) gave a 30-ppm decrease in the atmospheric CO$_2$, while the same doubling produced about a 30-ppm *increase* in a 13-box model (Keir, 1988). In the former model, the deep sea carbonate ion (CO$_3^=$) concentration is assumed to remain constant, while in the latter, economics of the carbonate throughput dictate that the deep CO$_3^=$ decrease.

18.2 Vertical Change in a Two-Box Ocean

A principal tenet of the ocean carbon cycle is that it is driven by the upward flux of dissolved nutrients into the photic zone, where in low-latitude surface waters nearly all of this flux is fixed into organic matter (Broecker, 1971). This determines the rate at which dissolved inorganic carbon is converted into the sinking particulate organic flux. Since both phosphate and carbon have relatively long residence times in the ocean, roughly on the order of 100,000 years, one can approximate the particulate rain of organic P and C as being totally reminer-

alized in the subsurface ocean, ignoring throughput by river input and depositional output.

The simplest carbon cycle model divides the ocean vertically into two boxes, and the circulation is parameterized as a single vertical exchange of water between them. If, in addition to organic carbon, the net biological production of CaCO$_3$ in the surface box is assumed to be proportional to the nutrient flux and the carbonate is completely redissolved in the deep box (i.e., no throughput), the atmospheric CO$_2$ concentration is independent of the vertical exchange transport. This is because the vertical gradients of total dissolved CO$_2$ and alkalinity remain the same regardless of the circulation rate, since the ratio of biological production to upwelling fluxes remains the same.

Complete recycling of particulate CaCO$_3$ fluxes has been assumed in more complex box models as well, but in general it is not a very good approximation of the carbon cycle on time scales longer than a few thousand years. Alkalinity is added to the ocean in the form of bicarbonate as a result of rock-weathering processes, and this input is removed as the long-term accumulation of CaCO$_3$ in deep sea sediments. (On glacial–interglacial time scales, erosion of carbonate reef platforms during low sea stands and buildup of reefs during interglacials constitute an additional variable source/sink of dissolved CO$_3^=$ to the ocean.) The carbonate sediment accumulation in the deep ocean is approximately 20–25% of the net biological production rate of CaCO$_3$ in the surface ocean. Therefore, unlike the deposition of organic carbon, the accumulation of CaCO$_3$ is a significant fraction of its rain rate, and changes in CaCO$_3$ production that might result from the ocean circulation may affect the atmospheric CO$_2$.

Two possible interactions of the ocean circulation with the carbonate system can be illustrated with a two-box model (Figure 18.1) including alkaline input and a dissolution flux that depends on the deep ocean dissolved carbonic acid system chemistry (Keir and Berger, 1983). A decrease in the thermohaline circulation would be accompanied by a decrease in both organic C and CaCO$_3$ production, if the latter is also controlled by the upwelling nutrient flux. Assuming the alkaline influx to the ocean from rivers remains relatively constant, the difference between carbonate production and alkaline influx from rivers would be reduced, and therefore a smaller proportion of the carbonate production would have to dissolve in order to rebalance the ocean's alkaline budget. After the reduction in carbonate production, the dissolution flux in the deep ocean would initially be in excess of the new steady-state value, causing the dissolved CO$_3^=$ of the deep ocean to rise until the lower dissolution rate was achieved. As a result, the surface ocean CO$_3^=$ concentration would also increase (Figure 18.2), causing a reduction in the atmospheric CO$_2$.

If, in contrast to the above situation, biological production of CaCO$_2$ is only weakly dependent on upwelling nutrient fluxes, change in the ocean circulation might alter atmospheric CO$_2$ by altering the ratio of organic C to CaCO$_3$ production. The basis of this possibility is the observation that in high upwelling areas, the organic C to CaCO$_3$ flux ratio observed in sediment traps is much larger than that in open ocean oligotrophic locations (Berger and Keir, 1984; Dymond and Lyle, 1985). If carbonate production as well as the river input simply remain con-

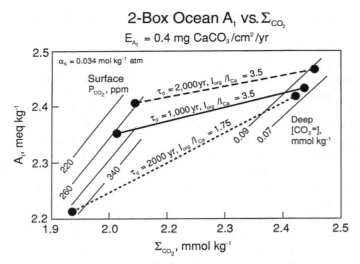

Figure 18.2: Change in the alkalinity (A$_t$) vs. total CO$_2$ (ΣCO$_2$) pattern of the two-box model when vertical exchange is reduced by 50% and thus the deep water residence time (τ_d) increases from 1,000 to 2,000 yr. Solid line shows the alkalinity vs. total CO$_2$ for the standard case ($\tau_d = 1,000$ yr) with atmospheric pCO$_2$ of 270 ppm. Dashed line shows pattern when circulation decreases by 50% ($\tau_d = 2,000$ yr) and the biological production of CaCO$_3$ also decreases by 50% (I$_{org}$/I$_{Ca}$ = 3.5). Dotted line shows pattern when the circulation decreases but the production of CaCO$_3$ remains constant (I$_{org}$/I$_{Ca}$ = 1.75). pCO$_2$ as a function of A$_t$ and ΣCO$_2$ corresponds to the solubility, α_s, at 20°C.

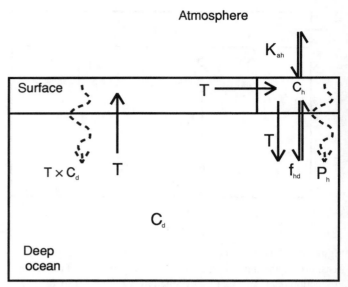

Figure 18.3: Three-box ocean model of Sarmiento and Toggweiler (1984), with meridional transport, T, and high-latitude vertical exchange, f$_{hd}$. The low-latitude production is determined by the upwelling nutrient flux, T × C$_d$. Biological production in the high-latitude surface, P$_h$, is a free parameter. K$_{ah}$ is the areally integrated piston velocity for CO$_2$ exchange between the atmosphere and high-latitude surface water.

stant after a reduction in thermohaline circulation, the difference between these carbonate fluxes would not change, while the organic carbon net production would be reduced in proportion to the circulation decrease. As a result, the deep sea CO$_3^=$ concentration would remain unchanged between the initial and final steady-state conditions; however, a greater surface-to-deep alkalinity gradient would develop under the reduced circulation, and this would increase the atmospheric CO$_2$, in contrast to the situation in which organic C and carbonate productions are proportional to each other (Figure 18.2).

The two examples above illustrate that the nature of the response of biological carbonate production to ocean circulation will affect the change in atmospheric CO$_2$. The response to a reduction in circulation could be either positive or negative, depending on the relative importance of deep ocean CO$_3^=$ concentration change versus that of the organic to carbonate rain ratio.

18.3 Solubility Pump Effects in a Three-Box Ocean

The process of poleward surface flow with cooling and subsequent sinking of cold water causes atmospheric CO$_2$ to be drawn into the sinking flux and reduces the atmospheric concentration from what it would be as a result of the biological pump alone. This occurs because the gas solubility increases with colder temperature, and therefore the CO$_2$ partial pressure (pCO$_2$) of the surface water decreases as it cools. Volk and

Hoffert (1985) analyzed this process in detail and termed it the "solubility pump." One notable conclusion from their analysis is that most of the variation of atmospheric CO$_2$ in the three-box ocean model with poleward surface flow and subsequent sinking (Figure 18.3) is primarily due to variation of the solubility pump. Fundamentally, the potentially low pCO$_2$ of the small area of cold water can exert a disproportionately large influence on the atmospheric CO$_2$.

In the special case where the conveyor belt transport, ζ, is the only ocean circulation parameter in the model, the degree of influence of warm versus cold surface water temperature on the CO$_2$ solubility that controls the atmospheric partial pressure (P_{atm}) is determined by the proportion of T versus the regional integrated piston velocity for CO$_2$ exchange between the cold, sinking polar water and the atmosphere, K_{ah} (Keir, 1993b). The approximate relationship is

$$P_{atm} = \frac{(TC_1^0 + \beta_1 K_{ah} C_2^0)}{(T\alpha_1 + \beta_1 K_{ah}\alpha_2)} \tag{18.1}$$

where C_1^0 and C_2^0 are the dissolved aqueous CO$_2$ gas concentrations that would exist in the warm and cold surface waters in a closed system with no gas exchange, α_1 and α_2 are the gas solubilities in warm and cold sea water, and β_1 is the partial change of aqueous dissolved CO$_2$ versus total dissolved inorganic carbon, equal to approximately 6%. Since the temperature effect on the closed system aqueous gas concentration per se is not very large (i.e., $C_1^0 \approx C_2^0$), one observes from Equation (18.1) that with high poleward water transport, the system tends toward warm temperature solubility control ($P_{atm} \approx C_1^0/\alpha_1$), whereas with rapid ventilation of the cold sinking water, the cold temperature solubility controls the atmospheric CO_2 content ($P_{atm} \approx C_2^0/\alpha_2$). The area and gas exchange flux of the warm surface ocean have little effect on the "effective" solubility, as long as this region is well ventilated relative to the upwelling of water into this surface region. More generally, it appears that the low-latitude biological pump tends to set the approximate concentration of total dissolved inorganic carbon and dissolved aqueous gas concentrations in the warm surface ocean. However, the overall solubility that influences the atmospheric CO_2 content appears to be determined by the dynamics governing rates of cooling and sinking in the upper ocean, together with the rate of gas exchange as this cooling occurs.

In the three-box ocean model, the concentration of deep ocean nutrients is very nearly constant over all circulation and biological changes in the system. Thus, the vertical gradients of nutrients and total CO_2 between the warm surface and deep ocean remain nearly the same regardless of the circulation, as in the two-box ocean model. Since the area of the polar box is small, the variation of high-latitude net production may be a small proportion of the total production over the whole ocean and still exert a large influence on atmospheric CO_2.

The effect of ocean circulation changes on the solubility pump is clearly seen from the model of Sarmiento and Toggweiler (1984), shown in Figure 18.3. In this model, 100% of the carbonate production is redissolved in the deep ocean, and changes in carbonate economics and in the rain ratio described in Section 18.2 do not affect atmospheric CO_2 when circulation varies. The model contains two transports, the meridional component (T) and a vertical exchange between the deep ocean and cold surface layer (f_{hd}). Figure 18.4 illustrates the model-calculated atmospheric CO_2 as a function of the two transport parameters, with the biological production of the high-latitude surface region held constant. The atmospheric CO_2 generally decreases as the high-latitude vertical exchange decreases; however, the CO_2 content does not vary in a monotonic fashion with changing meridional circulation, as indicated by the change in CO_2 along the dashed horizontal line in Figure 18.4. Starting from the model analog of the modern ocean, decreasing meridional circulation increases atmospheric CO_2. However, as the meridional circulation increases, the CO_2 content decreases to a minimum of approximately 240 ppm and then begins to increase. At constant cold surface-deep mixing rates that are lower than the modern analog (f_{hd}/f_{hd}^o <1), the position of the atmospheric CO_2 minimum shifts toward lower meridional transport.

The CO_2 minimum appears to be the result of two competing effects that occur when the meridional circulation changes. Increasing poleward flow tends to "wash out" nutrients and higher total CO_2 in the polar surface box. As a result, the pCO₂ of the cold surface water decreases. However, increasing poleward flow into the high latitudes decreases the residence time of this water in the polar region. Once this time becomes comparable to the time scale of the high-latitude surface water pCO₂ equilibration with the atmosphere, the system begins reverting to control by the higher-temperature solubility of the low-latitude surface. In other words, if the flow becomes fast enough, the low pCO₂ water is not in contact with the atmo-

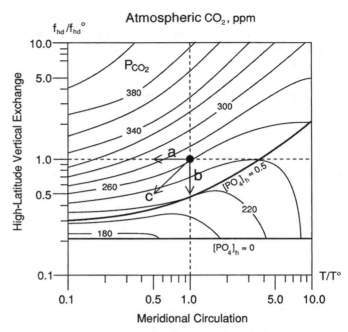

Figure 18.4: Atmospheric CO_2 as a function of meridional circulation and high-latitude vertical exchange, relative to standard ocean values (T^o and f_{hd}^o), when net biological production in the high latitude is constant at 2.3×10^6 mol C/s. Heavy contours show isophosphate concentrations in the high-latitude surface ocean of 0 and 0.5 μmol/kg. Arrows indicate effect of 50% reductions in the circulation components: (a) meridional circulation (T), (b) high-latitude vertical mixing (f_{hd}), and (c) both. (Data from Sarmiento and Toggweiler, 1984.) Reprinted with permission from *Nature*, J. L. Sarmiento and J. R. Toggweiler, *Nature* 308: 621–24. © 1984 MacMillan Magazines Ltd.

sphere long enough to remove any additional CO_2 effectively. In Figure 18.4, this is apparent at higher values of meridional circulation where, for example, the atmospheric CO_2 isolines no longer tend to parallel the 0.5 μmol/kg isoline of cold surface water phosphate (PO_4) concentration.

Since the process of deep water formation involves winter convection in the Greenland Sea, a change in the meridional circulation could have some associated alteration in the rate of high-latitude vertical exchange. The nature of the associated change in polar convection may well influence the atmospheric CO_2 change. For example, with constant high-latitude production in the model, the atmospheric CO_2 increases when the meridional component decreases by itself; however, the CO_2 concentration decreases when the meridional transport and polar convection decrease proportionately (Figure 18.4).

18.4 Vertical $CO_3^=$ Gradient

In addition to affecting the solubility pump, change in the thermohaline circulation could also alter the vertical distribution of nutrients and dissolved carbonate ion concentration. In the present ocean, maximum nutrient concentrations are usually found at intermediate depths of approximately 300–2,000 m, which is in part a consequence of re-solution of the particulate flux in the upper water column and the slow vertical advection that recycles the deep water formation. A decrease in the latter would cause the nutrient maximum together with metabolically produced CO$_2$ to be shifted to greater depths, actuating a temporal carbonate dissolution increase from the deep ocean's sediments. Boyle (1988) has proposed that the resulting increase in deep ocean alkalinity would mix into the surface waters and thereby decrease atmospheric CO$_2$.

In addition to decreasing the meridional circulation, other processes, such as increased upper-ocean mixing and deeper recycling of nutrients, could also shift nutrients to deeper depths. Boyle (1988) investigated the effect of these changes with a five-box ocean model (Figure 18.5A), and his results suggest that corresponding to a 0.4-μmol/kg decrease in intermediate water phosphate, the atmospheric CO$_2$ would decrease approximately 10–40 ppm, the result being somewhat dependent on the mechanism that causes the gradient change. In all cases it was assumed that the dissolution response would restore the dissolved CO$_3^=$ of the deep ocean to its prior steady-state value. As pointed out earlier, this might not be the case if a change in circulation alters the difference between biological production of CaCO$_2$ and alkalinity input from rivers;

however, Boyle's results do show the effect of the change in the vertical alkalinity and total CO$_2$ gradients by themselves.

The CO$_3^=$ concentration is proportional to the difference between total alkalinity (A_t) and total CO$_2$ (ΣCO$_2$), which is a mass-conserving quantity obtained directly from the model results. Figure 18.6 illustrates profiles of $A_t - \Sigma$CO$_2$ obtained from two of Boyle's five-box model calculations, one a standard analog of the present and the other a redistribution that results from a doubling of the upper-ocean exchange. In this example, the difference between surface and deep water $A_t - \Sigma$CO$_2$ increases. However, the nature of the vertical CO$_3^=$ gradient change in response to the circulation appears to be model-dependent. In the five-box analog of the present ocean shown in Figure 18.6, the upper-ocean mixing is much greater than that of intermediate and deep water. In order to produce a vertical $A_t - \Sigma$CO$_2$ distribution similar to that which exists, it must be assumed that the dissolution of the particulate carbonate rain takes place primarily in the upper ocean. Boyle (1988) notes that if carbonate dissolution takes place primarily in the deep ocean in this model, the nutrient downward shift of nutrients would be associated with an *increase* in the atmospheric CO$_2$.

A simple three-box vertical ocean model (Figure 18.5B) brings out an additional point concerning the structure of the change in the change in CO$_3^=$ gradient. The change in nutrient structure in this model has been described previously (Keir, 1991), and the distribution of $A_t - \Sigma$CO$_2$ can be calculated once the ratio of organic to carbonate production and the percentage of the particulate carbonate flux dissolving in the subsurface

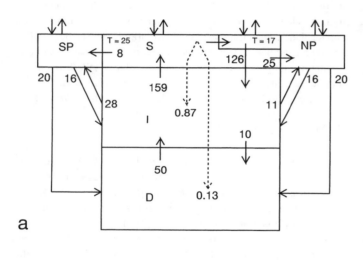

a

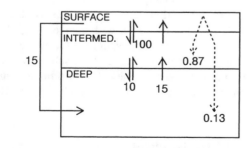

b

Figure 18.5: (A) Five-box model (IG-1) (Boyle, 1988), where the surface ocean is divided into warm (S) and cold regions (SP, NP), and (B) Vertical three-box ocean model (Keir, 1991). Whole numbers show transport in 10^6 m^3 C/s for standard ocean analogs. Dashed lines indicate fraction of particulate organic carbon reoxidized in intermediate and deep oceans. The organic carbon to CaCO$_3$ rain ratio is assumed to be 5 : 1 in both models.

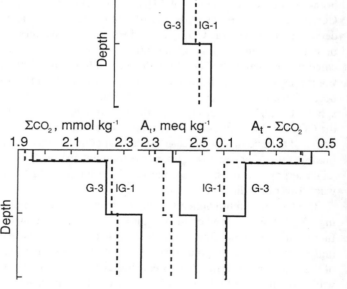

Figure 18.6: Vertical profiles of phosphate, total CO$_2$ (ΣCO$_2$), alkalinity (A_t), and $A_t - \Sigma$CO$_2$ in a five-box model. Dotted lines are for Boyle's IG-1 modern ocean; solid lines show profiles after surface-intermediate water exchange is doubled. The profiles are from the three vertical boxes in the model, the upper one being the warm surface ocean (S).

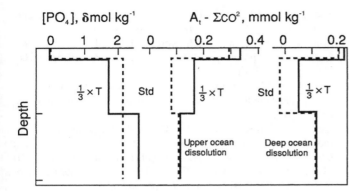

Figure 18.7: Vertical profiles of phosphate and $A_t - \Sigma CO_2$ from the three-box tandem model shown in Figure 18.5B. The dashed lines indicate profiles from the modern analog, and solid lines show the profiles when the meridional component is reduced by 62%. Two cases are shown, where the particulate carbonate flux is either redissolved mostly in the intermediate box or dissolved entirely in the deep box.

waters is specified. As in Boyle's model, it is assumed that the deep ocean value of $A_t - \Sigma CO_2$ is constant and the production of both organic carbon and $CaCO_3$ is proportional to the up-welling flux of nutrients. Because nutrients are shifted downward and their concentration in the intermediate layer decreases, the biological pump of carbon from the surface ocean becomes less effective; that is, less carbon per unit of upwelled water can be fixed according to nutrient limitation. Therefore the gradient of $CO_3^=$ in the upper part of the water column decreases (Figure 18.7). While the $CO_3^=$ gradient between intermediate and deep water always increases in concert with the downward nutrient shift, the increase in surface water $CO_3^=$ is less than the increase in intermediate water $CO_3^=$ concentration as a result of the weaker biological pump. As shown in Figure 18.7, it appears that the depth distribution of carbonate dissolution affects the tradeoff between an increased $CO_3^=$ gradient in deep water and decreased gradient in the upper ocean. The vertical redistribution of nutrients appears to be least effective in decreasing atmospheric CO_2 when all the carbonate flux dissolves in the deep ocean. It thus appears that the vertical distribution of carbonate dissolution together with the vertical structure of mixing in the ocean is critical to the nature of the so-called alkalinity effect that might result from thermohaline changes.

18.5 Proportional Circulation Changes

In ocean box models and the Hamburg general circulation model (GCM), experiments have been done where all transports are proportionately reduced (Sarmiento and Toggweiler, 1984; Broecker and Peng, 1987; Bacastow and Maier-Reimer, 1990; Heinze et al., 1991). These experiments produce a decrease in atmospheric CO_2, which appears to be caused by greater reduction in the upwelling total dissolved CO_2 flux than net carbon biological production in regions where the latter is not limited by low surface nutrient concentration. The simplest example is a three-box ocean model where both the meridional transport and the exchange of cold surface with

deep water are reduced. The increase in biological production over upwelling in the cold surface reduces its nutrient concentration and CO_2 partial pressure, drawing in a greater flux of CO_2 from the atmosphere into the sinking water. The accompanying reduction of the poleward flow of low-latitude surface water tends to oppose this reduction, but this effect is not as strong as the increase in the high-latitude biological pump, as shown by the individual vectors in Figure 18.4.

Greater reduction of the upwelling total dissolved CO_2 than of net carbon production occurs in the Hamburg GCM as well. Bacastow and Maier-Reimer (1990) found that a 40% reduction in the circulation in this model gives a decrease of only 24% in the net biological production. In their analysis of the results from the model computation, they found that the CO_2 reduction is enhanced by the solubility pump, which approximately doubles the effect that the biological pump would have produced by itself.

18.6 Time Scales of Circulation-Activated CO₂ Change

The discussion thus far has been limited to comparing one steady-state model atmospheric CO_2 concentration to another, without consideration of the time required to bring about the change. When a change in circulation immediately changes the total CO_2 or alkalinity budget of the surface ocean, the change in atmospheric CO_2 may be relatively rapid. For example, the atmospheric CO_2 transients produced by (1) a 50% decrease in the vertical exchange of water between the cold surface and deep water in the three-box ocean model (Wenk and Siegenthaler, 1985), and (2) the 40% decrease in all circulation in the Hamburg GCM (Bacastow and Maier-Reimer, 1990) appear similar. Both appear to reduce the atmospheric CO_2 by approximately 40 ppm overall, with a time constant of approximately 200 years. In the GCM transient, there also appears to be an initial decrease on the order of 10 ppm with a short time constant of approximately 10 years.

In contrast, the characteristic time scale for changes in deep sea carbonate dissolution appears to be approximately 3,000–6,000 years (Broecker and Peng, 1987; Keir, 1988; Sundquist, 1990). Two of the four circulation-induced effects, change in carbonate production and change in the vertical $CO_3^=$ gradient with restoration of the deep ocean concentration, depend entirely on the deep sea carbonate dissolution response to alter the atmospheric CO_2; one would expect a several-thousand-year time scale from these effects. The other two effects, change in the organic carbon to $CaCO_3$ rain ratio and change in the combined biological/solubility pump, would be expected to produce most of the atmospheric CO_2 change on a time scale of a few hundred years, because these mechanisms act directly on the balance of carbon and alkalinity fluxes into and out of the surface ocean.

18.7 Glacial–Interglacial CO₂ Change

Previous investigations of how the glacial–interglacial changes in CO_2 might have taken place have focused primarily on the interaction between the ocean's mixing and biological cycle of production and remineralization. Evidence for an increase in biological production, either globally owing to greater nutrient inventory or regionally in the Southern Ocean, has not been found (Boyle, 1992; Mortlock et al., 1991). In order to accomplish the 90-ppm lower atmospheric CO$_2$ with a decrease in ocean circulation, the associated decrease in upwelling total CO$_2$ flux in the high latitudes would have to be large, on the order of a factor of 5 or more (Knox and McElroy, 1984; Sarmiento and Toggweiler, 1984; Keir, 1988). Although it appears from Cd/Ca and changes in ^{13}C concentration (δ^{13}C) that the production of North Atlantic deep water decreased and the thermohaline circulation changed during glacial climates, measurements of the difference in radiocarbon age between planktonic and benthic foraminifera deposited during the last ice age indicate that the average age difference between surface and deep water has remained nearly constant (Broecker et al., 1990). Therefore, it appears that the deep ocean was on the average about as well ventilated during the last ice age as it is now. If the vertical circulation had greatly decreased, the vertical gradient in radiocarbon would have been much greater, and this would have produced a notable increase in the planktonic-benthic ^{14}C difference.

Thus it appears that something beside the mixing and biological cycle of the ocean may have contributed to the atmospheric CO$_2$ changes during the Pleistocene. One possibility is a change in the supply of alkalinity to the surface ocean, as a result of cycles of either coral reef buildup and erosion (Berger, 1982; Opdyke and Walker, 1992) or continental weathering (Munhoven and Francois, 1994). Another possibility is that the greater winds of the glacial climates cause greater air–sea gas exchange, which in turn enhances the uptake of CO$_2$ into the production of deep water (Keir, 1993a). Discussion of these effects is outside the scope of this chapter; however, both these possibilities do not seem likely to affect the ocean distribution of δ^{13}C. The distinction between, for example, a coral reef–driven CO$_2$ change and a wind-activated solubility pump change is that the former would change the long-term preservation of CaCO$_3$ in deep sea sediments, and the latter probably would not.

Acknowledgments

Wolfgang Berger, David Archer, Ed Boyle, and Martin Heimann provided helpful reviews of the original manuscript, and I am grateful for their comments. I would also like to acknowledge the support I have had from my wife, Angela, during the time this work was done, as well as in many other endeavors. She passed away suddenly in September 1993, after we returned to Kiel from the workshop in Colorado. I am grateful to Tom Wigley for the opportunity to attend the 1993 Global Change Institute.

References

Arrhenius, G. 1952. Sediment cores from the East Pacific. *Report of the Swedish Deep-Sea Expedition 1947–1948 Vol. 5*, 1–288.

Bacastow, R., and E. Maier-Reimer. 1990. Ocean circulation model of the carbon cycle. *Climate Dynamics 4*, 95–125.

Barnola, J. M., D. Raynaud, Y. S. Korotkevich, and C. Lorius. 1987. Vostok ice core provides 160,000-year record of atmospheric CO$_2$. *Nature 329*, 408–414.

Berger, W. H. 1982. Increase of carbon dioxide in the atmosphere during deglaciation: The coral reef hypothesis. *Naturwissenschaften 69*, 87–88.

Berger, W. H., and R. S. Keir. 1984. Glacial-Holocene changes in atmospheric CO$_2$ and the deep-sea record. In *Climate Processes and Climate Sensitivity*, J. E. Hansen and T. Takahashi, eds. AGU Monograph No. 29, Maurice Ewing Series No. 5, American Geophysical Union, Washington, D. C., pp. 337–351.

Boyle, E. A. 1988. The role of chemical fractionation in controlling late Quaternary atmospheric carbon dioxide. *Journal of Geophysical Research 93*, 15,701–15,714.

Boyle, E. A. 1992. Cd and C13 paleochemical ocean distributions during the stage 2 glacial maximum. *Annual Review of Earth and Planetary Science 20*, 245–287.

Boyle, E. A., and L. D. Keigwin. 1985. Comparison of Atlantic and Pacific paleochemical records for the last 215,000 years. *Earth and Planetary Science Letters 76*, 135–150.

Broecker, W. S. 1971. A kinetic model for the chemical composition of sea water. *Quaternary Research 1*, 188–207.

Broecker, W. S., and T.-H. Peng. 1987. The role of CaCO$_3$ compensation in the glacial to interglacial atmospheric CO$_2$ change. *Global Biogeochemical Cycles 1*, 15–29.

Broecker, W. S., T.-H. Peng, S. Trumbore, G. Bonani, and W. Wolfli. 1990. The distribution of radiocarbon in the glacial ocean. *Global Biogeochemical Cycles 4*, 103–117.

Crowley, T. J. 1985. Late Quaternary carbonate changes in the North Atlantic and Atlantic/Pacific comparisons. In *The Carbon Cycle and Atmospheric CO$_2$: Natural Variations Archean to Present*, E. T. Sundquist and W. S. Broecker, eds. AGU Monograph No. *32*, American Geophysical Union, Washington, D. C., pp. 271–284.

Duplessy, J.-C., N. J. Shackleton, R. G. Fairbanks, L. D. Labeyerie, D. Oppo, and N. Kallel. 1988. Deepwater source variations during the last climate cycle and their impact on the global deepwater circulation. *Paleoceanography 3*, 343–360.

Dymond, J., and M. Lyle. 1985. Flux comparisons between sediments and sediment traps in the eastern tropical Pacific: Implications for atmospheric CO$_2$ during the Pleistocene. *Limnology and Oceanography 30*, 699–712.

Heinze, C., E. Maier-Reimer, and K. Winn. 1991. Glacial pCO$_2$ reduction by the world ocean: Experiments with the Hamburg carbon cycle model. *Paleoceanography 6*, 395–430.

Keir, R. S. 1988. On the late Pleistocene ocean geochemistry and circulation. *Paleoceanography 3*, 413–445.

Keir, R. S. 1991. The effect of vertical nutrient redistribution of surface ocean δ^{13}C. *Global Biogeochemical Cycles 5*, 351–358.

Keir, R. S. 1993a. Are atmospheric CO$_2$ content and Pleistocene climate connected by wind speed over a polar Mediterranean Sea? *Global and Planetary Change 8*, 59–68.

Keir, R. S. 1993b. Cold surface ocean ventilation and its effect on atmospheric CO$_2$. *Journal of Geophysical Research 98*, 849–856.

Keir, R. S., and W. H. Berger. 1983. Atmospheric CO$_2$ content in the last 120,000 years: The phosphate extraction model. *Journal of Geophysical Research 88*, 6027–6038.

Knox, F., and M. B. McElroy. 1984. Changes in atmospheric CO$_2$: Influence of the marine biota at high latitudes. *Journal of Geophysical Research 89*, 4629–4637.

Mix, A. C., and R. G. Fairbanks. 1985. North Atlantic surface-ocean control of Pleistocene deep-ocean circulation. *Earth and Planetary Science Letters 73,* 231–243.

Mortlock, R. A., C. D. Charles, P. N. Froelich, M. A. Zibello, J. Saltzman, J. D. Hays, and L. H. Burckle. 1991. Evidence for lower productivity in the Antarctic Ocean during the last glaciation. *Nature 351,* 220–223.

Munhoven, G., and L. M. Francois. 1994. Glacial–interglacial changes in continental weathering: Possible implications for atmospheric CO$_2$. In *Carbon Cycling in the Glacial Ocean: Constraints on the Ocean's Role in Global Change,* R. Zahn, T. F. Pedersen, M. A. Kaminski, and L. Labeyrie, eds. Springer-Verlag, Berlin, Germany, pp. 39–58.

Neftel, A., H. Oeschger, T. Staffelbach, and B. Stauffer. 1988. CO$_2$ record in the Byrd core 50,000–5,000 years BP. *Nature 331,* 609–611.

Opdyke, B. N., and J. C. G. Walker. 1992. Return of the coral reef hypothesis: Basin to shelf partitioning of CaCO$_3$ and its effect on atmospheric CO$_2$. *Geology 20,* 733–736.

Sarmiento, J. L., and J. R. Toggweiler. 1984. A new model for the role of the oceans in determining atmospheric pCO$_2$. *Nature 308,* 621–624.

Siegenthaler, U., and T. Wenk. 1984. Rapid atmospheric CO$_2$ variations and ocean circulation. *Nature 308,* 624–626.

Sundquist, E. 1990. Influence of deep-sea benthic processes on atmospheric CO$_2$. *Philosophical Transactions of the Royal Society of London, Series A 331,* 155–165.

Volk, T., and M. I. Hoffert. 1985. Ocean carbon pumps: Analysis of relative strengths and efficiencies in ocean-driven atmospheric CO$_2$ changes. In *The Carbon Cycle and Atmospheric CO$_2$: Natural Variations Archean to Present,* E. T. Sundquist and W. S. Broecker, eds. AGU Monograph No. 32, American Geophysical Union, Washington, D. C., pp. 99–110.

Wenk, T., and U. Siegenthaler. 1985. The high-latitude ocean as a control of atmospheric CO$_2$. In *The Carbon Cycle and Atmospheric CO$_2$:* Natural Variations Archean to Present, E. T. Sundquist and W. S. Broecker, eds. AGU Monograph No. 32, American Geophysical Union, Washington, D. C., pp. 185–194.

19

Box Models of the
Terrestrial Biosphere

L. D. DANNY HARVEY

Abstract

We place box models within the hierarchy of terrestrial biosphere models used to assess atmosphere–biosphere carbon fluxes, develop the mathematical formulation of biosphere box models, and examine how gross and net fluxes resulting from land-use changes and CO_2 and temperature feedbacks can be separately and simultaneously incorporated into box models. We then summarize insights gained from sensitivity studies using a globally aggregated biosphere model, and close with a proposal for combining the box model approach with some of the simpler regionally disaggregated process models.

19.1 Introduction

Balancing the carbon cycle at the global scale requires that one account for the observed cumulative and year-to-year buildup in the amount of atmospheric CO_2 as the difference between emissions of CO_2 and its uptake by various sinks. It is widely assumed that the natural carbon cycle was sufficiently close to steady state prior to human influence that emissions were exactly balanced by removal processes. Hence, the task of balancing the carbon budget for the time interval since the beginning of human influence requires deriving estimates of the human-induced perturbation in both emission and sink terms such that the difference between total emission perturbations and total sink perturbations equals the observed buildup. The major emission terms in this balance are: emissions resulting from land-use changes, primarily deforestation; emissions resulting from combustion of fossil fuels and from the manufacture of cement; possible increases in plant respiration resulting from climatic warming; and emissions resulting from the oxidation of excess methane produced as a direct or indirect result of human activities. The major sink terms are: additional uptake of CO_2 by the oceans; likely enhanced photosynthesis resulting from the direct stimulatory effect of higher atmospheric CO_2 concentration; and possible additional burial of carbon in lake, reservoir, and marine sediments.

A hierarchy of models of the terrestrial biosphere has been used to assess the role of CO_2 photosynthesis and temperature-respiration feedback both in the buildup of CO_2 up to the present and in future atmospheric CO_2 increases. In addition to these two potential feedbacks, further emission or absorption of CO_2 will occur as a result of changes in the distribution of natural ecosystems as a result of climatic change. The net effect on the carbon budget of ecosystem shifts is likely to be different in the transient than in the steady state because of lags between the demise of a given ecosystem or species assemblage in a given area and its replacement with species or an ecosystem better adapted to the new climate. A further complicating factor is the impact of irreversible or at least prolonged deforestation by humans. This not only serves as a direct source of CO_2 to the atmosphere, but also reduces the future impact of possible CO_2-induced enhancements of photosynthesis by reducing the mass and extent of forest biomes.

In spite of the development of several more sophisticated terrestrial biosphere models, simple, regionally disaggregated box models can play a useful role as part of a larger, climate–carbon cycle modeling framework used to assess the impact of alternative future greenhouse gas emissions scenarios. Relatively simple biosphere models, coupled to models of ocean carbon uptake, are also needed in the analysis of atmospheric and oceanic ^{13}C and ^{14}C isotope data, in which exchange with long-lived soil carbon reservoirs is important.

19.2 A Hierarchy of Terrestrial Biosphere Models

At least four categories of terrestrial biosphere models have been developed that can be applied to questions of climate–biosphere interaction:

1. At the plant-to-hectare spatial scale and year-to-century time scale are ecophysiological models that explicitly parameterize gross photosynthesis, canopy respiration, and stem and root respiration at varying levels of detail, with coupling to modeled transpiration and canopy and possibly soil temperatures. An example is the FOREST-BGC model of Running and Coughlan (1988), which, in addition to the processes enumerated above, simulates C allocation above- and belowground, litterfall, decomposition and N mineralization, rainfall interception, evaporation, and transpiration at the forest canopy scale on a daily time step. Intermediate quantities that are calculated include stomatal resistance, foliage temperature, and soil moisture content. Bonan (1991) has developed a similar model and applied it to boreal forests. The Century model of Parton and colleagues (1987), which was initially developed for grasslands, adds further detail by distinguishing among structural, metabolic, and live plant C, N, and P, and between structural and nonstructural detritus, with the decay rate of structural detritus dependent on lignin content. FOREST-BGC has recently been expanded to cover a range of ecosystem types (BIOME-BGC; Running and Hunt, 1993). Unlike the gap models described next, species composition is not predicted but is a given.

2. Also at the tree-to-hectare spatial scale and year-to-century time scale are models that simulate the age distribution, species composition, and biomass of forest stands based on the relative and absolute temperature and moisture tolerances of individual species, mortality rate as a function of age, and frequency of disturbances, such as fires and wind throw, which open up space for competition. These forest gap models include a module that simulates plant demographics, and they contain biophysical modules that simulate the transfer of solar radiation based on the simulated vertical structure of the canopy, ground surface and subsurface temperature, and surface hydrology and evapotranspiration. Monthly to annual time steps are used. Plant growth is determined based on simple relationships involving accumulated degree days and drought days; photosynthesis and respiration are not explicitly represented. As discussed by Malanson (1993), the temperature–growth relationships for individual species are defined as a simple parabola, with end points at the present-day limits in growing degree days. This will lead to sig-

nificant errors in the predicted response to climatic change to the extent that present-day species range limits are determined by competition (or other factors such as soils) rather than by climate. Cook and Cole (1991) present tree ring evidence that at least some modeled species do not respond correctly to projected climatic change. In early applications of these models to climatic change, nutrient dynamics were not included (i.e., the FORET model of Solomon, 1986), while in later studies a feedback loop involving plant species, litter quality, soil nutrient availability, and plant growth was added (i.e., the LINKAGES model of Pastor and Post, 1988). This feedback loop has been found to alter significantly the modeled biomass response to changes in climate in some circumstances. The gap model paradigm has recently been extended to nonforest ecosystems, such as STEPPE for grasslands (Coffin and Lauenroth, 1990).

3. At the forest stand-to-biome spatial scale and century-to-millennium time scale are classification-based models in which the distribution of biomes for a given climate is determined based on the distribution of a number (typically two) of climatic variables. For each combination of climatic variables, a given biome is assumed to occur. Given the distribution of climatic variables used in the classification system for the present climate and for some future climate, the present and future distribution of biomes can be determined. The above- and belowground carbon density (kg C/m²) of each biome combined with the present and future areal extent of each biome can be used to deduce the change in total stored carbon. This approach has been used by a number of researchers. Prentice and Fung (1990), using the Goddard Institute for Space Studies general circulation model (GCM), concluded that the terrestrial biosphere would become a net sink for 235 Gt C under doubled atmospheric CO_2 conditions, primarily because of a projected increase in the extent of tropical forests resulting from precipitation increases. Schlesinger (1990) concluded that the terrestrial biosphere would release 45 Gt C based on the vegetation distribution under a doubled CO_2 climate given by Emanuel and colleagues (1985). Dixon and Turner (1991) concluded that soils could range from a net sink of 41 Gt C to a net source of 101 Gt C, based on doubled CO_2 climates as simulated by four GCMs. Using the same four GCM simulations of doubled CO_2, Smith and colleagues (1992) concluded that terrestrial carbon storage would increase by 8.5–180.5 Gt.

One obvious weakness of this approach is that it applies only to steady-state conditions, when the vegetation has fully adjusted to the new climate (assuming an unchanging climate persists long enough for this adjustment to occur), and therefore neglects potentially significant fluxes during the transient. Another weakness is that, in the above-mentioned studies, the influence of other factors, such as soil type and human interference, is ignored. The direct effects of higher CO_2 concentration in stimulating net photosynthesis and of higher temperature in increasing respiration rates are also ignored.

4. At the regional-to-global scale are models in which entire ecosystems or the entire terrestrial biosphere are either treated as a single box within a larger carbon cycle model (Oeschger et al., 1975; Siegenthaler and Oeschger, 1978; Laurmann and Spreiter, 1983), divided into two boxes with different turnover times (Machta, 1972; Bacastow and Keeling, 1973; Keeling and Bacastow, 1977; Revelle and Munk, 1977; Enting and Pearman, 1987), or divided into three to six boxes representing such components as woody tree parts, nonwoody tree parts, detritus, and soil carbon (Bolin, 1981; Svirezhev and Tarko, 1981; Emanuel et al., 1981, 1984; Goudriaan and Ketner, 1984; Kohlmaier et al., 1987; Harvey 1989a,b).

Simulation results obtained with the first three groups of models in the above hierarchy provide guidance in the formulation of feedback relationships for use in box models, whether globally aggregated or applied at individual grid points within a geographical distribution of vegetation types. Several studies indicate the importance of accounting for plant-nutrient-soil feedbacks in at least a rudimentary manner in order to simulate the effects on carbon storage of both temperature and atmospheric CO_2 concentration increases. For example, a temperature increase causes a decrease in soil organic matter content resulting from an increase in respiration rate. This in turn can cause a short-term increase in N availability and a transient stimulation of net primary productivity (NPP), which would partly offset the decrease of soil carbon. McGuire and colleagues (1992) conclude that N–C coupling can turn a decrease in NPP resulting from higher plant respiration following a 2°C temperature increase, which occurs in the absence of N cycle feedback, into an increase in NPP. Pastor and Post (1988) also conclude that increased N availability in a warmer climate would lead to increased productivity, if water is not limiting. On the other hand, Perry and colleagues (1991) argue that the potential increase in carbon storage resulting from a net release of N from more rapid decomposition of soil organic matter would not be fully realized because (1) there would be a shift from belowground to aboveground growth, as plants tend to adjust their shoot/root ratio to maintain the correct C/N balance, and (2) other factors would likely become limiting before all the extra N was used.

Turning from temperature feedback to CO_2 fertilization feedback, the stimulation of photosynthesis is believed to increase the C/N ration in plant litter, which in turn tends to reduce the availability of N for plants by reducing decomposition rates, thereby eventually reducing the photosynthetic response of plants to higher CO_2. Furthermore, Post and colleagues (1992) find, using a forest gap model, that the increase in whole ecosystem C storage that occurs if higher CO_2 is assumed to lead to increased fine root production is about half the increase that occurs if higher CO_2 is assumed to lead to greater aboveground growth. In an experimental field study in which yellow poplar saplings were grown at elevated CO_2 concentration for three years, Norby and colleagues (1992) found that there is no increase in whole-plant C storage despite a sustained increase in leaf-level photosynthesis and a decrease in foliar respiration. This is a result of an increase in the production of fine roots, which are short lived.

19.3 Formulation of Simple Box Models

Given a carbon reservoir i, containing an amount of carbon B_i, the flux of carbon from reservoir i to reservoir j as a result of a

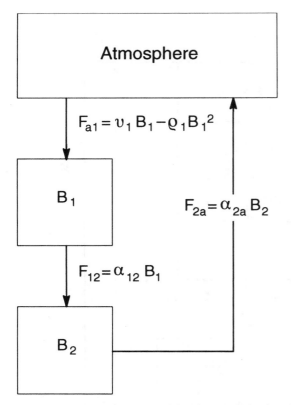

Figure 19.1: Illustration of a two-box biosphere model, where B's refer to carbon masses and F's refer to carbon fluxes. See text for explanation.

given process (such as respiration, or transformation to detritus or to soil carbon) is assumed in box models to be linearly proportional to the amount of carbon in reservoir i. That is, $F_{ij} = \alpha_{ij}B_i$. A slightly more sophisticated treatment is required in order to represent the carbon flux from the atmosphere to living biomass as a result of photosynthesis. In this case, a logistic-type equation is appropriate because it allows for initially rapid growth of biomass, but at a decreasing rate as the steady-state biomass is asymptotically approached. Using a logistic-type equation, the photosynthetic flux is given by $F_{ai} = v_iB_i - \rho_iB_i^2$, with subscript a referring to the atmosphere. The rate of change of carbon in each box is given by the sum of the fluxes into the box minus the sum of fluxes out of the box. This leads to a series of coupled ordinary differential equations.

To illustrate these concepts, consider a two-box model consisting of a living biomass box and a soil carbon and detritus box, as shown in Figure 19.1. The governing differential equations are

$$\frac{dB_1}{dt} = (v_1 - \alpha_{12})B_1 - \rho_1B_1^2 \tag{19.1}$$

$$\frac{dB_2}{dt} = \alpha_{12}B_1 - \alpha_{2a}B_2 \tag{19.2}$$

where B_1 is the biomass box and B_2 is the detritus/soil box. For small B_1, B_1 grows exponentially with a time constant of $(v_1 -$

$\alpha_{12})^{-1}$. Alternative formulations of the photosynthetic flux are possible, such as a saturation-type equation, whereby $F_{ai} = F_{max}(1 - \exp(-kB_i))$. For small kB_i, this gives the same governing equation as Equation 19.1, with $F_{max}k = v_i$ and $F_{max}k^2/2 = \rho_i$. The steady-state carbon contents can be easily inferred by setting the derivatives in Equations (19.1) and (19.2) to zero; they are

$$B_1 = \left(\frac{v_1 - \alpha_{12}}{\rho_1}\right) \tag{19.3}$$

$$B_2 = \left(\frac{\alpha_{12}}{\alpha_{2a}}\right) B_1 \tag{19.4}$$

Given estimates of the amounts of carbon in each box at present, or prior to human disturbance, and of the corresponding yearly carbon fluxes, the transfer coefficients α_{ij} can be readily computed as F_{ij}/B_i. v_i can then be chosen to give the desired time constant for biomass growth, and ρ_i can be chosen to give the desired steady-state biomass.

The assumption that the undisturbed terrestrial biosphere will tend to a steady state, as embodied in the above equations, is most likely not correct. Post and colleagues (1992) suggest that peat could be accumulating at a rate of 0.14 Gt C/yr and that an additional 0.45 Gt C/yr might be flowing from the terrestrial biosphere to oceans. These possible fluxes are small compared to the gross photosynthesis and respiration fluxes (on the order of 100 Gt C/yr) and could be handled by adding another, extremely long-lived carbon reservoir into which a small proportion of net photosynthesis flows.

Figure 19.2 illustrates a six-box biosphere model, consisting of boxes representing ground vegetation, nonwoody tree parts, woody tree parts, detritus, and rapidly and slowly overturning soil reservoirs. Since the transfer coefficients for box models are chosen such that the governing differential equations will converge to a steady-state solution with a predetermined mass in each box and predetermined fluxes between boxes, no new information is provided by the model steady-state solution. However, box models can be used to assess carbon fluxes following a disturbance (if any of the B_i are altered without changing transfer coefficients) or as a result of changes in the transfer coefficients as a result of changes in atmospheric CO_2 concentration or temperature.

19.4 Incorporation of CO_2 and Temperature Feedbacks

As indicated above, ecosystem models that include the coupling between nutrient dynamics and photosynthesis and/or that distinguish between aboveground and belowground production (which have different turnover rates) yield a much smaller relative increase in whole-plant or whole-ecosystem carbon storage than the relative increase in photosynthesis rate, even if the photosynthesis increase is sustained. Nutrient feedbacks might also partially reduce the effect on total carbon storage of temperature increases, at least during the transient. A decrease in soil moisture will also reduce the effect of warmer temperatures in increasing soil respiration. Evidence

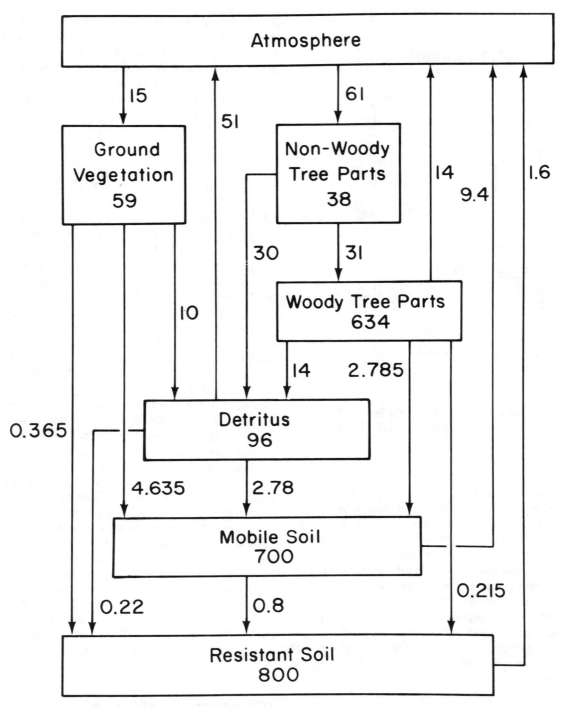

Figure 19.2: The six-box model used by Harvey (1989a,b). Numbers inside boxes refer to carbon masses (Gt), while numbers next to arrows refer to carbon fluxes (Gt C/yr). Reprinted with permission from L. D. D. Harvey, *Global Biogeochemical Cycles* 3: 137–53. © 1989, American Geophysical Union.

has recently emerged that elevated CO_2 also reduces dark respiration (Bunce, 1990). Since none of these factors is explicitly modeled in box models, they must be accounted for in the formulation of CO_2 and temperature feedbacks in the biosphere model and in the choice of the parameters that determine the strength of these feedbacks.

In the biosphere model of Harvey (1989b), a CO_2 fertilization effect is incorporated by multiplying v_i by a factor of

$$1 + \gamma = 1 + \beta \left(\frac{\ln CO_2(t)}{\ln CO_2(0)} \right) \qquad (19.5)$$

where $CO_2(t)$ is current CO_2 concentration, $CO_2(0)$ is the initial CO_2 concentration, and β is a direct fertilization factor for photosynthesis, estimated in various ways to be approximately 0.3–0.6 (Harvey, 1989b, 1991). The net effect on steady-state carbon storage in the corresponding box can be controlled through the manner in which ρ_i is altered in response to higher CO_2. If ρ_i is also multiplied by a factor of $1 + \gamma$, the steady-state biomass increases by a factor greater than $1 + \gamma$, as can be seen from Equation (19.3). If one postulates that steady-state biomass is multiplied by a factor R_B, then ρ_i must be multiplied by a factor of $1 + \zeta_i$, where

$$\xi_i = \frac{\upsilon_i(1 + \gamma - R_B) - (1 - R_B)\Sigma\alpha_{ij}}{R_B(\upsilon_i - \Sigma\alpha_{ij})} \tag{19.6}$$

If temperatures are below the optimum value for photosynthesis, then warming will lead to an increase in photosynthesis rate. As noted above, warming could also increase photosynthesis rates by causing a net release of N from soil organic matter. As discussed in Harvey (1989b), warming is also likely to increase the competitiveness of plant roots relative to microorganisms in sequestering a given supply of available soil N. Therefore, it is reasonable to postulate a temperature enhancement of photosynthesis rates, in addition to a CO_2 fertilization, although the magnitude of this effect is uncertain. It is also clear that the sign of the feedback must eventually reverse as temperatures increase above the optimal temperature for photosynthesis (which is species-dependent).

A common method of incorporating temperature effects on rate coefficients is to assume that the coefficient is multiplied by the factor

$$(Q_{10})^{(T - T_0)/10}$$

where Q_{10} is the factor by which the coefficient increases per $10°C$ warming, T is current temperature, and T_0 is initial temperature. In Harvey (1989b), the υ_i coefficient for ground vegetation is assumed to increase with temperature using a Q_{10} value of 1.4; υ_i for nonwoody tree parts is assumed to increase with a Q_{10} of 1.53. (This value, combined with a Q_{10} for respiration of woody tree parts of 2.0, gives a NPP Q_{10} of 1.4 for woody vegetation, thereby matching the value used for ground vegetation.) These temperature-dependent increases are in addition to CO_2 fertilization effects.

Reported Q_{10} values for plant maintenance and soil respiration range from 0.8 to 3.0, with most values equal to 2.0 or greater. Given possible CO_2 inhibition of maintenance respiration and water limitations on increases in soil respiration over large areas as soil moisture decreases in response to warming, values toward the low end of the reported range are likely to be most appropriate. In Harvey (1989a), a Q_{10} of 2.0 was adopted for both maintenance and soil respiration. The transformation of detritus to soil organic matter, and of transferring carbon from rapidly to slowly overturning soil reservoirs, is dependent at least in part on respiration by microorganisms and so should plausibly increase with increasing temperature as well. Finally, translocation from nonwoody to woody tree parts is assumed to increase with a Q_{10} of 2.0 in Harvey (1989a), based on a range of values reported in the literature.

The representation of feedbacks between temperature and photosynthetic and respiration fluxes by a Q_{10} formulation in box models is a simple parameterization that embodies a number of complex interactions, which are explicitly simulated to varying degrees by more complex models. Appropriate values must therefore be determined by comparison of results of box models and more complex models, and by comparison with observations. Furthermore, the derived values should be used with caution if applied to temperature changes or CO_2 concentrations that are significantly different from those used to derive the Q_{10} values.

To summarize, in the six-box model of Harvey (1989a), the transfer coefficients for fluxes from soil boxes to the atmosphere, between soil boxes, from detritus to the atmosphere and soil, and from woody tree parts to the atmosphere are all assumed to increase with temperature, while photosynthetic fluxes are assumed to increase with both temperature and CO_2 concentration. The only transfer coefficients not directly affected by CO_2 or temperature increases are those governing fluxes from biomass to soil and detritus, although these fluxes will increase if the biomass increases.

19.5 Incorporation of Fluxes Resulting from Land-Use Changes

The net CO_2 flux to the atmosphere in any given year as a result of deforestation involves the following components: emission from deforestation in that year, primarily as a result of biomass burning; emission from deforestation in previous years as a result of decay of detritus not consumed at the time of deforestation, accelerated oxidation of soil organic matter, and decomposition of products harvested from the forest; absorption of CO_2 from regrowth of vegetation on previously deforested land that has been abandoned; and absorption of CO_2 from the reaccumulation of soil organic matter after abandonment. Calculation of the net emission from deforestation in any given year requires a data base consisting of the areas deforested and abandoned in all preceding years that still affect emissions in the year in question, the initial above- and belowground carbon densities, the fate and turnover time of forest products, and response curves for carbon loss following deforestation and carbon accumulation following abandonment. Houghton and colleagues (1983) developed such a data base, with carbon contents and response curves for 10 different ecosystems and rates of deforestation from conversion to cropland, pastureland, or harvesting of forest products in 10 different regions. They computed a net emission in 1980 of 0.5–4.5 Gt C; however, in their most recent analysis (Houghton, 1991) the 1980 net emission was estimated to be 1.0–2.0 Gt C and to have grown to 1.50–3.0 Gt C by 1990. Harvey (1989b) analyzed the sources of uncertainty in the latest estimates available at the time and concluded that, for a net emission in 1980 of 1.0 Gt C the gross emission would have been 2.3 Gt C, while for a net 1980 emission of 2.5 Gt C the gross emission would have been 4.5 Gt C. Hence, gross emission is roughly twice the net emission, which implies that gross absorption is approximately equal to net emission. Thus, as discussed in Harvey (1989b), a given percent reduction in the gross emission rate will have approximately twice the relative effect on net emissions in the short term (a few decades), until the rate of carbon absorption resulting from regrowth on abandoned land itself decreases. Reducing the gross emission rate roughly in half would reduce net emissions to near zero.

Change in biomass resulting from land-use changes can be readily incorporated in the globally aggregated box model described above, either by directly specifying the net land-use flux or by specifying gross fluxes. The latter method was used by Harvey (1989b, 1990) so that scenarios could be constructed using input assumptions that are readily interpretable by policy makers. In particular, Harvey (1989b) specified changes in the gross fluxes resulting from conversion of forest to cropland, pastureland, and to harvested forest, the first two of which were regarded as permanent. A simple variation in the regrowth sink was assumed such that the regrowth sink equals the emission source that results from the harvesting of forests and decay of forest products 50 years after a constant flux that results from harvesting is reached. This implicitly assumes that forest harvesting is practiced on a sustainable basis, with forest biomass returning to the same amount prior to each successive harvest.

The gross emission fluxes obtained as described above are combined to give the net flux, which is then subtracted from each box in proportion to the amount of carbon already in the box; this is an admittedly crude procedure, but it is justified given the uncertainty concerning the current net flux. Since the carbon removed from the boxes already includes the effect of regrowth, where it is assumed to occur, the steady-state carbon content of the ground vegetation and nonwoody tree part boxes must be permanently reduced to prevent the masses in these boxes from rebounding. This requires dividing ρ_i by the same factor by which the box masses are multiplied as a result of the net land-use emission. The ρ_i are multiplied by additional factors to represent the effect of CO_2 fertilization, as described above.

19.6 Summary of Results Obtained with Globally Aggregated Box Models

Harvey (1989a) examined the effect of eight different model structures on the response of simple box models of the terrestrial biosphere to temperature and CO_2 fertilization feedbacks. Here, the differences obtained between Harvey's (1989a) Models 4, 6, and 7 (shown in Figure 19.3) are highlighted. These models differ in that there is a single soil reservoir in Model 4 and two soil reservoirs in Models 6 and 7, but with different pathways to the slowly overturning soil reservoir. The key insights obtained by comparing these three models are as follows:

- Dividing the soil reservoir into slowly and rapidly overturning components has no effect on the steady-state response to temperature-induced respiration increases if all the carbon that enters the slowly overturning reservoir enters directly from the biomass and detritus reservoirs (Model 7), whereas the soil carbon loss in response to increased temperature almost doubles if all the carbon entering the slowly overturning reservoir first passes through the rapidly overturning reservoir (Model 6) and if the soil-to-soil transfer coefficient is fixed.

- The transient response of total soil carbon in both models with two soil reservoirs is significantly slower than the transient response of Model 4.

- If the coefficient for transfer of carbon from rapidly to slowly overturning reservoirs increases in parallel with the increase of respiration coefficient, the choice of carbon pathway from biomass and detritus to slowly overturning soil carbon has no effect on the steady-state response, with the response of Model 6 becoming equal to that of Models 4 and 7. (Soil carbon loss decreases if the soil-to-soil coefficient increases with increasing temperature in Model 6.)

- For the case in which the soil-to-soil transfer coefficient increases with temperature, there is a small difference between the transient response of Models 6 and 7; however, this difference is small compared to the difference between either transient response and that of Model 4.

- Different combinations of temperature and CO_2 fertilization feedback that give comparable changes in total carbon storage up to the present can give dramatically different changes in the future.

Comparisons across the full set of models considered in Harvey (1989a) indicate that the effect of alternative model structures on the response of total carbon to temperature and CO_2 increases can be as large as the effect of varying β from 0.0 to 0.2 or varying respiration Q_{10} from 2.0 to 3.0. As summarized in Harvey (1989a), "Model structure, along with the functional relationships linking photosynthesis, respiration, and internal carbon transfers to atmospheric CO_2 and temperature, is an important source of uncertainty in carbon cycle models" (p. 150).

19.7 Hybrid Models and a Proposal for Future Research

Classification-based and box models can be linked, as illustrated by Esser (1987), Janecek and colleagues (1991), Raich and colleagues (1991), and Smith and Shugart (1993). Janecek and colleagues (1991) applied a simple box model consisting of two biomass boxes and a soil box to each grid point on a $1° \times 1°$ (latitude–longitude) global grid. One of 29 different vegetation types was assigned to each grid point. Carbon assimilation was computed as a function of light, temperature, and soil moisture; respiration was a function of temperature and amount of biomass; and decomposition rate was a function of temperature and soil moisture content. These functions differed between different vegetation types. The model has been used to simulate the present-day geographical distribution of seasonal carbon fluxes when forced by present-day seasonal temperatures and precipitation. Raich and colleagues (1991) used a similar approach for vegetation types on a $0.5° \times 0.5°$ grid covering South America. The model structure in their case consisted of two boxes, living vegetation and detritus/soil, but they computed the amount of organic nitrogen and carbon amounts in each box, along with the available inorganic nitrogen in the soil. This permitted coupling of the carbon and ni-

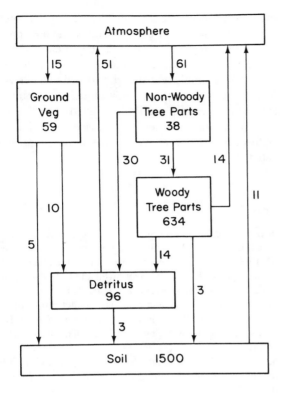

Model 4

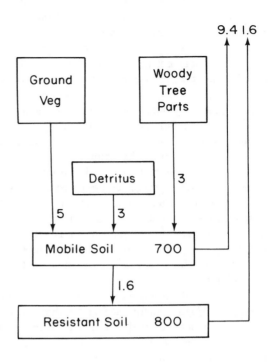

Model 6

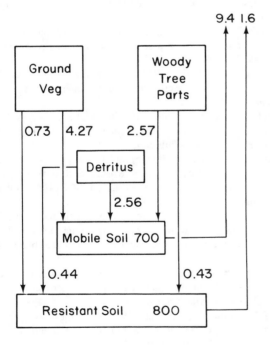

Model 7

Figure 19.3: (A) Model version 4 and the soil components and fluxes to and from soil components for (B) Model 6 and (C) Model 7 of Harvey (1989a). The other fluxes in Models 6 and 7 are the same as for Model 4. Reprinted with permission from L. D. D. Harvey, *Global Biogeochemical Cycles* 3: 137–53. © 1989, American Geophysical Union.

trogen cycles within a simple framework and on a continental scale.

Smith and Shugart (1993) attempted to overcome the inability of a classification-based approach to give transient carbon fluxes following a change of climate by assigning a two-box model to each grid point on a 0.5° × 0.5° global grid, where one box represented aboveground biomass and the other

box represented soil carbon. Steady-state vegetation distributions for current and future GCM-derived climates were determined based on the Holdridge classification, with steady-state above- and belowground carbon amounts assigned to each vegetation type. During the transition from one climate to the next, Smith and Shugart (1993) assumed the following changes for aboveground boxes: (1) that 0.02 of the initial carbon associated with each original vegetation type is lost per year as a result of dieback at grid points where dieback was considered likely to occur; (2) that 0.004 of the initial carbon is lost each year but is replaced by 0.004 of the carbon associated with the new steady-state vegetation type as a result of successional replacement at those grid points where successional replacement was not expected to be limited by migration rates; and (3) that 0.004 of the initial carbon is lost each year but is replaced by 0.001 of the carbon associated with the new steady-state vegetation type as a result of successional replacement at those grid points where successional replacement was expected to be limited by migration rates. For belowground carbon, the rate of loss of soil carbon at grid points where the steady-state carbon content decreases was set at 0.02 per year of the difference between initial and perturbed steady-state carbon amounts, while the rate of increase associated with new vegetation communities was set at the corresponding vegetation replacement rate (either 0.001 or 0.004 per year), thus ignoring likely lags between vegetation and soil carbon development.

When the hybrid Holdridge-box model was applied to GCM-derived climatic change scenarios, the terrestrial biosphere served as a net source of carbon to the atmosphere, with cumulative emissions of up to 200 Gt C 100–150 years after the climatic change was applied.

These preliminary results need to be refined in the following ways: (1) by assuming a gradual shift in the "target" steady-state vegetation distributions, corresponding to a gradual rather than step function change in climate; (2) by allowing for hypothetical CO_2 fertilization effects; and (3) by allowing for a hypothetical temperature–respiration feedback. All these effects could be readily incorporated by applying Harvey's (1989a) six-box biosphere model at individual grid points. For each vegetation type an appropriate steady-state partitioning of carbon among the various boxes, flux coefficients, and appropriate time constants for photosynthetic growth would be applied. A continuously changing steady-state biomass at each grid point could be incorporated by continuously changing ρ; CO_2 and temperature feedbacks could be incorporated as described above with an additional constraint on soil temperature–respiration feedback based on model-projected (or -deduced) changes in soil moisture. As a further step, N dynamics could be added based on the formulation of Raich and colleagues (1991).

19.8 Conclusion

Vegetation climate feedbacks could play a potentially important role in determining future atmospheric CO_2 concentrations. These feedbacks include (1) transient emissions (of up to 2 Gt C/yr) resulting from forest dieback and the lag between decline of maladapted species and establishment of species better adapted to the new (but changing) climate; (2) direct CO_2 effects on photosynthesis and respiration; and (3) temperature effects on photosynthesis and respiration, both by plants and in soils. As well, land-use changes could continue to contribute a significant CO_2 flux to the atmosphere.

Simple box models of the globally aggregated biosphere have provided a number of useful insights and are appropriate for coupling to simple models of the complete carbon cycle. Incorporation in box models of the major nutrient feedbacks can be done in a simple and straightforward manner. A useful next step would be to use multibox models with nutrient dynamics within a geographically distributed classification system to compute transient net fluxes associated with all three sets of feedbacks listed above for scenarios of transient climatic change.

References

Bacastow, R., and C. D. Keeling. 1973. Atmospheric carbon dioxide and radiocarbon in the natural carbon cycle: II. Changes from A. D. 1700 to 2070 as deduced from a geochemical model. In *Carbon and the Biosphere,* G. M. Woodwell and E. V. Pecan, eds. U.S. Atomic Energy Commission, Washington, D.C., pp. 86–135.

Bolin, B. 1981. Steady state and response characteristics of a simple model of the carbon cycle. In *Carbon Cycle Modelling,* B. Bolin, ed. SCOPE 16, John Wiley and Sons, Chichester, U.K., pp. 315–331.

Bonan, G. B. 1991. Atmosphere–biosphere exchange of carbon dioxide in boreal forests. *Journal of Geophysical Research 96,* 7301–7312.

Bunce, J. A. 1990. Short and long term inhibition of respiratory carbon dioxide efflux by elevated carbon dioxide. *Annals of Botany 65,* 637–642.

Coffin, D. P., and W. K. Lauenroth. 1990. A gap dynamics simulation model of succession in a semiarid grassland. *Ecological Modelling 49,* 229–266.

Cook, E. R., and J. Cole. 1991. On predicting the response of forests in eastern North America to future climatic change. *Climatic Change 19,* 271–282.

Dixon, R. K., and D. P. Turner. 1991. The global carbon cycle and climate change: Responses and feedbacks from below ground systems. *Environmental Pollution 73,* 245–262.

Emanuel, W. R., G. G. Killough, and J. S. Olson. 1981. Modelling the circulation of carbon in the world's terrestrial ecosystems. In *Carbon Cycle Modelling,* B. Bolin, ed. SCOPE 16, John Wiley, and Sons Chichester, U.K., pp. 335–353.

Emanuel, W. R., G. G., Killough, W. M. Post, and H. H. Shugart. 1984. Modeling terrestrial ecosystems in the global carbon cycle with shifts in carbon storage capacity by land use change. *Ecology 65,* 970–983.

Emanuel, W. R., H. H. Shugart, and M. P. Stevenson. 1985. Climatic change and the broad-scale distribution of terrestrial ecosystem complexes. *Climatic Change 7,* 457–460.

Enting, I. G., and G. I. Pearman. 1987. Description of a one-dimensional carbon cycle model calibrated using techniques of constrained inversion. *Tellus 39B,* 459–476.

Esser, G. 1987. Sensitivity of global carbon pools and fluxes to human and potential climatic impacts. *Tellus 39B,* 245–260.

Goudriaan, J., and P. Ketner. 1984. A simulation study for the global carbon cycle, including Man's impact on the biosphere. *Climatic Change 6,* 167–192.

Harvey, L. D. D. 1989a. Effect of model structure on the response of terrestrial biosphere models to CO_2 and temperature increases. *Global Biogeochemical Cycles 3,* 137–153.

Harvey, L. D. D. 1989b. Managing atmospheric CO_2. *Climatic Change 15,* 343–381.

Harvey, L. D. D. 1990. Managing atmospheric CO_2: Policy implications. *Energy 15*, 91–104.

Harvey, L. D. D. 1991. Comments on "The aerial fertilization effect of CO_2 and its implications for global carbon cycling and maximum greenhouse warming," by S. B. Idso *Bulletin of the American Meteorological Society 72*, 1905–1907.

Houghton, R. A. 1991. Tropical deforestation and atmospheric carbon dioxide. *Climatic Change 19*, 99–118.

Houghton, R. A., J. E. Hobbie, J. M. Melillo, B. Moore, B. J. Peterson, G. R. Shaver, and G. M. Woodwell. 1983. Changes in the carbon content of terrestrial biota and soils between 1860 and 1980: A net release of CO_2 to the atmosphere. *Ecological Monographs 53*, 235–262.

Janecek, A., M. K. B. Lüdeke, J. Kindermann, T. Lang, A. Klaudius, R. D. Otto, F.-W. Badeck, G. H. Kohlmaier. 1991. A global high resolution mechanistic and prognostic model for the seasonal and long term CO_2 exchange between the terrestrial ecosystems and the atmosphere: The Frankfurt Biosphere Model (FBM). Presented at the 3rd International Environmental Chemistry Congress in Brazil, Salvador-Bahia, September 30–October 4, 1991 (submitted to *Science of the Total Environment*).

Keeling, C. D., and R. B. Bacastow. 1977. Impact of industrial gases on climate. In *Energy and Climate* (Geophysics Study Committee), National Academy of Sciences, Washington, D.C., pp. 72–95.

Kohlmaier, G. H., H. Brohl, E. O. Sire, M. Plochl, and R. Revelle. 1987. Modelling stimulation of plants and ecosystem response to present levels of excess atmospheric CO_2. *Tellus 39B*, 155–170.

Laurmann, J. A., and J. R. Spreiter. 1983. The effects of carbon cycle model error in calculating future atmospheric carbon dioxide levels. *Climatic Change 5*, 145–181.

Machta, L. 1972. The role of the oceans and the biosphere in the carbon dioxide cycle. In *The Changing Chemistry of the Oceans*, D. Dryssen and D. Jagner, eds. Almqvist and Wiksell, Stockholm, Sweden, pp. 121–146.

Malanson, G. P. 1993. Comment on modeling ecological response to climatic change. *Climatic Change 23*, 95–109.

McGuire, A. D., J. M. Melillo, L. A. Joyce, D. W. Kicklighter, A. L. Grace, B. Moore III, and C. J. Vorosmarty. 1992. Interactions between carbon and nitrogen dynamics in estimating net primary productivity for potential vegetation in North America. *Global Biogeochemical Cycles 6*, 101–124.

Norby, R. J., C. A. Gunderson, S. D. Wullschleger, E. G. O'Neill, and M. K. MacCracken. 1992. Productivity and compensatory response of yellow-poplar trees in elevated CO_2. *Nature 357*, 322–324.

Oeschger, H., U. Siegenthaler, U. Schotterer, and A. Gugelmann. 1975. A box diffusion model to study the carbon dioxide exchange in nature. *Tellus 27*, 168–192.

Parton, W. J., D. S. Schimel, C. V. Cole, and D. S. Ojima. 1987. Analysis of factors controlling organic matter levels in Great Plains grasslands. *Soil Science Society of America 51*, 1173–1179.

Pastor, J., and W. M. Post. 1988. Response of northern forests to CO_2-induced climatic change: Dependence on soil water and nitrogen availabilities. *Nature 334*, 55–58.

Perry, D. A., J. G. Borchers, D. P. Turner, S. V. Gregory, C. R. Perry, R. K. Dixon, S. C. Hart, B. Kauffman, R. P. Neilson, and P. Sollins. 1991. Biological feedbacks to climate change: Terrestrial ecosystems as sinks and sources of carbon and nitrogen. *The Northwest Environmental Journal 7*, 203–232.

Post, W. M., J. Pastor, A. W. King, and W. R. Emanuel. 1992. Aspects of the interaction between vegetation and soil under global change. *Water, Air, and Soil Pollution 64*, 345–363.

Prentice, K. C., and I. Y. Fung. 1990. The sensitivity of terrestrial carbon storage to climate change. *Nature 346*, 48–51.

Raich, J. W., E. B. Rastetter, J. M. Melillo, D. W. Kicklighter, P. A. Steudler, B. J. Peterson, A. L. Grace, B. Moore III, and C. J. Vorosmarty. 1991. Potential net primary productivity in South America: Application of a global model. *Ecological Applications 1*, 399–429.

Revelle, R., and W. Munk. 1977. The carbon dioxide cycle and the biosphere. In *Energy and Climate* (Geophysics Study Committee), National Academy Press, Washington, D.C., pp. 140–158.

Running, S. W., and J. C. Coughlan. 1988. A general model of forest ecosystem processes for regional applications: I. Hydrological balance, canopy gas exchange and primary production processes. *Ecological Modelling 42*, 125–154.

Running, S. W., and E. R. Hunt. 1993. Generalization of a forest ecosystem process model for other biomes, BIOME-BGC, and an application for global-scale process models. In *Scaling Physiological Processes: Leaf to Globe*, J. R. Ehleringer and C. B. Fields, eds. Academic Press, San Diego, Calif., pp. 141–258.

Schlesinger, W. H. 1990. Evidence from chronosequence studies for a low carbon-storage potential of soils. *Nature 348*, 232–234.

Siegenthaler, U., and H. Oeschger. 1978. Predicting future atmospheric carbon dioxide levels. *Science 199*, 388–395.

Smith, T. M., R. Leemans, and H. H. Shugart. 1992. Sensitivity of terrestrial carbon storage to CO_2-induced climate change: Comparison of four scenarios based on general circulation models. *Climatic Change 21*, 367–384.

Smith, T. M., and H. H. Shugart. 1993. The transient response of terrestrial carbon storage to a perturbed climate. *Nature 361*, 523–526.

Solomon, A. M. 1986. Transient response modelling of forests to CO_2-induced climate change: Simulation modeling experiments in eastern North America. *Oecologia 68*, 567–579.

Svirezhev, Y. M., and A. M. Tarko. 1981. The global role of the biosphere in the stabilization of atmospheric CO_2 and temperature. In *Carbon Cycle Modelling*, B. Bolin, ed. SCOPE 16, John Wiley and Sons, Chichester, U.K., pp. 355–364.

20

Impacts of Climate and CO_2 on the Terrestrial Carbon Cycle

F. I. WOODWARD AND T. M. SMITH

Abstract

The characteristics of a new vegetation production model, integrated with a novel approach to nitrogen uptake from the soil, is described and assessed. The model shows an effective capacity to predict plant and vegetation processes scaling up from the leaf, through the canopy to the biome scale. When the model is used to assess the impact of CO$_2$ enrichment and global climate change, it becomes clear that global warming dominates the net primary production (NPP) responses of major biomes. In general terms, a global warming of 3°C and a 10% increase in precipitation is likely to cause reductions in NPP. However, this reduction might be offset by a doubling in atmospheric CO$_2$.

20.1 Introduction

The qualitative and quantitative natures of the past and current terrestrial carbon pools have been described by other contributors to this volume (Houghton, Gifford, Schlesinger, Hall, and Keeling) and recently by Sundquist (1993). This chapter aims to describe a global-scale use of plant physiological mechanisms for investigating the influences of changes in climate and CO$_2$ on photosynthesis, transpiration, and nitrogen uptake. These mechanisms can be scaled to indicate the changes in carbon fluxes to and from vegetation with a changing climate and atmospheric concentration of CO$_2$. These predictions are a necessary part of the future predictions of climate derived from general circulation models (GCMs), through the direct effects of vegetation on the fluxes of CO$_2$ and water vapor into the atmosphere, and the subsequent effects on the radiation balance of the Earth (Shine et al., 1990; Wigley and Raper, 1992; Schimel et al., 1996).

The described model is an incomplete reflection of nature; a number of important feedbacks that determine plant responses are not yet included. For example, not included at this stage is a treatment of the fluxes of litter and soil carbon and nitrogen associated with decomposition and immobilization, although nitrogen uptake to the plant is explicitly included. The absence of a complete nitrogen cycle may lead to greater sensitivities of plant responses to CO$_2$ enrichment than in a model with a more complete nitrogen cycle (e.g., Rastetter et al., 1992). However, models with well-represented soil processes often possess poor plant physiological mechanisms, indicating problems in interpreting predictions. The model presented here assumes that during model runs, the only significant change in soil processes occurs through the impact of climate on nitrogen uptake.

The strong influence of vegetation on climate has been clearly demonstrated (Shukla and Mintz, 1982), through changes in evapotranspiration; however, the influence of changes in the CO$_2$ fluxes from vegetation has not been considered with the same mechanistic detail. Much of the early carbon cycle modeling (Bolin, 1981) was based on the box model approach (Harvey, 1989), in which processes were defined as little more than inspired guesses of the likely mechanisms and rates. The need for quick and complete products for inclusion in GCMs has encouraged the proliferation of empirical, semiempirical, and correlational models of vegetation – the quick-fix approach – yet such approaches cannot be used in new conditions of climate and CO$_2$ (Woodward, 1987). The fact that many of these models can be defined by a small set of simple equations (e.g., Lieth, 1975; Uchijima and Seino, 1985; Gifford, 1993) makes them seductively attractive, but careful assessment of the approaches and the many simplifying assumptions should alert the user to significant problems (Gifford, 1993).

A problem in all the correlational models is the lack of any capacity to include directly the effects of future CO$_2$ enrichment on plant processes. These effects can, in turn, have a significant influence on the carbon and water cycles of vegetation (Woodward et al., 1991); therefore, a climate feedback is expected. This chapter describes a mechanistic approach to predicting the effects of climatic change and CO$_2$ enrichment on the CO$_2$ fluxes by plants and major vegetation types. These mechanisms are based, primarily, upon responses measured under controlled experiments.

Correlational and mechanistic approaches to flux predictions have different advantages and disadvantages. The former are generally based on field observations and have the major limitations of being unable to include the effects of CO$_2$ enrichment and to differentiate between the effects of different climatic variables. However, they have the benefit that complex ranges of processes and scales can be dealt with by small sets of simple relationships. This contrasts with a mechanistic approach, which would need to simulate a wide range of processes, such as plant photosynthesis, respiration, senescence, leaf fall, and leaf decomposition by mycorrhizas, saprophytic fungi, bacteria, and animals. This modeling also needs to be achieved for the major vegetation types of the world, for which mechanistic understanding and parameter values are often very restricted. Clearly, neither approach is satisfactory. However, the emphasis in this chapter is on a mechanistic, process-based approach where possible, on the grounds that this is the only technique for adequately incorporating the effects of elevated CO$_2$ on vegetation.

20.2 Modeling Approach

The processes of plant photosynthesis and respiration drive the biospheric carbon cycle. Therefore it makes logical sense to start with models of these processes, in order to predict the effects of changes in temperature and CO$_2$. The process of photosynthesis (in plants with the C$_3$ biochemical pathway) is well characterized, and one model (Farquhar et al., 1980) is generally used and very well validated against observations (Harley et al., 1992; McMurtrie and Wang, 1993; Wullschleger, 1993). The subsequent steps in modeling to reach a global coverage are less obvious, and also less well understood. One approach to this goal follows a sequence in which all steps have explicit responses to CO$_2$ and temperature:

1. Leaf photosynthesis and stomatal conductance
2. Nitrogen uptake from the soil

3.   Nitrogen partitioning within the leaf canopy
4.   Leaf photosynthesis and respiration through the leaf canopy
5.   Annual net primary productivity (NPP) and response to precipitation
6.   Major biome NPP
7.   Modeling NPP with whole-plant respiration

20.2.1 Leaf Photosynthesis and Stomatal Conductance

The Farquhar model (Farquhar et al., 1980) provides a well-defined mechanistic model for predicting photosynthetic and stomatal conductance responses of leaves to temperature and to CO_2. Recent papers by Harley and colleagues (1992) and McMurtrie and Wang (1993) provide all the necessary equations for coding the model for computer simulations. Wullschleger (1993) provides information from a wide range of species for investigating the responses of species from different vegetation types.

Actual measurements of the response of wheat leaf gas exchange to CO_2 (Figure 20.1) indicate the marked nonlinear response of photosynthesis to changes in the intercellular CO_2 partial pressure (i.e., the mean CO_2 partial pressure in the air spaces of the leaf). The slope of the line labeled $-g_s$ is the stomatal conductance times -1. The intercepts of the two $-g_s$ lines with the x-axis indicate the ambient CO_2 partial pressure of the particular measurement. The slope is steeper for the approximately current ambient partial pressure of 35 Pa than for a partial pressure of 56 Pa (the partial pressure predicted for around the year 2050 by Wigley and Raper, 1992). The steeper slope at 35 Pa indicates that the stomatal conductance decreases with increasing CO_2, while the photosynthetic rate increases (26%). Therefore leaves under CO_2 enrichment are predicted to be more water use–efficient than at the current

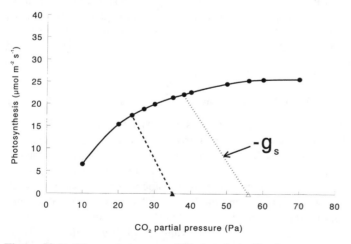

Figure 20.1: Observed responses (filled circles) of leaf photosynthesis to the intercellular CO_2 partial pressure. Ambient CO_2 partial pressure (filled triangle, open triangle) intercepts (35 and 56 Pa) connect to main curve (solid line) by lines with slopes that are -1 times the stomatal conductance (g_s). In all cases throughout the chapter, actual observation points are indicated by symbols, and modeled output is shown as continuous lines.

CO_2 concentration (Morison, 1985). The increase in photosynthesis and water use efficiency and the decrease in stomatal conductance with CO_2 partial pressure are generally observed in experiments (Bazzaz, 1990; Woodward et al., 1991; Woodward, 1992).

The influence of increasing temperature on photosynthesis depends on the absolute temperature (Slatyer, 1977), such that at suboptimal temperatures a 3°C increase in temperature could lead to as much as a 40% increase in photosynthetic rate. In contrast, at high temperatures (greater than approximately 25–30°C) an increase in temperature can cause a reduced rate of photosynthesis, because the photorespiratory rate has a higher temperature coefficient than photosynthesis (Jordan and Ogren, 1984; Brooks and Farquhar, 1985). This negative effect of temperature is reduced by an increase in CO_2 partial pressure, which enhances the rate of photosynthesis relative to that of photorespiration (Harley et al., 1992; McMurtrie and Wang, 1993).

20.2.2 Nitrogen Uptake from the Soil

The response of leaf CO_2 exchange to temperature and CO_2 is well understood and characterized. However, the leaf is considered in isolation with no account taken of the rate of nutrient uptake from the soil, although it is very clear that the concentration of leaf nitrogen (taken as a surrogate for the photosynthetic enzyme rubisco) positively influences the maximum rate of photosynthesis (Field and Mooney, 1986; Evans, 1989). Therefore, in order to model the CO_2 exchange of individual plants and plants in canopies it is necessary to calculate, or model, the uptake of nitrogen from the soil. Woodward and Smith (1994a,b) have described and defined the close relationship between the maximum rate of photosynthesis (Amax) and the rate of nitrogen supply from soils across the whole global range of soil types.

It is also necessary to predict the influence of temperature on the uptake of nitrogen across the spectrum – from soils that are strongly organic, with a very large soil carbon content (up to 40,000 g/m²), to those with a low soil carbon content (up to approximately 500 g/m²). The mobility and method of nitrogen uptake into the plant varies considerably, but in a predictable manner determined by the soil carbon and nitrogen contents (Woodward and Smith, 1994b) over this range of soils. For plants in high carbon soils, the route of uptake is dominantly through mycorrhizal hyphae (Read, 1991); however, the plant plays an increasing role in uptake as the soil carbon levels decrease.

The influence of soil carbon concentration and temperature on nitrogen uptake has been described elsewhere (Woodward and Smith, 1994a,b). The uptake kinetics follow those generally observed in experimental systems (Clarkson and Warner, 1979; Bravo and Uribe, 1981; Clarkson, 1985). These responses of nitrogen uptake to temperature have been incorporated with the responses of leaf gas exchange to temperature (Harley et al., 1992; McMurtrie and Wang, 1993) in order to predict the response of soil nutrient uptake to temperature and the ensuing response of photosynthesis (Figure 20.2). An increase in 3°C is selected to be similar to the current range of

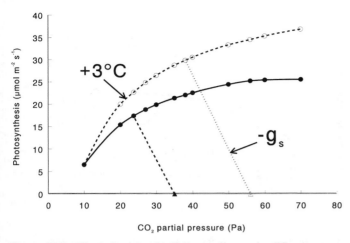

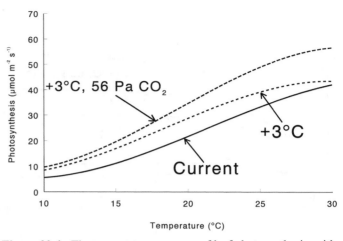

Figure 20.2: The influence of a 3°C warming on the CO_2 response of leaf photosynthesis, incorporating an increase in the uptake rate of nitrogen from the soil (dashed curve). Details are provided in Figure 20.1.

Figure 20.4: The temperature response of leaf photosynthesis, with sensitivities to a 3°C warming and with CO_2 enrichment (from 35 to 56 Pa).

GCM projections for future climates toward the end of the next century (Wigley and Raper, 1992).

A 3°C warming and an increase in the CO_2 partial pressure from 35 to 56 Pa leads to a 70% increase in photosynthetic rate, compared to the 26% predicted for the increase in CO_2 at constant temperature (Figure 20.1). This increase reflects the increased supply of nitrogen with increasing temperature. An interesting feature of the temperature response, with nitrogen uptake, is that stomatal conductance remains constant with increasing CO_2 (the slopes of the $-g_s$ lines on Figure 20.2 are equal), unlike the case at constant temperature (Figure 20.1). This effect is shown more clearly (Figure 20.3) by comparing the responses of g_s at the two temperatures, where it can be seen that stomatal conductance increases more with a temperature increase of 3°C at a CO_2 partial pressure of 56 Pa than at 35 Pa.

The response of photosynthesis to a warming of 3°C and an increase of CO_2 partial pressure from 35 to 56 Pa varies with

temperature (Figure 20.4). The relative response to warming decreases with temperature, from 50% at 10°C to 4% at 30°C. This demonstrates the greater increase in photorespiration (oxygen fixation) than photosynthesis (CO_2 fixation) with increasing temperature, in spite of the general increase in nutrient uptake with increasing temperature (Lawlor, 1987; Lawlor et al., 1987). This negative effect of photorespiration on photosynthetic rate is overridden by the increase in CO_2 partial pressure to 56 Pa.

20.2.3 Nitrogen Partitioning within the Leaf Canopy

Once nitrogen is taken up by roots, it is partitioned within the plant to different growing and living organs. In the approach described here, nitrogen is partitioned to the different layers of leaves in a plant canopy in a direct and positive response to the average irradiance of the leaf, a response that maximizes photosynthetic CO_2 fixation (Woodward, 1990; Thornley and Johnson, 1990). The maximum rate of photosynthesis (Amax) by a particular leaf is directly defined from the uptake rate of nitrogen into that leaf (Woodward and Smith, 1994a,b).

20.2.4 Leaf Photosynthesis and Respiration through the Leaf Canopy

Once the nitrogen taken up into the plant is partitioned between the leaves of the canopy, it is then possible to calculate the photosynthetic and dark respiratory rates of all the leaves in the plant canopy. The parameters for the photosynthesis model are first calculated from the Amax value (Woodward and Smith, 1994a) and published correlations (Wullschleger, 1993). The dark respiration rate is determined from the nitrogen supply to the leaf (Charles-Edwards et al., 1986; Ryan, 1991).

The influence of 23°C warming and an increase in CO_2 partial pressure to 56 Pa is predicted (Figure 20.5) for a range of leaf area indexes (LAI) between 1 and 9 and for two temperatures (10 and 20°C). In addition, the influence of soil charac-

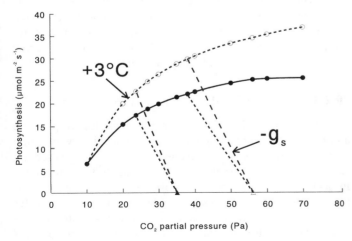

Figure 20.3: The influence of a 3°C warming on the response of stomatal conductance (g_s) to CO_2.

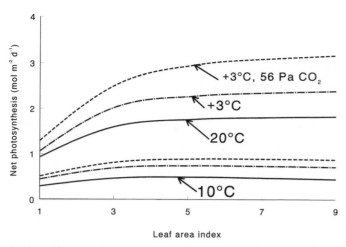

Figure 20.5: Canopy net CO_2 exchange, on a low carbon (1,000 g/m²) soil, at 10°C and 20°C and with warming and CO_2 enrichment. The curves at 10°C follow the same series as those at 20°C.

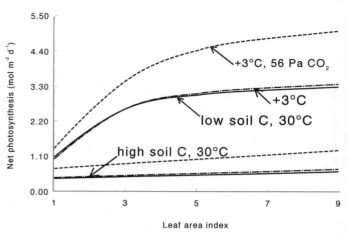

Figure 20.7: Canopy net CO_2 exchange on low and high carbon soils and with warming and CO_2 enrichment. The curves on the high carbon soil follow the same series as those on the low carbon soil.

teristics – low soil carbon (1,000 g/m²) and high soil carbon (30,000 g/m²) – are also investigated (Figures 20.5 and 20.6). For a low carbon soil with a high capacity for nutrient uptake, canopy net photosynthesis (day photosynthesis minus night respiration) increases from an LAI of 1 to approximately 4, at which point the response reaches an asymptote (Figure 20.5). At both temperatures, an increase in temperature and in CO_2 partial pressure increase the daily rate of net photosynthesis. At an LAI of 4, the relative increase with warming and CO_2 enrichment is greater at 10°C (80%) than at 20°C (62%).

The responses to temperature and CO_2 on a high carbon soil (Figure 20.6), which is highly organic and provides nitrogen at a slow rate, are rather different from those on the soil with low carbon (Figure 20.5), where N is supplied at a higher rate (Woodward and Smith, 1994b). The rates of net photosynthesis are much lower, as a consequence of the low supply rate of carbon – a feature that is characteristic of boreal forest species (Woodward and Smith, 1994a,b). However, at 20°C, both warming and CO_2 enrichment lead to a marked stimulation in

photosynthesis, primarily driven through the increased rate of nutrient supply. At an LAI of 4, and at 20°C, the relative increase in canopy photosynthesis with warming and CO_2 enrichment is 133%. At the lower temperature of 10°C, any increase in LAI above unity leads to a decline in net photosynthesis. This reflects the low supply rate of nitrogen from the soil and the low photosynthetic rate at this temperature. Warming and CO_2 enrichment slightly ameliorate the effect, but not to a great extent. This response mirrors the response seen by tundra vegetation during experimental studies (Oechel and Strain, 1985; Woodward et al., 1991).

Increasing the temperature of both soil types to 30°C causes further increases in canopy photosynthesis (Figure 20.7); however, the effect of a 3°C increase is very small, and reflects the asymptotic nature of the uptake of nutrients at high temperatures (Clarkson, 1985). With both low and high carbon soils there is a marked response to CO_2 enrichment, and this is strongly determined by the reduced photorespiratory rate. The responses at this high temperature indicate that when water is not limiting photosynthesis, then a warming of 3°C will have little effect on photosynthesis, unlike at 20°C for both soil types and at 10°C for the low carbon soil. In all cases CO_2 enrichment is predicted to increase the rate of canopy photosynthesis; however, these responses will be strongly modified in the field when water supply is taken into account (Woodward et al., 1991).

20.2.5 Annual NPP and Response to Precipitation

In this context, NPP is photosynthesis minus leaf respiration, and in this section the respiratory costs associated with living sapwood and the synthetic respiratory costs of constructing leaves and sapwood are not included in the model simulations. In Section 20.2.6, the synthetic costs of constructing sapwood and leaves are assumed equal for all biomes. Biomes differ only in the amount of leaf tissue (LAI, which is controlled primarily by climate) and sapwood to supply water to this leaf area. No other variations in dry matter partitioning have been incorporated.

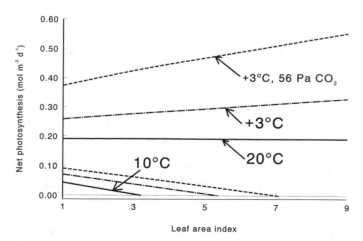

Figure 20.6: Canopy net CO_2 exchange, on a high carbon soil (30,000 g/m²) soil. Details are as explained in Figure 20.5.

The NPP model developed so far also includes a prediction of stomatal conductance, which is influenced by temperature and CO₂ through photosynthesis, and directly by the humidity of the air (Ball et al., 1987). It is clear that the LAI of a canopy is strongly influenced by the balance between canopy transpiration and the supply of water as precipitation (Woodward, 1987). Thus the realized NPP and LAI of the canopy will vary with precipitation. This effect on NPP has therefore been investigated for different temperatures, soil carbon concentrations, and precipitations for canopies of different LAI (Figures 20.8 and 20.9). In this stage of the model there is an additional feedback from the soil to the leaf, by the impact of soil water status on stomatal conductance (Gollan et al., 1992). In the model, precipitation throughfall through the plant canopy infiltrates into the soil. This increases the soil water content. Evapotranspiration decreases the soil water content, and the balance of these two processes is used to derive a soil water content which controls stomatal conductance. At the global scale, plant rooting depth and soil water holding capacity in this zone may well vary. However, this extent is not known and so, at this stage, the same maximum plant available water is assumed for all sites.

At a mean temperature of 10°C, the NPP is highest on soils with a low soil carbon, and therefore a high rate of nutrient supply (Figure 20.8A). This is characteristic of thin Alpine soils where herbaceous canopies show very high photosynthetic rates (Körner and Diemer, 1987). The domed response surface is a measure of increased leaf self-shading, and therefore reduced photosynthetic capacity with increasing LAI, and the reduction in NPP with increasing precipitation indicates a cloudier and lower irradiance environment.

The NPP of canopies on high soil carbon is low, increasing slightly at low LAI, when leaf self-shading is at a minimum.

Such a response is likely to be stable even when taking account of the full nutrient cycle, as leaf litter in this system will be slow to decompose, maintaining the high carbon levels in the soil and slow rates of nitrogen uptake. These low rates of NPP are characteristic of plants from the tundra (Woodward and Smith, 1994a,b). Increasing the temperature by 3°C and the CO₂ to 56 Pa increases the NPP by a maximum of 92% for canopies with an LAI of unity and for plants on high soil carbon (Figure 20.8B). The increase is about 85% for plants on low soil carbon, with an LAI of 7 and at an annual precipitation of 500 mm.

The response surfaces of the canopies on the two soil types change markedly when the temperature is increased to 20°C. At this temperature, characteristic of warm temperate and subtropical climates, precipitation is likely to be the major limitation to NPP (Figure 20.9A), and no marked decline in NPP is observed at the highest precipitation, even though irradiance is lower. At low precipitation, and for LAIs greater than 4, NPP is greatest on the soils with high soil carbon. This results from the low photosynthetic rates and the correlated low stomatal conductances, which define lower canopy conductances than for plants on low soil carbon. On low soil carbon, stomatal conductances are higher and therefore, with a shorter growing season, restricted by the more rapid use of the limited water supply. Once the precipitation exceeds about 500 mm/yr, the NPP on the low soil carbon is greater, with adequate water to supply higher rates of transpiration. Increasing the temperature by 3°C, even with an increase in CO₂ partial pressure, markedly influences NPP for canopies at LAIs greater than 4 and on low soil carbon (Figure 20.9A). In this case the increased transpiration of canopies with increasing LAI exceeds the rate of supply by the precipitation. NPP for the canopies on high soil carbon increases with temperature and CO₂, with no obvious decline in NPP as LAI increases.

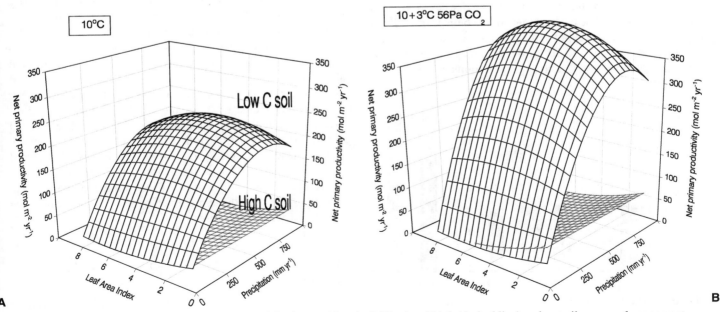

Figure 20.8: The influence of leaf area index, precipitation, and low (solid line) and high (dashed line) carbon soil on annual canopy net primary productivity (photosynthesis minus leaf respiration): (A) at a mean temperature of 10°C and a CO₂ partial pressure of 35 Pa; (B) with a warming of 3°C and a CO₂ partial pressure of 56 Pa.

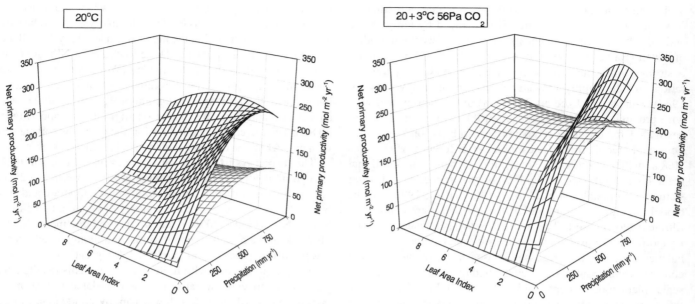

A   B

Figure 20.9: The influence of leaf area index, precipitation, and low (solid line) and high (dotted line) carbon soil on annual canopy net primary productivity (photosynthesis minus leaf respiration): (A) at a mean temperature of 20°C and a CO_2 partial pressure of 35 Pa; (B) with a warming of 3°C and a CO_2 partial pressure of 56 Pa.

20.2.6 Major Biome NPP

The influence of a 3°C global warming, an increase in CO_2 partial pressure to 56 Pa, and an annual increase in precipitation of 10% on the NPP of the major biome types of the world can now be simulated using the described model. The climatic changes are in the ranges investigated by Smith and colleagues (1992) for estimating the effect of future climatic change on the distribution of vegetation. The areal extents of four major biomes – tundra, grassland, dry forest, and mesic wet forest (Smith et al., 1992) – have been used to calculate their global NPP (photosynthesis minus leaf respiration) with the model described in the previous sections. The mean climate of the biomes has been calculated from Müller (1982) and Smith and colleagues (1992).

The NPPs (Figure 20.10) of the four major biomes, taking into account their areal extents, agree quite well with pub-

lished data (Melillo et al., 1993). These two approaches do differ significantly in some respects. Most notable is that the approach taken by Melillo and colleagues (1993) is to characterize the NPP of an established and geographically distributed global suite of vegetation. The model presented here also predicts NPP globally, but there are no given characteristics of vegetation that constrain the model projections.

The β-factor (Bacastow and Keeling, 1973) is also shown (Figure 20.10) for the simulation with an increase in CO_2 partial pressure from 35 to 56 Pa. The greatest response of NPP to CO_2 enrichment, as measured by the β-factor, is for the dry forest, in which CO_2 not only directly stimulates the photosynthetic rate, but also leads to a reduction in the canopy stomatal conductance, effectively increasing canopy water use efficiency in this seasonally dry environment. The β-factors are also high for grassland and wet mesic forests. For grassland the explanation is the same as for dry forest. In contrast, for wet forest the increase is more the result of an increase in photosynthetic rate through the canopy of high LAI. The reduced rate of photorespiration with CO_2 enrichment is especially marked in warm climates and for shaded leaves (Woodward and Smith, 1994a) but is much less so in the cool tundra, where the β-factor is lowest. These biome responses are closely in line with those predicted by Melillo and colleagues (1993) for extant vegetation.

20.2.7 Modeling Biome NPP with Whole-Plant Respiration

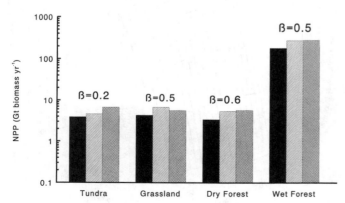

Figure 20.10: Predicted net primary productivity of four major biome types in (black histogram) current climate; (single hatch) current climate with CO_2 enrichment to 56 Pa, with β-factor; (double hatch) current climate with a warming of 3°C and a CO_2 partial pressure of 56 Pa. 45% of annual biomass is carbon.

The NPP estimated thus far accounts only for leaf respiration. Other significant respiratory costs also need to be accounted for, in particular those of the sapwood – the living parts of the water- and sap-conducting pathways of the plant. Waring and colleagues (1982) and Friend (1993) describe methods for cal-

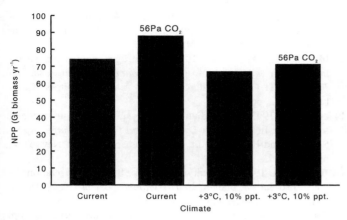

Figure 20.11: Estimated net primary productivity (photosynthesis minus whole-plant respiration) for the wet forest biome, at current climate; current climate with CO_2 enrichment to 56 Pa; with warming by 3°C and an increase in precipitation by 10% at 35 Pa CO_2; and with warming by 3°C and an increase in precipitation by 10% at 56 Pa CO_2.

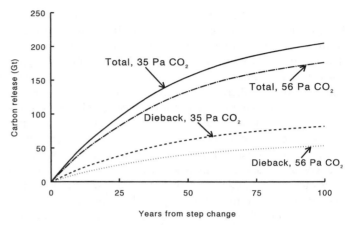

Figure 20.12: Estimated trend in global release of carbon from the terrestrial biosphere following a step increase in temperature of 3°C (from Smith and Shugart, 1993). The influence of a CO_2 enrichment to 56 Pa is modeled for biosphere dieback (dotted line) and for the total carbon release (dashed/dotted line).

culating the amount of sapwood, and its respiratory costs, in relation to the height to which water must be conducted (the canopy height) and the LAI. These relationships have been used to calculate the maintenance costs of the sapwood with a marked temperature sensitivity, and the synthetic costs with no temperature sensitivity (McCree, 1974; Charles-Edwards et al., 1986; Thornley and Johnson, 1990). The respiratory costs must also be added to the synthetic costs of constructing the leaves in the canopy, assuming that the whole leaf canopy is constructed annually.

These accountable costs can be readily incorporated in the prediction of NPP. Losses through other processes such as carbohydrate leaching and the influence of senescence cannot be included at this time, through lack of a quantitative description of the processes. The current and future NPP of the wet mesic forest (Figure 20.11) are now considerably less than for the incomplete respiratory budget (Figure 20.10), and the β-factor is reduced to 0.3. A 3°C warming and a 10% increase in precipitation reduce NPP; however, this reduction is almost offset when the climatic change also includes the increase in CO_2 partial pressure.

20.2.8 Biome Changes with Climatic Change

Smith and Shugart (1993) clearly demonstrated that after a step increase in temperature, there would be a net release of CO_2 from the biosphere to the atmosphere, primarily as a consequence of increased rates of soil decomposition and the dieback of current vegetation. The fluxes involved are large (Figure 20.12) and would exert a strong positive feedback on the climate. The time for the biosphere to reset the balance by increasing the uptake of carbon through new successional vegetation and through new migrating vegetation is likely to be at least 10 times longer than the time scale of dieback and decomposition (Figure 20.12). Therefore, in the short-term period of centuries, the movement of species around the landscape to new geographical positions and the development of

new soils under a new and developing vegetation will not counteract the large expected releases of CO_2 (Figure 20.12). However, the rate of vegetation dieback is primarily due to drought, so it is possible that the extensive dieback predicted by Smith and Shugart (1993) might be limited by a direct effect of increasing CO_2.

The NPP model described in this chapter was applied to the climates and vegetation types described by Smith and Shugart (1993), with a 3°C warming and 10% increase in precipitation, to determine the degree of effectiveness of CO_2 enrichment in reducing these impacts. The increase in CO_2 concentration diminished dieback by about 35% (Figure 20.12), through a reduced biome transpiration under CO_2 enrichment and a consequent amelioration of the effects of drought. The total release of carbon was reduced only by 14%, because it has been assumed that the rate of soil decomposition is not changed by CO_2.

20.3 Conclusion

Our current understanding of the processes of photosynthesis and respiration and the optical characteristics of canopies makes it possible to predict with some confidence the effects of future CO_2 enrichment on plants. Across a range of soil types, and with adequate water supply, an increase in the CO_2 partial pressure from 35 to 56 Pa, the range of increase expected around the year 2050 (Wigley and Raper, 1992), causes a modest 26% increase in photosynthetic rate. This response is increased to 70% by warming, particularly through an increased rate of nutrient supply. These responses are carried through to CO_2 exchange for the whole canopy (including leaf respiration), as long as the water supply is maintained and with some variation in response to temperature. Small increments of warming (3°C) will have the greatest effect in warm (not hot or cold) climates, although the influence of CO_2 may be large in hot climates, through significant reductions in photorespiration.

The inclusion of water supply in the models exerts large effects on the responses of canopy NPP. In warm climates, a warming of 3°C is predicted to have a large effect on canopies that are on soils with low carbon and high nutrient supply rates. In these situations – for example, grasslands and agricultural lands – there is a predicted trend to canopies of lower LAI but higher productivities at these levels. For vegetation on soils with high carbon content and low nutrient supply rates – for example, forests – warming and CO_2 enrichment is predicted to increase productivity. However, the impact of changes in nutrient cycling as a consequence of differences in litter quality (e.g., C/N ratio) is not included in the model, a feature that may influence the degree of response.

Calculation of the total NPP of the major biomes indicates that all biomes respond positively to increases in temperature and CO_2. However, when a more complete assessment of NPP is attempted it becomes clear that a warming of 3°C, even with a 10% increase in precipitation, is likely to cause a reduction in NPP, which might be offset through an increase in the partial pressure of CO_2. The continued increase in CO_2 may also serve to reduce the rate of biome dieback as the climate warms, a response that may reduce the net release of carbon to the atmosphere compared with that, as predicted by Smith and Shugart (1993) for the transition to a warmer world.

References

Bacastow, R., and C. D. Keeling. 1973. Atmospheric carbon dioxide and radiocarbon in the natural carbon cycle: II. Changes from 1700 to 2070 as deduced from a geochemical model. In *Carbon and the Biosphere,* G. M. Woodwell and E. V. Pecan, eds. U.S. Atomic Energy Commission, Washington, D.C., pp. 86–135.

Ball, J. T., I. E. Woodrow, and J. A. Berry. 1987. A model predicting stomatal conductance and its contribution to the control of photosynthesis under different environmental conditions. In *Progress in Photosynthesis Research: IV. Proceedings of the VIIth International Congress on Photosynthesis,* I. Biggins, ed. Martinus-Nijhoff, Dordrecht, The Netherlands, pp. 221–224.

Bazzaz, F. A. 1990. The response of natural ecosystems to the rising global CO_2 levels. *Annual Review of Ecology and Systematics 21,* 167–196.

Bolin, B. (ed.). 1981. *Carbon Cycle Modelling.* SCOPE Volume 16, John Wiley and Sons, Chichester, U.K.

Bravo, F. P., and E. G. Uribe. 1981. Temperature dependence on concentration kinetics of absorption of phosphate and potassium in corn roots. *Plant Physiology 67,* 815–819.

Brooks, A., and G. D. Farquhar. 1985. Effect of temperature on the CO_2/O_2 specificity of ribulose–1,5–bisphosphate carboxylase/oxygenase and the rate of respiration in the light: Estimates from gas-exchange experiments on spinach. *Planta 165,* 397–406.

Charles-Edwards, D. A., D. Doley, and G. M. Rimmington. 1986. *Modelling Plant Growth and Development.* Academic Press, Sydney, Australia.

Clarkson, D. T. 1985. Factors affecting mineral nutrient acquisition by higher plants. *Annual Review of Plant Physiology 36,* 77–115.

Clarkson, D. T., and A. Warner. 1979. Relationships between root temperature and the transport of ammonium and nitrate ions by Italian and perennial ryegrass (*Lolium multiflorum* and *Lolium perenne*). *Plant Physiology 64,* 557–561.

Evans, J. R. 1989. Photosynthesis and nitrogen relationships in leaves of C_3 plants. *Oecologia 78,* 9–19.

Farquhar, G. D., S. von Caemmerer, and J. A. Berry. 1980. A biochemical model of photosynthetic CO_2 assimilation in leaves of C_3 species. *Planta 149,* 78–90.

Field, C., and H. A. Mooney. 1986. The nitrogen–photosynthesis relationship in wild plants. In *On the Economy of Plant Form and Function,* T. Givnish, ed. Cambridge University Press, Cambridge, U.K., pp. 25–55.

Friend, A. D. 1993. The prediction and physiological significance of tree height. In *Vegetation Dynamics and Global Change,* A. M. Solomon and H. H. Shugart, eds. Chapman and Hall, New York, pp. 101–115.

Gifford, R. M. 1993. Implications of CO_2 effects on vegetation for the global carbon budget. In *The Global Carbon Cycle,* Martin Heimann, ed. Proceedings of the NATO ASI at Il Ciocco, Italy, September 8–20, 1991. Springer-Verlag, Berlin, Germany, pp. 165–205.

Gollan, T., U. Schurr, and E. D. Schulze. 1992. Stomatal response to drying soil in relation to changes in the xylem sap composition of *Helianthus annuus:* I. The concentration of cations, anions, amino acids in, and pH of, the xylem sap. *Plant, Cell and Environment 15,* 551–559.

Harley, P. C., R. B. Thomas, J. F. Reynolds, and B. R. Strain. 1992. Modelling photosynthesis of cotton grown in elevated CO_2. *Plant, Cell and Environment 15,* 271–282.

Harvey, L. D. D. 1989. Effect of model structure on the response of terrestrial biosphere models to CO_2 and temperature increases. *Global Biogeochemical Cycles 3,* 137–153.

Jordan, D. B., and W. L. Ogren. 1984. The CO_2/O_2 specificity of ribulose 1,5-bisphosphate carboxylase/oxygenase: Dependence on ribulose-bisphosphate concentration, pH and temperature. *Planta 161,* 308–313.

Körner, C., and M. Diemer. 1987. *In situ* photosynthetic responses to light, temperature and carbon dioxide in herbaceous plants from low and high altitude. *Functional Ecology 1,* 179–194.

Lawlor, D. W. 1987. *Photosynthesis: Metabolism, Control and Physiology.* Longman, Harlow, U.K.

Lawlor, D. W., F. A. Boyle, A. T. Young, A. J. Keys, and A. C. Kendall. 1987. Nitrate nutrition and temperature effects on wheat: Photosynthesis and photorespiration of leaves. *Journal of Experimental Botany 38,* 393–408.

Lieth, H. 1975. Modelling the primary productivity of the world. In *Primary Productivity of the Biosphere,* H. Lieth and R. B. Whittaker, eds. Springer-Verlag, New York, pp. 237–263.

McCree, K. J. 1974. Equations for the rate of dark respiration of white clover and grain sorghum, as functions of dry weight, photosynthetic rate, and temperature. *Crop Science 14,* 509–514.

McMurtrie, R. E., and Y. P. Wang. 1993. Mathematical models of the photosynthetic response of tree stands to rising CO_2 concentrations and temperatures. *Plant, Cell and Environment 16,* 1–13.

Melillo, J. M., A. D. McGuire, D. W. Kicklighter, B. Moore III, C. J. Vorosmarty, and A. L. Schloss. 1993. Global climate change and terrestrial net primary production. *Nature 363,* 234–240.

Morison, J. I. L. 1985. Sensitivity of stomata and water use efficiency to high CO_2. *Plant, Cell and Environment 8,* 467–474.

Müller, M. J. 1982. *Selected Climatic Data for a Global Set of Standard Stations for Vegetation Science.* W. Junk, The Hague, The Netherlands.

Oechel, W. C., and B. R. Strain. 1985. Native species responses to increased carbon dioxide concentration. In *Direct Effects of Increasing Carbon Dioxide on Vegetation,* B. R. Strain and J. D. Cure, eds. DOE/ER-0238, U.S. Department of Energy, Washington, D.C., pp. 117–154.

Rastetter, E. B., R. B. McKane, G. R. Shaver, and J. M. Melillo. 1992. Changes in C storage by terrestrial ecosystems: How C–N interactions restrict responses to CO$_2$ and temperature. *Water, Air, and Soil Pollution 64,* 327–344.

Read, D. J. 1991. Mycorrhizas in ecosystems. *Experientia 47,* 376–391.

Ryan, M. G. 1991. Effects of climate change on plant respiration. *Ecological Applications 1,* 157–167.

Schimel, D. S., D. Alves, I. G. Enting, M. Heimann, F. Joos, D. Raynaud, and T. M. L. Wigley. 1996. CO$_2$ and the carbon cycle. In *Climate Change 1995: The Science of Climate Change, Contribution of Working Group I to the Second Assessment Report of the Intergovernmental Panel on Climate Change,* J. T. Houghton, L. G. Meira Filho, B. A. Callander, N. Harris, A. Kattenberg, and K. Maskell, eds. Cambridge University Press, New York, pp. 65–86.

Shine, K., R. G. Derwent, D. J. Wuebbles, J.-J. Morcrette, A. J. Apling, J. P. Blanchet, R. J. Charlson, D. Crommelynck, H. Grassl, N. Husson, G. J. Jenkins, I. Karol, M. D. King, V. Ramanathan, H. Rodhe, G.-Y. Shi, G. Thomas, W.-C. Wang, T. M. L. Wigley, and T. Yamanouchi. 1990. Radiative forcing of climate. In *Climate Change: The IPCC Scientific Assessment,* J. T. Houghton, G. J. Jenkins, and J. J. Ephraums, eds. Cambridge University Press, Cambridge, U.K., pp. 41–68.

Shukla, J., and Y. Mintz. 1982. Influence of land-surface evapotranspiration on the Earth's climate. *Science 215,* 1498–1501.

Slatyer, R. O. 1977. Altitudinal variation in the photosynthetic characteristics of Snow Gum, *Eucalyptus pauciflora* Sieb. ex Spreng: III. Temperature response of material grown in contrasting thermal environments. *Australian Journal of Plant Physiology 4,* 301–312.

Smith, T. M., and H. H. Shugart. 1993. The transient response of terrestrial carbon storage to a perturbed climate. *Nature 363,* 523–526.

Smith, T. M., R. Leemans, and H. H. Shugart. 1992. Sensitivity of terrestrial carbon storage to CO$_2$-induced climate change: Comparison of four scenarios based on general circulation models. *Climatic Change 21,* 367–384.

Sundquist, E. T. 1993. The global carbon dioxide budget. *Science 259,* 934–941.

Thornley, J. H. M., and I. R. Johnson. 1990. *Plant and Crop Modelling: A Mathematical Approach to Plant and Crop Physiology.* Oxford Science Publications, Oxford, U.K.

Uchijima, Z., and H. Seino. 1985. Agroclimatological evaluation of net primary productivity of natural vegetation: I. Chikugo model for evaluating net primary productivity. *Journal of Agricultural Meteorology 40,* 343–352.

Waring, R. H., P. E. Schroeder, and R. Oren. 1982. Application of the pipe model theory to predict canopy leaf area. *Canadian Journal of Forest Research 12,* 556–560.

Wigley, T. M. L., and S. C. B. Raper. 1992. Implications for climate and sea level of revised IPCC emissions scenarios. *Nature 357,* 293–300.

Woodward, F. I. 1987. *Climate and Plant Distribution.* Cambridge University Press, Cambridge, U.K.

Woodward, F. I. 1990. Global change: Translating plant ecophysiological responses to ecosystems. *Trends in Ecology and Evolution 5,* 308–311.

Woodward, F. I. 1992. Predicting plant responses to global environmental change. *New Phytologist 122,* 239–251.

Woodward, F. I., and T. M. Smith. 1994b. Predictions and measurements of the maximum photosynthetic rate, Amax, at the global scale. In *Ecophysiology of Photosynthesis,* E. D. Schulze and M. M. Caldwell, eds. Springer-Verlag, Berlin, Germany, pp. 491–509.

Woodward, F. I., and T. M. Smith. 1994a. Global photosynthesis and stomatal conductance: Modelling the controls by soil and climate. *Advances in Botanical Research 20,* 1–41.

Woodward, F. I., G. B. Thompson, and I. F. McKee. 1991. The effects of elevated concentrations of carbon dioxide on individual plants, populations, communities and ecosystems. *Annals of Botany 67, Supplement 1,* 23–38.

Wullschleger, S. D. 1993. Biochemical limitations to carbon assimilation in C$_3$ plants – A retrospective analysis of the A/C$_i$ curves from 109 species. *Journal of Experimental Botany 44,* 907–920.

21

Stabilization of CO$_2$ Concentration Levels

TOM M. L. WIGLEY

Abstract

The various CO_2 concentration stabilization profiles used in Intergovernmental Panel on Climate Change (IPCC) analyses are described, and their methods of construction explained. Padé approximant coefficients are given to allow readers to recalculate the profiles precisely. Forward and inverse initialization strategies are discussed, and industrial emissions requirements for the "S" and "WRE" profiles are compared and evaluated. Details of the post-1990 carbon budget breakdown for the specific case of WRE550 are given. Uncertainties in the industrial emissions required for stabilization following any given profile are quantified. Uncertainties considered are those resulting from the effects of: CO_2 fertilization formulation (logarithmic versus rectangular hyperbolic); the prescribed future "history" of net deforestation; terrestrial sink specification and the IPCC restriction of this sink to CO_2 fertilization only; and ocean flux uncertainties. Sink specification is the greatest source of uncertainty, leading to potential errors in implied emissions of up to ±2 Gt C/yr for WRE550. Errors of this magnitude are equivalent to a misspecification of the stabilization level of approximately ±50 ppmv.

21.1 Introduction

The stabilization of future atmospheric CO_2 concentration is one of the primary aims of the United Nations Framework Convention on Climate Change (UNFCCC). Article 2 of this convention gives the following objective: "To achieve stabilization of greenhouse gas concentrations . . . at a level that would prevent dangerous anthropogenic interference with the climate system. . . ." What is required to stabilize CO_2 concentration? To provide insight into this question, Working Group I of the Intergovernmental Panel on Climate Change (IPCC) initiated an extensive inter-model comparison exercise in which profiles of future concentration changes, stabilizing at different levels from 350 to 750 ppmv, were used in inverse carbon cycle model calculations to determine the corresponding emissions requirements. Full details of this exercise are given in Enting and colleagues (1994). Selected results are presented in Schimel and colleagues (1995), with updates in Schimel and colleagues (1996). The present chapter describes these analyses in more detail and addresses some of the uncertainties involved in the calculations.

21.2 Modeling Strategy

To carry out these inverse calculations in a consistent and well-defined manner, it was necessary to specify a procedure for initializing the various carbon cycle models in the year 1990 in a way that ensured a credible 1980s-mean carbon budget. Standard data sets also had to be developed or adopted for past changes in CO_2 concentrations ($C(t)$), industrial emissions ($I(t)$), net emissions resulting from land-use changes ($D_n(t)$), and future concentrations.

21.2.1 Initialization

Two possible initialization procedures were specified: an inverse and a forward method. The inverse initialization procedure followed the method of Wigley (1993), based on earlier work by Siegenthaler and Oeschger (1987), Enting and Mansbridge (1987), and Enting and Pearman (1987). The forward procedure is described below.

To explain these two initialization procedures, it is necessary first to write down the carbon budget, as follows:

$$2.123 \, dC/dt = dM/dt = I + D_n - F_{oc} - X \qquad (21.1)$$

where C is concentration (ppmv), M is the atmospheric carbon mass, I is industrial emissions, D_n is net emissions from land-use changes (mainly deforestation), F_{oc} is the atmosphere-to-ocean flux, and X is an additional (terrestrial) sink term. M, F_{oc}, X, I, and D_n are in Gt C/yr. The need for initialization arises because the 1980s-mean budget does not balance if only dM/dt, I, D_n, and F_{oc} are considered. The initialization procedure determines the additional term $X(t)$ in order to ensure a realistically balanced budget.

For the calculations described in Enting and colleagues (1994) and Schimel and colleagues (1995), the following 1980s-mean budget values were used: $\Delta M(80s) = 3.38$ Gt C/yr, $I(80s) = 5.45$ Gt C/yr, and $D_n(80s) = 1.60$ Gt C/yr (see, e.g., Schimel et al., 1996, table 2.1). $F_{oc}(80s)$ values varied from model to model, with most values around the current "best-guess" value of 2.0 Gt C/yr. These numbers require that X be nonzero, specifically $X(80s) = 1.67$ Gt C/yr. As noted in Schimel and colleagues (1995), these budget figures are not (and were not at the time) the most up-to-date values. In Schimel and colleagues (1996; see table 2.1), revised budget details were used: $\Delta M(80s) = 3.28$ Gt C/yr, $I(80s) = 5.46$ Gt C/yr, and $D_n(80s) = 1.10$ Gt C/yr. With $F_{oc}(80s) = 2.0$ Gt C/yr, these require $X(80s) = 1.28$ Gt C/yr. These are still (early 1997) the most up-to-date figures. The calculations below use the more recent budget figures.

The X term has been referred to as the "missing sink," a misnomer because there are many candidates for X. The sink is certainly not missing; what is missing is a reliable quantification of its components. For the stabilization calculations, we took the simplest credible way of quantifying X – by assuming it was solely due to the effects of CO_2 fertilization. Using CO_2 fertilization alone is a credible method because the implied value of the fertilization factor is consistent with small-scale experimental data. In other words, if small-scale CO_2 fertilization results are applied at the global scale, then the carbon budget can be realistically balanced using this factor alone. Nevertheless, using CO_2 fertilization alone is still clearly an oversimplification, since $X(t)$ must also involve other processes, such as climate effects and nitrogen fertilization. The effect of this simplified treatment is discussed in Section 21.5.3.

For forward initialization, the model is run in forward mode to 1990, forced with both industrial and land-use emissions. This gives $C(t)$ as an output. The fertilization factor is then adjusted to optimize the fit between the modeled and observed $C(t)$ values.

For inverse initialization, $D_n(t)$ is not required. In this case, the model is run in inverse mode to 1990, forced with observed $C(t)$ and industrial emissions. For any given value of the fertilization factor, this gives $D_n(t)$ as an output. The fertilization factor is then adjusted to give an appropriate 1980s-mean value of D_n, here 1.1 Gt C/yr.

21.2.2 Data Specification

To complete the problem specification, historical data sets for $C(t)$, $I(t)$, and $D_n(t)$ are required. For $C(t)$, a smoothed concentration history was produced based on ice core data (Friedli et al., 1986) and directly observed Mauna Loa data (Keeling and Whorf, 1991). (Although not used here, a revision of this has been produced using more recent ice core data from Etheridge et al., 1996; see Appendix 21.1.) Industrial emissions data were taken from Marland and Boden (1991) and Keeling (1991). The data were linearly extrapolated back to 1844 to avoid a step jump at the start of the published record. Recent data were updated by G. Marland (personal communication). The industrial emissions values used in Schimel and colleagues (1995) are given in Enting and colleagues (1994; their Table 21.3A2). For the latest IPCC report (Schimel et al., 1996) and for the work reported here, these were further revised (G. Marland, personal communication). Over 1980–90, these latest values differ only very slightly from those published in Trends '93 (Keeling, 1994; Marland et al., 1994; mean difference, 0.003 Gt C/yr). Net land-use emissions data were from Houghton (1991; see Enting et al., 1994; their Table A1), extended linearly back from 1850 to $D_n(t) = 0.2$ Gt C/yr in 1764 based on discussions with R. A. Houghton. The latter data are needed only for the forward approach to initialization.

The next requirement was to produce future concentration stabilization "profiles." These were determined as follows. First, a series of stabilization levels were chosen (350, 450, 550, 650, and 750 ppmv). Then, stabilization dates and pathways to stabilization were determined iteratively in order to ensure that the implied emissions changes were smoothly varying and not unrealistically rapid. This required choosing different stabilization dates for each profile; the dates chosen were 2150, 2100, 2150, 2200, and 2250, respectively.

The initial stabilization profiles produced in this way were meant to be "illustrative rather than prescriptive" (Enting et al., 1994, p. 6). They were designed primarily to give an indication of the likely range of emissions required to stabilize CO_2 concentrations, to provide a methodological framework for future emissions estimates, and to give an indication of the accuracy with which such calculations can be carried out at present. Essentially, the pathways were meant to be "policy-neutral." They were not meant to be interpreted in policy or economic terms.

Nevertheless, the resulting emissions pathways were criticized on economic grounds because they showed, in general, immediate departures from the standard "existing policies" (or "business-as-usual," BAU) emissions scenarios developed by the IPCC in 1992 (Leggett et al., 1992). Because of this, an alternative set of concentration profiles were developed (Wigley et al., 1996), as described in Section 21.3. To distinguish these two set of profiles, they are referred to as Sxxx and WRExxx,

where xxx is the stabilization level. There are, in fact, two sets of Sxxx profiles, those originally constructed by Enting and colleagues (Sxxxold) based on the old 1980s-mean carbon budget and used in Enting and colleagues (1994) and Schimel and colleagues (1995), and a new set of profiles (Sxxxnew) based on the most recent 1980s-mean budget and used in Schimel and colleagues (1996). Only the Sxxxnew profiles are considered here.

To ensure that these various concentration profiles led to smoothly varying emissions, an initial set of plausible profiles were specified, and these were refined by an iterative adjustment process. Final profiles based on these calculations were obtained by fitting Padé approximants to concentrations and/or derivatives at specific points (Enting et al., 1994, p. 76). Padé approximants are ratios of polynomials: a form equivalent to the following was used:

$$\tilde{C}(x) = (ax + bx^2 + cx^3)/(1 + dx + ex^2)$$

where

$$\tilde{C} = (C - C_0)/(C_s - C_0)$$

$$x = (t - t_0)/(t_s - t_0)$$

where C_0 is the "initial" concentration at $t = t_0 = 1990.5$ and C_s is the "final" concentration (i.e., the stabilization level) at $t = t_s$. For the monotonically increasing cases (achieving stabilization at 450, 550, 650, or 750 ppmv), c may be taken as zero. The fit is obtained by specifying C and dC/dt at $t = t_0$ and at $t = t_s$ (where dC/dt must be zero), and by specifying C at one other point (t_1) between t_0 and t_s. For S350 and WRE350, where concentration rises to a maximum before declining to 350 ppmv, an additional zero derivative constraint must be imposed at the point of maximum concentration, requiring c to be nonzero. Padé coefficient details are given in Appendix 21.1.

The need to match the profiles smoothly to observations in 1990 explains why new profiles had to be produced when the history of CO_2 concentration changes was revised, as between the 1994 (Schimel et al., 1995) and 1995 (Schimel et al., 1996) IPCC reports.

Finally, for the inverse calculations forced with the stabilization profiles, it is necessary to specify the future "history" of $D_n(t)$. To do this, it was assumed that $D_n(t)$ followed the IS92a scenario of Leggett and colleagues (1992) up to 2075, declined to zero over 2075–2100, and stayed at zero subsequently. The need to specify $D_n(t)$ creates a matching problem in 1990, since the 1990 value is not only determined by the initialization procedure but also specified in the IS92a scenario. In general, these two values differ, so it was necessary to choose a set of transition values. This was done by taking the initialization value in 1990 and interpolating this linearly over 1990–2000 to the IS92a scenario value in 2000 (viz., 1.3 Gt C/yr).

21.3 Revised Stabilization Profiles

Subsequent to the specifications made for the exercise in Enting and colleagues (1994), described above, changes were made to the 1980s-mean carbon budget details. First, the "best-guess" value for $D_n(80s)$ was reduced to 1.1 Gt C/yr (Schimel et al., 1996, table 2.1). Second, the 1980s-mean atmospheric mass buildup was changed to 3.220Gt20C/yr (Schimel et al., 1995), and this was subsequently altered again to 3.2820Gt20C/yr (based on Kohmyr et al., 1985; Conway et al., 1994; and personal communication from P. Tans). This required the specification of a new concentration history (changes were required only over the decade of the 1980s), which, as noted above, required all the stabilization profiles to be respecified (in order to ensure a smooth match of C and dC/dt in 1990). Should the more recent (Etheridge et al., 1996) ice core concentration data be used in future calculations, this will not require the stabilization profiles to be recalculated, since the 1990.5 values of C and dC/dt have not been changed (see Appendix 21.1).

The simplest way to update the original (Enting et al., 1994) stabilization profiles is to use the same Padé approximant method as previously, with revised constraints on C and dC/dt at 1990.5 and maintaining the original "pinning points" at $t = t_1$ (see Appendix 21.1, Table 21.A2). This, unfortunately, leads to new profiles that differ substantially from the original. To maintain as much consistency with the original profiles as possible, new pinning points were defined, minimizing the overall difference between the old (Sxxxold) and new (Sxxxnew) profiles (see Appendix 21.1, Table 21.A2).

In addition to the Sxxx profiles, a radically different set of profiles has been devised and analyzed by Wigley and colleagues (1996): the WRExxx profiles. In devising these profiles, economic considerations were taken into account, albeit only qualitatively. Basically, it was assumed that the immediate (1990) departure of emissions from BAU expectations implied by the Sxxx profiles was highly unlikely, and that a slow (but increasing with time) departure was more likely. The economic arguments behind this are given in Wigley and colleagues (1996) and elaborated on in Appendix 21.2. As an idealization of the "slow departure from BAU" assumption, the WRExxx profiles follow the IPCC IS92a BAU emissions scenario (Leggett et al., 1992) for 10–30 years, depending on the chosen stabilization level. This does not mean that no policies need be introduced to reduce emissions, or that no action should be taken until 2000–2020 (see Appendix 21.2). Rather, it means that much of the "action" is in the planning, that is, in defining and evaluating strategies to reduce CO_2 emissions within a time frame commensurate with the economic and technical realities of the global energy system.

To devise these new profiles, the following procedure was used. First, concentration projections were made using the Wigley's (1993) model for the central IPCC92 emissions scenario, IS92a (Leggett et al., 1992). The new profiles were then constrained to follow these to 2000, 2005, 2010, 2015, and 2020 for stabilization levels of 350, 450, 550, 650, and 750 ppmv, respectively. (The earlier departure from IS92a concentrations for lower stabilization values reflects the greater difficulty of attaining lower stabilization levels.) These concentrations are model-dependent. Therefore, if inverted with a different model (as in Schimel et al., 1996), the equivalent emissions differ (slightly) from IS92a.

The same stabilization dates were used as before. To derive pathways from the points of departure from IS92a to the stabilization points, Padé approximants were used. The constraints were as before, matching C and dC/dt at the end points and fixing an intermediate value (the point of maximum concentration for the 350-ppmv stabilization case). Details are given in Appendix 21.1.

In response to the "needs" of IPCC to consider a more extreme case, an additional WRE profile was added after the publication of Wigley and colleagues (1996), one in which BAU emissions were followed to 2050 and the eventual stabilization level was taken as 1,000 ppmv (WRE1000; see Appendix 21.1). This extreme case was, in fact, added to the set of profiles analyzed by IPCC in response to a request from industry. Although the WRE1000 profile may be considered to be within the range of possibilities allowable under Article 2 of the UNFCCC, given that we do not yet know how to quantify the constraint of "dangerous interference" to the climate system, to follow such a profile would, in my view, be foolhardy. Apart from the much larger changes in climate that would ensue, relative to stabilization targets of 750 ppmv and below, the long-term implications for sea level verge on the catastrophic (see, e.g., Wigley, 1995; Warrick et al., 1996; and Raper et al., 1996). Furthermore, the effect of a 1,000-ppmv atmospheric CO_2 level on ocean chemistry and biology is, while highly uncertain, potentially disastrous – ocean pH could fall to levels that may make it difficult for many shell-producing organisms to produce their carbonate shells, upsetting the entire food chain.

21.4 Results

The purpose of developing the set of concentration stabilization profiles described above was to estimate the range of CO_2 emissions required to achieve stabilization at different levels. The primary results, therefore, are the estimates of post-1990 emissions determined by running a carbon cycle model in inverse mode. For the original stabilization profiles (Sxxxold), these calculations have been carried out using a large number of models; the results are described in Enting and colleagues (1994) and in Schimel and colleagues (1995). For the revised "S" profiles (Sxxxnew), results have been calculated using the models of Siegenthaler and Joos (1992), Jain and colleagues (1995), and Wigley (1993); the former results are used in the primary IPCC calculations illustrated in Schimel and colleagues (1996). Schimel and colleagues (1996) also give results for the WRE profiles. The sensitivity of emissions to the pathway toward a given stabilization level has been considered by Wigley and colleagues (1996) and Schimel and colleagues (1996) by comparing WRExxx and Sxxxnew results, and by Wigley and colleagues (1996, fig. 2) by considering perturbations of the WRE550 case. A somewhat different pathway sensitivity analysis is given in Enting and colleagues (1994, pp. 30, 31, and 49). Here we carry out a range of additional sensitivity and uncertainty analyses using the Wigley (1993) model.

21.4.1 Forward Initialization

For forward initialization, $I(t)$ and $D_n(t)$ are specified, $F_{oc}(t)$ and $X(t)$ are generated by the model (with $X(t)$ involving a free

Table 1. *Optimized results for forward initialization*

C(1765.0)	β	r	WRMS	C(1990.5)	dC/dt(1990.5)	ΔM(80s)
276.0	0.243	1.160	4.05	359.07	1.97	3.75
277.0	0.272	1.179	3.48	358.37	1.92	3.67
*278.0	0.303	1.198	2.95	357.72	1.88	3.60
279.0	0.337	1.219	2.49	357.06	1.84	3.52
280.0	0.374	1.242	2.15	356.42	1.80	3.44
*281.0	0.417	1.268	1.99	355.69	1.75	3.36
282.0	0.464	1.296	2.06	354.98	1.71	3.28
283.0	0.515	1.326	2.32	354.33	1.67	3.20
*284.0	0.573	1.360	2.74	353.65	1.62	3.12
Observed				353.64	1.54	3.28
Units			ppmv	ppmv	ppmv/yr	Gt C/yr

Notes: WRMS = minimum value of the weighted root mean square error; rows = cases illustrated in Figure 21.1; β and r = logarithmic and rectangular hyperbolic CO₂ fertilization factors, respectively.

parameter, the fertilization factor), and $C(t)$ is output by the model. The fertilization factor is then varied to optimize the fit between observed and modeled $C(t)$. It is essential with this method to add a further constraint: Ideally, C and dC/dt in 1990.5 should match the observed values. (This is a rigorous constraint for $C(t)$, but less so for dC/dt.) One way to accommodate this constraint is to allow the initial value of the $C(t)$ history (i.e., the 1764–65 value) to be a free parameter as well as the fertilization factor, and to optimize with respect to both this and the fertilization factor. How well this can be done depends on the particular model used. Here, using the Wigley (1993) model, we show that it is impossible to obtain a satisfactory "constrained" fit for the new budget.

For forward initialization, it is necessary to define a metric to minimize for optimization. Rather than use a standard root mean square (RMS) difference between modeled and observed results, we use a weighted RMS (WRMS) with a weight of 1 up to 1957 and 3 for 1958–90 (to reflect the greater accuracy of the more recent data).

A range of forward initialization calculations were carried out varying the initial concentration and optimizing with respect to the fertilization factor. Results are shown in Table 21.1. If the initial concentration $C(1765.0)$ is 278 ppmv (i.e., similar to that in the $C(t)$ data set used in inverse initialization), then the best fit is obtained for a fertilization factor (r) of 1.20, equivalent to a 20% enhancement in net primary productivity (NPP) for a CO₂ doubling.[1] Although such a value for r

is a realistic result for the fertilization factor, this particular forward initialization result is difficult to apply directly to future forward or inverse calculations because of inconsistencies with observations over 1980–90. As Table 21.1 shows, the model-based concentration change over the 1980s differs considerably from observations ($\Delta M(80s) = 3.60$ Gt C/yr, much higher than the observed value), and the 1990.5 concentration value (357.7 ppmv) and its rate of change (1.92 ppmv/yr) are both much higher than the observed values (see also Figure 21.1). The error in $\Delta M(80s)$ is outside the range of uncertainty for this quantity, conflicting with our need to have a budget balance that is within observational uncertainty. The high 1990.5 C and dC/dt values would lead to spurious concentrations for the early part of any stabilization profile that joined smoothly to them, or would require a step change in concentrations in 1990, which is clearly inadmissible.

Furthermore, using $C(1765.0) = 278$ ppmv gives a suboptimal fit to the observed $C(t)$: The weighted RMS value of 3.60 ppmv is considerably higher than the lowest possible value of 1.99 ppmv obtained for $C(1765.0) \approx 281$ ppmv. Even this fully optimized fit is unsatisfactory: $C(1990.5)$ and dC/dt in 1990.5 are both noticeably higher than observed. In fact, as Table 21.1 and Figure 21.1 show, it is impossible to match simultaneously C and dC/dt in 1990.5 and $\Delta M(80s)$. Matching $C(1990.5)$ (using $C(1765.0) \approx 284$ ppmv) gives a value of dC/dt in 1990.5 that is too high and a value of $\Delta M(80s)$ that is too low. Matching dC/dt (using $C(1765.0) \approx 286$ ppmv; the results are not shown here) leads to values that are too low for both $C(1990.5)$ and $\Delta M(80s)$, while matching $\Delta M(80s)$ (using $C(1765.0) \approx 282$ ppmv) leads to values of $C(1990.5)$ and dC/dt in 1990.5 that are both too high. Furthermore, the initial (1765) concentration values required in these partial matches are 4–8 ppmv above the ice core value, and probably outside the range of observational uncertainty in this value. Because of these problems, we abandon the forward initialization method and concentrate on results obtained using inverse initialization. Most calculations reported in Enting and

---

[1]The model used here employs either a logarithmic or a rectangular hyperbolic form for CO₂-fertilized NPP enhancement (the latter following Gifford, 1993). In all calculations here, except those comparing the two fertilization forms, we use the rectangular hyperbolic formulation. The relationship between this and the logarithmic form (for which the parameter is β) is described in detail in Wigley (1993). Using Gifford's formulation, r is the ratio between NPP at 680 ppmv and 340 ppmv. The β value corresponding to $r = 1.20$ is 0.30.

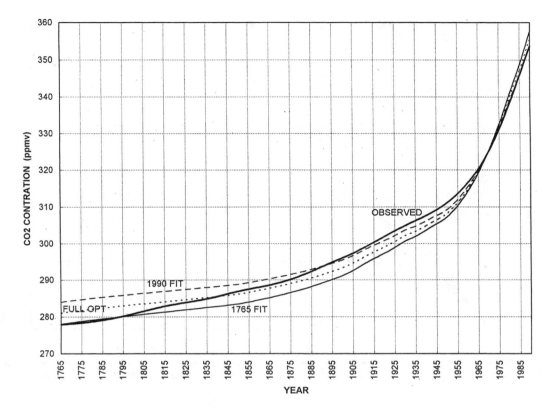

Figure 21.1: Optimized forward fits to observed concentration changes based on different fitting constraints. The bold solid curve gives the observations; the thin solid curve denotes modeled concentration fixed at the observed value in 1765 (27820ppmv), the long-dashed curve (C(1765) = 284 ppmv) denotes modeled concentration fixed to match the observed value in 1990; and the short-dashed curve (C(1765) = 281 ppmv) is fully optimized with no constraints.

colleagues (1994, table 5.1) used the inverse initialization method.

21.4.2 Inverse Initialization

With inverse initialization, $I(t)$ and $C(t)$ are specified, $F_{oc}(t)$ and $X(t)$ are calculated by the model, and $D_n(t)$ is computed by an inverse calculation. The value of $D_n(t)$ depends on the chosen value of the fertilization factor; consequently, by varying this factor, one can satisfy one additional constraint – in this case, that $D_n(80s)$ should equal a specified value (1.6 Gt C/yr in the original calculations mode by Enting and colleagues 1.1 Gt C/yr in the revised calculations presented here).

By specifying $C(t)$ and estimating $D_n(t)$, this method takes advantage of the fact that the $D_n(t)$ uncertainties are much larger than the $C(t)$ uncertainties. Inverse initialization has another advantage in that the magnitude of uncertainties in future emissions for any given stabilization profile may be assessed by varying $D_n(80s)$ over the range of likely values (following the method of Wigley, 1993; see below). For any given $D_n(80s)$ value, matching the observed $C(t)$ record by inverse modeling leads to a corresponding value for r (or β). The range of r values obtained in this way more than spans the range of values implied by the forward initialization exercise (see Table 21.1), so a full set of inverse initialization results will encompass those that would have been obtained with forward initialization.

Figure 21.2 shows the Sxxxnew and WRExxx concentration profiles, while the required emissions, using the Wigley (1993) model, are shown in Figure 21.3. The results in Figure 21.3 are, of course, model-dependent. The range of intermodel differences in emissions for any given profile is similar to the range of differences among different profiles (see Enting et al.,

1994). Four important results are independent of this uncertainty, however. First, in all cases the final emissions level is substantially below the current level. Second, for the Sxxxnew profiles (as is the case for the old "S" profiles used in Enting et al., 1994), there is, for all stabilization levels, an immediate drop in emissions below the IS92a BAU case. Third, when comparing the "S" and "WRE" profile results, the maximum emissions levels are considerably higher in the latter. This follows directly from the fact that the WRE emissions are constrained to follow a BAU trajectory for the first 10–30 years. To compensate for this, WRE emissions begin to decline sooner and decline for a longer period after maximizing, eventually becoming less than those for the S profiles. Finally, for any given stabilization level, integrated emissions over periods longer than the time to stabilization are relatively insensitive to the pathway to stabilization. For earlier times, the WRE profiles allow substantially higher cumulative emissions.

A more detailed analysis of the WRE550 case is presented in Figures 21.4 and 21.5. Figure 21.4 shows changes over time of the various carbon budget components. The annual mass buildup (corresponding to dC/dt) maximizes in 2020; industrial emissions peak slightly later because both the ocean and fertilization fluxes continue to increase until around 2050. Even though the stabilization level (550 ppmv) is substantially above the present concentration level (360 ppmv), the fertilization flux increases by only approximately 0.9 Gt C/yr (to 2.3 Gt C/yr, relative to the 1990 level of 1.4 Gt C/yr). The ocean flux also increases as a result of increasing atmospheric concentration, but by a larger amount (from 2.2 Gt C/yr in 1990 to a maximum of 4.320Gt C/yr).

After peaking around 2050, redistributions of carbon between the terrestrial biosphere reservoirs and the slow ap-

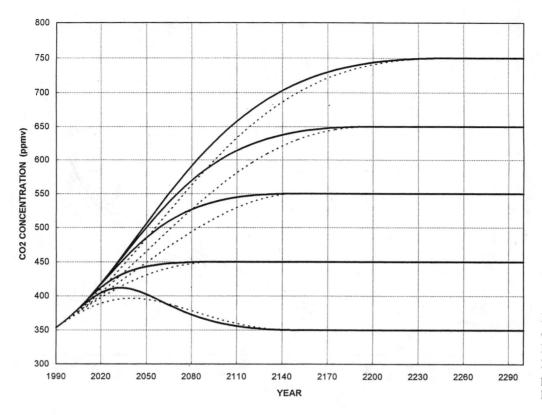

Figure 21.2: Stabilization concentration profiles used by IPCC (Schimel et al., 1996). Dashed lines denote revised "S" profiles (Sxxxnew in text); solid lines denote "WRE" profiles.

proach to a new steady state require the fertilization flux to decline slowly over time. Similarly, the continuing flux of carbon into the ocean even after the atmospheric concentration stabilizes reduces the atmosphere–ocean partial pressure difference, requiring the ocean flux to decline slowly with time. These two declining fluxes in turn require that the anthro-pogenic flux continue to decrease with time after the point of concentration stabilization.

Changes in the model's terrestrial carbon reservoirs are shown in Figure 21.5. Total storage increases with time as the result of net deforestation (a loss) and the continuing fertilization flux (a gain). Over the period 1991–2300, the integrated

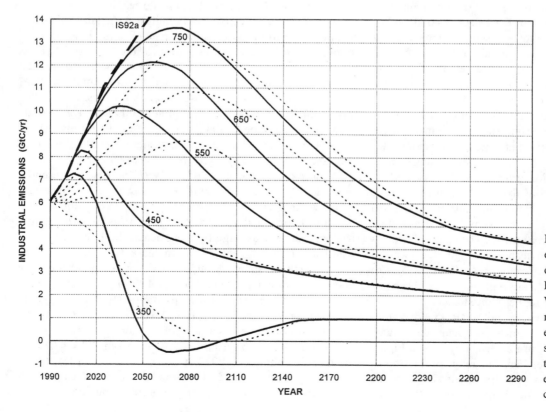

Figure 21.3: Industrial emissions corresponding to the concentration profiles shown in Figure 21.2, based on the Wigley (1993) carbon cycle model. Dashed lines denote emissions for the "S" profiles; solid lines denote emissions for the "WRE" profiles. IS92a emissions are shown for comparison.

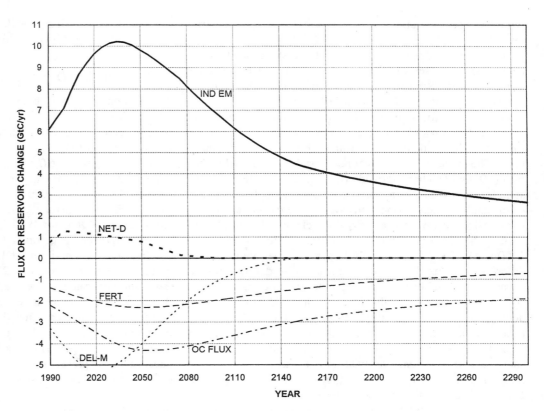

Figure 21.4: Time series of annual carbon budget breakdown for the WRE550 concentration stabilization profile: IND EM = industrial emissions; NET-D = net deforestation (prescribed); OC FLUX = atmosphere-to-ocean flux; FERT = fertilization flux into terrestrial biosphere; DEL-M = atmospheric mass increment. Note that the signs of each component have been chosen so that the sum of all components is zero.

net deforestation flux is 77 Gt C, while the integrated fertilization flux (from Figure 21.4) is 470 Gt C. Note that the latter is equivalent to an average flux of 1.5 Gt C/yr, similar to the current (1990) value. Assuming a constant terrestrial sink would therefore give results very similar to those obtained here. The net gain by the terrestrial biosphere, approximately

390 Gt C over 1991–2300, is distributed 52% into the soil carbon pool, 45% into the living vegetation pool, and 3% into the litter pool (Figure 21.5). The increase in the living vegetation pool of approximately 175 Gt C (approximately 25% relative to 1990) appears to be reasonable, given the relatively small integrated net deforestation effect and the prolonged period

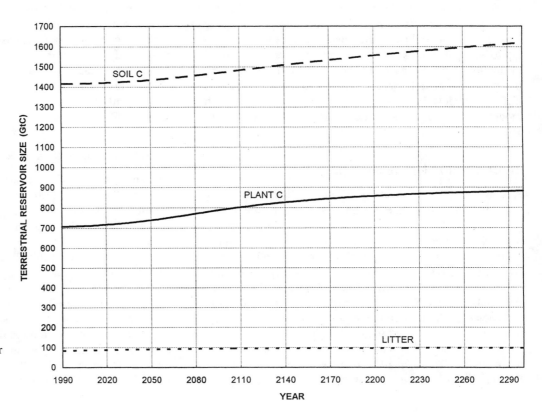

Figure 21.5: Terrestrial reservoir changes for the WRE550 concentration stabilization profile.

with CO_2 concentrations close to double the preindustrial level.

These results depend on the way deforestation and terrestrial biosphere reservoir changes are modeled. The influence of deforestation is far more complex than its simulation in the present model, which does not account for the particular use made of any deforested area. This could reduce the potential for a fertilization sink. Changes in the biomass pool available for NPP enhancement are determined by gross deforestation (D_g). The present model assumes a stable relationship between D_g and D_n characterized primarily by a single time constant for regrowth. When land use is changed, and particularly when land is taken permanently out of the carbon cycle, the relationship between D_g and D_n must change. In the present model, the integrated gross deforestation flux over 1991–2300 is 410 Gt C, substantially larger than the integrated net deforestation flux. If a significant fraction of deforested land became unavailable for regrowth, the potential exists for a significant reduction in the fertilization sink. However, reducing this sink substantially does not alter the implied industrial emissions in a major way for any given stabilization level; the uncertainty range is less than the range of values associated with different stabilization pathways. The potential for change is, in fact, clear from Figure 21.5. Since $D_n(t)$ and $C(t)$ are fixed, and since $C(t)$ determines $F_{oc}(t)$, changes in $X(t)$ must be directly reflected in $I(t)$. Thus, reducing $X(t)$ to zero can reduce the peak value of $I(t)$ only by approximately 2 Gt C/yr.

21.5 Uncertainties

Two types of uncertainties may be distinguished: those resulting from errors in model structure and those resulting from model parameter uncertainties. The two must, of course, overlap – altering model parameters is a standard way of attempting to account for possible model structure errors or oversimplifications. Enting and colleagues (1994) illustrated some of the structure-related uncertainties by comparing results from different models. Here, we consider uncertainties arising from model parameter uncertainties (for the model of Wigley, 1993).

21.5.1 Fertilization Formulation

A priori, one would expect the rectangular hyperbolic formulation for NPP enhancement used in the above calculations to give lower emissions than the logarithmic formulation. For both, NPP increases as concentration increases. At high concentrations, however, the growth rate of NPP with increasing concentration is significantly less in the rectangular hyperbolic case. For the cases considered here, however, the differences are small. For the WRE profiles, the maximum effect (larger emissions for the logarithmic case) occurs around the time of maximum emissions and varies from less than 0.1 Gt C/yr for WRE350, through 0.20 Gt C/yr for WRE550, to 0.46 Gt C/yr for WRE750. These small differences are partially due to the fact that both cases are constrained by the initialization to give

the same mean fertilization flux over the 1980s. This allows less scope for any subsequent fertilization flux differential.

21.5.2 Post-1990 $D_n(t)$ Scenario

The next uncertainty is that associated with the $D_n(t)$ scenario. Clearly, if $D_n(t)$ were larger, then $I(t)$ must be smaller. This follows directly from Equation (21.1), since the atmospheric mass increment (dM/dt) is fixed, and since $F_{oc}(t)$ and $X(t)$ are controlled largely by $C(t)$; thus, $I(t) + D_n(t)$ must be almost independent of $D_n(t)$ for any given $C(t)$. We therefore expect only small changes in total emissions to arise from any change in $D_n(t)$. To demonstrate this, Figure 21.6 shows total emissions ($D_n(t)$) + $I(t)$) for WRE550 using the standard (IS92a-based) $D_n(t)$ scenario compared with results using a scenario in which $D_n(t)$ stays constant at 1.3 Gt C/yr to 2100 before declining linearly to zero over 2100–2200. Over the period to 2200, the integrated net deforestation flux difference between these two cases is quite large, 127 Gt C.

Figure 21.6 gives a somewhat counterintuitive result: The total emissions allowed is actually slightly higher in the case where $D_n(t)$ is larger. This is a model-specific result. It arises because, in the present model, the larger net deforestation flux depletes the terrestrial carbon reservoirs more. Hence, their fluxes into the atmosphere through the oxidation of organic matter are less, allowing a higher flux from other (anthropogenic) sources. This is, however, a relatively minor effect: The difference in the sum of $I(t)$ and $D_n(t)$ (i.e., in total anthropogenic emissions) is only approximately 0.02 Gt C/yr at the time of maximum emissions, increasing slowly to approximately 0.2 Gt C/yr by 2300.

These results may be critically dependent on (and therefore more sensitive to) the current model's simple formulation of the relationships between $D_n(t)$, $D_g(t)$, and the fertilization sink – more specifically, to the use assumed for the deforested land. Since the larger $D_n(t)$ case corresponds to a much larger integrated gross deforestation flux, there is considerable potential in this case for a more significant reduction in the fertilization flux if land previously available for NPP were removed permanently from production or replaced by biomass of much lower NPP.

21.5.3 Additional Sink Specification

The main source of uncertainty is that arising from the assumption that CO_2 fertilization is the only terrestrial sink, through the magnitude of the fertilization effect implied by this assumption. This, in turn, is determined by the assumed value of $D_n(80s)$.

The use of inverse initialization with fertilization tuned to match a specific value of $D_n(80s)$ is a more powerful method than might first appear to be the case. By varying $D_n(80s)$, one can investigate the effect of not only uncertainties in the net deforestation component of the budget, but also uncertainties related to the way the additional sink ($X(t)$) is specified and uncertainties arising from our imprecise knowledge of ocean uptake (as determined by $F_{oc}(80s)$). A brief explanation of this is given in Wigley (1993), which we elaborate on here.

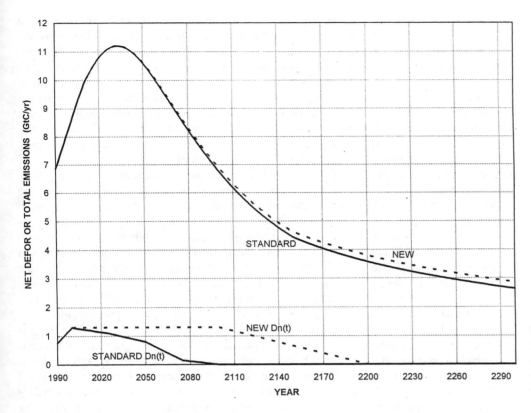

Figure 21.6: Effect of assumed future changes in net deforestation on total emissions (net deforestation plus industrial) under stabilization profile WRE550.

To see how this more general interpretation arises, we first write Equation (21.1) in the form

$$dM/dt = I + D_n - F_{oc} - X_{fert} - Y \qquad (21.2)$$

which separates the sink term (X) into a fertilization component (X_{fert}) and an unspecified component (Y). We now define an effective net deforestation term by

$$D_n^* = D_n - Y \qquad (21.3)$$

which allows us to rewrite Equation (21.2) as

$$dM/dt = I + D_n^* - F_{oc} - X_{fert} \qquad (21.4)$$

Equation (21.4) has the same form as Equation (21.1) in that X_{fert} is the only explicitly specified sink term; however, the net deforestation term must now be interpreted as combining the true net deforestation and the unspecified sink. In our analyses thus far, we have identified D_n^* with D_n, which is equivalent to assuming $Y \equiv 0$ or $X \equiv X_{fert}$; however, this is not necessary.

A more general interpretation is to use Equation (21.4) for initialization and to revert to the $D_n^* \equiv D_n$ case for future projections. This is equivalent to assuming that a nonzero Y term exists up to 1990, and that this term drops linearly to zero over 1990–2000. On the assumption that the unspecified sink term (Y) is always positive, setting this term to zero after 1990 is a way of maximizing the error associated with its nonspecification. This, in turn, means that for any specified D_n^* value, the inverse-calculated $I(t)$ is a lower bound.

A specific example illustrates this. In the calculations above, we have used $D_n(80s) = 1.1$ Gt C/yr in initialization. More generally, we can begin by assuming $D_n^*(80s) = 1.1$ Gt C/yr, which leads to the same fertilization term as previously,

as is clear from Equation (21.4). This is equivalent to assuming that $D_n(80s)$ is, for example, $(1.1 + \delta)$ Gt C/yr, and that the 1980s-mean value of the unspecified sink is δ Gt C/yr. The actual value of $D_n(80s)$ does not need to be specified; for $\delta > 0$, it must exceed 1.1 Gt C/yr in this particular example.

For the future projections, however, the true value of $D_n(t)$ *is* specified. These calculations are performed by using Equation (21.2) with $Y \equiv 0$. The method therefore assumes a smooth transition from the D_n^* interpretation up to 1990 (with nonzero Y) to the D_n interpretation from 2000 onward (with Y necessarily zero). If $Y > 0$ up to 1990, it is unlikely to drop to zero over 1990–2000; therefore, so the calculated $I(t)$ results corresponding to $D_n^*(80s) = 1.1$ Gt C/yr must be lower bound values.

Another way to view this is to note that the specification of $D_n^*(80s)$ for initialization is equivalent to specifying both $X_{fert}(80s)$ and $Y(80s)$. The value of $X_{fert}(80s)$ that is effectively being specified is determined uniquely by $D_n^*(80s)$ through Equation (21.4) as

$$X_{fert}(80s) = I(80s) + D_n^*(80s) - F_{oc}(80s) - \Delta M(80s) \quad (21.5)$$

while the assumed value of $Y(80s)$ is determined only if $D_n(80s)$ is known, through

$$Y(80s) = D_n(80s) - D_n^*(80s) \qquad (21.6)$$

For an uncertainty analysis, we need to know the likely range of values for $D_n^*(80s)$. This may be estimated using information given in Schimel and colleagues (1996). For the various budget components, we have, as central values, $\Delta M(80s) = 3.28$ Gt C/yr, $I(80s) = 5.46$ Gt C/yr, and $F_{oc}(80s) = 2.0$ Gt C/yr. If we take $D_n^*(80s) = 1.1 \pm 0.7$ Gt C/yr together with these values, the balanced-budget constraint leads to

X_{fert} = 0.6 (1.3) 2.0 Gt C/yr, corresponding to fertilization factor values of r = 1.06 (1.18) 1.33. In Schimel and colleagues (1996), the stated uncertainty range for X_{fert} is 0.5–2.0 Gt C/yr, so the range for $D_n^*(80s)$ assumed above is fully consistent with this.

Note that $D_n^*(80s)$ = 1.1 ± 0.7 Gt C/yr, together with $D_n(80s)$ = 1.1 ± 1.1 Gt C/yr (Schimel et al., 1996, Table 21.1), implies $Y(80s)$ = 0.0 ± 1.3 Gt C/yr. While the error bars on this inferred Y value are reasonable, the central value is less so. If the central estimates of the other budget components are reasonable, then a Y value close to zero would require, for example, the nitrogen fertilization and climate effects on the budget of the 1980s to balance approximately, which is unlikely given the estimates provided in Schimel and colleagues (1996) for these terms. Alternatively, for Y to be noticeably positive (as would be the case if nitrogen fertilization were the dominant component of Y), we would require, for example, an ocean flux term somewhat less than 2.0 Gt C/yr and/or a net deforestation term somewhat larger than 1.1 Gt C/yr.

We now assess the implications of the uncertainty in $D_n^*(80s)$ for a particular case, that of the WRE550 stabilization profile. The inverse calculations are the same as previously, except that the input value of $D_n(80s)$, which determines the CO$_2$ fertilization factor, is now interpreted more generally as $D_n^*(80s)$. The three values considered, $D_n^*(80s)$ = 0.4 (1.1) 1.8 Gt C/yr, correspond to additional sinks (i.e., sinks not directly related to CO$_2$ fertilization) of $Y(80s)$ = − 1.3 (0.0) 1.3 Gt C/yr. As noted above, the negative $Y(80s)$ values (which lead to higher industrial emissions because they require a larger CO$_2$ fertilization effect to balance the 1980s budget) seem, a priori, less likely than positive $Y(80s)$ values. We retain the full range,

however, since we have insufficient quantitative information on the full budget to make a sound judgment regarding the magnitude of Y. A negative $Y(80s)$ value could easily be eliminated by choosing different central values for $F_{oc}(80s)$ and/or $D_n(80s)$, for example. The crucial aspect here is that the range of CO$_2$ fertilization values used lies within and is consistent with a priori determined uncertainty bounds. No matter what $Y(80s)$ actually is, the present methodology assumes that $Y(t)$ declines to zero by 2000, leaving CO$_2$ fertilization as the only terrestrial sink from then onward.

Varying $D_n^*(80s)$ to assess uncertainties is not straightforward for the WRE stabilization profiles. This is because the profiles themselves are dependent on the value of $D_n^*(80s)$, which is a key factor in determining future concentrations under the IS92a emissions scenario (which WRE550 follows to 2010). Thus, if we change $D_n^*(80s)$, the concentration values to 2010 will change. The concentration profile after 2010 must therefore change to provide a smooth match in 2010.

Results are shown in Figures 21.7 and 21.8. Figure 21.7 shows the modified WRE550 concentration profiles adjusted to be consistent with the different $D_n^*(80s)$ values, while the implied industrial emissions are shown in Figure 21.8. The emissions differences in Figure 21.8 arise from two effects. First, the concentration profile differences lead to slight emissions differences even for fixed $D_n^*(80s)$ (i.e., fixed fertilization factor). The dominant effect, however, is that which results from running the inverse calculations with the different values of $D_n^*(80s)$. This acts in the opposite direction to (and overwhelms) the first effect. Overall, the resulting emissions uncertainty is (necessarily) zero to 2010, increases to approximately ±2 Gt C/yr by around 2050, and declines slowly thereafter.

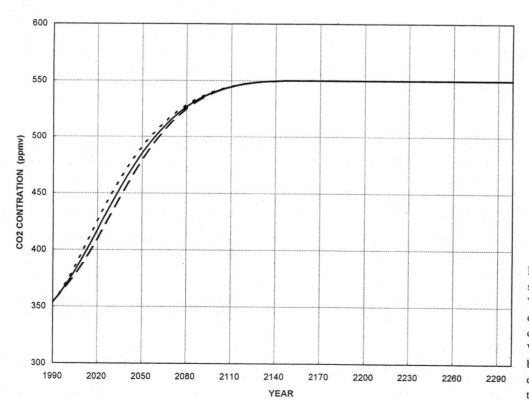

Figure 21.7: Concentration stabilization profiles equivalent to WRE550 for three different values of (0.4, 1.1, and 1.8 Gt C/yr). The central curve is the standard WRE550 case. Profiles differ because of the constraint that emissions should follow IS92a out to 2010.

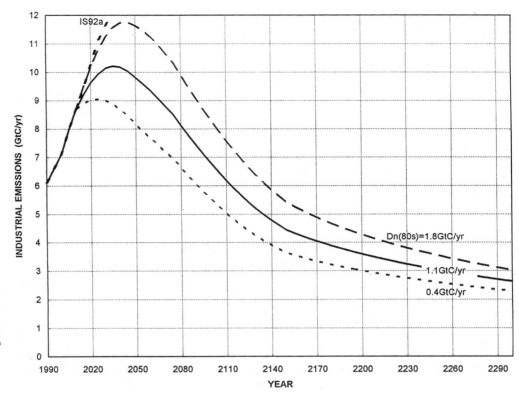

Figure 21.8: Industrial emissions uncertainties for the WRE550 case arising from uncertainties in the specification of the terrestrial sink, corresponding to three values of D_n^* (0.4, 1.1, and 1.8 Gt C/yr). The corresponding concentration profiles are shown in Figure 21.7.

By comparing Figures 21.8 and 21.3, it can be seen that this uncertainty (i.e., the uncertainty resulting from the way the terrestrial sink term is represented) is similar in magnitude to the emissions differences between the S550new and WRE550 cases, and is equivalent to a misspecification of the stabilization level by approximately ±50 ppmv.

The possibility of a bias in the results of Figure 21.8 has already been noted. The implicit assumption that $Y(t)$ drops to zero by 2000 means that the set of $I(t)$ projections shown in the figure may all be too low – an additional sink after 2000 would clearly allow higher industrial emissions. Compensating for this, however, is the possibility that the negative values of $Y(80s)$ implied by $D_n^*(80s) = 1.1 \pm 0.7$ Gt C/yr are unrealistic. Since these values correspond to the higher values of $D_n^*(80s)$ and, hence, to the higher fertilization factors, this in turn would imply that the lower values of $I(t)$ in Figure 21.8 are more likely. On balance, therefore, and within the constraints of the Wigley (1993) carbon cycle model used here, Figure 21.8 provides a reasonably unbiased view of the effect of terrestrial sink uncertainties on future values of $I(t)$.

21.5.4 Ocean Flux Effects

With the present methodology, it is not possible to separate out the effects of ocean flux uncertainties, even though these are large and they must clearly affect the trajectory of future emissions corresponding to any given concentration stabilization profile. There are two reasons for this: there is compensation between the ocean and terrestrial biosphere sinks in "balancing" the 1980s-mean budget; and (in the model) these two sinks vary similarly in the future as concentrations change (see Figure 21.4). When the budget is written as in Equation (21.4),

with the 1980s-mean values of dM/dt, I, and D_n^* all specified, the sum of F_{oc} and X_{fert} must be fixed. Choosing a lower value of $F_{oc}(80s)$ therefore requires a higher value of $X_{fert}(80s)$ and of the fertilization factor, r. For the future, the effect of a lower ocean flux is then largely offset by the higher fertilization effect. This is illustrated in Figure 21.9, which shows little difference in the implied industrial emissions for WRE550 for a range of $F_{oc}(80s)$ values, 2.0 ± 0.8 Gt C/yr. Other profiles show a similarly low sensitivity.

21.6 Conclusions

This chapter describes the methods used to construct profiles that stabilize atmospheric CO₂ concentration at different levels, and the inverse carbon cycle modeling procedure used to calculate the emissions required to follow these different profiles. Two different types of pathways toward each stabilization level were considered (Figure 21.2). The emissions requirements vary greatly according to the pathway (depicted in Figure 21.3). In all cases, however, emissions must eventually drop to well below current levels in order to achieve and maintain concentration stabilization. Even after stabilization is attained, emissions must continue to drop as the fluxes into the ocean and the terrestrial biosphere continue to decrease (Figure 21.4).

Uncertainties involved in the calculation of emissions were quantified. The method of formulating CO₂ fertilization has only a small influence – tenths of a Gt C at most. The choice of scenario for future net deforestation ($D_n(t)$, $t > 1990$) must clearly affect the implied industrial emissions ($I(t)$); however, it has only a small effect on total emissions ($D_n(t) + I(t)$) (Fig-

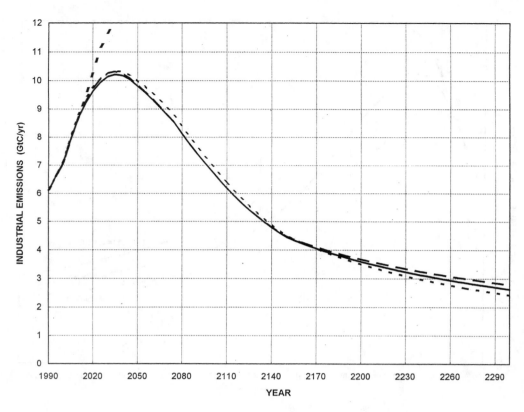

Figure 21.9: Effect of different 1980s-mean ocean fluxes on emissions requirements for the WRE550 stabilization profile. The effect is small because of compensating changes in the fertilization flux.

ure 21.6). With the present methodology, ocean flux uncertainties have only a minor influence on implied emissions (Figure 21.9) because such errors are compensated for in tuning the CO_2 fertilization flux.

The most important source of uncertainty is the result of the method used to quantify the residual (or so-called missing) sink. A general analysis is presented that suggests that misspecification of this sink, either quantitatively or qualitatively, could lead to errors of up to ±2 Gt C/yr in the implied emissions for any given stabilization profile (Figure 21.8). Such an error is equivalent to a misspecification of the stabilization level by approximately ±50 ppmv. Other calculations support this conclusion. For example, details of the future carbon budget in the WRE550 case (Figure 21.4) show that the residual (fertilization) sink in the future will remain of comparable magnitude to today's sink (1.4 Gt C/yr). If this sink were removed instantaneously in 1990, the resulting industrial emissions allowed would have to be smaller, leading to a lower bound for such future emissions of order 1–2 Gt C below the original results. Conversely, if the fertilization sink shown in Figure 21.4 were to be supplemented in the future by, for example, a nitrogen fertilization sink of magnitude comparable to its estimated current strength, this would allow substantially higher industrial emissions in the future.

Acknowledgments

The work reported here was supported by the U.S. Department of Energy under grant number DE-FG02–86ER60397. Many individuals have provided useful comments on this work, and discussions with Richard Richels have been particularly valuable.

References

Bruce, J. P., H. Lee, and E. F. Haites. 1996. Summary for policymakers. In *Climate Change 1995: Economic and Social Dimensions of Climate Change, Contribution of Working Group III to the Second Assessment Report of the Intergovernmental Panel on Climate Change,* J. P. Bruce, H. Lee, and E. F. Haites, eds. Cambridge University Press, Cambridge, U.K., pp. 1–16.

Conway, T. J., P. Tans, L. S. Waterman, K. W. Thoning, D. R. Buanerkitzis, K. A. Maserie, and N. Zhang. 1994. Evidence for interannual variability of the carbon cycle from the NOAA/CMDL global air sampling network. *Journal of Geophysical Research 99,* 22831–22855.

Enting, I. G., and J. V. Mansbridge. 1987. Inversion relations for the deconvolution of CO_2 data from ice cores. *Inverse Problems 3,* L63–L69.

Enting, I. G., and G. I. Pearman. 1987. Description of a one-dimensional carbon cycle model calibrated using techniques of constrained inversion. *Tellus 39B,* 459–476.

Enting, I. G., T. M. L. Wigley, and M. Heimann. 1994. *Future Emissions and Concentrations of Carbon Dioxide: Key Ocean/Atmosphere/Land Analyses.* Technical Paper No. 31, Division of Atmospheric Research, CSIRO, Australia.

Etheridge, D. M., L. P. Steele, R. L. Langenfelds, R. J. Francey, J.-M. Barnola, and V. I. Morgan. 1996. Natural and anthropogenic changes in atmospheric CO_2 over the last 1000 years from air in Antarctic ice and firn. *Journal of Geophysical Research 101,* 4115–4128.

Friedli, H., H. Loetscher, H. Oeschger, U. Siegenthaler, and B. Stauffer. 1986. Ice core record of the $^{13}C/^{12}C$ ratio of atmospheric CO_2 in the past two centuries. *Nature 324,* 237–238.

Gifford, R. M. 1993. Implications of CO_2 effects on vegetation for the global carbon budget. In *The Global Carbon Cycle,* M. Heimann, ed. Proceedings of the NATO ASI at Il Ciocco, Italy,

September 8–20, 1991. Springer-Verlag, Berlin, Germany, pp. 165–205.

Houghton, R. A. 1991. The role of forests in affecting the greenhouse gas composition of the atmosphere. In *Global Climate Change and Life on Earth*, R. L. Wyman, ed. Chapman and Hall, New York, pp. 43–55.

Jain, A. K., H. S. Kheshgi, M. I. Hoffert, and D. J. Wuebbles. 1995. Distribution of radiocarbon as a test of global carbon cycle models. *Global Biogeochemical Cycles 9*, 153–166.

Keeling, C. D. 1991. CO$_2$ emissions – Historical record, global. In *Trends '91: A Compendium of Data on Global Change*, T. A. Boden, R. J. Sepanski, and F. W. Stoss, eds. Carbon Dioxide Information Analysis Center, Oak Ridge National Laboratories, Oak Ridge, Tenn., pp. 382–385.

Keeling, C. D. 1994. Global historical CO$_2$ emissions. In *Trends '93: A Compendium of Data on Global Change*, T. A. Boden, D. P. Kaiser, R. J. Sepanski, and F. W. Stoss, eds. Carbon Dioxide Information Analysis Center, Oak Ridge National Laboratories, Oak Ridge, Tenn., pp. 501–504.

Keeling, C. D., and T. P. Whorf. 1991. Atmospheric CO$_2$–Modern record, Mauna Loa. In *Trends '91: A Compendium of Data on Global Change*, T. A. Boden, R. J. Sepanski, and F. W. Stoss, eds. Carbon Dioxide Information Analysis Center, Oak Ridge National Laboratories, Oak Ridge, Tenn., pp. 12–14.

Kohmyr, W. D., R. H. Gammon, T. B. Harris, L. S. Waterman, T. J. Conway, W. R. Taylor, and K. W. Thoning. 1985. Global atmospheric CO$_2$ distribution and variations from 1968–82 NOAA/GMCC CO$_2$ flask sample data. *Journal of Geophysical Research 90*, 5567–5596.

Leggett, J. A., W. J. Pepper, and R. J. Swart. 1992. Emissions scenarios for IPCC: An update. In *Climate Change 1992: The Supplementary Report to the IPCC Scientific Assessment*, J. T. Houghton, B. A. Callander, and S. K. Varney, eds. Cambridge University Press, Cambridge, U.K., pp. 69–95.

Marland, G., R. J. Andres, and T. A. Boden. 1994. Global, regional, and national CO$_2$ emissions. In *Trends '93: A Compendium of Data on Global Change*, T. A. Boden, D. P. Kaiser, R. J. Sepanski, and F. W. Stoss, eds. Carbon Dioxide Information Analysis Center, Oak Ridge National Laboratories, Oak Ridge, Tenn., pp. 505–508.

Marland, G., and T. A. Boden. 1991. CO$_2$ emissions: Modern record, global. In *Trends 91: A Compendium of Data on Global Change*, T. A. Boden, R. J. Sepanski, and F. W. Stoss, eds. Carbon Dioxide Information Analysis Center, Oak Ridge National Laboratories, Oak Ridge, Tenn., pp. 386–389.

Raper, S. C. B., T. M. L. Wigley, and R. A. Warrick. 1996. Global sea level rise: Past and future. In *Sea-Level Rise and Coastal Subsidence: Causes, Consequences and Strategies*, J. D. Milliman and B. U. Haq, eds. Kluwer Academic Publishers, Dordrecht, The Netherlands, pp. 11–45.

Schimel, D., D. Alves, I. Enting, M. Heimann, F. Joos, D. Raynaud, and T. M. L. Wigley. 1996. CO$_2$ and the carbon cycle. In *Climate Change 1995: The Science of Climate Change, Contribution of Working Group I to the Second Assessment Report of the Intergovernmental Panel on Climate Change*, J. T. Houghton, L. G. Meira Filho, B. A. Callander, N. Harris, A. Kattenberg, and K. Maskell, eds. Cambridge University Press, Cambridge, U.K., pp. 65–86.

Schimel, D., I. Enting, M. Heimann, T. M. L. Wigley, D. Raynaud, D. Alves, and U. Siegenthaler. 1995. CO$_2$ and the carbon cycle. In *Climate Change 1994: Radiative Forcing of Climate Change and an Evaluation of the IPCC IS92 Emissions Scenarios*, J. T. Houghton, L. G. Meira Filho, J. Bruce, H. Lee, B. A. Callander, E. Haites, N. Harris and K. Maskell, eds. Cambridge University Press, Cambridge, U.K., pp. 35–71.

Siegenthaler, U., and F. Joos. 1992. Use of a simple model for studying oceanic tracer distributions and the global carbon cycle. *Tellus 44B*, 186–207.

Siegenthaler, U., and H. Oeschger. 1987. Biospheric CO$_2$ emissions during the past 200 years reconstructed by deconvolution of ice core data. *Tellus 39B*, 140–154.

Warrick, R. A., J. Oerlemans, P. Woodworth, M. Meier, and C. Le Provost. 1996. Changes in sea level. In *Climate Change 1995: The Science of Climate Change, Contribution of Working Group I to the Second Assessment Report of the Intergovernmental Panel on Climate Change*, J. T. Houghton, L. G. Meira Filho, B. A. Callander, N. Harris, A. Kattenberg, and K. Maskell, eds. Cambridge University Press, Cambridge, U.K., pp. 358–405.

Wigley, T. M. L. 1991. Could reducing fossil-fuel emissions cause global warming? *Nature 349*, 503–506.

Wigley, T. M. L. 1993. Balancing the carbon budget: Implications for projections of future carbon dioxide concentration changes. *Tellus 45B*, 409–425.

Wigley, T. M. L. 1995. Global-mean temperature and sea level consequences of greenhouse gas concentration stabilization. *Geophysical Research Letters 22*, 45–48.

Wigley, T. M. L., R. Richels, and J. A. Edmonds. 1996. Economic and environmental choices in the stabilization of atmospheric CO$_2$ concentrations. *Nature 379*, 242–245.

Appendix 21.1: Historical and Stabilization Concentration Data

Historical concentration data affect the results of stabilization calculations in two ways: the total change over the decade of the 1980s is a key component of the 1980s-mean budget used for inverse initialization; the specific values of $C(1990.5)$ and dC/dt in 1990.5 (taken as $C(1991.0)–C(1990.0)$) determine the stabilization profiles either through the need to match these profiles smoothly to past data (for the Sxxx profiles) or as initial conditions for future concentrations under the IS92a emissions scenario (used in the WRExxx profiles).

For the most recent IPCC calculations (Schimel et al., 1996), the concentration history data used are given in Table 21.A1, denoted "IPCC." More recent ice core results from Etheridge and colleagues (1996) have allowed an improvement of these data (denoted "Etheridge" in the Table 21.A1). The ice core data have been spline-smoothed (D. Etheridge, personal communication) and joined smoothly to the same historical atmospheric observed data used for the IPCC data set. This new blended data set is provided here for future reference.

The newer data show interesting, shorter time-scale fluctuations, not evident in the IPCC data: most noticeably, slight declines in concentration over the 1820s and 1830s and around 1940. Since both data sets employ the same directly observed data in recent years, they do not differ over the decade of the 1980s – they are identical from 1966 onward. Therefore, their use in future analyses would not require re-calculation of the concentration stabilization profiles.

The stabilization profiles for future concentration changes may be calculated simply from the Padé approximant formula given in the text, using the coefficients presented in Table 21.A3. These coefficients are based on values of C and dC/dt listed in Table 21.A2. It is possible to determine the various

Table 21.A1. *Historical concentration data sets (midyear values in ppmv)*

Year	Midyear IPCC	Etheridge	Year	Midyear IPCC	Etheridge	Year	Midyear IPCC	Etheridge	Year	Midyear IPCC	Etheridge
1764	277.938	277.430	1821	283.551	285.036	1878	290.696	289.797	1935	306.280	309.410
1765	277.965	277.547	1822	283.648	285.067	1879	290.909	290.255	1936	306.546	309.742
1766	277.995	277.669	1823	283.743	285.084	1880	291.129	290.730	1937	306.815	310.012
1767	278.028	277.797	1824	283.836	285.083	1881	291.355	291.210	1938	307.087	310.210
1768	278.063	277.929	1825	283.928	285.058	1882	291.587	291.683	1939	307.365	310.328
1769	278.102	278.067	1826	284.020	285.011	1883	291.824	292.135	1940	307.650	310.368
1770	278.143	278.209	1827	284.112	284.940	1884	292.066	292.558	1941	307.943	310.348
1771	278.187	278.357	1828	284.203	284.847	1885	292.313	292.940	1942	308.246	310.288
1772	278.234	278.509	1829	284.296	284.735	1886	292.562	293.277	1943	308.560	310.213
1773	278.284	278.665	1830	284.391	284.608	1887	292.815	293.570	1944	308.887	310.148
1774	278.337	278.824	1831	284.487	284.467	1888	293.071	293.818	1945	309.228	310.110
1775	278.393	278.987	1832	284.584	284.312	1889	293.328	294.023	1946	309.584	310.115
1776	278.452	279.153	1833	284.686	284.145	1890	293.586	294.193	1947	309.956	310.175
1777	278.514	279.321	1834	284.792	283.972	1891	293.843	294.333	1948	310.344	310.297
1778	278.579	279.490	1835	284.900	283.808	1892	294.098	294.450	1949	310.749	310.487
1779	278.648	279.660	1836	285.014	283.665	1893	294.350	294.557	1950	311.172	310.756
1780	278.719	278.830	1837	285.132	283.548	1894	294.598	294.660	1951	311.614	311.114
1781	278.794	280.001	1838	285.257	283.458	1895	294.842	294.767	1952	312.077	311.551
1782	278.872	280.174	1839	285.387	283.403	1896	295.082	294.895	1953	312.561	312.050
1783	278.953	280.350	1840	285.524	283.390	1897	295.320	295.052	1954	313.068	312.598
1784	279.038	280.530	1841	285.665	283.420	1898	295.558	295.250	1955	313.599	313.189
1785	279.126	280.715	1842	285.809	283.490	1899	295.797	295.492	1956	314.154	313.815
1786	279.217	280.905	1843	285.953	283.597	1900	296.038	295.780	1957	314.737	314.465
1787	279.311	281.098	1844	286.097	283.735	1901	296.284	296.107	1958	315.347	315.134
1788	279.409	281.293	1845	286.238	283.892	1902	296.535	296.465	1959	315.984	315.823
1789	279.510	281.491	1846	286.378	284.060	1903	296.794	296.845	1960	316.646	316.526
1790	279.615	281.689	1847	286.515	284.232	1904	297.062	297.242	1961	317.328	317.240
1791	279.722	281.886	1848	286.649	284.407	1905	297.337	297.650	1962	318.025	317.965
1792	279.833	282.082	1849	286.780	284.577	1906	297.620	298.060	1963	318.742	318.707
1793	279.948	282.275	1850	286.908	284.734	1907	297.910	298.470	1964	319.489	319.474
1794	280.067	282.465	1851	287.032	284.877	1908	298.204	298.877	1965	320.282	320.278
1795	280.188	282.651	1852	287.154	285.010	1909	298.504	299.280	1966	321.133	321.133
1796	280.313	282.832	1853	287.272	285.142	1910	298.806	299.672	1967	322.045	322.045
1797	280.442	283.007	1854	287.386	285.280	1911	299.111	300.050	1968	323.021	323.021
1798	280.573	283.175	1855	287.497	285.425	1912	299.419	300.412	1969	324.059	324.059
1799	280.708	283.334	1856	287.605	285.579	1913	299.728	300.757	1970	325.155	325.155
1800	280.845	283.483	1857	287.710	285.742	1914	300.040	301.087	1971	326.299	326.299
1801	280.984	283.624	1858	287.815	285.907	1915	300.352	301.410	1972	327.484	327.484
1802	281.124	283.754	1859	287.920	286.070	1916	300.666	301.730	1973	328.698	328.698
1803	281.266	283.875	1860	288.025	286.227	1917	300.980	302.052	1974	329.933	329.933
1804	281.408	283.987	1861	288.131	286.375	1918	301.294	302.380	1975	331.194	331.194
1805	281.550	284.091	1862	288.239	286.510	1919	301.608	302.712	1976	332.498	332.498
1806	281.693	284.187	1863	288.350	286.637	1920	301.923	303.052	1977	333.853	333.853
1807	281.835	284.275	1864	288.465	286.757	1921	302.237	303.402	1978	335.254	335.254
1808	281.976	284.354	1865	288.585	286.872	1922	302.550	303.765	1979	336.690	336.690
1809	282.116	284.425	1866	288.709	286.990	1923	302.862	304.145	1980	338.150	338.150
1810	282.253	284.488	1867	288.839	287.110	1924	303.172	304.545	1981	339.628	339.628
1811	282.389	284.545	1868	288.974	287.235	1925	303.478	304.965	1982	341.126	341.126
1812	282.522	284.599	1869	289.116	287.372	1926	303.779	305.400	1983	342.650	342.650
1813	282.652	284.650	1870	289.263	287.527	1927	304.075	305.845	1984	344.206	344.206
1814	282.779	284.700	1871	289.416	287.702	1928	304.366	306.302	1985	345.797	345.797
1815	282.901	284.750	1872	289.577	287.897	1929	304.651	306.770	1986	347.397	347.397
1816	283.020	284.800	1873	289.744	288.117	1930	304.930	307.237	1987	348.980	348.980
1817	283.134	284.850	1874	289.919	288.372	1931	305.206	307.700	1988	350.551	350.551
1818	283.243	284.900	1875	290.103	288.668	1932	305.478	308.157	1989	352.100	352.100
1819	283.349	284.949	1876	290.293	289.000	1933	305.746	308.605	1990	353.636	353.636
1820	283.451	284.995	1877	290.491	289.375	1934	306.013	309.027			

Note: All calculations in the main text use the IPCC data set. If these data are used, the present chapter should be cited as the source reference. If the "Etheridge" data are used, the present chapter and Etheridge and colleagues (1996) should both be cited.

coefficients analytically from the given values of C and dC/dt. In all cases, there are some simple algebraic constraints on the coefficients. First, the chosen form (see the main chapter text) has $\tilde{C} = 0$ when $x = 0$, which automatically ensures that $C = C_0$ when $t = t_0$. Second, since $C = C_s$ when $t = t_s$ (i.e., $\tilde{C} = 1$ when $x = 1$), we must have

$$a + b + c = 1 + d + e$$

Third, since $dC/dt = 0$ when $t = t_s$ (i.e., $d\tilde{C}/dx = 0$ at $x = 1$), we must also have

$$e = 1 + b + 2c$$

The above two equations lead to

$$a = 2 + c + d$$

Thus, instead of five coefficients, only three are independent. In the monotonic cases (i.e., all profiles bar S350 and WRE350), where $c \equiv 0$, only two coefficients are independent.

Appendix 21.2: Additional Implications of the WRExxx/Sxxx Profile Comparisons

The derivation of the WRExxx profiles is described in Wigley and colleagues (1996), where a discussion of the economic and climate implications of these profiles is given. Here we give a review of this discussion, elaborating on points that are covered only briefly in the original work. The stabilization

profiles originally derived by Enting and colleagues (1994), the Sxxx profiles, were designed as a standard data set for a large intermodel comparison exercise carried out as input into the 1994 IPCC report (Schimel et al., 1995). These profiles were meant to be "illustrative rather than prescriptive" (p. 6). The WRE pathways were generated to widen the playing field, and to account for the fact that the rapid and immediate departure from BAU emissions implied by the IPCC (Enting et al.) profiles may be unrealistic (for technical, economic, and/or policy reasons).

The work of Wigley and colleagues (1996) falls squarely in the domain of the chemical and physical sciences in developing and concentrating on results from carbon cycle and climate models. There is no economic modeling involved; however, their paper reviews recent economic literature relevant to the concentration stabilization issue and interprets the assumptions and results qualitatively in economic terms.

The WRE profiles were designed so that the implied emissions would follow BAU for 10–30 years (depending on the stabilization level). It is important to note that the "follow BAU" assumption was used only as an approximation to (or idealization of) trajectories that departed initially only slowly from BAU. *Slow initial departure from BAU* is the primary underlying assumption of the paper by Wigley and colleagues.

The emissions results (see main chapter text, Figure 21.3), as is obvious a priori, show that considerably larger initial emissions are allowable for the "slow departure" (WRE) cases

Table 21.A2. *Fixed points and derivatives used in determining Padé approximant coefficients*

Profile[a]	t_0	C_0	$dC/dt\,_0$	t_1	C_1	$dC/dt|_1$	t_s	C_s	$dC/dt|_s$
S350 old	1990.5	354.170	1.700	2041.0	396.50	0.0	2150.5	350.0	0.0
S450 old	1990.5	354.170	1.700	2031.0	411.91	—	2100.5	450.0	0.0
S550 old	1990.5	354.170	1.700	2031.0	420.78	—	2150.5	550.0	0.0
S650 old	1990.5	354.170	1.700	2031.0	429.23	—	2200.5	650.0	0.0
S750 old	1990.5	354.170	1.700	2051.0	489.17	—	2250.5	750.0	0.0
S350 new	1990.5	353.636	1.542	2041.0	396.50	0.0	2150.5	350.0	0.0
S450 new	1990.5	353.636	1.542	2051.0	431.028	—	2100.5	450.0	0.0
S550 new	1990.5	353.636	1.542	2071.0	480.206	—	2150.5	550.0	0.0
S650 new	1990.5	353.636	1.542	2091.0	546.910	—	2200.5	650.0	0.0
S750 new	1990.5	353.636	1.542	2111.0	634.139	—	2250.5	750.0	0.0
WRE350	2000.5	371.270	1.970	2033.5	412.0	0.0	2150.5	350.0	0.0
WRE450	2005.5	381.662	2.186	2050.5	443.0	—	2100.5	450.0	0.0
WRE550 (0.4)	2010.5	398.582	2.735	2050.5	491.0	—	2150.5	550.0	0.0
WRE550 (1.1)	2010.5	393.014	2.357	2050.5	485.0	—	2150.5	550.0	0.0
WRE550 (1.8)	2010.5	387.714	2.006	2050.5	479.0	—	2150.5	550.0	0.0
WRE650	2015.5	405.234	2.532	2050.5	500.0	—	2200.5	650.0	0.0
WRE750	2020.5	418.384	2.725	2050.5	506.0	—	2250.5	750.0	0.0
WRE1000	2050.5	508.560	3.242	2100.5	675.0	—	2375.5	1000.0	0.0

[a]Parenthetical numbers for WRE550 give the value of D^*_n (80s) (in Gt C/yr) used in computing the IS92a portions of the profiles. For other WRExxx cases, 1.1 Gt C/yr was used.

Table 21.A3. *Padé approximant coefficients*

Profile[a]	a	b	c	d	e
S350old	−65.22761	133.37010	−66.66247	−0.56514	1.04516
S450old	1.95137	−0.83601	—	−0.04863	0.16399
S550old	1.38896	−0.99943	—	−0.61104	0.00057
S650old	1.20677	−0.30213	—	−0.79323	0.69787
S750 old	1.11664	0.82918	—	−0.88336	1.82918
S350new	−67.85504	138.36700	−68.69380	−1.16124	1.97940
S450new	1.76020	−0.46961	—	−0.23980	0.53039
S550new	1.25644	−0.72857	—	−0.74356	0.27134
S650new	1.09264	−0.05750	—	−0.90736	0.94250
S750new	1.01149	1.06222	—	−0.98851	2.06222
WRE350	−13.89281	32.99947	−13.71046	−2.18235	6.57855
WRE450	3.03887	4.40261	—	1.03887	5.04261
WRE550 (0.4)	2.52876	0.93938	—	0.52876	1.93938
WRE550 (1.1)	2.10197	1.48791	—	0.10197	2.48791
WRE550 (1.8)	1.73053	1.97890	—	−0.26947	2.97890
WRE650	1.91375	1.48848	—	−0.08625	2.48848
WRE750	1.88999	1.46930	—	−0.11001	2.46930
WRE1000	2.14401	1.55569	0.14401	2.55569	

[a]Parenthetical numbers for WRE550 give the value of D^*_n (80s) (in Gt C/yr) used in computing the IS92a portions of the profiles. For other WRExxx cases, 1.1 Gt C/yr was used.

compared with the IPCC (Enting et al.) cases. For stabilization at 550 ppmv, larger industrial emissions are allowable in the WRE case compared with the IPCC case to around 2070, beyond which emissions in the WRE case drop below those in the IPCC case. For reasons associated with the carbon cycle, cumulative emissions are *always* more in the WRE cases, even out to and beyond the point of concentration stabilization – this is the carbon cycle "bonus" identified by Wigley and colleagues.

The main results of the paper by Wigley and colleagues, and explanations of these results, are as follows:

1. The analysis shows that there is a wide range of emissions pathways by which CO_2 concentration stabilization at any given time and level may be achieved. Specifically, two concentration profiles are considered for each stabilization level (the "S," or IPCC, and "WRE" profiles – see the main chapter text, Figure 21.2), resulting in wide differentials in emissions (see the main chapter text, Figure 21.3).

Determining the emissions required to follow a given concentration pathway is an inverse calculation. In any inverse calculation, small changes in the input (here, concentration) generally lead to large changes in output (here,

emissions). Thus, small differences in concentration pathway lead to large differences in required emissions.

2. Because the emissions trajectories differ substantially, the potential emissions-reduction *costs* may differ quite widely between concentration profiles with the same stabilization level.

This requires no explanation

3. Because, for any given stabilization level, the "S" and "WRE" concentration profiles are similar (see the main chapter text, Figure 21.2), the attendant climate and sea-level change differences are relatively small, generally less than 0.2°C and 4 cm for global-mean temperature and sea level, respectively (for central estimates of the key climate and sea-level model parameters) – see Wigley and colleagues (1996; their figure 3).

Small concentration differences imply small differences in radiative forcing. Because the climate and sea-level systems have very large inertia (associated with the size of the oceans and the large ice masses), they respond slowly to external forcing. Hence, small forcing differences lead to even smaller response (global-mean temperature and sea-level) differences.

4. Climate change differences between profiles with the same stabilization level are sufficiently small that they

lie within the uncertainties in climate change predictions that arise from aerosol uncertainties.

Sulfate aerosols (derived primarily from anthropogenic sulfur dioxide (SO_2) emissions) are major players in the climate change issue, second only to CO_2. We know that future climate will be sensitive to aerosol effects and, hence, to changes in SO_2 emissions. However, there are large uncertainties in these emissions and their radiative forcing effects, leading to significant uncertainties in future climate change. These uncertainties exceed the climate change differentials between different stabilization pathways leading to any given stabilization level.

5. Because of aerosol effects, the climate change differentials between different profiles in early decades may even be counterintuitive – in other words, it is possible that the pathway to stabilization at a given concentration level that has higher CO_2 emissions could, at least initially, have *lower* climate damage costs (i.e., smaller climate change) than a lower-emissions pathway.

This effect arises from the very different time scales for atmospheric responses to changes in CO_2 emissions (slow) and SO_2 emissions (rapid); see Wigley (1991). Wigley and colleagues show that, if future CO_2 and SO_2 emissions changes were to occur in parallel, then the SO_2 effect would cause *higher* CO_2 emissions pathways to have *smaller* (and, hence, slower) global-mean temperature and sea-level changes for a number of decades.

Some important implications of these results are as follows:

*   If, *no matter what the reason* (technological, economic, or political), future emissions were to depart initially only slowly from BAU, this would not preclude stabilization of the concentration of CO_2 in the atmosphere.

*   Since large emissions differences between profiles with the same stabilization level arise from and correspond to small concentration and climate change differences, mitigation cost differences could be larger than the differences in the benefits of averted damages.

*   It is difficult to estimate the differences in emissions reduction costs between different stabilization profiles; consequently, such differential cost estimates are uncertain. However, differential damage estimates are even more uncertain – indeed, because of aerosol uncertainties related to SO_2 emissions, we cannot even be confident of the *sign* of these differentials.

The economic discussion in Wigley and colleagues focuses on the mitigation costs associated with different emissions pathways. A number of factors contribute to the expectation that emissions pathways that initially depart slowly from BAU will involve lower mitigation costs than those that depart rapidly from BAU. Apart from the carbon cycle bonus noted above, these factors are as follows:

*   Positive marginal productivity of capital. Continuing economic growth (which, in simple terms, requires us to

"discount" the future in economic analyses) implies that, to finance a future burden, the further in the future this burden lies the smaller is the set of resources that must be set aside now. This, of course, requires us to make appropriate investment decisions today.

*   Capital stock. Stock for energy production and use is typically long-lived (e.g., power plants, housing, and transport). The current system is configured based upon a set of expectations about the future. Unanticipated changes will be costly. Time is therefore needed to re-optimize the capital stock. As noted by IPCC Working Group III, "Implementing emission reductions at rates that can be absorbed in the course of normal stock turnover is likely to be cheaper than enforcing premature retirement (of capital stock) now" (Bruce et al., 1996, p. 12).

*   Technical progress. There is ample evidence for past and potential future improvements in the efficiency of energy supply, transformation, and end-use technologies. Thus, the availability of low-carbon substitutes is likely to improve and their costs are likely to be reduced over time. *This does not mean that we should not attempt to accelerate appropriate technological change.*

Although mitigation costs and cost differentials are uncertain, it is clear that they are less uncertain in the near term than are estimates of the benefits and benefit differentials accruing from averted climate change damages. The latter involve huge uncertainties associated with the following factors: aerosol forcing uncertainties; estimation of regional climate change even, if we could quantify aerosol (and other) forcings reliably; and determination of damages, even if we could estimate regional climate changes reliably. The paper by Wigley and colleagues focuses on the mitigation-cost side of the cost–benefit equation. The paper can in no way be construed as a cost–benefit analysis.

Wigley and colleagues have been misrepresented or misinterpreted concerning whether or not they advocate a delay in responding to the climate change issue. They do *not*. They state quite clearly that their results *cannot be used to justify a delay in implementing policies to reduce future anthropogenic climate change.* At some time in the not-too-distant future, global CO_2 emissions must drop substantially below what they might be under a BAU scenario. For this to be possible, planning must begin now. This is essential if society is to reduce its current (and projected) heavy dependence on a fossil fuel–intensive energy infrastructure, and to ensure that appropriate and cost-effective technology is available to facilitate this change.

The primary result of the analysis by Wigley and colleagues is that if, *no matter what the reason*, future emissions were initially to depart only slowly from BAU, this would not preclude stabilization of the concentration of CO_2 in the atmosphere. This should be reassuring to everyone.

A secondary, but still important, result is that there may well be CO_2 concentration and emissions pathways that are economically advantageous (i.e., involving significantly lower mitigation costs) yet that may be only minimally more expensive in terms of damages. In other words, with the proper policies, the future may be a win–win situation, in which both the differential economic costs (relative to BAU) and the absolute environmental costs are small. Much more work is required (in the areas of carbon cycle modeling, climate modeling, impacts assessment, and economic analysis) before these latter ideas can be satisfactorily quantified and tested, particularly on the damages side.

What Wigley and colleagues have shown is that there is a wider range of emissions control options than was previously realized. They do not present any final judgments among the possibilities, in terms of either stabilization target or pathway, nor do they say anything about how any particular future emissions trajectory might be realized. Making these judgments and determining how any emissions control target might be achieved are challenges for the future for both scientists and policy makers.

Tom M. L. Wigley
UCAR/NCAR
P.O. Box 3000
Boulder, CO 80307

March 1996

Part V

Appendixes

Organizing Committee Members

William R. Emanuel
Inez Fung
Michael B. McElroy
Berrien Moore III
David S. Schimel
William H. Schlesinger
Ulrich Siegenthaler, GCI Co-director
Robert T. Watson, GCI Co-director
Tom M. L. Wigley, GCI Co-director

Working Group Members

Group A	*Group B*	*Group C*
Missing Carbon Sink	Paleo-CO_2 Variations	Modeling Strategies
Chair	*Chair*	*Co-chairs*
W. Schlesinger	U. Siegenthaler	W. Emanuel
		B. Moore
Rapporteur	*Rapporteur*	*Rapporteur*
D. Lashof	K. Caldeira	D. Harvey
Members	*Members*	*Members*
B. Bolin	D. Barnes	R. Braswell
S. Craig	P. Friedlingstein	J. Edmonds
I. Fung	J. Harden	I. Enting
R. Gifford	R. Keir	E. Hausman
D. Hall	J. Lynch-Stieglitz	M. Heimann
K. Harrison	B. Opdyke	M. Hoffert
E. Holland	D. Raynaud	P. S. Liss
R. A. Houghton	M. Stuiver	R. Richels
C. D. Keeling	E. Sundquist	J. Sarmiento
R. Keeling	T. Volk	D. Schimel
G. Marland	T. Wigley	B. Semtner
T.-H. Peng		R. Watson
L. Ver		I. Woodward
A. Watson		

Reviewers

Referees from Outside

David Archer, University of Chicago
A. N. D. Auclair, Science and Policy Associates
Michael Bender, University of Rhode Island
Wolfgang Berger, Scripps Institution of Oceanography
Robert Berner, Yale University
E. Boyle, Massachusetts Institute of Technology
J. di Primio
J. Goudriann, Wageningen
Atul Jain, Lawrence Livermore National Laboratory
A. P. Jaques, Environment Canada
Paul Jarvis, University of Edinburgh

Haroon Kheshgi*, Exxon
T. G. F. Kittel, National Center for Atmospheric Research
Rik Leemans*, RIVM
Austin Long, University of Arizona
R. J. Luxmoorek, Oak Ridge National Laboratory
Fred Mackenzie, University of Hawaii
Pam Matson, University of California at Berkeley
Wim Mook, Netherlands Institute for Sea Research
James Morison, University of Reading
James Murphy, Hadley Centre for Climate Prediction and Research
Nebojsa Nakicenova, International Institute for Applied Systems Analysis
Michael Oppenheimer, Electricite de France
Joyce Penner, Lawrence Livermore National Laboratory
Niel Plummer, U.S. Geological Survey
W. M. Post*, Oak Ridge National Laboratory
Michael Prather, University of California at Irvine
Joel Swisher, RISO National Laboratory
Robbie Toggweiler*, Geophysical Fluid Dynamics Laboratory
Susan Trumbore, University of California at Irvine
John Weyant, Stanford University
Tim Whorf, Scripps Institution of Oceanography
Gary Yohe, Wesleyan University

GCI Participants/Authors Who Reviewed Papers

Robert Dixon, U.S. Department of Energy
Bert Drake, Smithsonian Environmental Research Center
William Emanuel*, University of Virginia
Ian Enting, CSIRO
Inez Fung, University of Victoria
Jennifer Harden, U.S. Geological Survey
L. D. Danny Harvey, University of Toronto
Martin Heimann**, Max-Planck-Institut für Meterologie
Elisabeth Holland, National Center for Atmospheric Research
Peter Liss, University of East Anglia
Gregg Marland, Oak Ridge National Laboratory
Jorge Sarmiento*, Princeton University
David Schimel, National Center for Atmospheric Research
William Schlesinger, Duke University
Allen Solomon, Environmental Protection Agency
Eric Sundquist, U.S. Geological Survey
Andrew Watson
Tom Wigley, University Corporation for Atmospheric Research
Ian Woodward, University of Sheffield

*Reviewed two papers.
**Reviewed three papers.

Index